Functional Nanomaterials and Nanotechnologies
Applications for Energy and Environment

Functional Nanomaterials and Nanotechnologies
Applications for Energy and Environment

Dr. Vikas Mittal
Editor and Lead Author

$\mathcal{CWP}$
Central West Publishing

Disclaimer
Every effort has been made by the publisher, editor and authors while preparing this book, however, no warranties are made regarding the accuracy and completeness of the content. The publisher, editor and authors disclaim without any limitation all warranties as well as any implied warranties about sales, along with fitness of the content for a particular purpose. Citation of any website and other information sources does not mean any endorsement from the publisher and authors. For ascertaining the suitability of the contents contained herein for a particular lab or commercial use, consultation with the subject expert is needed. In addition, while using the information and methods contained herein, the practitioners and researchers need to be mindful for their own safety, along with the safety of others, including the professional parties and premises for whom they have professional responsibility. To the fullest extent of law, the publisher, editor and authors are not liable in all circumstances (special, incidental, and consequential) for any injury and/or damage to persons and property, along with any potential loss of profit and other commercial damages due to the use of any methods, products, guidelines, procedures contained in the material herein.

NATIONAL
LIBRARY
OF AUSTRALIA

A catalogue record for this book is available from the National Library of Australia

ISBN (print): 978-0-6482205-3-4
ISBN (e-book): 978-0-6482205-2-7

Contents

Preface

Nanotechnology and nanomaterials have impacted all engineering, science and technology sectors owing to their superior properties and performance profiles as compared to the conventional materials. Energy and environment are the two main sectors among these which have experienced widespread application of these nanotechnologies and nanomaterials. Specifically, the applications of such materials and technologies in enhanced oil recovery, gas adsorption, gas separation, corrosion protection, fuel cell components and chemical reactions have been achieved. To elaborate these in detail, the present volume provides fundamental aspects of these materials and technologies, along with significant insights into their development for particular applications.

In this respect, Chapter 1 reviews the application of nanotechnology for enhanced oil recovery. The use of metal organic frameworks and metal oxides for the gas adsorption and other environmental applications is reviewed in Chapter 2. The potential of ionic liquids for gas separation processes is also discussed in Chapter 3. Chapter 4 presents the recent developments in the generation of self-healing coatings for anti-corrosion applications. The use of nano-carbon based polymer composite materials as membranes in fuel cells is described in Chapter 5. The developments in deep eutectic solvents for gas separation and EOR applications are reviewed in Chapter 6. Chapter 7 specifically discusses the effectiveness of graphene for corrosion protection applications. Theoretical insights into the gas molecular adsorption on graphene are presented in Chapter 8, with special focus on gas sensing applications. Chapter 9 also presents a related DFT study of the N_2O adsorption on graphene and heteroatom-doped graphene. Nano-catalysis is the focus of Chapter 10, where the advances in the synthesis and application of nano-catalysts in the field of energy have been demonstrated. Various advanced carbon based coatings have been presented in Chapter 11. Finally, Chapter 12 is the exhaustive review of the various adsorbents for the adsorption of different gases.

The support of all the chapter contributors is greatly acknowledged towards the successful accomplishment of the book. The book is dedicated to my family for their constant support.

Vikas MITTAL
Australia

1

Application of Nanotechnology for Enhanced Oil Recovery

1.1 Introduction

Nanotechnology refers to the technology where at least one of the constituents is sized at the nanometer scale, generally from a few nanometres to a few hundred nanometers. One of the fundamental concepts encompassed by nanotechnology is that of the nanoparticle. A nanoparticle is generally defined as an aggregation of between 10 and 100,000 atoms, with a radius of 1-100 nm [1]. The unique benefit of nanotechnology, in contrast to micro- and macro-scale technologies, is the change in properties of matter that takes place at the nanometre scale. Scaling the dimension of a macroscopic material down in size initially does not affect the material's properties, but eventually causes minor property changes, and finally causes major property changes once the size reduces below 100 nm [1]. Due to this unique change in properties, the potential offered by the materials designed at the nanometer scale is significant. In this respect, the application and potential offered by nanotechnology specifically for enhanced oil recovery has developed manifold in the last few decades.

Enhanced oil recovery (EOR) is defined by Morrow and Heller [2] as oil recovery brought about through means beyond that of the natural forces present in the reservoir. This definition includes tertiary oil recovery, which is oil recovery that takes place after the completion of primary (natural reservoir forces) and secondary (water or gas flooding) oil recovery methods. There are a variety of physical, chemical, and biological processes that have been theorised and tested for tertiary recovery methods. Researchers have explored, among other methods, microbial bio-surfactants, aqueous chemical flooding, polymer flooding, and CO_2 injection for tertiary oil recovery [3-6]. The goal of tertiary recovery methods is to increase the amount of oil that can be recovered economically. As mentioned earlier, this is the area where nanotechnology has a significant benefit. It has been mentioned that primary and secondary oil recovery generally leave between 50% and 65% of the original oil in place

Arjun Dhillon*, Ali Elkamel and Vikas Mittal**, The Petroleum Institute (part of Khalifa University of Science and Technology), Abu Dhabi, UAE
*Current address: University of Waterloo, Canada; **Current address: Bletchington, Wellington County, Australia

(OOIP) [7]. Nanoparticle-based technologies can be used to replace or augment existing EOR methods for greater economic recovery efficiencies. This chapter explores some promising prospects for such applications, as well as some other methods for improving oil recovery.

1.2 Improved CO_2 Foams

A popular and established form of tertiary oil recovery is carbon dioxide flooding. Carbon dioxide is miscible in many hydrocarbons present in crude oil, particularly those with between five and twenty carbon atoms, at sufficiently high pressures, as noted by Holm [8]. Also, upon dissolution of carbon dioxide in crude oil, the oil reduces its viscosity by a factor of 5 to 10 as it swells by 10% to 60% [8]. By reducing the viscosity of the oil, the carbon dioxide improves the oil's mobility and enables higher oil recovery, as a result. Martin and Taber also reported that the vaporizing-gas-drive mechanism enables miscibility, although carbon dioxide solubility at reservoir pressures also helps to swell oil volume at an earlier point [9].

A main weakness of gas injection EOR techniques such as CO_2 flooding is said to be the relatively high mobility of gases [10]. According to Aronofsky and Ramey, the efficiency and success of oil extraction depends on the mobility ratio, which is the ratio of the mobility of the displacing phase (CO_2) to that of the displaced phase (oil/water) [11]. A lower mobility ratio implies a greater sweep efficiency and, thus, is desirable when comparing injection processes. Since the mobility ratio is directly proportional to the mobility of the displacing phase (CO_2), the high mobility of carbon dioxide results in a higher mobility ratio, which has an adverse effect on the sweep efficiency of this oil recovery method. Heller has identified the cause of this to be the formation of viscous fingers, where the carbon dioxide is able to move faster than fluids such as oil and water, with the given pressure gradient, which reduces the sweep efficiency [12]. A method known as water-alternating-gas (WAG) has been employed frequently to try to address the issue of mobility-related problems arising from the carbon dioxide. This method involves alternating between injection of water and carbon dioxide. Although this method is preferable to carbon dioxide injection alone, it is also reported to be inadequate and justified only in the absence of an acceptable alternative [13]. It is for this reason that CO_2 foams are being explored as a potential solution for tertiary oil recovery.

It has been frequently reported that injecting the carbon dioxide gas as foam reduces its mobility [10,14-16]. Making the gas into foam causes it to lose much of its mobility, while its efficiency remains unaffected. The viscosity of the carbon dioxide is increased, which results in a mobility ratio that is more favourable for oil recovery. Experimentation has shown that carbon dioxide foams created using a surfactant known as alpha olefin sulfonate (AOS) were successful in improving oil recovery and efficient exploitation of the carbon dioxide [10]. As shown in Figure 1.1, the comparison of miscible gas and

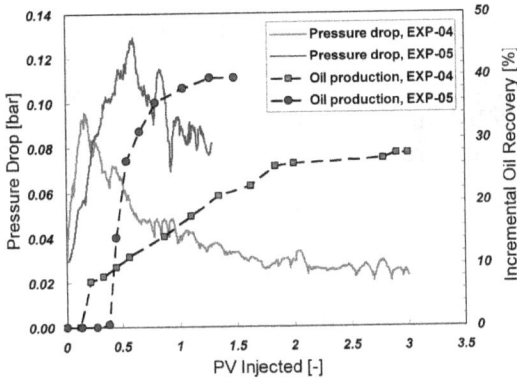

Figure 1.1 Pressure drop and oil production profiles for the miscible gas (EXP-04) and miscible foam (EXP-05) experiments. Reproduced from Reference 10 with permission from American Chemical Society.

miscible foam indicated a significant increase in the oil production when foam was used.

The application of nanotechnology in this area exists in the form of improving carbon dioxide foams with nanoparticles. Worthen *et al.* [17] identified the potential for nanoparticle reinforcement of carbon dioxide foams for improved foam stability and longevity. The authors successfully created carbon dioxide foam using 50% SiOH silica (SiO_2) nanoparticles in the place of surfactants or polymers. When compared to carbon dioxide foams formed using nanoparticles coated with polyethylene glycol (PEG), as well as those formed by the surfactant Tergitol™ 15-S-20, the carbon dioxide foams created using the silica nanoparticles were observed to be much more stable. Espinosa *et al.* [18] used 5 nm silica nanoparticles to form a carbon dioxide foam in water. The supercritical foam underwent PEG surface treatment and was found to be very stable, even at high reservoir temperatures. The foam was produced at temperatures between 50 °C and 95 °C successfully, demonstrating its tolerance of high temperatures. There are currently a large number of ongoing research efforts in this area with a goal of augmenting carbon dioxide foam EOR methods with nanoparticles [19,20]. In summary, carbon dioxide foam flooding is a tertiary oil recovery method with a lot of potential offered as a result of nanotechnology.

1.3 EOR Surfactants

Surfactant flooding, a tertiary oil recovery process that involves the addition of surfactants to waterflooding formulations to improve its overall oil recovery, has been investigated under various conditions as an EOR technique for many years [21-24]. A surfactant is a chemical whose molecular structure is such that it is capable of reducing interfacial tension (IFT). In the context of oil recovery, surfactants are typically amphiphilic molecules that reduce the oil/water IFT. Uren and Fahmy [25] stated that reducing the interfacial tension between oil and injected water should generally improve the oil recovery brought about through waterflooding. Surfactants possess both hydrophilic and hydrophobic components, indicating that the molecules are able to interact with both oil and water, thereby, reducing the IFT that is present between the two liquid phases. As stated by Levitt *et al.* [26], lowering IFT allows more oil to be released from the formation's capillary forces that would otherwise prevent the oil from being extracted.

The utility of surfactants for EOR comes from their molecular properties, which can be modified at the nanometer scale. Kong and Ohadi [27] mentioned that nanotechnology has the techniques and methods necessary to specifically tailor surfactants for necessary conditions or functions. The authors further reported that there is potential for a more controlled addition of these surfactants to the oil reservoir, implying a better overall recovery. Amanullah and Al-Tahini [28] described the additional benefit of nano-sized surfactants, referencing the fact that nanoscale particles have a much greater accessible surface area for a given mass or volume, compared to their microscale counterparts. The fact that nanoparticles have a much higher specific surface area enables them to interact to a much greater extent than larger particles. The result of this increased reactivity is a reduced required concentration for nano-sized surfactants to achieve the same result as a higher concentration of larger surfactants. Kong and Ohadi [27] also mentioned that surfactant flooding, like carbon dioxide flooding, can suffer from a poor mobility ratio. The authors suggested this issue can be resolved using nanoparticles to modify the viscosity of the injected solution. Dexter and Middelberg [29] investigated biological surfactants derived from peptides (proteins) that are capable of lowering IFT at the oil/water interface. These nanoscale surfactants are an attractive alternative to synthetic surfactants for oil recovery.

The benefit of nanotechnology to improvement of surfactant flooding exists primarily in the form of enhanced surfactants. Using nanoscale chemical and biological synthesis methods, it becomes possible to create surfactants that can drastically reduce the oil/water interfacial tension in low concentrations. The ongoing research and experimentation in this field has the potential to make surfactant flooding more efficient and economical.

1.4 Nano-fluids for Improving Oil Recovery

Nano-fluids are defined by Hendraningrat *et al.* [30] as suspension mixtures of solid nanoparticles contained within water, oil, ethylene glycol, or other fluids. It has been reported that nano-fluids exhibit unique properties such as high thermal conductivity and rapid heat transfer [31]. These enhanced thermal properties of nano-fluids can be useful in various applications throughout the upstream oil and gas industry, including enhanced oil recovery. Another significant advantage of nano-fluids is the fact that these can act as a mechanism for delivering nanoparticles into oil reservoirs to perform specific functions. These important features of nano-fluids enable their usage for a variety of different applications in enhanced oil recovery.

An important physicochemical characteristic of reservoir rock is its wettability. Abdallah *et al.* [32] defined this property as the affinity of a particular fluid to a given solid surface. This property is the result of the intermolecular forces acting at the surfaces and interface of the fluid and solid. A fluid that is considered wetting with respect to a surface will increase its contact with the surface by spreading out, and will displace a non-wetting fluid that will reduce its own contact with the surface. In oil recovery, the focus is on the relative wettabilities of water and oil, with respect to the formation rock. Water-wet rock will be hydrophilic, while oil-wet rock will be hydrophobic, implying that water-wet and oil-wet conditions are opposite extremes on the wettability continuum. Laboratory experiments have indicated that oil recovery is most successful when wettability is neutral, meaning that the rock does not preferentially attract water over oil, or vice versa [33]. It was also stated that reservoir rock wettability influences the recovery factor, which has a role in determining how commercially feasible an oil field is [34]. These findings have led to numerous studies related to wettability alteration procedures under various conditions with the goal of improving oil recovery [35-38]. Hendraningrat *et al.* [30] created a nano-fluid by suspending silica nanoparticles (7 nm average size) in synthetic brine and tested its oil recovery potential with Berea sandstone. In this experiment, it was found that the nanoparticles reduced the interfacial tension and increased oil recovery in both the secondary and tertiary phases of EOR. The additional recovery was nearly 2% of the original oil in place (OOIP) for tertiary nano-fluid flooding, and was nearly 8% OOIP for secondary nano-fluid flooding (compared to brine flooding). Li *et al.* [39] also reported that nano-fluids containing silica nanoparticles reduced interfacial tension at the oil/water interface, and further noted that the nano-fluid was able to alter the rock wettability, making it more water-wet. Ju *et al.* [40] reported the mechanism for this phenomenon to be the adsorption of nanoparticles onto the pore walls, resulting from Brownian motion, which made their surface more hydrophilic. It was also discovered by Li *et al.* [39] that increasing the concentration of silica nanoparticles in the fluid caused an increase in oil recovery until a maximum point of 0.05 wt%, after which an increase in nanoparticle concentration resulted in a decrease in oil recovery.

Nanoparticles can be added to fluids that are already in use in oil recovery to form nano-fluids with enhanced properties. Amanullah and Al-Tahini [28] emphasized the utility that specifically functionalised and tailored nanoparticles could potentially have in oil recovery fluids such as drilling fluids, completion fluids, and stimulation fluids. The authors suggested that the addition of nanoparticles to these various fluids will be of importance when facing production and environmental obstacles. According to Amanullah and Al-Tahini, the quantity of solid particles in drilling fluids impacts the level of formation damage, the level of productivity, and the rate of penetration (ROP). Due to the high specific surface area, and high resulting reactivity of nanoparticles, low concentrations of nanoparticles in fluids can achieve similar results to those brought about by higher concentrations of their larger-scale counterparts. The use of nanoparticles would, therefore, reduce the problems associated with the solids content in drilling fluids. The authors further identified drilling fluid filtrate loss as an important additional issue faced by existing drilling fluids that do not incorporate nanotechnology. Amanullah [41] described the process of filtrate loss, also known as fluid invasion, as the result of a pressure differential within the oil reservoir. When drilling fluids apply a significantly higher amount of pressure on the reservoir than the internal reservoir pressure, such a differential exists and results in fluid filtrate and particles forcibly entering the formation. As stated by Donaldson and Chernoglazov [42], filtrate migration occurs radially through the walls of the wellbore. It occurs to varying degrees depending on formation porosity, and displaces water and hydrocarbons along its path. Donaldson and Chernoglazov [42] also specified that fluid invasion can span between 5 and 10 cm in high porosity formations, and up to 3 m in low porosity formations. According to Amanullah [41], the consequences of fluid invasion can be severe, including alteration of the permeability of the formation and reduction or termination of the flow of reservoir fluids toward the production well. These consequences are somewhat prevented by the drilling fluid through the formation of a filter cake. Elkatatny *et al.* [43] explained that a filter cake forms on the face of the formation as particles of different sizes adhere to it, reducing its permeability. The filter cake formation can help to reduce fluid invasion into the reservoir formation, whereas the filter cake formed by standard drilling fluids does not completely prevent this problem. It is for this reason that nanotechnology has the potential to improve upon existing drilling fluid technologies. Amanullah and Al-Tahini [28] also suggested that the inclusion of nanoparticles in drilling fluids has the potential to solve the problem of fluid invasion as a result of unique nanoparticle properties. A drilling fluid incorporating nanotechnology would be thin, yet essentially impermeable and resistant to erosion. This lack of permeability would help to prevent the adverse effects of fluid invasion into the formation such as formation damage and loss of productivity. An additional benefit to such drilling fluids, as noted by Amanullah and Al-Tahini [28], is the ease with which their filter cake can be removed. The high reactivity resulting from the high specific surface area of nanoparticles enables efficient chemical removal of the filter cake from the wellbore.

Research has recently been conducted on a nano-fluid made up of hydrophobic and lipophilic polysilicon (HLP) nanoparticles in ethanol. It has been reported that the nanoparticles caused an alteration in rock wettability from a water-wet condition to a more neutral wettability [44]. In addition to the wettability alteration, a decrease in oil/water interfacial tension was noted as a result of flooding with this HLP nano-fluid. In the core-flood experiments conducted on water-wet sandstone, the authors observed a reduction in oil/water IFT by a factor of ten, a decrease in water-wetness denoted by a 24° decrease in contact angle, and a resulting 19.31% increase in oil recovery after injection of the HLP nano-fluid. Core-flood experiments conducted by Onyekonwu and Ogolo [45] also demonstrated an increase in oil recovery from ethanol-based HLP nanofluid, as well as an increase in oil recovery from neutrally-wet polysilicon nanoparticles dispersed in ethanol.

Some nano-fluids are also capable of improving oil recovery directly through a mechanism known as disjoining pressure. Wasan and Nikolov [46] described this as the formation of a wedge-like transition area that exists at the interface between an aqueous nano-fluid, a phase such as oil, and a solid.

This is the result of nanoparticles assembling at the interface between the solid and oil phases, as shown in Figure 1.2 [47]. As stated by McElfresh *et al.* [48], significant quantities of nanoparticles in nano-fluids have a tendency to move toward the phase boundary between an oil droplet and the surrounding water, and exert a force on the oil droplet. This ordering of nanoparticles is favourable due to increased overall entropy, according to Kondiparty *et al.* [47], as it increases the freedom of nanoparticles in the bulk solution. As nanoparticles fill the area between the droplet and the solid it is adhered to, this force acts at the vertex (point of contact) of the droplet and solid to push the droplet away from this surface. In this way, oil droplets can be forced away from the solid rock of the formation by nanoparticles, leading to a greater amount of recovered oil. Hendraningrat *et al.* [30] also observed that when using hydrophilic silica nanoparticles in nano-fluid flooding, the disjoining pressure mechanism in conjunction with the reduction of interfacial tension resulted in improved oil recovery.

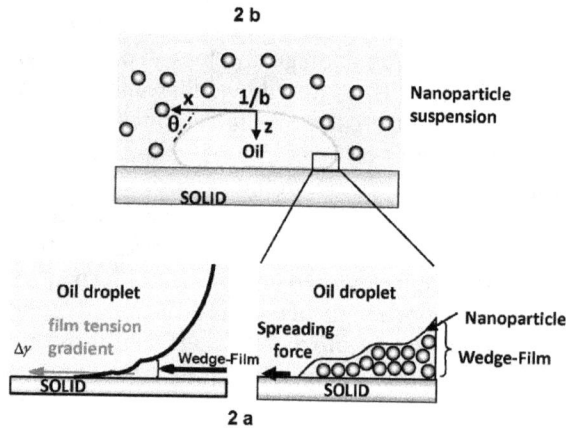

Figure 1.2 Disjoining pressure mechanism caused by nano-fluid mobilizing adsorbed oil droplet from solid. Reproduced from Reference 47 with permission from American Chemical Society.

Zirconium oxide nano-fluids have also been investigated for EOR purposes. Karimi *et al.* [49] added zirconium oxide (ZrO_2) nanoparticles to non-ionic surfactant fluid mixtures to form a nano-fluid and used this nano-fluid to conduct EOR tests on carbonate rock. This rock was initially found to be strongly oil-wet, having a great wetting preference for oil over water, but was changed to water-wet after exposure to the zirconium oxide nano-fluid. The authors stated that the wettability alteration occurred as a result of adsorption of the zirconium oxide nanoparticles on the surface of the rock. The authors further mentioned that the adsorption process was slow, requiring a minimum of two days to complete. Figure 1.3 also shows the scanning electron microscopy (SEM) images of surface of the rock aged in fluids 1-3. Fluid 1 had no nanoparticles, whereas fluid 2 and 3 differed in nanoparticle content. For fluid 1, the morphology of the rock surface resembled that of clean surface. In the case of fluids 2 and 3, formation of nanostructures on the rock surface was observed. The authors identified these nanostructures as nano-sized ribbons (fluid 2) and nano-flower (fluid 3). The nanostructures formed on the rock surface were suggested to have altered the wettability of an oil-wet rock to significantly water-wet nature. Thus, nanoparticle concentration was observed to have significant effect on the performance of nano-fluid. The formation of nanostructures was also confirmed through X-ray diffraction. The obtained nano-fluids also resulted in significantly enhanced oil recovery.

Figure 1.3 SEM images of oil-wet carbonate rock aged in fluid 1 (A), fluid 2 (B), and fluid 3 (C). Reproduced from Reference 49 with permission from American Chemical Society.

Similar to zirconium oxide nanoparticles, alumina (Al_2O_3) nanoparticles have also been shown to change rock wettability from strongly oil-wet to strongly water-wet in nano-fluid form. Giraldo *et al.* [50] created nano-fluids using anionic surfactants and alumina nanoparticles, and tested their wettability alteration potential on sandstone that had been previously made strongly oil-wet. The nano-fluid was able to change the wettability of the sandstone to strongly water-wet, which is also demonstrated in Figure 1.4.

In summary, nano-fluids offer a variety of EOR applications, especially as research continues on functionalising nanoparticles in more complex and useful ways. Nanoparticles can be used both to create new fluids and to enhance conventional fluids, making nano-fluids research quite expansive.

Figure 1.4 Contact angles for the treatment with a nanoparticle concentration of 100 ppm: (a) oil/air/rock system before treatment, (b) oil/air/system after treatment, (c) water/air/system before treatment, (d) water/air/system after treatment. Reproduced from Reference 50 with permission from American Chemical Society.

1.5 Emulsions and Nano-emulsions for EOR

Emulsions and nano-emulsions have been the topics of numerous studies for the purpose of enhancing oil recovery. Mandal *et al.* [51] defined emulsions as suspension mixtures involving droplets of one liquid in another, where the two liquids are immiscible and the droplet size is greater than 0.1 µm. For EOR purposes, the two main immiscible liquids that are of concern are oil and water. Emulsions of oil and water, both oil-in-water and water-in-oil, have been investigated for EOR. For instance, Figure 1.5 demonstrates the performance of emulsion with 20% oil content.

Oil-in-water emulsions are used in the process of emulsion flooding, a process that involves the usage of these emulsions as displacing fluid for oil recovery. The mechanism for this EOR process, as described by Thomas and Farouq Ali [52], is the blocking of permeable paths for the fluids injected into the reservoir. The oil droplets dispersed in the emulsion can block these preferred fluid paths, thereby, forcing the injection fluids to areas that have not been contacted. This result can be made possible under the right circumstances, making emulsion flooding a potential EOR technique. Another benefit to oil-in-water emulsions is their altered viscosity. As previously mentioned, the success of oil recovery is partially dependent on mobility ratio, which in turn is partially dependent on the viscosity of the displacing fluid, making this viscosity an important factor for EOR. Pal [53] concluded that there was a relationship between the viscosity of an emulsion and the size of its constituent droplets. It was found that decreasing the droplet size resulted in an increase in the emulsion viscosity. This further adds to the potential of emulsion flooding as an effective EOR method.

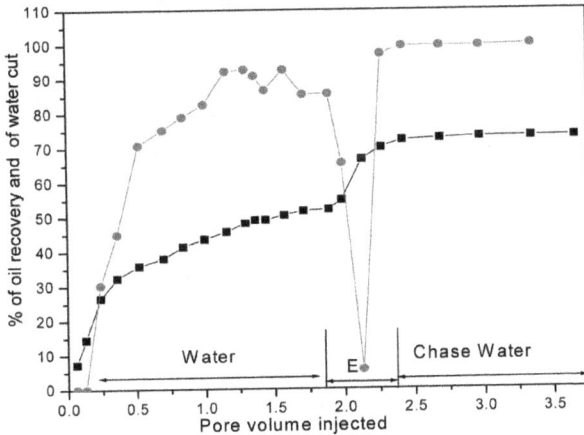

Figure 1.5 Production performance of emulsion (20% oil) flooding: ■, % oil recovery; ●, % water cut. Reproduced from Reference 51 with permission from American Chemical Society.

A major issue faced by emulsion flooding is the inherent instability of emulsions. Since oil and water are immiscible, droplets of oil within water tend to coalesce in order to form a separated phase, thereby minimising the contact surface between the oil and water phases. It is for this reason that certain conditions must be met for emulsions to be stable. This condition, according to Chen and Tao [54], is the presence of an emulsifying agent or emulsifier. The authors defined such an emulsifier as a surfactant that is capable of film formation at the oil/water interfaces in order to prevent coalescence. Nanoparticles have been investigated as potential alternatives to conventional surfactants for stabilising emulsions. Zhang *et al.* [55] observed that it was possible to stabilise oil/water emulsions with silica nanoparticles which had undergone surface functionalization. The emulsions generated with these nanoparticles were 2-4 μm in size on average, and were reported to reduce permeability of the rock with respect to the injected water. The uniformity of the nanoparticles is reflected in the tight, organised interfacial barrier they form at the oil/water interface. According to Zhang *et al.* [55], this results in significantly greater emulsion stability, especially at high-temperatures, which is an excellent quality for an EOR agent. In the experiments performed by Zhang *et al.* [55], the emulsions formed using nanoparticles remained stable for weeks under conditions involving both ambient and elevated temperatures. Further research by Zhang *et al.* [56] used functionalized silica 5 nm nanoparticles to stabilise emulsions for several months after their formation.

Peptide surfactants mentioned previously in this paper also offer a unique advantage to the creation and manipulation of oil/water emulsions for oil recovery. Dexter and Middelberg [29] stated that proteins can be modified with hydrophilic and hydrophobic groups along their polyamide chain. This enables the creation of protein-based surfactants, which could be used to lower IFT at oil/water interfaces. According to Dexter and Middelberg [29], the peptides adsorb at the oil/water interface and cross-link with one another to stabilize an emulsion. Unlike typical surfactants, these surfactants arrange themselves parallel to the interface rather than perpendicular, as their hydrophilic and hydrophobic groups are along their chain rather than at the ends of it. The authors further stated that the cross-linking can be deactivated and reactivated chemically with a change in pH or a change in the concentration of certain metal ions present, as seen in Figure 1.6. Since it is possible to chemically manipulate these peptide surfactants in order to promote or hinder the formation of interfacial films around emulsion droplets, it is possible to create or break down emulsions. Another attractive feature of peptide surfactants is the potential for inexpensive production, as it has been shown by Morreale *et al.* [57] that such peptides could be produced biologically through a process known as bioprocess-centred molecular design. Furthermore, according to

Malcolm *et al.* [58], these surfactants, as biomolecules, are biodegradable, making them more environmentally-friendly option for stabilising emulsions.

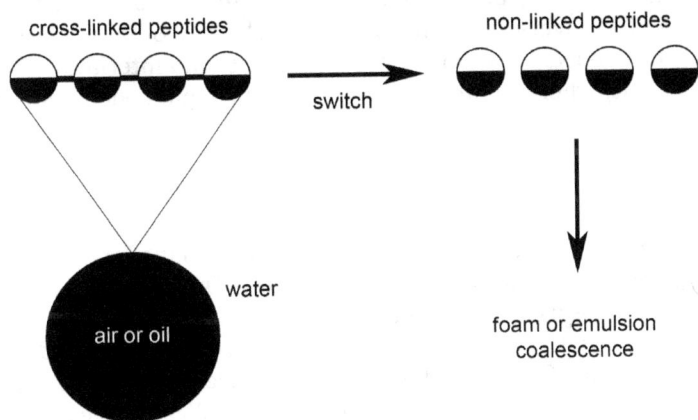

Figure 1.6 Switchable cross-linking of peptide surfactants to control formation and collapse of emulsions. Reproduced from Reference 29 with permission from American Chemical Society.

The potential applications for nano-emulsions for EOR are similar to those of emulsions. Nano-emulsions are smaller droplets than typical emulsions, with droplet size varying within the range of 50-500 nm, according to Mandal *et al.* [59]. These nano-emulsions, similarly to other emulsions, can be used as displacing fluid for emulsion flooding. The comparative benefit of nano-emulsion flooding over traditional emulsion flooding comes from the unique properties held by nano-emulsions. Nano-emulsions can be transported and stored without loss of utility due to processes such as sedimentation or creaming. Robins [60] defined creaming as the buoyancy-driven movement and separation of droplets from an emulsion to the top of a container. This process is mentioned to be similar to sedimentation, but in the opposite direction. According to Mandal *et al.* [59], it is due to the small droplet size of nano-emulsions that separation driven by density differences does not occur significantly, resulting in the prevention of both sedimentation and creaming. Additionally, the small size of the droplets allows them access to the pores within reservoir rock, which should be expected to improve oil recovery when nano-emulsion flooding is used. Research studies have also reported the preparation of oil-in-water nano-emulsions from mineral oil using non-ionic surfactants [59]. Flooding tests performed using these nano-emulsions exhibited an increase of more than 30% OOIP recovered in comparison to waterflooding alone.

Thus, nanotechnology offers numerous ways to develop emulsion flooding as a tertiary oil recovery technique. Nanoparticles and other nano-engineered technology can be used to tailor better surfactants and interfacial films to stabilize emulsions, whereas nano-emulsions offer unique benefits for oil recovery.

1.6 Ferro-fluids

Ferro-fluids are nano-fluids that are composed of nanoparticles possessing magnetic properties, according to Odenbach [61]. Odenbach [61] further stated that the particles used in ferro-fluids are normally magnetite (Fe_3O_4) nanoparticles with an average size of 10 nm which have also been modified with a surfactant. The role of the surfactant is said to preven coagulative aggregation, which would otherwise be the result of the van der Waals forces that cause interactions between particles. The author further mentioned that it is, therefore, possible to produce ferro-fluids with adequate stability, in liquids such as water, oil, and kerosene. The author attributed the unique superparamagnetic characteristic of ferro-fluids on the macroscopic scale to the nanoscale magnetism of their constituent nanoparticles. This property is what makes the ferro-fluids unique in their behaviour, especially in the presence of magnetic fields.

Kothari *et al.* [62] identified the magneto-rheological effect as the applicable capability of ferro-fluids that is the result of their magnetic properties. This effect gives rise to the ability of ferro-fluids to alter their viscosity as a consequence of exposure to an external magnetic field. Since viscosity is a significant factor that influences fluid mobility in enhanced oil recovery, the potential for viscosity alteration using ferro-fluids is promising. The authors described an application that manipulated this capability of ferro-fluids in surfactant flooding. Ferro-fluids used in addition to surfactant flooding reduced the viscosity of reservoir fluids after coming into contact with them, once a magnetic field was applied. The authors opined that this decrease in reservoir fluid viscosity can help to improve fluid flow [62]. Improved flow of the fluids within the reservoir can be expected to have a benefit on the oil recovery process of surfactant flooding, making ferro-fluid-enhanced surfactant flooding EOR more effective than conventional surfactant flooding.

Pramana *et al.* [63] explored the use of ferro-fluids for heating heavy oil. To successfully recover heavy oil, the viscosity must be reduced, and this is normally achieved through thermal methods. The authors investigated the potential of electromagnetic induction heating as a method of thermally reducing the viscosity of heavy oil in the oil reservoir. Ferro-fluids were used as part of this process because they have been found to heat up more rapidly under these circumstances than fluids that are not magnetic. According to the authors, this is due to a phenomenon known as hysteresis, which results in hysteresis heating. The ferro-fluid used in this experiment was found to heat at a faster rate than both graphite fluid and brine, making it preferential as a fluid for this type of thermal EOR method.

In another study, Melle *et al.* [64] also investigated the application of ferro-fluids for the stabilization of oil/water emulsions, also as demonstrated in Figure 1.7. The mechanism for emulsion stabilization involving ferro-fluids is essentially the same as the mechanism used by silica nanoparticles for the same purpose. The magnetic nanoparticles in ferro-fluids are able to adsorb at the interface between the oil and water phases, allowing a sufficient reduction in interfacial tension and the resulting stabilization of the oil/water emulsion. The benefit to using ferro-fluids as opposed to standard nano-fluids for stabilizing emulsions is the ability to manipulate the stability and motion of droplets within the emulsion using an external magnetic field. The authors observed that specific manipulation of the magnetic field caused the separation of the emulsion into its constituent phases. As with peptide surfactants that were mentioned previously, these ferro-fluids are able to destabilize and stabilize oil/water emulsions at will. Additionally, it was found that manipulating the magnetic field in a different way enabled transportation of the emulsion droplets through the continuous phase, while the emulsions remained stable. The movement of these droplets was controllable through further manipulation of the magnetic field. It was also observed that small changes in the magnetic field caused the emulsion droplets to increase their length to a small degree. These features of ferro-fluid-stabilized emulsions make them very attractive as EOR agents for emulsion flooding. These emulsions can be controlled in various ways using magnetic fields, making them much more versatile than typical oil/water emulsions used in conventional emulsion flooding.

1.7 Metallic Nanoparticles for Heavy Oil Recovery

Metallic nanoparticles have been the subject of interest for some investigations of EOR involving heavy oil. The United States Geological Survey defines heavy oil as a class of crude oil that is heavier than light crude oil, possessing an API gravity of no more than 22° and a viscosity of no less than 100 cP [65].

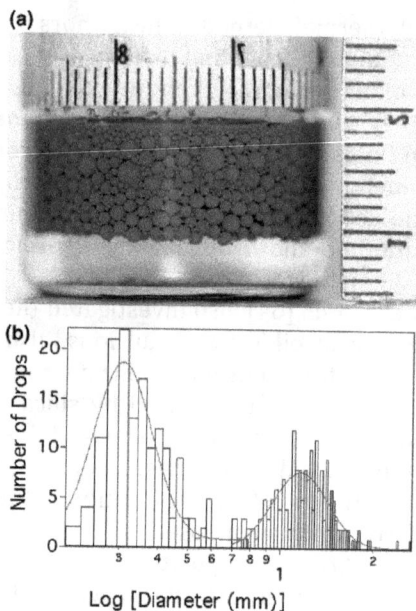

Figure 1.7 (a) Oil-in-water emulsion stabilized with carbonyl iron paramagnetic particles (4 wt%). and (b) histogram of the distribution of drops sizes corresponding to the emulsion in Figure 1.7a. Reproduced from Reference 64 with permission from American Chemical Society.

The high density and viscosity of heavy oil make its extraction and transportation much less economical, in comparison to light oil. Due to the challenging economics of heavy oil extraction, there is a lot of heavy oil that is not being exploited in oil reservoirs [65]. It is for this reason that the oil industry has been looking into novel nanotechnology solutions to this issue.

Presently, heavy oil extraction is performed mainly through thermal methods that involve heating the oil to reduce its viscosity. Shokrlu and Babadagli [66] identified the injection of air or steam as conventional thermal methods, and electrical or electromagnetic heating as unconventional thermal methods. The authors tested several different types of metal nanoparticles, and observed that they helped to enhance the property of heat transfer within the heavy oil. By improving the heat transfer, thermal methods of viscosity reduction can take place more efficiently, thereby making heavy oil recovery more economical. Furthermore, the authors observed that viscosity reduction of the heavy oil at various temperatures, which occurred through a mechanism known as aquathermolysis, was catalyzed by the same metallic nanoparticles that improved heat transfer. According to the authors, this catalytic effect of the nanoparticles can be expected to reduce the amount of thermal energy input required to extract heavy oil, thereby making the process even more economical.

Hashemi *et al.* [67] have also performed experiments to explore the catalytic capabilities of metallic nanoparticles. The authors observed that trimetallic nanoparticles, composed of tungsten, nickel, and molybdenum, were able to catalyze chemical reactions that upgraded heavy oil, reducing its overall viscosity. This concept was called an "underground refinery" approach. Using the nanoparticles in the form of a hot vacuum gas oil (VGO) nano-fluid, the heavy oil was broken into lighter components through catalytic hydrocracking. The resulting components had lower viscosities and densities than heavy oil, making them easier to extract. The benefits to oil recovery are summarized in Figures 1.8. Additionally, by hydrocracking heavy oil within the reservoir, less processing was required during oil refining, making the economics of heavy oil recovery even more favourable.

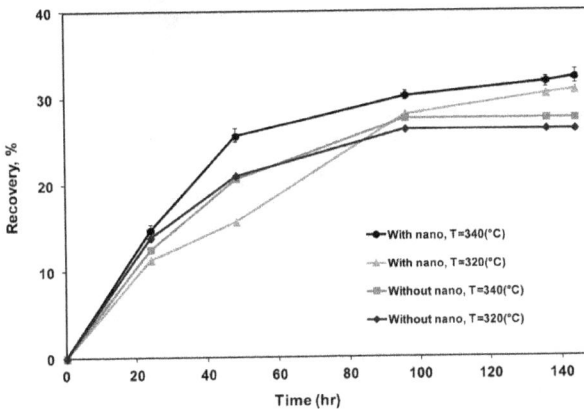

Figure 1.8 Increase in heavy oil recovery due to nano-fluid catalyzed hydrocracking. Reproduced from Reference 67 with permission from American Chemical Society.

In another study, Ehtesabi *et al.* [68] studied the recovery of heavy oil in sandstone cores using titanium dioxide nanoparticles. Oil recovery was increased to 80% with a very low nanoparticle content, as compared to 49% achieved by using only water for coreflooding. Rock surface was observed to change in wettability for oil-wet to water-wet after treatment with nanoparticles. As also seen in Figure 1.9, nanoparticles formed homogenous deposition on the surface of core plug and nanorods with average diameter of 60 nm were observed.

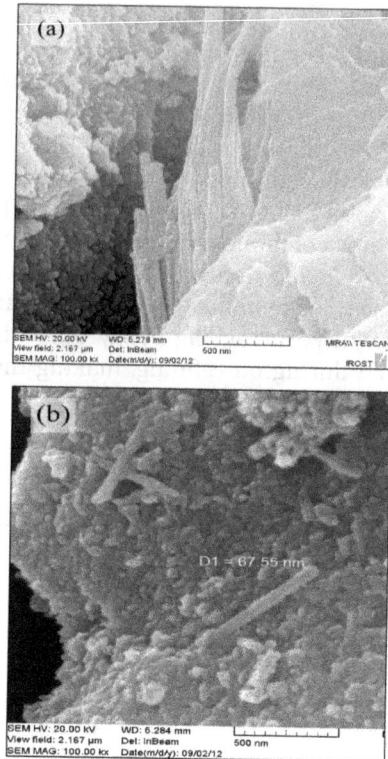

Figure 1.9 SEM images of the core plug after the flooding with a 1% concentration of TiO_2 nanoparticles. Reproduced from Reference 68 with permission from American Chemical Society.

1.8 Conclusions

Nanotechnology has been explored in recent years as a potential solution to a variety of problems involving oil recovery, particularly with respect to enhanced oil recovery. Nanotechnology involves manipulation of matter at the nanoscale level, generally dealing with particles 1-100 nm in size. This review has investigated a variety of applications of nanotechnology in the upstream petroleum industry, particularly in the area of enhanced oil recovery. The unique properties of nanoparticles as a result of their small size enable a wide range of potential applications within a number of different EOR techniques. For instance, carbon dioxide flooding has been made significantly more practical with the addition of nanoparticles, which can reduce mobility-related issues of CO_2 foams. Surfactant flooding can benefit from nanoscale manipulation of EOR surfactants as well. Nano-fluids containing various functional nanoparticles are versatile and can be used for a variety of EOR-related purposes. Nanotechnology also helps stabilise emulsions of oil and water for emulsion flooding. Superparamagnetic ferro-fluids have uses in EOR due to their behaviour in magnetic fields. Metallic nanoparticles have significant catalytic abilities, which is useful in heavy oil EOR. Thus, engineering at the nanoscale level can result in a variety of novel materials with enhanced properties that are optimized for various purposes. With nanotechnology, it is possible to tailor the properties of nanoparticles and nano-fluids to perform optimally under reservoir conditions, making them promising as a way of improving EOR methods. It is likely that nanotechnology will remain of interest to the oil and gas industry, as new applications of nanomaterials for both the upstream and downstream petroleum industry are continually revealed.

References

1. Bhushan, B. (2007) *Springer Handbook of Nanotechnology*, 2nd edition, Springer, New York.
2. Morrow, N. R., and Heller, J. P. (1985) Fundamentals of enhanced recovery. In: *Enhanced Oil Recovery, I, Fundamentals and Analyses*, Donaldson, E. C., Chilingarian, G. V., and Yen, T. F. (eds.), Elsevier Science Publishers B. V., The Netherlands, pp. 47-74.
3. Kamath, K., Zammerilli, A. M., Comberiati, J. R., Taylor, B. D., and Slagle, F. D. (1982) Mechanism of Tertiary Oil Recovery by Aqueous Chemical Flooding. *Society of Petroleum Engineers Oilfield and Geothermal Chemistry Symposium*, USA. Online: https://www.onepetro.org/conference-paper/SPE-10599-MS (assessed 15th March 2017).
4. Lang, W., and Rehage, G. (1980) Recent Results on the Use of Polymers in Tertiary Oil Recovery in Brines of High Salinity. *Society of Petroleum Engineers Oilfield and Geothermal Chemistry Symposium*, USA. Online: https://www.onepetro.org/conference-paper/SPE-8983-MS (assessed 16th March 2017).

5. Maudgalya, S., Knapp, R. M., McInerney, M. J., Folmsbee, M., and Nagle, D. P. (2005) Tertiary Oil Recovery with Microbial Biosurfactant Treatment of Low-Permeability Berea Sandstone Cores. *Society of Petroleum Engineers Production Operations Symposium*, USA. Online: https://www.onepetro.org/conference-paper/SPE-94213-MS (assessed 21st March 2017).

6. Wang, G., and Locke, D. (1980) A laboratory study of the effects of CO_2 injection sequence on tertiary oil recovery. *SPE Journal*, **20**(4), 278-280.

7. Rao, D., and Hughes, R. (2011) Current research and challenges pertaining to CO_2 flooding and sequestration. *The Way Ahead*, **7**(2), 17-19.

8. Holm, L. W. (1982) CO_2 flooding: Its time has come. *Journal of Petroleum Technology*, **34**(12), 2739-2745.

9. Martin, F. D., and Taber, J. J. (1992) Carbon dioxide flooding. *Journal of Petroleum Technology*, **44**(4), 396-400.

10. Farajzadeh, R., Andrianov, A., and Zitha, P. L. (2010) Investigation of immiscible and miscible foam for enhancing oil recovery. *Industrial & Engineering Chemistry Research*, 49(4), 1910-1919.

11. Aronofsky, J. S., and Ramey, H.,J., Jr. (1956) Mobility ratio - Its influence on injection or production histories in five-spot water flood. *Journal of Petroleum Technology*, **8**(9), 205-210.

12. Heller, J. P. (1984) Reservoir Application of Mobility Control Foams in CO_2 Floods. *Society of Petroleum Engineers Enhanced Oil Recovery Symposium*, USA. Online: https://www.onepetro.org/conference-paper/SPE-12644-MS (assessed 10th March 2017).

13. Kulkarni, M., and Rao, D. (2005) Experimental investigation of miscible and immiscible Water-Alternating-Gas (WAG) process performance. *Journal of Petroleum Science and Engineering*, **48**(1), 1-20.

14. Bernard, G. G., Holm, L. W., and Harvey, C. P. (1980) Use of surfactant to reduce CO_2 mobility in oil displacement. *Society of Petroleum Engineers Journal*, **20**(4), 281-292.

15. Khalil, F., and Asghari, K. (2006) Application of CO_2-foam as a means of reducing carbon dioxide mobility. *Journal of Canadian Petroleum Technology*, **45**(5), 37-42.

16. Lee, H. O., and Heller, J.,P. (1990) Laboratory measurements of CO_2-foam mobility. *SPE Reservoir Engineering*, **5**(2), 193-197.

17. Worthen, A., Bagaria, H., Chen, Y., Bryant, S. L., Huh, C., and Johnston, K. P. (2012) Nanoparticle Stabilized Carbon Dioxide in Water Foams for Enhanced Oil Recovery. *Society of Petroleum Engineers Improved Oil Recovery Symposium*, USA. Online: https://www.onepetro.org/conference-paper/SPE-154285-MS (assessed 21st March 2017).

18. Espinosa, D. A., Johnston, K. P., Bryant, S. L., and Huh, C. (2010), 'Nanoparticle-Stabilized Supercritical CO_2 Foams for Potential Mobility Control Applications' in *Society of Petroleum Engineers Improved Oil Recovery Symposium*. Tulsa, Oklahoma. 24-28 April 2010. Texas: Society of Petroleum Engineers.

19. National Energy Technology Laboratory (2013) *Nanoparticle-stabilized CO_2 Foam for CO_2-EOR Application*. Online: http://www.netl.doe.gov/technologies/oil-gas/Petroleum/projects/EP/ImprovedRec/FE0005979-NMIMT.html (Accessed 1st August 2015).

20. National Energy Technology Laboratory (2013) *Use of Engineered Nanoparticle-stabilized CO2 Foams to Improve Volumetric Sweep of CO2 EOR Processes*. Online: http://netldev.netl.doe.gov/research/oil-and-gas/project-summaries/enhanced-oil-recovery/fe0005917-uta (Accessed 1st August 2015).
21. Adams, W. T., and Schievelbein, V. H. (1987) Surfactant flooding carbonate reservoirs. *SPE Reservoir Engineering*, **2**(4), 619-626.
22. Hirasaki, G. J., Miller, C. A., and Puerto, M. (2008), Recent Advances in Surfactant EOR. *Society of Petroleum Engineers Annual Technical Conference and Exhibition*, USA. Online: https://www.onepetro.org/journal-paper/SPE-115386-PA (assessed 27th March 2017).
23. Mattax, C. C., Blackwell, R. J., and Tomich, J. F. (1983) Recent Advances in Surfactant Flooding. *11th World Petroleum Conference*, UK. Online: https://www.onepetro.org/conference-paper/WPC-20220 (assessed 13th March 2017).
24. Ziegler, V. M. (1988) Laboratory investigation of high temperature surfactant flooding. *SPE Reservoir Engineering*, **3**(2), 586-596.
25. Uren, L. C., and Fahmy, E. H. (1927) Factors influencing the recovery of petroleum from unconsolidated sands by waterflooding. *Transactions of the AIME*, **77**(1), 318-335.
26. Levitt, D., Jackson, A., Heinson, C., Britton, L. N., Malik, T., Dwarakanath, V., and Pope, G. A. (2006) Identification and Evaluation of High-Performance EOR Surfactants. *Society of Petroleum Engineers/Department of Energy Symposium on Improved Oil Recovery*, USA. Online: https://www.onepetro.org/journal-paper/SPE-100089-PA (assessed 10th March 2017).
27. Kong, X., and Ohadi, M. (2010) Applications of Micro and Nano Technologies in the Oil and Gas Industry – Overview of the Recent Progress. *Abu Dhabi International Petroleum Exhibition and Conference*, UAE. Online: https://www.onepetro.org/conference-paper/SPE-138241-MS (assessed 19th March 2017).
28. Amanullah, M., and Al-Tahini, A. (2009) Nano-Technology - Its Significance in Smart Fluid Development for Oil and Gas Field Application. *Society of Petroleum Engineers Saudi Arabia Section Technical Symposium*, Saudi Arabia. Online: https://www.onepetro.org/conference-paper/SPE-126102-MS (assessed 10th March 2017).
29. Dexter, A., and Middelberg, A. (2008) Peptides as functional surfactants. *Industrial & Engineering Chemistry Research*, **47**(17), 6391-6398.
30. Hendraningrat, L., Li, S., and Torsæter, O. (2013) A Coreflood Investigation of Nanofluid Enhanced Oil Recovery in Low-Medium Permeability Berea Sandstone. *Society of Petroleum Engineers International Symposium on Oilfield Chemistry*, USA. Online: https://www.onepetro.org/conference-paper/SPE-164106-MS (assessed 9th March 2017).
31. Murshed, S., Leong, K., and Yang, C. (2008) Thermophysical and electrokinetic properties of nanofluids – A critical review. *Applied Thermal Engineering*, **28**(17), 2109-2125.

32. Abdallah, W., Buckley, J. S., Carnegie, A., Edwards, J., Herold, B., Fordham, E., Graue, A., Habashy, T., Seleznev, N., Signer, C., Hussain, H., Montaron, B., and Ziauddin, M. (2007) Fundamentals of wettability. *Oilfield Review*, 19(2), 44-61.
33. Morrow, N. R. (1990) Wettability and its effect on oil recovery. *Journal of Petroleum Technology*, **42**(12), 1476-1484.
34. Aziz, A.K. (2011) Impact of Wettability Alteration on Recovery Factor. *SPE/DGS Saudi Arabia Section Technical Symposium and Exhibition*, Saudi Arabia. Online: https://www.onepetro.org/conference-paper/SPE-149044-MS (assessed 21st March 2017).
35. Chen, P., and Mohanty, K. K. (2014) Wettability Alteration in High Temperature Carbonate Reservoirs. *Society of Petroleum Engineers Improved Oil Recovery Symposium*, USA. Online: https://www.onepetro.org/conference-paper/SPE-169125-MS (assessed 29th March 2017).
36. Gupta, R., and Mohanty, K. K. (2008) Wettability Alteration of Fractured Carbonate Reservoirs. *Society of Petroleum Engineers/Department of Energy Improved Oil Recovery Symposium*, USA. Online: https://www.onepetro.org/conference-paper/SPE-113407-MS (assessed 21st March 2017).
37. Nasralla, R. A., Bataweel, M. A., and Nasr-El-Din, H.A. (2011) Investigation of Wettability Alteration by Low Salinity Water in Sandstone Rock. *Society of Petroleum Engineers Offshore Europe Oil and Gas Conference and Exhibition*, UK. Online: https://www.onepetro.org/conference-paper/SPE-146322-MS (assessed 11th March 2017).
38. Tang, G., and Kovscek, A. R. (2002) Wettability Alteration of Diatomite Induced by Hot-Fluid Injection. *Society of Petroleum Engineers Annual Technical Conference and Exhibition*, USA. Online: https://www.onepetro.org/conference-paper/SPE-77461-MS (assessed 19th March 2017).
39. Li, S., Hendraningrat, L., and Torsaeter, O. (2013) Improved Oil Recovery by Hydrophilic Silica Nanoparticles Suspension: 2-Phase Flow Experimental Studies. *International Petroleum Technology Conference*, China. Online: https://www.onepetro.org/conference-paper/IPTC-16707-MS (assessed 19th February 2017)
40. Ju, B., Fan, T., and Ma, M. (2006) Enhanced oil recovery by flooding with hydrophilic nanoparticles. *China Particuology*, **4**(1), 41-46.
41. Amanullah, M. (2003) A Novel Method of Assessment of Spurt and Filtrate Related Formation Damage Potential of Drilling and Drilling-in Fluids. *Society of Petroleum Engineers Asia Pacific Oil and Gas Conference and Exhibition*, Indonesia. Online: https://www.onepetro.org/conference-paper/SPE-80484-MS (assessed 20th March 2017).
42. Donaldson, E. C., and Chernoglazov, V. (1987) Characterization of drilling mud fluid invasion. *Journal of Petroleum Science and Engineering*, **1**(1), 3-13.
43. Elkatatny, S., Mahmoud, M., and Nasr-El-Din, H. (2012) Characterization of filter cake generated by water-based drilling fluids Using CT scan. *SPE Drilling & Completion*, 27(2), 282-293.
44. Shahrabadi, A., Bagherzadeh, H., Roustaei, A., and Golghanddashti, H. (2012) Experimental Investigation of HLP Nanofluid Potential to Enhance Oil Recovery: A Mechanistic Approach. *Society of Petroleum Engineers International Oilfield*

Nanotechnology Conference, The Netherlands. Online: https://www.onepetro.org/conference-paper/SPE-156642-MS (assessed 19th March 2017).

45. Onyekonwu, M., and Ogolo, N. (2010) Investigating the Use of Nanoparticles in Enhancing Oil Recovery. *34th Annual Society of Petroleum Engineers International Conference and Exhibition*, Nigeria. Online: https://www.onepetro.org/conference-paper/SPE-140744-MS (assessed 29th March 2017).

46. Wasan, D., and Nikolov, A. (2003) Spreading of nanofluids on solids. *Nature*, 423(6936), 156-159.

47. Kondiparty, K., Nikolov, A., Wu, S., and Wasan D. (2011) Wetting and spreading of nanofluids on solid surfaces driven by the structural disjoining pressure: Statics analysis and experiments. *Langmuir*, 27(7), 3324-3335.

48. McElfresh, P., Holcomb, D., and Ector, D. (2012), Application of Nanofluid Technology to Improve Recovery in Oil and Gas Wells. *Society of Petroleum Engineers International Oilfield Nanotechnology Conference*, The Netherlands. Online: https://www.onepetro.org/conference-paper/SPE-154827-MS (assessed 19th February 2017).

49. Karimi, A., Fakhroueian, Z., Bahramian, A., Khiabani, N. P., Darabad, J. B., Azin, R., and Arya, S. (2012) Wettability alteration in carbonates using zirconium oxide nanofluids: EOR implications. *Energy & Fuels*, **26**(2), 1028-1036.

50. Giraldo, J., Benjumea, P., Lopera, S., Cortes, F. B., and Ruiz, M. A. (2013) Wettability alteration of sandstone cores by alumina-based nanofluids. *Energy & Fuels*, **27**(7), 3659-3665.

51. Mandal, A., Samanta, A., Bera, A., and Ojha, K. (2010) Characterization of oil-water emulsion and its use in enhanced oil recovery. *Industrial & Engineering Chemistry Research*, 49(24), 12756-12761.

52. Thomas, S., and Farouq Ali, S.M. (1989) Flow of emulsions in porous media, and potential for enhanced oil recovery. *Journal of Petroleum Science and Engineering*, **3**(1-2), 121-136.

53. Pal, R. (1998) A novel method to correlate emulsion viscosity data. *Colloids and Surfaces A: Physicochemical and Engineering Aspects*, **137**(1-3), 275-286.

54. Chen, G., and Tao, D. (2005) An experimental study of stability of oil-water emulsion. *Fuel Processing Technology*, **86**(5), 499-508.

55. Zhang, T., Roberts, M., Bryant, S. L., and Huh, C. (2009) Foams and Emulsions Stabilized with Nanoparticles for Potential Conformance Control Applications. *Society of Petroleum Engineers International Symposium on Oilfield Chemistry*, USA. Online: https://www.onepetro.org/conference-paper/SPE-121744-MS (assessed 19th March 2017).

56. Zhang, T., Davidson, D., Bryant, S. L., and Huh, C. (2010) Nanoparticle-Stabilized Emulsions for Applications in Enhanced Oil Recovery. *Society of Petroleum Engineers Improved Oil Recovery Symposium*, USA. Online: https://www.onepetro.org/conference-paper/SPE-129885-MS (assessed 21st March 2017).

57. Morreale, G., Lee, E. G., Jones, D. B., and Middelberg, A. P. (2004) Bioprocess-centred molecular design (BMD) for the efficient production of an interfacially active peptide. *Biotechnology and Bioengineering*, **87**(7), 912-923.

58. Malcolm, A., Dexter, A., and Middelberg, A. (2007) Peptide surfactants (Pepfactants) for switchable foams and emulsions. *Asia-Pacific Journal of Chemical Engineering*, 2(5), 362-367.
59. Mandal, A., Bera, A., Ojha, K., and Kumar, T. (2012) Characterization of Surfactant Stabilized Nanoemulsion and Its Use in Enhanced Oil Recovery. *Society of Petroleum Engineers International Oilfield Nanotechnology Conference*, The Netherlands. Online: https://www.onepetro.org/conference-paper/SPE-155406-MS (assessed 28th March 2017).
60. Robins, M. (2000) Emulsions - creaming phenomena. *Current Opinion in Colloids and Interface Science*, 5(5-6), 265-272.
61. Odenbach, S. (2003) Ferrofluids - magnetically controlled suspensions. *Colloids and Surfaces A: Physicochemical and Engineering Aspects*, 217(1-3), 171-178.
62. Kothari, N., Raina, B., Chandak, K. B., Iyer, V., and Mahajan, H. P. (2010) Application of Ferrofluid for Enhanced Surfactant Flooding in EOR. *Society of Petroleum Engineers EUROPEC/EAGE Annual Conference and Exhibition*, Spain. Online: https://www.onepetro.org/conference-paper/SPE-131272-MS (assessed 19th March 2017).
63. Pramana, A. A., Abdassah, D., Rachmat, S., and Mikrajuddin, A. (2010) Electromagnetic induction heat generation of nano-ferrofluid and other stimulants for heavy oil recovery. *AIP Conference Proceedings*, 1284(1), 163-166.
64. Melle, S., Lask, M., and Fuller, G. (2005) Pickering emulsions with controllable stability. *Langmuir*, 21(6), 2158-2162.
65. United States Geological Survey (2013). *Heavy Oil and Natural Bitumen - Strategic Petroleum Resources*. Online: http://pubs.usgs.gov/fs/fs070-03/fs070-03.html. (Accessed 21st Aug 2016).
66. Shokrlu, Y., and Babadagli, T. (2010) Effects of Nano-Sized Metals on Viscosity Reduction of Heavy Oil/Bitumen During Thermal Applications. *Canadian Unconventional Resources & International Petroleum Conference*, Canada. Online: https://www.onepetro.org/conference-paper/SPE-137540-MS (assessed 19th March 2017).
67. Hashemi, R., Nassar, N., and Almao, P. (2013) Enhanced heavy oil recovery by in-situ prepared ultradispersed multimetallic nanoparticles: A study of hot fluid flooding for athabasca bitumen recovery. *Energy & Fuels*, 27(4), 2194-2201.
68. Ehtesabi, H., Ahadian, M. M., Taghikhani, V., and Ghazanfari, M. H. (2014) Enhanced heavy oil recovery in sandstone cores using TiO_2 nanofluids. *Energy & Fuels*, 28, 423-430.

MOFs and Metal Oxides for Gas Adsorption and Environmental Applications

2.1 Introduction

Acid gases like carbon dioxide (CO_2) and hydrogen sulfide (H_2S) represent two most abundant harmful gases in the natural gas streams which need to be removed for achieving usefulness of natural gas. Similarly, environmental pollution due to the release of harmful gases from different industrial and household processes leads to severe degradation of the quality of air and significant efforts are required to minimize pollution level as well as to remove the pollutants from the environment. Some of the harmful materials that are present in our environment are H_2S, CO_x, SO_x, NO_x, nitrogen containing compounds (NCCs), sulfur containing compounds (SCCs), volatile organic compounds (VOCs), dyes, pharmaceuticals and personal care products (PPCPs).

2.1.1 Classification of Hazardous Materials

Depending on the sources, the most plentiful hazardous materials can be categorized into two classes, namely naturally originating hazardous materials and anthropogenic. A substantial quantity of naturally occurring materials are available underground as well as in water, soil, air and minerals. However, the anthropogenic materials generate from chemical reactions, combustion or from the effluent of toxic materials. The global energy need is satisfied by the naturally originating materials such as coal, crude oil and natural gas. As mentioned earlier, natural gas contains harmful gases like CO_2 and H_2S among others, which require removal. In addition, these energy sources when burnt generate high amounts of harmful gases to the atmosphere. The major toxic gases causing environmental air pollution are CO_2, H_2S, SO_x, NH_3, NO_x, VOCs and other hydrocarbons [1]. The gases like SO_2, CH_4, O_3 and N_2O are considered as greenhouse gases and the ejection of these gases enhances the ozone levels present in the troposphere. Vehicles related pollutants such as CH_4, CO, NO_2, SO_2 as well as carbon black also cause global warming. Currently, the carbon balance of the world is considered as one of the critical environmental

Haleema Saleem and Vikas Mittal*, The Petroleum Institute (part of Khalifa University of Science and Technology), Abu Dhabi, UAE
*Current address: Bletchington, Wellington County, Australia

problem, and hence the reduction of anthropogenic CO_2 release has now become a very significant issue. VOCs are chemicals having high vapor pressure, and are usually discharged from adhesives, resins, paints, solvents, etc. [2]. Benzene, toluene, xylene and phenolics are some of the common hazardous VOCs. Nitrogen or sulfur containing organic compounds are naturally originating species, and are available in fossil fuels as well as oils like jet fuel, gasoline, crude oil, heating oil and diesel. The combustion of fossil fuels is a large source of toxic releases, which causes global warming, greenhouse effect, dangerous impact on living species and hazardous air pollution [3-5].

Further, due to the global industrial advancement, the severity of water pollution has become critical, thus, posing significant challenges to the health of living beings [6]. Dissolved heavy metal salts or organic pollutants are often present in the polluted water [7,8]. The organic pollutants existing in the natural water resources cause a great damage to the aquatic habitats and also to the human health. Some of the toxic pollutants existing in the waste water are benzene, phenols, nitro-benzenes, chlorinated benzenes, phthalates and chlorinated phenols. Dye materials are commonly used in plastics, textiles, paper and leather industries, and a large amount of these materials ends up in water as harmful pollutant [9,10]. Separation of these dyes from waste water is very crucial, but challenging. PPCPs are chemical contaminants that are present in water, and are also unsafe for living organisms and have to be removed [11]. The disposal of heavy metal ions like chromium (Cr), lead (Pb), copper (Cu), arsenic (As), mercury (Hg), antimony (Sb), cadmium (Cd) and manganese (Mn) in processed water is also highly toxic to human beings as well as ecological systems [12,13].

2.1.2 Adsorption of Hazardous Materials

In the last few decades, the porous materials have evolved as good adsorbents that exhibit significant potential for the purification of air, gas and water streams by adsorptive separation of the contaminants, and hence a great deal of research focus has been devoted for the examination and analysis of the advanced porous materials [14]. For the decontamination, adsorption has been regarded as the preferable technique due to its low cost, no or harmless secondary products, design simplicity, ease of operation and efficient adsorbent regeneration. This method is based on the porous adsorbent ability to selectively adsorb specific materials from either refinery streams or from the atmosphere. The adsorbent materials can be structurally tuned to generate suitable pore shape and size, thus, providing efficient contact with the adsorbate materials, thus, leading to effective separation of pollutants during the adsorption process. Depending on the interaction between the porous sorbents and the adsorbates, the adsorbents can be classified as physical and chemical adsorbents [15]. With respect to physical adsorbents, the adsorbates are trapped inside the adsorbent pores through weak van der Waals forces. Hence, the adsorbent material can be easily regenerated by physical treatments or simple solvent exchange. However, in the chemical adsorbents, absorption takes place by the chemical bond formation between the adsorbent and the adsorbate. Therefore, chemical treatments are required for the regeneration of the spent adsorbents, which may also occasionally lead to permanent loss of adsorbent performance. The capability of the adsorption process depends on the selectivity for specific compounds, regenerability of adsorbents, durability and adsorption ability of the adsorbents. For the adsorptive removal of harmful compounds, different porous adsorbents like zeolites [16,17], activated carbons [18,19] , mesoporous materials [20,21] and metal organic frameworks (MOFs) [22,23] have been studied. For the effective adsorption, specific pore geometry, porosity and adsorption sites are needed [24]. In addition to this, certain active species such as different functional groups (basic or acidic), metal oxides, metal ions, polyoxometalates and metal salts are generally included into the adsorbents for the selective removal of harmful compounds by interactions such as hydrogen bonding, acid-base and π- π interactions.

In this chapter, the main focus is to elucidate the role of MOFs and metal oxides for the adsorptive removal of hazardous gases like H_2S, NH_3, CO_2, and other hydrocarbons present in the atmosphere as well as the dissolved heavy metal salts or the organic pollutants present in the polluted water. Beneficial characteristics of MOFs and metal oxides have been discussed that ensure their effective role in the adsorption of specific toxic compounds. The effect of linkers, open metal sites, MOF modification/functionalization, central metal ions and the adsorbent/adsorbate interactions are also explained. Further, the analysis of adsorption mechanisms such as electrostatic interactions, hydrogen bonding and acid-base interactions for the effective adsorption is also presented. Thus, this review presents an up-to-date study of theoretical predictions, experimental results as well as developing concepts on the adsorption ability of different MOFs and metal oxides, with specific emphasis on the removal of harmful gases like CO_2, H_2S, NH_3, and water contaminants like the heavy metals and organic contaminants, among other contaminants. In addition to this, observations in terms of benefits and disadvantages of employing different metal oxide frameworks as well as metal oxides have also been pointed out, along with discussing in detail the thermodynamic and kinetic analysis.

2.2 Metal Organic Frameworks

MOFs consists of two major components, i.e., an organic molecule (linker) and a metal ion or a group of metal ions. The linkers are generally di-, tri- or tetradendate ligands [25]. Examples of some MOFs are Cu-BTC, MOF-5 and CPO-27. The porous MOFs are very attractive because of their versatile applications like CO_2 storage/adsorption [26], vapor adsorption [27], hydrogen storage[28], chemical separation [29], drug delivery [30] and so on. These materials are preferred than other porous materials due to their pore functionality, high and tunable porosity, pore compositions and open metal sites. For the adsorption related applications, MOFs are promising materials due to their controlled tuning of the pore surface, which causes the selective adsorption of certain guest molecules which have specific functional groups. The properties of the MOFs can be tuned for obtaining the desired performance, by orderly and intentionally tuning the functionalities and structures. Hence, many research studies have been carried out on the adsorption of different liquid and gaseous components using MOF, due to their beneficial features of pore geometry and porosity [31]. Depending on the adsorbates, both modified as well as virgin MOFs can be used for the adsorptive removal of different toxic gases and liquids.

2.2.1 Adsorptive Removal of Harmful Gases

Using different MOFs, adsorption of various harmful gases/vapors like H_2S [32,33], NH_3 [4], CO_2 [34,35], SO_2 [36], benzene [37], NO [38] have been studied. Most of the aforementioned gases/vapors are either required to be removed from products or are ejected from different industries as waste by-products. For a clean and safe environment, efficient removal of these harmful gases is very important. As mentioned earlier, MOFs have advantages like high porosity as well as easy tunability of pore shape and size, from mesoporous to microporous scale, by altering the nature of organic molecule and connectivity of inorganic component [39]. The characteristic examples include zirconium based MOFs namely UiO-68(Zr- terphenyldicarboxylate, Zr-TPDC), UiO-67 (Zr-BPDC, Zr- biphenyldicarboxylate) and UiO-66 (Zr- benzenedicarboxylate, Zr-BDC) having high surface areas 4170, 3000 and 1187 m^2/g respectively [40]. The large enhancement in the surface area is obtained by varying the organic linker. The following sections describe the specific use of MOFs for the removal of various harmful gases.

CO_2 Adsorption

The adsorption as well as desorption of CO_2 in MOFs are attributed to the variations happening in the framework structures, and have also been described using the "gate effect mechanism" or a "breathing type mechanism". In a further study, Walton *et al.* [41] added through the comparison of Monte Carlo simulation and experimental adsorption of CO_2 for MOF IRMOF-1 that attractive electrostatic interactions between the CO_2 molecules resulted in the unusual shape of the adsorption isotherms (Figure 2.1).

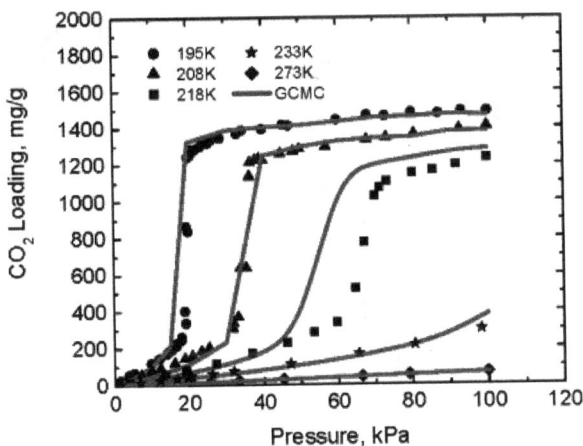

Figure 2.1 Comparison of Monte Carlo simulations (GCMC) and experimental adsorption isotherms for CO_2 in IRMOF-1. Reproduced from Reference 41 with permission from American Chemical Society.

For the application in flue streams, MOF adsorbents for the CO_2 capture should meet the requirements such as high CO_2 adsorption ability, corrosion resistance, high thermal stability and good selectivity for CO_2 over several other components present in flue gas. It should also be noted that most of these adsorbents exhibit little adsorption in the low-pressure conditions i.e. 0.1–0.2 bars, which is the realistic pressure range employed for the CO_2 adsorption from flue gas stream [42]. Further, a reduction in their adsorption capacity is seen when it exposed to gaseous mixtures under dynamic conditions.

Hence, significant efforts are required for the enhancement in the CO_2 adsorption capacity under equilibrium conditions. In this study, the modifications in MOFs was achieved in three aspects namely (1) metal ions, (2) organic linkers and (3) novel consolidation of both. As shown in Figure 2.2, the functionalized MOF was subjected to cyclic gas treatments. A small 0.75 wt% change in the weight of the sample was observed over repeated cycles, which confirmed that the material was able to withstand cyclic exposure to the mixed gas stream. The adsorption performance was maintained over the cycles and the material could be regenerated effectively.

For the metal ions, generating open metal sites is an effective method, as the vacant coordination sites at metal ions act as primary sites for the guest molecules [43]. Britt et al. [43] stated that MOF loaded with open Mg sites, Mg-MOF-74 ($Mg2(2,5$-dioxidoterephthalate)) had high selectivity, easy regeneration and excellent performance for the CO_2 removal. Mg-MOF-74 was exposed to a gas stream consisting of 20% CO_2 in CH_4, and the adsorbent was

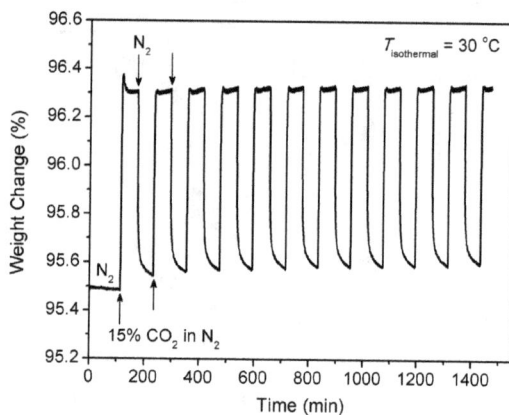

Figure 2.2 Gas cycling experiment for the functionalized MOF at 30°C, with 15% CO_2 in N_2 followed by a flow of pure N_2. Reproduced from Reference 42 with permission from American Chemical Society.

observed to capture CO_2 alone, without CH_4. The adsorbent retained 89 g of CO_2/kg of material prior to breakthrough (2.23 mmol.g^{-1}). Dietzel et al. [44] studied the outcome of the metal center on the CO_2 adsorption ability and selectivity using a series of iso-structural MOFs namely M-CPO-27s (M=Ni, Mn, Co, Mg, Zn). It was observed that at high pressure (50 bar) and 298 K, the CO_2 adsorption was 63 wt% for MOF CPO-27(Mg) and 51 wt% for the MOF CPO-

27(Ni). In another research by Caskey *et al.* [45], it was observed that at 1 atm, the CPO-27(Co) exhibited CO_2 adsorption of 30.6 wt%, whereas CPO-27(Mg) showed an adsorption of 35.2 wt%. The CO_2 adsorption ability of CPO-27(Mg) was found to be superior as compared to the other members of the series, thereby representing a major development in the CO_2 adsorption ability of the MOFs. In addition to this, for enhancing the performance, the unsaturated metal sites could be functionalized using groups like amine. Couck *et al.* [46] proved the efficient separation of CH_4 and CO_2 gases using amino functionalized MIL-53(Al). Due to the presence of quadrupole moment in CO_2, these CO_2 molecules have a greater affinity for the NH_2 groups. Breakthrough experiments confirmed that the CH_4, which was weakly adsorbed, was able to transmit through the MOF packed column during the adsorption of CO_2. Thus, at ambient conditions, effective separation of CO_2 and CH_4 could be achieved using the aforementioned MOF adsorbent.

For the organic linkers modification, Bae *et al.* [47] analyzed the feasibility of utilizing MOFs which are carborane based or coordinated with ligand mixture. A higher selectivity for CO_2 over CH_4 was noticed in these MOF materials. Carboranes have several beneficial material properties like thermal stability, rigidity and chemical stability. In the pursuit of preparing novel MOFs having high CO_2 adsorption capacity as well as stability, the capability to generate MOFs having different organic linkers as well as metal joints is considered to contribute significant flexibility, along with particular physical behavior and chemical functionalities. Recently, Yaghi *et al.* [48] prepared a new class of MOFs called zeolitic imidazole frameworks (ZIFs), where metal atoms like Zn were bonded with nitrogen atoms by ditopic imidazolate or functionalized Im links to generate the neutral frameworks. The materials had good chemical as well as thermal stabilities (until 500 °C). In another study by Phan *et al.* [49], it was stated that at 0 °C, 1 L of ZIF-69 had the ability to store 82.6 L of CO_2. Further, ZIFs also exhibited higher selectivity than other MOF types for the adsorption of CO_2 from flue gases. It was found that only CO_2 could slip into the ZIF cages, whereas other gas molecules could just pass through them without any obstruction.

Hu *et al.* [50,51] studied the effect of addition of alkaline metal cations like Na^+, Li^+ and K^+ into a previously studied Zr-MOF having excellent CO_2 adsorption properties, by facile neutralization reactions. The CO_2 adsorption capacity of the MOFs was analyzed employing breakthrough experiments and sorption isotherms. From the binary mixed gas breakthrough experiment, it was seen that out of the different Zr-MOFs, the MOF UiO-66 (Zr)-$(COONa)_2$ exhibited high CO_2/N_2 selectivity value of 99.6. This was considered to be highest among all the previously obtained values under analogous testing conditions. The MOFs were reported to have good CO_2 adsorption capacity, high water stability and excellent CO_2/N_2 selectivity. Further, their preparation could be scaled up very efficiently by utilizing commercially accessible reagents with batch reactors as well as environmental friendly solvent like water. The results obtained in the study confirmed that the synthesized MOFs are very efficient adsorbents for the removal of CO_2 from flue gas. Yang *et al.* [52] observed enhanced CO_2 adsorption with adsorbents attained by sonication or microwave heating. At 298 K, the small sized and homogenous particles of CPO-27(Mg), generated by sonication method, adsorbed almost 350 mg.g^{-1} of CO_2. In addition to this, the materials also exhibited large isosteric heat of adsorption. Table 2.1 also demonstrates the CO_2 adsorption capacities of some of the MOFs.

Table 2.1 Comparison of CO_2 adsorption capacities of some of the MOF adsorbents

Sl. No.	Adsorbent	Temperature	Pressure	CO_2 adsorption capacity (wt%)	Reference
1	CPO-27(Mg)	298 K	50 bar	63 wt%	131
2	CPO-27(Ni)	298 K	50 bar	51 wt%	131
3	CPO-27(Co)	Ambient temperature	1 atm	30.6 wt%	132
4	CPO-27(Mg)	Ambient temperature	1 atm	35.2 wt%	132

Adsorption of Sulfur Containing Compounds

The use of fossil resources, like petroleum, natural gas and coal, generates sulfur containing species like SO_2, H_2S, mercaptan, thiophene and thioether. These sulfur containing compounds are harmful to human health and the environment and hence the desulphurization processes are needed. As mentioned earlier, MOFs are also promising absorbents for the removal of sulfur containing comounds. In a study by Achmann *et al.* [53], it was noted that the adsorbent MOF-199 possessed high efficiency for the removal of sulfur from low sulfur model fuels and oil. The MOF also exhibited the maximal sorption rates as well as rapid sorption characteristics. The minor pores in the MOF-199 provided better stabilization for Cu- adsorbed sulfur species by the van der Waals force, and thereby caused an excellent sulfur adsorption performance. In another study, Chavan *et al.* [54] analyzed the H_2S adsorption performance on the adsorbent Ni-MOF-74. The adsorbent exhibited maximal uptake and higher H_2S adsorption enthalpy, when compared to other adsorbents, while preserving its original porosity and structure (Figure 2.3).

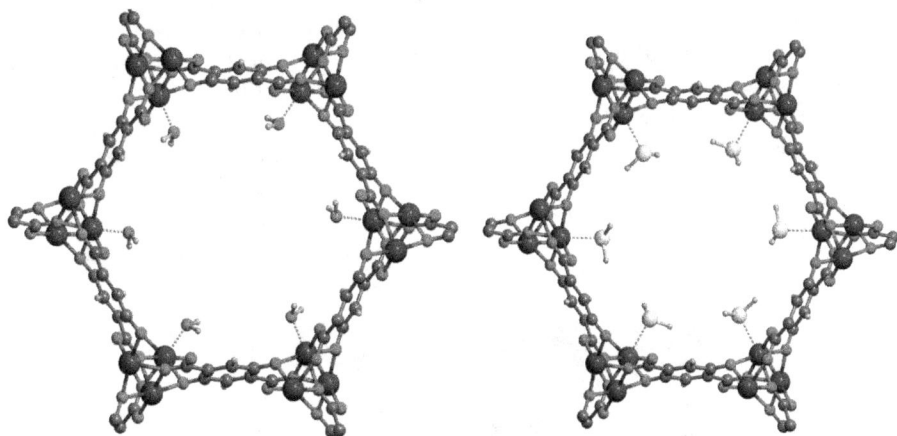

Figure 2.3 H_2O (left) and H_2S (right) molecules as arranged in the CPO-27-Ni channels. Reproduced from Reference 54 with permission from American Chemical Society.

In a study reported by de Voorde *et al.* [55], DFT analysis on the adsorption behavior of S-heterocyclic compounds on MOF-74 series (Mg, Cu, Ni, Zn and Co) was carried out. It was confirmed that the metal ion had a remarkable effect on the MOFs adsorption ability and S-heterocyclic species affinity. It was suggested that the adsorbent Ni-MOF-74 offered the best desulfurization behavior. Hamon *et al.* [56] examined the efficiency of MIL-series e. g. MIL-100(Cr), MIL-53(Cr, Fe and Al), MIL-101(Cr) and MIL-47 for the adsorption of H_2S at 303 K. The study suggested that the –OH groups present in the MIL-53 adsorbent promoted the H_2S adsorption. In a different study by Hamon *et al.*, [57] it was reported that the –OH groups existing in the MOFs performed as proton donor, thereby increasing the H_2S adsorption. The authors observed that on H_2S adsorption, the MIL-47(V) structure remained rigid up to a pressure of 1.8 MPa. On the other hand, the MIL-53(Cr) initially present in the large pore form (LP) switched to its narrow pore version (NP) at very low pressure, and exhibited a second structural transition from the NP to the LP at higher pressure. The authors observed that experimental and simulated adsorption enthalpies for H_2S decreased in the sequence: MIL-53(Cr) NP > MIL-47(V) > MIL- 53(Cr) LP, also depicted in Figure 2.4.

Several previous studies have showed that the efficiency of MOFs can be influenced by the factors such as metals ions, substituent groups and coordinatively unsaturated sites (CUS). Further, most of the studies were based on the adsorption of thiophene species from liquid fuel. Chen *et al.* [58] performed research on the desulfurization ability of MOFs for the H_2S, mercaptan and thioether species. A systematic DFT analysis on the influence of MOFs metal centre (structures as well as metal ions) and organic ligands on the adsorption of H_2S, CH_3SCH_3 and CH_3CH_2SH was carried out. It was noted that the desulfurization ability of MOF-74 and MOF-199 with open metal site, was superior to the other MOFs without coordinatively unsaturated sites. The observation confirmed the remarkable influence of CUS on the sulfur compounds adsorption. Here, IRMOF-3 containing amino groups exhibited the maximal sulfur

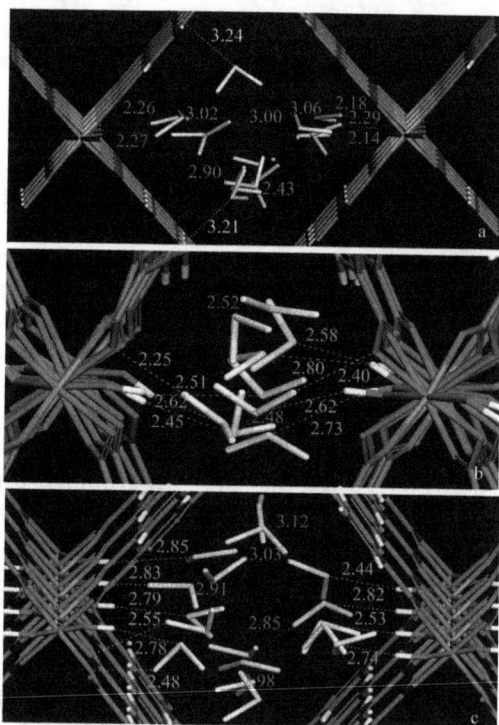

Figure 2.4 Illustrations of the preferential arrangements for H_2S simulated at 303 K in MIL-47(V) (a), MIL-53(Cr) NP (b) and MIL-53(Cr) LP (c) solids. Reproduced from Reference 57 with permission from American Chemical Society.

capacity, succeeded by IRMOF-8. However, it was noted that the density of CUS sites present in the Zn-MOF-74 (6.16×10^{-3} mol/g) was greater than Cu-MOF-199 (4.96×10^{-3} mol/g), and hence the sulfur removal ability of Zn-MOF-74 was confirmed to be lesser than that of Cu-MOF-199. Fernandez *et al.* [59] studied the adsorptive removal of toxic acid gases by using fluorinated MOF (FMOF-2), which was prepared using zinc nitrate hexahydrate and 2,2'-bis(4-carboxyphenyl) hexafluoropropane. The MOF was observed to be stable for the H_2S and SO_2 adsorption. With FMOF-2, the estimated weight capacities for H_2S and SO_2 were 8.3% and 14% respectively, at 1 bar and room temperature. Dathe *et al.* [36] studied the SO_2 adsorption using $Ba(CH_3COO)_2$ (or $Ba(NO_3)_2$ and $BaCl_2$)-impregnated Cu-BTC. The impregnation accelerated the formation of small micro-crystals of barium salts in Cu-BTC pores. Also, the sectional destruction of the host structure took place in $BaCl_2$. It was also noticed that at greater temperature, the SO_2 adsorption surpassed the stoichiometric ability based on Ba^{2+} concentration. Accordingly, the excess SO_2 adsorption was because of the chemical bonding between the SO_2 and metal cations, thereby resulting in the generation of Cu-sulfates. Hence, at lower temperature, Cu-BTC performed as an excellent host material for obtaining very well dispersed barium salts. However, at higher temperature, the adsorbent Cu-BTC decomposed and generated unique Cu species that performed as SO_x storage sites, thus producing Cu-sulfates finally. Thus, the main limitation of these bariums salts impregnated Cu-BTC adsorbent is the irreversible SO_x storage in their structure.

Adsorption of Other Harmful Gases

For the adsorption of various toxic gases, the open metal sites of MOFs have been stated repeatedly as the active sites. Further, a large number of studies stated that the properties of MOFs could be controlled by the loading of active metals (Pd, Cu or Ag) or active species (sulfated zirconia or polyoxymetalates) or by generating composites [60]. Glover *et al.* [61] demonstrated the adsorption of various toxic gases like SO_2, NH_3, octane vapor and CNCl using M-CPO-27 (M=Zn, Ni, Mg or Co) under humid and dry conditions. The breakthrough results confirmed that only under the dry conditions, the adsorbents with open metal sites were able to adsorb the aforementioned toxic gases. In humid conditions, the adsorptive ability decreased significantly as the aggressive adsorption of water started dominating. In the case of NH_3, the reduction in the adsorption capacity was not identified, as it was adsorbed similarly under both humid and dry conditions. Hence, it could be confirmed that the presence of water vapour had a negative influence on the adsorption of different toxic gases using the CPO-27 type adsorbents.

Petit and Bandosz [62] studied different composites of MOFs (MIL-100(Fe), MOF-5 or Cu-BTC) and graphitic compound (GO, graphite or graphite oxide), under ambient conditions, for the adsorption of NO_2, H_2S and NH_3. It was observed that the open metal sites of MOFs coordinated with oxide groups of GO, which caused the generation of a new pore space in the interface between the MOF units and carbon layers, to form composites having definite properties. With the GO/Cu-BTC composites, more than 50% (for H_2S), 12% (for NH_3) and 4% (for NO_2) enhancement in the adsorption ability was observed. The increase in the adsorption ability for the toxic gases seen in the composites was attributed to the porosity generated in the interface where the dispersive forces were substantial. Britt *et al.* [63] studied the possible suitableness of six MOFs, consisting of 2-amino terephthalate (IRMOF-3), diacetylene-1,4-bis(4-benzoic acid) (IRMOF-62), $Zn_4O(CO_2)_6$ group linked by terephthalate (MOF-5), benzene-1,3,5-tris(4- benzoate) (MOF-177), Cu-BTC and $Zn_2O_2(CO_2)_2$ chains linked by 2,5-dihydroxyterephthalate (CPO-27), for the adsorption of different hazardous gases/vapors including NH_3, chlorine, SO_2, benzene, tetrahydrothiophene and ethylene oxide. The obtained results were correlated with BPL carbon. It was seen that different factors like the active adsorption site having specific functional group and open metal sites of Cu-BTC or CPO-27 played a critical role in deciding the dynamic adsorption ability of these MOFs. Compared to the other MOFs, CPO-27 exhibited excellent performance, and the adsorption ability was almost 6 times higher than the BPL carbon. The improved adsorption was because of the existence of an immensely reactive 5 coordinate Zn species, together with the reactive oxo-group present in CPO-27. However, in the case of all other gases except Cl_2, the MOF Cu-BTC exhibited higher efficiency than BPL carbon. For NH_3 adsorption in IRMOF-3, the existence of NH_2 in the MOF greatly increased the adsorption ability, when compared to MOF-5 or IRMOF-1. Prior to the breakthrough, the IRMOF-3 adsorbed NH_3 approximately 71 times, when compared to the BPL carbon, due to the ability of NH_3 to easily form the hydrogen bonds.

CO, another toxic gas in the environment, has good binding capacity with the metal sites and MOFs with CUS can easily adsorb CO from gaseous mixture. Karra and Walton [64] conducted molecular simulation analysis and observed that at 298 K, Cu-BTC was a good adsorbent for CO over N_2 and H_2. It was proposed that the electrostatic interaction between the CO dipole and partial charge of CUS of Cu-BTC controlled the adsorptive behavior. Different MOF materials can also be used successfully to adsorb NO by adsorption processes [65]. The gravimetric NO adsorption on the open metal sites of Cu-BTC at 196 K and 1 bar was examined by Xiao *et al.* [65]. It was observed that by using 1 g of Cu-BTC almost 9 mmol of NO was adsorbed, which was remarkably greater than any other porous solids for the NO adsorption (Figure 2.5). The adsorption of benzene (both liquid and vapor phase) was also analyzed by Jhung *et al.* [66] using MIL-101(Cr), chromium terephthalate adsorbent. 16.7 mmol.g^{-1} of benzene (vapor phase) was adsorbed by the adsorbent MIL-101(Cr). The obtained value was approximately 8.7, 5.5 and two times higher than the H-ZSM-5, SBA-15 and a commercial activated carbon, respectively. Due to the high porosity of MIL-101(Cr), it displayed higher benzene adsorption capacity than the pitch based activated carbon (12.4 mmol.g^{-1}) with a higher surface area of 2600–3600 m^2 g^{-1}. The potential utilization of MIL-101(Cr) in the vapor phase adsorption of VOCs like p-xylene and ethyl acetate was also studied by Shi *et al.* [67].

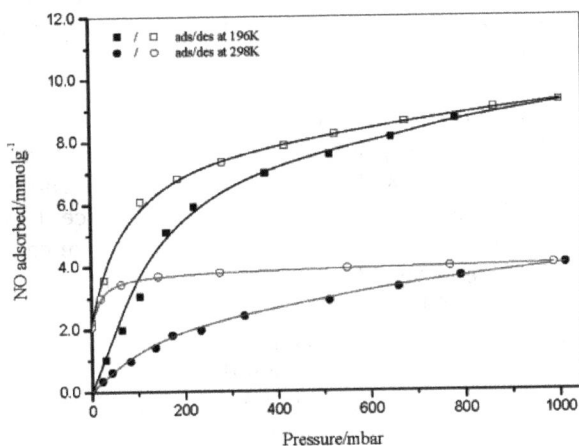

Figure 2.5 Adsorption (filled symbols) and desorption (open symbols) isotherms at 196 K (squares) and 298 K (circles) for nitric oxide on HKUST-1. Reproduced from Reference 65 with permission from American Chemical Society.

2.2.2 Adsorptive Elimination of Contaminants from Wastewater

Various MOFs have been employed for the adsorption of different hazardous contaminants from water. The adsorption occurs through different mechanisms like the electrostatic interactions, hydrogen bonding, acid-base interactions, hydrophobic interactions and π-π interactions. One of the substantial barrier for the application of several types of MOFs is the water instability (both in vapour and liquid phase). This instability of MOFs in water is a critical problem and many research studies have been carried out to increase their water stability [68]. Computational as well as experimental studies showed that several MOFs like MIL-101-V, MOF-5 experienced degradation by ligand displacement, when these materials were exposed to water [69,70]. On the other hand, MOFs like UiO-66 [71], ZIF-8 [72] MIL-101-Cr [73], MIL-100 (Fe, Al, Cr) [74,75] had good water stability and were suitable for the adsorption application in water or water purification. Several methods like incorporation of water repellent functional groups [76], composite formation [77] and fluorination [78] have been achieved to improve the water stability [79].

Haque *et al.* [80] reported the adsorptive separation of dye using MOFs. The authors utilized two types of porous Cr based MOFs, namely MIL-53-Cr and MIL-101-Cr, for the adsorption of methyl orange (MO), an anionic hazardous dye, from the aqueous solution. It was noticed that the performance of the MOFs was superior to the activated carbon. The adsorption kinetic constant and the adsorption capacity of MIL-101-Cr were higher than that of MIL-53-Cr, thereby portraying the significance of pore size as well as porosity for the adsorption. On the other hand, a better adsorption was seen after the functionalization of MIL-101-Cr with ethylenediamine (ED) and protonated ED (PED). Generally, MO exists in the sulfate form, and hence it exhibits greater electrostatic interaction with an adsorbent having positive charge. It was observed that the positive charge dispersion enhanced in the order MIL-101-Cr < ED-MIL-101-Cr < PED-MIL-101-Cr. On the other hand, the positive charge on PED-MIL-101-Cr reduced with enhancing the solution pH, due to the deprotonation of the protonated adsorbent. In another study, Haque *et al.* [81] analyzed the adsorption of methylene blue (MB-cationic) and MO from aqueous solution, using NH_2-MIL-101-Al, an NH_2-functionalised MOF. The ultimate adsorption ability of NH_2-MIL-101-Al, for the removal of MB was observed to be 762±12 mg/g at 30 °C, which was higher than most other porous materials and MOFs. However, by using non-amine functionalized framework (MIL-101-Al), lower adsorption capacity was noticed (195 mg/g). Thus, it suggested that the electrostatic interactions between the cationic MB and the amino groups of the MOF favored the adsorption. Haque *et al.* [82] also reported that the iron terephtahalate MOF (MOF-235) could adsorb both the cationic (MB) as well as anionic (MO) dyes from the waste water. Here, the adsorption kinetic constant and adsorption capacity of MOF-235 were higher as compared to that of the activated carbon. The higher adsorption of MB or MO was explained on the basis of the electrostatic interactions between the adsorbents and dyes. MB existed in positive form and MO existed in negative form, and hence electrostatic interactions were observed with the adsorbents having negative (charge balancing anion) charges as well as positive framework respectively. After the equilibration with MOF-235, MB and MO adsorptions at varying pH values were determined. By increasing the pH of the MO solution, it was observed that the quantity of adsorbed MO reduced, as the positive charge density reduced with increasing pH. However, the quantity of adsorbed MB increased with pH increase, as the negative charge concentration increased with pH increase. In another study, Leng *et al.* [83] analyzed the adsorption behavior of uranine using MIL-101-Cr and observed the aforementioned type of interactions. The adsorptive removal ability of uranine using MIL-101-Cr was found to be 126.9 mg/g, which was much higher than the activated carbon (17.5 mg/g). This was due to the electrostatic interactions, which were explained by the zeta potential analysis. The zeta potential of MIL-101-Cr was observed to be around 21.1 mV, and zeta potential reduced to 12.0 mV after the uranine interaction. Chen *et al.* [84] studied the xylenol orange (XO) adsorption on MIL-101-Cr and observed significant adsorption ability of the material. It was observed that the enhancement in the pH of the solution decreased the amount of adsorbed XO. No adsorption was observed, when the pH of the solution attained 12. Thus, after the analysis of zeta potential, the authors concluded that the interactions between the adsorbent and dye played a key role in the adsorption process.

Hydrogen bonding also has a major role in adsorption of harmful materials using MOFs. Liu *et al.* [85] examined the adsorption of phenol and para-nitrophenol (PNP) from aqueous solution, using different MOFs (NH_2-MIL-101-Al and MIL-100-Fe, Cr). During the phenol adsorption, limited as well as similar adsorption ability was exhibited by all the three MOFs. On the other hand, NH_2-MIL-101-Al exhibited very high adsorption of PNP, which was about 1.9 times and 4.3 times greater than the MIL-100-Cr and MIL-100-Fe, respectively. The significant adsorption ability of NH_2-MIL-101-Al was because of the hydrogen bonding between the PNP and NH_2 groups in NH_2-MIL-101-Al, as illustrated in Figure 2.6. Xie *et al.* [86] examined nitrobenzene adsorption from waste water using two types of aluminum based MOFs, namely MIL-68-Al (1130 ± 10 mg/g) and CAU-1 (970 ± 10 mg/g). The adsorption values were greater than the experimental values reported for other porous materials. The two MOFs had μ-OH groups in Al-O-Al units, which played a crucial role in attaining higher adsorption by the generation of hydrogen bonding between the nitrogen atom in nitrobenzene and μ-OH groups of MOFs.

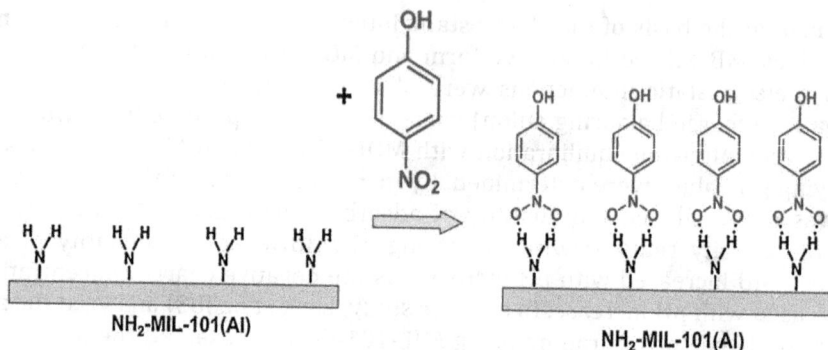

Figure 2.6 Mechanism of PNP adsorption on NH2-MIL-101-Al by hydrogen bonding. Reproduced from Reference 85 with permission from American Chemical Society.

Khan *et al.* [87] studied the adsorption of phthalic acid from water using ZIF-8 and other MOFs like NH$_2$-UiO-66, MIL-53-Cr, UiO-66, NH$_2$-MIL-100-Cr, MIL-100-Cr and MIL-100-Fe. When compared to other MOFs, the ZIF-8 exhibited higher adsorption at high pH. This might be due to the electrostatic interactions between the ZIF-8 (positive charged surface) and PA^{2-} as well as H-PA$^-$ ions. Further, it was also obserevd that the amine functionalized MOFs exhibited higher adsorption of phthalic acid at lower pH, than the MOFs without NH$_2$ group. At low pH, the phthalic acid did not deprotonate, and the acid-base interaction between the basic sites and phthalic acid exhibited dominancy. Tong *et al.* [88] studied the effect of framework metal ions in the adsorption of dyes using MIL-100-Fe/Cr. The adsorbents MIL-100-Cr and MIL-100-Fe had identical pore volumes (0.75 cm^3/g and 0.76 cm^3/g) as well as surface areas (1760 m^2/g and 1770 m^2/g). On the other hand, it was noticed that both the MOFs exhibited different behavior for the adsorption of MB and MO from wastewater. For the adsorption of MO, the absorbent MIL-100-Fe displayed excellent adsorption (1045.2 mg/g), when compared to the adsorbent MIL-100-Cr (211.8 mg/g). Due to the greater binding energy of water molecules with the metal sites of MIL-100-Cr, the aggressive adsorption of water and MO was more outstanding at the MIL-100-Cr surface. The adsorption of water molecules occurred at the apertures of hexagonal and pentagonal windows, which limited the accessibility of MIL-100-Cr cages for MO to a higher degree. Size selective adsorption is also noticed during adsorption using MOFs. Wang *et al.* [89] analyzed the size selective adsorption of dyes having different sizes, where neutral MOFs were used for the adsorption. The MOFs had face centered cubic topologies with 3-dimensional frameworks, and were used for the adsorption of three different dyes, i.e, MB, MO and rhodamine (RB). It was noticed that MB and MO were adsorbed, however, RB was not adsorbed due to its larger size that blocked the transmission through the smaller windows of the MOFs.

Presence of uranium in water is one of the most serious means of hazard to human health [90]. Adsorption is considered as the best method for the uranium removal from water due to its design flexibility, low cost and toxic pollutant insensitivity. Feng *et al.* [91] examined uranium adsorption capacity of the MOF HKUST-1 from aqueous solution. The most advantageous adsorption conditions were fixed along with the adsorbent dosage (0.03g), solution pH (at 6) as well as the agitation time (2 h). It was noticed that when the initial concentration of uranium was less than 200 mg/g, the temperature had no visible effect on the adsorption. However, when the initial concentration of uranium exceeded 200 mg/g, a greater temperature favored the adsorption process. At 318 K, the adsorbent HKUST-1 displayed an excellent adsorption capacity, at initial uranium concentration of 800 mg/L and solution pH value of 6. Further, the endothermic and impulsive nature of the adsorption process were indicated by the thermodynamics of HKUST-1/uranium system.

2.3 Metal Oxides

Transitional metal oxide nanoparticles have been adequately researched because of the attractive behavior as advanced nanomaterials in environmental as well as energy areas due to their excellent adsorption performance [92,93]. As mentioned earlier, harmful gases pose serious challenges to the process industries as well as environment. Wastewater from food processing, pharmaceutical and textile industries has constantly been a crucial environmental issue, as the contaminants have adverse effects on aquatic life as well as human health. Several dyes as organic contaminants are stable and difficult to degrade in conventional wastewater treatment methods. Thus, establishing effective and beneficial adsorption methods are needed for removing the organic contaminants present in the wastewater [94]. Metal oxide based adsorbents confirmed to be an efficient solution for the removal of a wide variety of contaminants. These oxides are available in various shapes and sizes, in the form of micro-particles, nanoparticles, nanocomposites and granules, thereby demonstrating distinct characteristics which are critical to their sorption properties. Several research studies have been carried out during the last two decades for developing various optimum methods of preparation of metal oxide adsorbents. Out of the different methods, thermal decomposition and reduction, sol–gel processes and hydrothermal preparation [95] are the most promising ones.

2.3.1 Adsorption of CO_2

In addition to the removal of CO_2 from natural gas, the removal of CO_2 from the large CO_2 emission sources like thermal power plants has also a significant role for preventing the global warming [96]. For the efficient removal of CO_2 using solid adsorbents, higher selectivity of CO_2 against other gases like nitrogen is very important. Chemical adsorption of CO_2 is considered as one of the methods to enhance the selectivity of the adsorbents. The metal oxides display improved selectivity due to the chemical adsorption between the CO_2 molecules and their surfaces. Chemical adsorption also allows the metal oxides for CO_2 adsorption in the existence of water vapor. H_2O undergoes physical adsorption and hence it has limited effect on the CO_2, which is adsorbed chemically. This behavior is different from zeolites, which are noted to have higher loss of CO_2 in the presence of water vapor. Together with the high selectivity for CO_2, the high surface area of the adsorbent is also very important for the efficient adsorption. The advancements in nanostructure control have facilitated the preparation of oxides with higher surface area by generating nanoparticles and porous structures. The utilization of these methods allows added enhancement in the CO_2 adsorption ability of the metal oxides.

Cai *et al.* [97] explored the synthesis of mesoporous alumina (MA) and MA-supported metal oxides. Aluminum isopropoxide, aluminum chloride, and aluminum nitrate nonahydrate were used as aluminum precursors, whereas nickel, magnesium, iron, chromium, copper, cerium, lanthanum, yttrium, calcium, tin chlorides, or nitrates were metal precursors. The authors observed that the use of aluminum nitrate nonahydrate for the synthesis led to large mesopores with size ranging from ~7 nm to 16 nm, improved ordering of the oxides as well as improved adsorption affinity toward CO_2. Figure 2.7 shows the temperature-programmed desorption (TPD) profiles of the <A and MA supported metal oxide materials for CO_2. The profiles exhibited a broad peak, which indicated the presence of a wide range of basic sites on the surface of these materials.

In another study, Leon *et al.* [98] studied magnesium-aluminum double oxides derived from the thermal treatment of layered hydroxides for CO_2 adsorption. The authors analyzed the effects of various factors, such as the incorporated cation (K or Na), mode of addition of precursors, sonication, and calcination temperature, and related these with the adsorption capacity by the

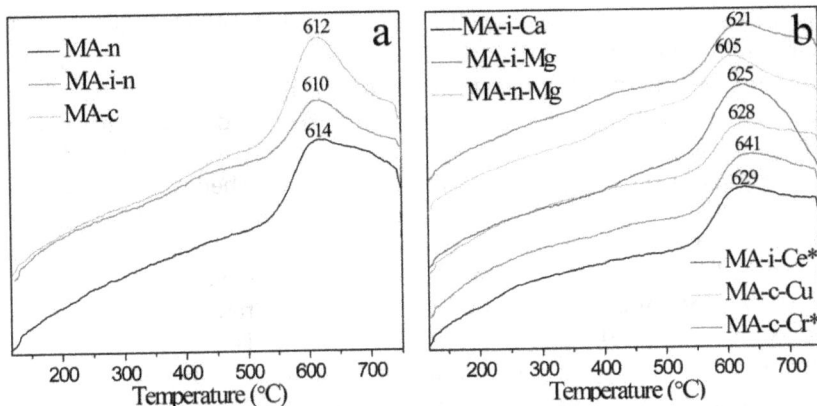

Figure 2.7 CO_2-TPD profiles for (a) MAs and (b) MA supported-metal oxides. Reproduced from Reference 97 with permission from American Chemical Society.

use of thermogravimetry and calorimetry. As observed in Figure 2.8, the equilibrium adsorption isotherms at 6.7 kPa and 323 K of the mixed oxides, recorded using microcalorimetry, exhibited the presence of combined chemisorption and physisorption behavior. It was observed that Langmuir transformation fit of the curves was achieved at low coverages, however, did not fit the data at higher extent of coverage, thus, indicating multilayer adsorption.

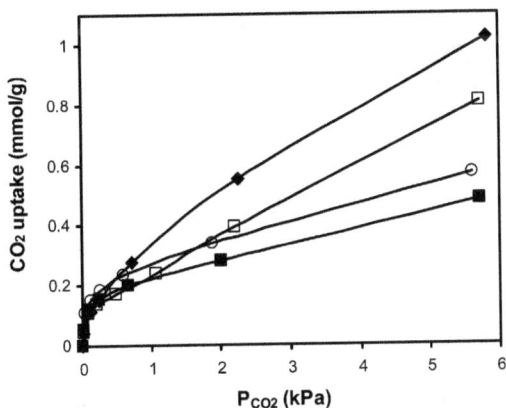

Figure 2.8 CO_2 adsorption isotherms at 323 K and 6.7 kPa, measured by microcalorimetry for (♦) 1K723F, (□) 2K723F, (■) 1KUS723F, and (○) 2KUS723F. Reproduced from Reference 98 with permission from American Chemical Society.

Yoshikawa *et al.* [99] also studied the CO_2 adsorption behavior of single metal oxides. CeO_2, Al_2O_3, SiO_2 and ZrO_2 were generated, and the chemically adsorbed CO_2 amount was measured using CO_2 pulse injection. From the adsorption analysis, it was observed that the CeO_2 based adsorbents exhibited the highest CO_2 adsorption capacity. The temperature programmed desorption investigation of CO_2 confirmed that the CO_2, adsorbed on the CeO_2, got desorbed mostly at temperatures from 323K to 473 K, with the peak at 393 K. Further, a very small amount of desorption remained till the temperature reached 723 K. For analyzing the CO_2 adsorption as well as desorption mechanisms, the amount of CO_2 adsorbed on CeO_2 and the temperature dependence were examined with the help of FTIR spectroscopy. The FTIR analysis results confirmed that the hydrogen carbonate and bidentate carbonate species were essentially formed on the CeO_2 adsorbent surface. The lowest decomposition temperature was exhibited by the hydrogen carbonate species, which revealed that this species could be utilized to reduce the energy consumption for the capture of CO_2. It was also noted that for increasing the hydrogen carbonate, treatment with water vapor proved to be efficient. As the exhaust gases from the thermal power plants consist of CO_2 and water vapor, it has the ability to help the generation of hydrogen carbonate species. Hence, CeO_2 can be considered as an efficient adsorbent for the CO_2 adsorption in thermal power plants. Wang *et al.* [100] also studied high-temperature adsorption of CO_2 on mixed oxides, which were derived from hydrotalcite-like compounds. The oxides were characterized with various physicochemical techniques such as x-ray diffraction analysis, Fourier transform infrared spectroscopy, thermogravimetric analysis and BET. It was observed that the generated oxides were of either periclase or spinel phase, with an interparticle pore diameter of 9.6-15.4 nm. The oxides were observed to exhibit high CO_2 adsorption capability at 350 °C. For instance, mixed oxide CaCoAlO was observed to adsorb 1.39 mmol/g of CO_2 (6.12 wt%) from a gas mixture (8% CO_2 in N2) at 350 °C and 1 atm. The range of adsorption for other mixed oxides was 0.87–1.28 mmol/g (3.83–5.63 wt%) of CO_2. Figure 2.9 also demonstrates the CO2 adsorption data in the fixed bed reaction, in comparison with the simulated breakthrough curves.

In other studies for CO_2 adsorption, Zhao *et al.* [101] synthesized MgO nano/micro-particles with various morphologies as well as porous structures and studied the CO_2 adsorption ability of these porous materials. Song *et al.* [102] analyzed the CO_2 adsorption kinetics of MgO adsorbent under various CO_2 partial pressures and adsorption temperatures. The adsorption behaviors of MgO were successfully predicted by the pseudo-second order models. The

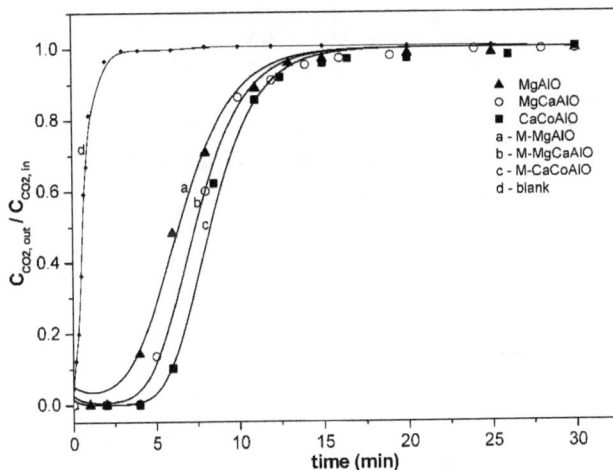

Figure 2.9 CO_2 sorption data in the fixed-bed reactor and simulated breakthrough curves. "M-" represents predicated curves. Reproduced from Reference 100 with permission from American Chemical Society.

material exhibited two stage process of adsorption, having a fast initial step succeeded by a sluggish second step. The first step was due to the diffusion of film from the bulk gas phase to the MgO exterior, whereas the second slow step was attributed to the strong inter-particle diffusion resistance due to the large number of narrow pores, thus, reducing the diffusion rate of CO_2. From the pore size and the pore area distribution, it was noted that the surface area of the generated MgO primarily located in small pores under 3 nm, where the CO_2 molecules underwent strong diffusion resistance. The study confirmed that the slow diffusion of CO_2 in the narrow pores reduced the MgO adsorption kinetics in spite of the fact that numerous pores contributed to larger surface area. Thus, it was revealed that an outstanding chemical adsorbent should have greater surface area for chemisorption as well as suitable pore size distribution for promoting the diffusion of CO_2 in the adsorbent porous structure. Kumar et al. [103] analyzed the CO_2 adsorption ability of mesoporous MgO, which acted as a promising pre-combustion CO_2 adsorbent. It was observed that at 350 °C temperature and CO_2 pressure of 10 bar, almost 96.96% of MgO changed to $MgCO_3$. In another study by Vu et al. [104], mesoporous MgO adsorbent was synthesized using aerogel technique and analysis of the CO_2 adsorption behavior at intermediate temperature range of 250–400 °C was carried out. At 325 °C and 120 min, the MgO sample exhibited higher adsorption capacity of 13.9 wt%. Thus, it was confirmed that MgO exhibited higher CO_2 adsorption ability after modification and had the potential of being an excellent adsorbent for CO_2 removal.

Yu-Dong *et al.* [105] examined the CO_2 adsorption efficiency of MgO adsorbent in the presence of water vapor in a fixed bed. Initially, the analysis of CO_2 adsorption capacity of MgO adsorbent under dry condition was carried out. At 100 mL/min flow rate, it took almost 14 min for the CO_2 concentration of the outlet to reach 14.5%. A decrease in the time to reach the equilibrium was seen with the increased flow rate and it took just 4 min for that of 300 mL/min to obtain equilibrium. It was found that at 50 °C, with the increase in relative humidity from 30% to 70%, the CO_2 adsorption ability of MgO adsorbent enhanced from 0.82 mol/kg to 3.42 mol/kg. At a concentration of water vapor of 8.547% v/v and at an adsorption temperature of 75 °C, the CO_2 adsorption capacity reached a maximum value of 3.54 mol/kg. It was reported that the CO_2 adsorption efficiency of MgO was greatly dependent on the presence of water vapor. However, the time to obtain equilibrium increased because of the controlled diffusion of CO_2 arising due to the condensation of water vapor. The results obtained from the cycle experiments confirmed that the condensed water increased the aggregation of MgO particles. Further, it was also noted that at higher temperature, i.e., 100 °C, no water vapor condensation took place and the porous structure remained stable during the cycling adsorption. Elimination of the porous structure degradation and enhancement in the CO_2 adsorption capacity was observed at adsorption temperature between 100 °C and 110 °C. The study confirmed that MgO adsorbent could be developed as a promising material for CO_2 adsorption from wet flue gas.

2.3.2 Adsorption of H_2S

As mentioned earlier, H_2S is a toxic, corrosive gas and is one of the commonly seen contaminants present in several industrially essential feedstocks like coal gas, natural gas, liquefied petroleum gas, jet fuel and naphtha. H_2S is also a dominant source of acid rain, as the oxidation of H_2S causes the formation of sulfur oxide. Further, it can also poison several industrial catalysts like the supported nickel catalyst, which is employed in hydrocarbon steam reforming. Hence, the separation of H_2S from these liquid or gaseous streams is very essential. H_2S adsorption using different metal oxides or their mixtures has been explored in great details to overcome these challenges.

Metal oxides such as iron oxides are widely used for H_2S adsorption. These oxides separate H_2S by forming iron sulfides which are insoluble in nature.

The different chemical reactions occurring in this process are as described as follows:

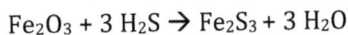

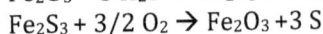

$$Fe_2O_3 + 3\ H_2S \rightarrow Fe_2S_3 + 3\ H_2O$$
$$Fe_2S_3 + 3/2\ O_2 \rightarrow Fe_2O_3 + 3\ S$$

For the adsorption process, iron oxide is commonly employed in a form known as "iron sponge". The iron sponge consists of iron oxide (Fe_2O_3 and Fe_3O_4) impregnated wood chips, and can be used in batch as well as continuous systems. After the saturation, the sponge can be regenerated. However, after each regeneration cycle, the activity generally gets decreased by one-third [106]. In continuous system operation, the air is progressively added to the gaseous stream, and the regeneration of iron sponge occurs simultaneously. The H_2S removal rate is generally very high i.e., 2500 mg H_2S/g Fe_2O_3. However, the iron sponge has some disadvantages like high operating cost, chemically intensive process and production of continuous waste stream that might be either disposed of as dangerous waste or regenerated expensively. Some commercially generated iron oxide systems like Sulfatreat 410 HPR are available, which can produce non-hazardous waste. The adsorption capacity of Sulfatreat 410 HPR was reported to be 150 mg H_2S/g adsorbent [106]. Carnes et al. [107] reported that the metal oxide reactivities depend on intrinsic crystallite reactivity, crystallite size and the surface area. Further, it was also seen that when compared to the micro-crystalline structures, the nano-crystalline structures exhibited enhanced reactivity with H_2S. This is because of the higher surface area that promotes adsorption. Further, the presence of Fe_2O_3 enhanced the reaction due to the formation of iron sulfides which acted as the catalyst. Rodriguez and Maiti [108] reported that the H_2S adsorption ability of a metal oxide depended on the electronic band gap energy. More H_2S adsorption took place, when electronic band gap energy was less. This is due to the fact that the electronic bad gap is negatively related to oxide chemical reactivity. However, the reactivity depends on the mixing of oxide's bands with H_2S orbitals. In the case of better mixing, the oxides react well with sulfur containing molecules, thus forming metal sulfides. This causes the dissociation of H_2S molecules and sulfur immobilization in metal sulfides. In spite of this, the application of metal oxides for H_2S removal has some drawbacks like high cost, low selectivity, low separation effectiveness and lesser sorption/desorption rate.

At higher temperatures (200 °C - 400 °C), zinc oxide (ZnO) based materials are employed for removing H_2S traces. This is because, ZnO have better selectivity for sulfides than the iron oxides [109]. Davidson *et al.* [109] conducted research on the H_2S removal capacity of ZnO, and it was observed that the ZnO surface reacted with H_2S to form a zinc sulfide insoluble layer, thus, separating the H_2S from the gas stream. Almost 40% of available H_2S was converted and the reaction depicted in the equation below led to the H_2S removal.

$$ZnO + H_2S \rightarrow ZnS + H_2O$$

Many commercial products have been developed employing ZnO, and the ultimate sulfur loading on these oxides materials has been observed in the range 300-400 mg S/g sorbent. Majority of research studies carried out till date for the advancement of solid H_2S adsorbents focused on the materials which are acceptable for higher temperatures (> 300 °C). Ca- and Fe-containing materials are among the most researched solid materials for H_2S removal from coal gas. This is because of their low cost and higher reactivity [110]. However, due to the presence of large amount of H_2 and large CO/CO_2 ratio in the coal gas, these materials have some disadvantages. For example, the Fe_3O_4 is reduced to FeO and Fe, however, the Fe_3C formed during the process decreases the sulfur uptake of the solid [111]. Lew *et al.* [112] also observed that ZnO was superior to iron oxide due to its agreeable sulfidation thermodynamics. However, the sulfidation kinetics of ZnO was slower than iron oxide. The major disadvantage of using ZnO based solid adsorbents for the application of hot gas cleaning is the damage which occurs because of the reduction of ZnO to elemental Zn which is volatile. Thus, a number of metal oxides have the potential for efficient use as gas adsorbents, however, many challenges are required to be resolved.

Several mixed metal oxides like Cu-Fe-O, Zn-Fe-V-O, Zn-V-O, Zn-Fe-Ti-O, Zn-Ti-O, Cu-Mo-O and Cu-Mn-O have been examined for the H_2S adsorption applications [113-116]. It was observed that the regeneration ability as well as reactivity got enhanced when these materials were deposited on an appropriate support. The mixed metal oxide $ZnFe_2O_4$ also exhibited good H_2S adsorption ability at higher temperature range of 500 °C – 700 °C [117]. For high temperature fuel gas desulfurization, Kobayshi *et al.* [118] examined the zinc ferrite silicon dioxide composites. Zinc ferrite was observed to undergo sulfidation in the reducing environment to form iron sulfides, zinc blende and wurtzite in the presence of 500 ppm of H_2S.

In another study, Huang *et al.* [119] studied the adsorption and reaction of H_2S on TiO_2 rutile (110) and anatase (101) surfaces using periodic density functional theory (DFT). The authors observed that H2S, HS, S, and H preferentially adsorbed at the Ti_{5c}, O_{2c}, $(Ti_{5c})_2$, and O_{2c} sites, respectively, on the rutile surface, whereas these sites were Ti_{5c}, $(Ti_{5c})_2$, $(-O_{2c})(-Ti_{5c})$, and O_{2c}, respectively, on the anatase surface. Figure 2.10 also shows the adsorbed H_2S and the fragments HS, S and H on the TiO_2 rutile surface. Novochinskii *et al.* [120]

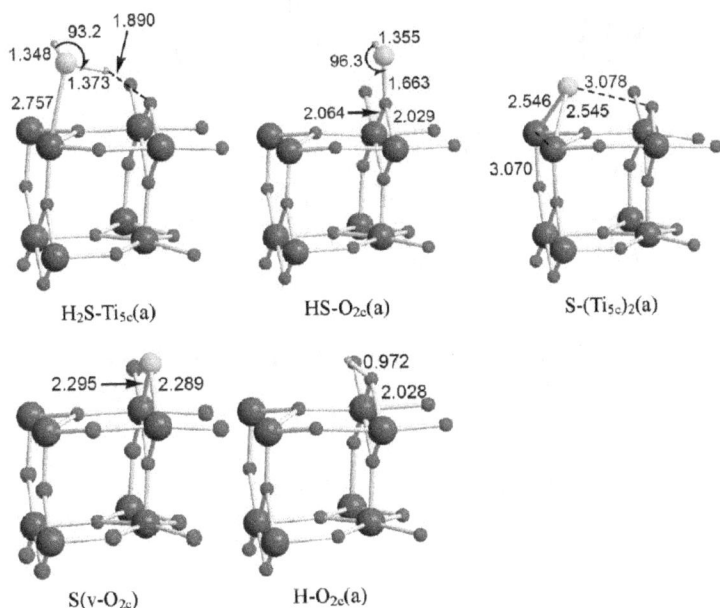

H_2S-Ti_{5c}(a) HS-O_{2c}(a) S-$(Ti_{5c})_2$(a)

$S(v$-$O_{2c})$ H-O_{2c}(a)

Figure 2.10 Geometries of adsorbed H_2S and its fragments, HS, S, and H, on the TiO_2 rutile (110) surface. Reproduced from Reference 119 with permission from American Chemical Society.

also reported a functional ZnO based adsorbent for low-temperature H_2S removal from steam-containing gas mixtures [120]. The authors observed very low H_2S outlet concentrations (20 parts per billion by volume) over the modified ZnO adsorbent. The authors observed the sulfur-trap capacity (which is the amount of H_2S trapped before breakthrough) to be dependent on factors such as space velocity, temperature, steam concentration, CO_2 concentration, and particle size. Figure 2.11 also demonstrates the performance of a ZnO-based trap at different concentrations of water vapor and H_2S. The authors observed higher trap capacity when the H_2S inlet concentration was higher. The trap capacity was also observed to decrease as a function of temperature. Presence of steam was noticed to inhibit the capture of H_2S by the adsorbent due to the shifting of the equilibrium of the reaction. Furthermore, increasing the CO_2 concentration in the feed up to 12 vol % was also observed to decreases the capacity of ZnO for H_2S adsorption.

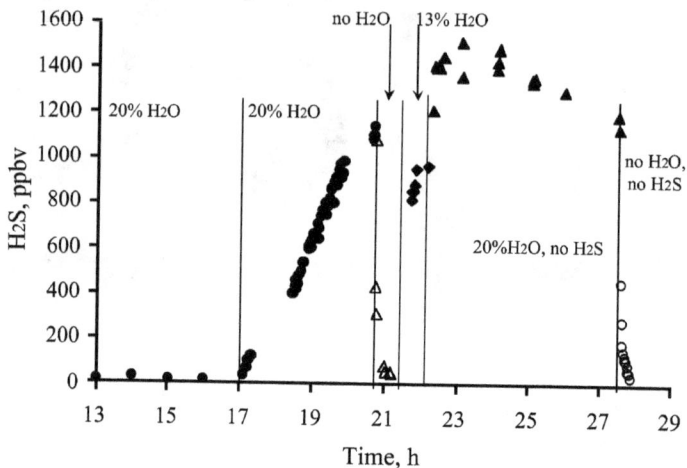

Figure 2.11 Performance of a ZnO-based trap at different concentrations of water vapor and H_2S. Reproduced from Reference 120 with permission from American Chemical Society.

For low temperature H_2S removal applications, several other studies have also been reported [107,109,121,122]. Carnes and Klabunde [107] examined the nano-crystalline metal oxides which were generated using sol gel technique. It was confirmed that at lesser temperature range (25 °C - 100 °C), the activity order was observed to be ZnO > CaO > Al2O3 >> MgO. Davidson *et al.* [109] examined the adsorption ability of H_2S using doped and high surface area ZnO at lower temperatures of 25 °C - 45 °C, and observed about 40% conversion of H_2S. Baird *et al.* [123] examined the ZnO doped with different oxides of Fe, Co and Cu, mixed Co–Zn oxides, Co-Zn-Al oxides, Co–Fe oxides to generate H_2S adsorbents at lower temperature. Out of the mixed metal oxides, Co-Zn oxide was reported to have the best H_2S adsorption ability at room temperature. Xue *et al.* [124] generated a series of pure metal oxides as well as mixed metal oxides from different hydrous oxides and hydroxy-carbonate precursors. It was observed that adsorbents like CuO, Zn/Co, Zn/Ti/Zr, Zn/Al and Zn/Mn mixed oxides exhibited good H_2S adsorption capacity at room temperature. The benefits of using hydroxycarbonate precursors included good dispersion and a synergy of metal components which were expected to be continued in the mixed metal oxide succeeding calcination [125]. The high degree of dispersion of metal oxides is one of the most attractive feature of metal oxide based adsorbents for the low temperature performance. Some studies have also focused on the advancement of the mixed metal oxide adsorbents like Cu-Zn and Cu-Zn-Al for the H_2S adsorption at lower temperatures. These Cu based mixed metal oxides are considered to effective adsorbents for low temperature applications because of their low cost than the Zn–Co mixed metal-oxide adsorbents. However, the Cu based mixed oxides, which are generally reduced by hydrogen before being used as adsorbents, were also analyzed for the H_2S adsorption and some other organic sulfur compounds for middle or higher temperatures also [126,127]. Jiang *et al.* [128] generated a series of Cu-Zn hydroxy-carbonate precursors by co-precipitation method, with changing Cu/Zn molar ratios. In addition to this, two series of Cu-Zn-Al hydroxy-carbonate precursors were prepared with varying metal molar ratio, by co-precipitation as well as multi-precipitation methods. The high surface area Cu-Zn-Al metal oxides displayed remarkable sulfur removal capacities at the temperature range 25 °C - 100 °C. It was observed that for H_2S adsorption at low temperature, the Cu rich adsorbents were more suited than the Zn rich adsorbents. The probable reason for this behavior could be due to the fast sulfidation rate of CuO than ZnO because of the lesser rearrangement of anion lattice.

Polychronopoulou *et al.* [129] also studied the efficiency of mixed metal oxides Fe-Mn-Zn-Ti-O of differing composition, generated using sol gel technique, for the removal of H_2S from a gaseous mixture consisting of 7.5 vol % CO_2, 1-3 vol % H_2O, 25 vol % H_2 and 0.06 vol % H_2S, analyzed at temperature range 25 °C - 100 °C. The chemical interaction between the mixed metal oxide and H_2S was studied and it was observed that the molar ratio of Fe/Mn employed in the preparation of aforementioned metal oxides mostly determined the size and morphology of the generated solid particles. Comparison of the temperature effect on the H_2S adsorption for the sol gel generated 5Fe-15 Mn-40 Zn-40 Ti-O solid and the commercial nickel based solid was made in the presence of 1 vol % and 3 vol % H_2O in the feed stream. In the 1 vol % H_2O system, the commercial solid appeared to be better than the sol get generated mixed metal oxide at 25 °C - 100 °C temperature range. For the commercial solid, the H_2S adsorption enhanced by 141% by the increase in adsorption temperature from 25 °C - 100 °C. The presence of 1-3 vol % H_2O in the 7.5% CO_2/25% H_2/0.06% H_2S/He gas mixture remarkably increased the H_2S adsorption at 25 °C for 5 Fe-15 Mn-40 Zn-40 Ti-O solid.

In a recent study, in-situ formed graphene/ZnO nanostructured composites were employed for low temperature hydrogen sulfide removal from natural gas [130]. Nanostructured composites of graphene and highly dispersed sub-20 nm sized ZnO nanoparticles (TRGZ) were successfully prepared via by combining freeze-drying and thermal annealing processes. A series of compositions with different weight ratios of ZnO nanoparticles were prepared and used as a reactive sorbent in low temperature hydrogen sulfide (H_2S) removal from natural gas. The composite sorbent having a ZnO mass ratio of 45.1 wt% exhibited a significantly greater H_2S adsorption capacity (3.46 mmol.g^{-1}) than that of pure ZnO (1.06 mmol.g^{-1}), indicating that hybridization of ZnO with graphene significantly improved the H_2S removal ability. Figure 2.12 also shows the transmission elecrom micrographs (TEM) of (a) pristine ZnO nanoparticles, (b) TRGZ-1, (c) TRGZ-2, and (d) TRGZ-3 graphene/ZnO nanohybrids. The authors suggested that the possible mechanism involved initial physisorption of H_2S molecules with oxygenated functional groups of graphene which later can reach to the finely dispersed active ZnO nanoparticles and get chemisorbed through reactive adsorption. The EDX spectra of H_2S treated samples exhibited a strong peak for sulfur element indicating the reactive adsorption of H_2S on the sorbent.

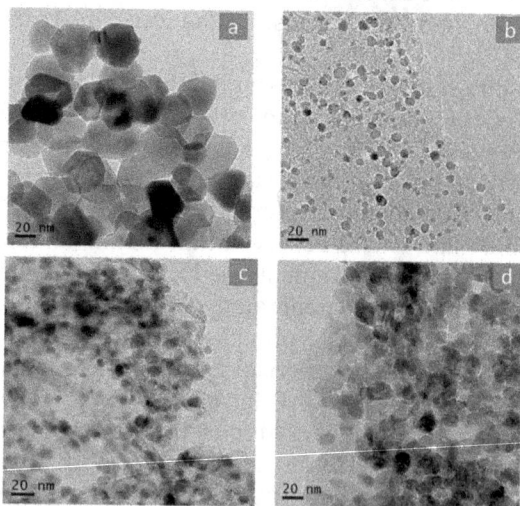

Figure 2.12 TEM images of (a) pristine ZnO nanoparticles, (b) TRGZ-1, (c) TRGZ-2, and (d) TRGZ-3 graphene/ZnO nanohybrids. Reproduced from Reference 130 with permission from Royal Society of Chemistry.

2.3.3 Adsorption of Organic Contaminants

Among the different physical methods employed for dye removal, adsorption is widely used method and considerable attention has been paid for the application of nano-sized adsorbent materials because of their higher surface area. In recent years, the utilization of magnetic particle technology for solving environmental issues has also gained attention. Different types of iron oxides have been studied for the dye adsorption, due to their reactive surface. These materials enable faster magnetic separation just after the adsorption. The magnetic particles can be employed for adsorbing the contaminants from gaseous or aqueous effluents and can subsequently be easily removed from the medium.

Luiz *et al.* [131] generated magnetic adsorbents by combining the magnetic properties of iron oxides and adsorption properties of the activated carbon. The iron oxide/activated carbon composites were used for the removal of volatile organic compounds like phenol, chloroform, drimaren red dye and chlorobenzene from the aqueous solution. It was noted that the existence of iron oxide did not influence the adsorption efficiency of activated carbon. In another study by Khosravi *et al.* [132], iron oxide nano-spheres were prepared using solvo-thermal method and were employed as an efficient adsorbent for anionic dye separation from the aqueous solution. Here, efficiency of iron oxide nano-spheres for the adsorption of anionic dyes namely reactive orange (RO) and reactive yellow (RY) was analyzed. The maximal adsorption capacity of RY and RO was observed to be 25.0 mg/g and 32.5 mg/g, respectively. The kinetic analysis confirmed rapid adsorption rate onto nano-spheres, which took place within 10 min. The adsorption process was exothermic and the optimum pH was about 4, i.e., acidic solution. At the final stage, due to the magnetic nature of the iron oxide nano-spheres, smooth separation from solution by a magnet could be achieved.

2.3.4 Heavy Metal Removal from Water/Wastewater

Heavy metal exposure, even at a very small level, is considered to be a danger for the living beings [133]. For the heavy metal separation from the water/wastewater, adsorption methods have become prominent techniques. Out of the various adsorbents available, nano-sized metal oxides (NMOs) are regarded as promising ones for the heavy metal removal from aqueous systems [134] .This is due to their greater surface areas as well as higher activities generated by the size quantization effect. The NMOs offer numerous advantages like preferable sorption, higher capacities and faster kinetics towards the heavy metals present in the water as well as wastewater. The examples for NMOs include nano-sized ferric oxides, aluminum oxides, magnesium oxides, manganese oxides, titanium oxides and cerium oxides. These materials exist in different forms such as tubes, particles, etc. On the other hand, when the metal oxide size decreases from the micrometer level to nanometer level, the enhancement in the surface energy causes poor stability. Hence, NMOs are prone to agglomeration because of the presence of van der Waals forces and other interactions, and thereby the high adsorption capacity as well as selectivity of NMOs can significantly reduce or even disappear [135]. Thus, the shape and size of these adsorbents are critical factors influencing their adsorption behavior.

The utilization of iron oxide based nano-materials is considered to be very effective for the separation of metal contaminants from water, due to their remarkable features such as small size, magnetic property, reusability and high surface area [136]. For instance, iron oxide nanomaterials exhibit excellent adsorption ability for the separation of arsenic from water. Feng *et al.* [137] generated super magnetic Fe_3O_4 coated with ascorbic acid, by hydrothermal process. The material had a surface area of $179 \, m^2/g$ and diameter < 10 nm. At room temperature, the material exhibited super paramagnetic behavior and the saturation magnetization attained $40 \, emug^{-1}$, which indicated that these materials were suitable for use as an adsorbent for the separation of arsenic from waste water. The ultimate adsorption capacity of As(III) and As(V) was found to be $46.06 \, mg/g$ and $16.56 \, mg/g$ respectively.

The excessive application of copper in the industries leads to the accumulation in the environment, thereby polluting the water. Hao *et al.* [138] generated a magnetic nano-adsorbent (MNP-NH$_2$) by covalent binding of 1,6-hexadiamine on the Fe_3O_4 nanoparticles surface, and analyzed its adsorption capacity for the removal of copper ions from the aqueous solution. Different factors influencing the adsorption performance such as salinity, initial Cu^{2+} concentration, temperature, amount of MNP-NH$_2$, contact time and pH were examined. The magnetic nanoparticles were able to separate 98% of copper from tap water as well as the polluted water. It was noticed that the adsorption occurred faster and the equilibrium was attained within 5 min. The chemical sorption was noticed to be the rate limiting step of the sorption mechanism, and the ultimate adsorption ability was observed to be $25.77 \, mg.g^{-1}$ at 298 K and pH value 6. The adsorbent MNP-NH$_2$ exhibited good reusability as well as higher stability under the experimental conditions. Complete desorption of the Cu^{2+} ions was achieved by using $0.1 \, mol.L^{-1}$ HCl solution for < 1 min and the regenerated MNP-NH$_2$ was able to retain the original metal discharge level. It was observed that the adsorption ability of the nano-adsorbent MNP-NH$_2$ remained constant, and no change in the desorption ability occurred during 15 adsorption - desorption cycles. In another study by Nashaat *et al.* [139], the adsorption mechanics, equilibria, thermodynamics and kinetics of Pb(II) ions onto Fe$_3$O$_4$ nano-adsorbents were analyzed. It was found that Fe$_3$O$_4$ nano-adsorbents had greater removal efficiency as Pb(II) was adsorbed from waste water in a very brief time and the equilibrium was accomplished within 30 min. Further, the adsorption was dependent on temperature, pH and the Pb(II) initial concentration. At pH value 5.5, the maximal removal of Pb(II) was obserevd. As the temperature and Pb(II) initial concentration was increased, the adsorption was also observed to enhance. The adsorption isotherms were determined using the Freundlich and Langmuir models. The thermodynamics of the adsorption of Pb(II) revealed the impulse nature, endothermic behavior and physisoprtion process. The regeneration as well as the desorption analysis confirmed the repeated use of nano-adsorbents without influencing the adsorption capacity. This confirmed the usefulness of Fe$_3$O$_4$ nano-adsorbents as inexpensive and effective adsorbents for the fast removal and recovery of the metal ions from wastewater.

The adsorptive behavior of nano-sized manganese oxides (NMnOs) are considered to be superior when compared to its bulk counterpart due to its greater surface area as well as polymorphic structures [140]. In the past several decades, NMnOs have been used for the sorption of anionic and cationic contaminants like arsenate, phosphate and heavy metal ions from natural water. Alumina (Al$_2$O$_3$) is also a conventional adsorbent for the removal of heavy metals and γ- Al$_2$O$_3$is regarded to be superior to α- Al$_2$O$_3$ in adsorption performance [141]. Nano-sized γ- Al$_2$O$_3$ adsorbents are generated using sol-gel method, and the materials have been used as solid phase extraction materials for the pre-concentration/separation of the trace metal ions. In the case of TiO$_2$, it was observed that the bulk and nano-TiO$_2$ differed in surface acidity, catalytic reactivity and chemical behavior, based on their various surface planes. From the BET analysis, it was seen that the nominal particle sizes were 8.3 and 329.8 nm, for nano-sized and bulk particles respectively [142]. The specific area of bulk particles was found to be 9.5 m^2/g, whereas the nanoparticles' specific area was 185.5 m^2/g. In a research study by Liang *et al.* [143], the nano-TiO$_2$ adsorbent, with BET surface area 208 m^2/g and diameter 10–50 nm exhibited adsorption capacity of Cd and Zn as 7.9 mg/g and 15.3 mg/g respectively at pH value 9. The effects of contact time, pH, interfering ions and elution solution on the adsorption performance of nano-TiO$_2$ for the removal of Zn and Cd were examined. The presence of common anions and cations (in the concentration range of 100-5000 mg/L) was found to have no significant effect on the metal adsorption (Cd^{2+} ions and Zn^{2+} of 1.0 mg/mL) under the specified conditions. Many research studies have also confirmed the role of nano-structured ZnO to effectively remove the heavy metals from various media [144]. In a related study, Lee *et al.* [145] generated NZnOs powder using the solution combustion method. The authors observed that as compared with two TiO_2 powders, P25 and the other one generated by homogenous precipitation at lower temperature, the NZnO powder displayed greater Cu^{2+} adsorption from the solution.

Table 2.2 summarizes the heavy metals adsorption capacities of different metal oxides reviewed in this study.

Table 2.2 Heavy metal adsorption capacities of some of the metal oxide adsorbents

Sl. No.	Adsorbent	Tempera-ture	pH value	Heavy metal	Adsorption capacity (mg/g)	Refer-ence
1	Super-magnetic Fe_3O_4 coated with ascorbic acid	room tempera-ture	-	As(III)	46.06 mg/g	115
2	Super-magnetic Fe_3O_4 coated with ascorbic acid	room tempera-ture	-	As(V)	16.56 mg/g	115
3	Magnetic nano-adsorbent (MNP-NH_2)	298 K	6	Cu	25.77 mg/g	116
4	Nano-TiO_2 adsorbent		9	Cd	7.9 mg/g	103
5	Nano-TiO_2 adsorbent		9	Zn	15.3 mg/g	103

2.4 Summary and Outlook

Adsorption represents an immensely useful technology for the removal of harmful gases like H_2S, CO_2, SO_2, NH_3 present in the gas stocks and air as well as separation of organic contaminants and heavy metals from the wastewater. It has attracted remarkable consideration in both scientific exploration as well as commercial utilizations. Further, the development of state-of-the-art porous materials for the adsorption process is required to enhance the adsorption capacity or to overcome the existing disadvantages such as small pore volume or pore size of adsorbents. In this chapter, a comprehensive assessment on the use of MOFs and metal oxides for gas/vapor and liquid phase adsorptions has been presented. Further, focus has also been devoted to the separation of a variety of contaminants like dyes and various heavy metals present in the water using the MOFs and metal oxides. As mentioned before, MOFs are promising adsorbents for the aforementioned applications due to their unique properties like higher surface area as well as tunable porosities. In addition to these properties, MOFs are able to consolidate functionalities by means of grafting or loading active or functional species and thereby increasing the adsorption capacity. Several modifications in MOFs, which were done in three aspects namely (1) metal ions, (2) organic liners and (3) novel consolidation of both, for enhancing the adsorption ability were also analyzed. The metal oxides exhibit enhanced selectivity because of the presence of chemical adsorption between the gas molecules and their surfaces. Both the surface area as well as pore size of the adsorbent greatly affects the gas diffusion in the adsorbent porous structure. Their reactivity depends on the crystallite size, intrinsic crystallite reactivity as well as the surface area. Nano sized metal oxides (NMOs) are considered to be an excellent adsorbent for the heavy metal removal from aqueous systems. In spite of the high adsorption capacity of the MOFs and metal oxides, there are several economical and technical barriers to promote the usage of these materials for gas adsorption and environmental applications. Most of these challenges are associated with material stability, remigration, high temperature performance, durability, loss of adsorption and selectivity, etc. However, with the employment of advanced nanotechnologies, it is envisaged that many of the currently existing limitations would be overcome in due course of time, which will pave the way for the large scale application of these MOFs and metal oxide materials for industrial and commercial applications.

References

1. Pawelec, B., Navarro, R. M., Campos-Martin, J. M., and Fierro, J. L. G. (2011) Towards near zero-sulfur liquid fuels: a perspective review. *Catalysis Science and Technology*, **1**, 23-42.
2. Vandenbroucke, A. M., Morent, R., Geyter, N. D., and Leys, C. (2011) Non-thermal plasmas for non-catalytic and catalytic VOC abatement. *Journal of Hazardous Materials*, **195**, 30-54.
3. Garces, H. F., Galindo, H. M., Garces, L. J., Hunt, J., Morey, A., and Suib, S. L. (2010) Low temperature H_2S dry-desulfurization with zinc oxide. *Microporous and Mesoporous Materials*, **127**, 190-197.
4. Petit, C., and Bandosz, T. J. (2010) Enhanced adsorption of ammonia on metal-organic framework/graphite oxide composites: analysis of surface interactions. *Advanced Functional Materials*, **20**, 111-118.

5. Colvile, R. N., Hutchinson, E. J., Mindell, J. S., and Warren, R. F. (2001) The transport sector as a source of air pollution. *Atmospheric Environment*, **30**, 1537-1565.
6. Bao, C., and Fang, C.-L. (2012) Water resources flows related to urbanization in china: challenges and perspectives for water management and urban development. *Water Resources Management*, **26**, 531-552.
7. Gupta, V. K., and Suhas (2009) Application of low-cost adsorbents for dye removal - a review. *Journal of Environmental Management*, **90**, 2313-2342.
8. Gong, Y., Zhao, X., Cai, Z., O'Reilly, S. E., Hao, X., and Zhao, D. (2014) A review of oil, dispersed oil and sediment interactions in the aquatic environment: influence on the fate, transport and remediation of oil spills. *Marine Pollution Bulletin*, **79**, 16-33.
9. Crini, G. (2006) Non-conventional low-cost adsorbents for dye removal: a review. *Bioresource Technology*, **97**, 1061-1085.
10. Chen, S., Zhang, J., Zhang, C., Yue, Q., Li, Y., and Li, C. (2010) Equilibrium and kinetic studies of methyl orange and methyl violet adsorption on activated carbon derived from phragmites australis. *Desalination*, **252**, 149-156.
11. Hasan, Z., Jeon, J., and Jhung, S. H. (2012) Adsorptive removal of naproxen and clofibric acid from water using metal-organic frameworks. *Journal of Hazardous Materials*, **209-210**, 151-157.
12. Ke, F., Qiu, L.-G., Yuan, Y.-P., Peng, F.-M., Jiang, X., Xie, A.-J., Shen, Y.-H., and Zhu, J.-F. (2011) Thiol-functionalization of metal-organic framework by a facile coordinationbased postsynthetic strategy and enhanced removal of Hg^{2+} from water. *Journal of Hazardous Materials*, **196**, 36-43.
13. Zhu, B.-J., Yu, X.-Y., Jia, Y., Peng, F.-M., Sun, B., Zhang, M.-Y., Luo, T., Liu, J.-H., and Huang, X.-J. (2012) Iron 1,3,5-benzenetricarboxylic metal-organic coordination polymers prepared by solvothermal method and their application in effcient As(V) removal from aqueous solutions. *Journal of Physical Chemistry C*, **116**, 8601–8607.
14. Li, J.-R., Sculley, J., and Zhou, H.-C. (2012) Metal- organic frame works for separations. *Chemical Reviews*, **112**, 869-932.
15. Babich, I. V., and Moulijn, J. A. (2003) Science and technology of novel processes for deep desulfurization of oil refinery streams: a review. *Fuel*, **82**, 607-631.
16. Velu, S., Ma, X., and Song, C. (2003) Selective adsorption for removing sulfur from jet fuel over zeolite-based adsorbents. *Industrial and Engineering Chemistry Research*, **42**, 5293-5304.
17. Choi, S., Drese, J.,H., and Jones, C. W. (2009) Adsorbent materials for carbon dioxide capture from large anthropogenic point sources. *ChemSusChem*, **2**, 796-854.
18. Zhou, A., Ma, X., and Song, C. S. (2006) Liquid-phase adsorption of multi-ring thiophenic sulfur compounds on carbon materials with different surface properties. *Journal of Physical Chemistry B*, **110**, 4699-4707.
19. Deliyanni, E., Seredych, M., and Bandosz, T. J. (2009) Interactions of 4,6-dimethyldibenzothiophene with the surface of activated carbons. *Langmuir*, **25**, 9302-9312.
20. Haque, E., Jun, J. W., Talapaneni, S. N., Vinu, A., and Jhung, S. H. (2010) Superior adsorption capacity of mesoporous carbon nitride with basic CN framework for phenol. *Journal of Materials Chemistry*, **20**, 10801-10803.

21. Wang, Y., Yang, R.,T., and Heinzel, J. M. (2008) Desulfurization of jet fuel by π-complexation adsorption with metal halides supported on MCM-41 and SBA-15 mesoporous materials. *Chemical Engineering Science*, **63**, 356-365.

22. Li, J.-R., Kuppler, R. J., and Zhou, H.-C. (2009) Selective gas adsorption and separation in metal-organic frameworks. *Chemical Society Reviews*, **38**, 1477-1504.

23. Jiang, H.-L., and Xu, Q. (2011) Porous metal-organic frameworks as platforms for functional applications. *Chemical Communications*, **47**, 3351-3370.

24. Seredych, M., and Bandosz, T .J. (2011) Removal of dibenzothiophenes from model diesel fuel on sulfur rich activated carbons. *Applied Catalysis, B: Environmental*, **106**, 133-141.

25. Ferey, G. (2008) Hybrid porous solids: past, present, future. *Chemical Society Reviews*, **37**, 191-214.

26. Sumida, K., Rogow, D. L., Mason, J. A., McDonald, T. M., Bloch, E. D., Herm, Z. R., Bae, T.-H., Long, J. R. (2012) Carbon dioxide capture in metal-organic frameworks. *Chemical Reviews*, **112**, 724-781.

27. Wu, H., Gong, Q., Olson, D. H., and Li, J. (2012) Commensurate adsorption of hydrocarbons and alcohols in microporous metal organic frameworks. *Chemical Reviews*, **112**, 836-868.

28. Suh, M. P., Park, H. J., Prasad, T. K., and Lim, D.-W. (2012) Hydrogen storage in metal-organic frameworks. *Chemical Reviews*, **112**, 782-835.

29. Li, J.-R., Sculley, J., and Zhou, H.-C. (2012) Metal-organic frameworks for separations. *Chemical Reviews*, **112**, 869-932.

30. Horcajada, P., Gref, R., Baati, T., Allan, P. K., Maurin, G., Couvreur, P., Ferey, G., Morris, R. E., and Serre, C. (2012) Metal-organic frameworks in biomedicine. *Chemical Reviews*, **112**, 1232-1268.

31. Furukawa, H., Ko, N., Go, Y. B., Aratani, N., Choi, S. B., Choi, E., Yazaydin, A. O., Snurr, R. Q., O'Keeffe, M., Kim, J., and Yaghi, O. M. (2010) Ultrahigh porosity in metal-organic frameworks. *Science*, **329**, 424-428.

32. Hamon, L., Serre, C., Devic, T., Loiseau, T., Millange, F., Ferey, G., and Weireld, G. D. (2009) Comparative study of hydrogen sulfide adsorption in the MIL-53(Al, Cr, Fe),MIL-47(V), MIL-100(Cr), and MIL-101(Cr) metal organic frameworks at room temperature, *Journal of American Chemical Society*, **131**, 8775-8777.

33. Hamon, L., Leclerc, H., Ghoufi, A., Oliviero, L., Travert, A., Lavalley, J.-C., Devic, T., Serre, C., Ferey, G., Weireld, G. D., Vimont, A., and Maurin, G. (2011) Molecular insight into the adsorption of H_2S in the flexible MIL-53(Cr) and rigid MIL-47(V) MOFs:infrared spectroscopy combined to molecular simulations. *Journal of Physical Chemistry C*, **115**, 2047-2056.

34. Liu, J., Thallapally, P. K., McGrail, B. P., and Brown, D. R. (2012) Progress in adsorption-based CO_2 capture by metal-organic frameworks. *Chemical Society Reviews*, **41**, 2308-2322.

35. Sumida, K., Rogow, D. L., Mason, J. A., McDonald, T. M., Bloch, E. D., Herm, Z. R., Bae, T.-H., and Long, J. R. (2012) Carbon dioxide capture in metal-organic frameworks. *Chemical Reviews*, **112**, 724-781.

36. Dathe, H., Peringer, E., Roberts, V., Jentys, A., and Lercher, J. A. (2005) Metal organic frameworks based on Cu^{2+} and benzene-1,3,5-tricarboxylate as host for SO_2 trapping agents. *C. R. Chimie*, **8**, 753-763.

37. Planchais, A., Devautour-Vinot, S., Giret, S., Salles, F., Trens, P., Fateeva, A., Devic, T., Yot, P., Serre, C., Ramsahye, N,m and Maurin, G. (2013) Adsorption of benzene in the cation-containing MOFs MIL-141. *Journal of Physical Chemistry C*, **117**(38), 19393-19401.

38. Huxford, R. C., Rocca, J. D., and Lin W. (2010) Metal-organic frameworks as potential drug carriers. *Current Opinion in Chemical Biology*, **14**(2), 262-268.

39. Kitagawa, S., Kitaura, R., and Noro, S.-I. (2004) Functional porous coordination polymers. *Angewandte Chemie International Edition*, **43**, 2334-2375.

40. Cavka, J. H., Jakobsen, S., Olsbye, U., Guillou, N., Lamberti, C., Bordiga, S., and Lillerud, K. P. (2008) A new zirconium inorganic building brick forming metal organic frameworks with exceptional stability. *Journal of American Chemical Society*, **130**, 13850-13851.

41. Walton, K. S., Millward, A. R., Dubbeldam, D., Frost, H., Low, J. J., Yaghi, O. M., and Snurr, R. Q. (2008) Understanding inflections and steps in carbon dioxide adsorption isotherms in metal-organic frameworks. *Journal of American Chemical Society*, **130**, 406-407.

42. Demessence, A., D'Alessandro, D. M., Foo, M. L., and Long, J. R. (2009) Strong CO_2 binding in a water-stable, triazolate-bridged metal-organic framework functionalized with ethylenediamine. *Journal of American Chemical Society*, **131**, 8784-8786.

43. Britt, D., Furukawa, H., Wang, B., Glover, T. G., and Yaghi, O. M. (2009) Highly efficient separation of carbon dioxide by a metal-organic framework replete with open metal sites. *Proceedings of the National Academy of Sciences of the USA*, **106**, 20637-20640.

44. Dietzel, P. D. C., Besikiotis, V., and Blom, R. (2009) Application of metal-organic frameworks with coordinatively unsaturated metal sites in storage and separation of methane and carbon dioxide. *Journal of Materials Chemistry*, **19**, 7362-7370.

45. Caskey, S. R., Wong-Foy, A. G., and Matzger, A. J. (2008) Dramatic tuning of carbon dioxide uptake via metal substitution in a coordination polymer with cylindrical pores. *Journal of American Chemical Society*, **130**, 10870-10871.

46. Couck, S., Denayer, J. F. M., Baron, G. V., Remy, T., Gascon, J., and Kapteijn, F. (2009) An amine-functionalized MIL-53 metal-organic framework with large separation power for CO_2 and CH_4. *Journal of American Chemical Society*, **131**, 6326-6327.

47. Bae, Y. S., Farha, O. K., Spokoyny, A. M., Mirkin, C. A., Hupp, J. T., and Snurr, R. Q. (2008) Carborane-based metal-organic frameworks as highly selective sorbents for CO_2 over methane. *Chemical Communications*, **37**, 4135-4137.

48. Hayashi, H., Cote, A. P., Furukawa, H., O'Keeffe, M., and Yaghi, O. M. (2007) Zeolite A imidazolate frameworks. *Nature Materials*, **6**, 501-506.

49. Phan, A., Doonan, C. J., Uribe-romo, F. J., Knobler, C. B., O'Keeffe, M., and Yaghi, O. M. (2010) Synthesis, structure, and carbon dioxide capture properties of zeolitic imidazolate frameworks. *Accounts of Chemical Research*, **43**, 58-67.

50. Hu, Z., Khurana, M., Seah, Y. H., Zhang, M., Guo, Z., and Zhao, D. (2015) Ionized Zr-MOFs for highly efficient post-combustion CO_2 capture, *Chemical Engineering Science*, **124**, 61-69.

51. Hu, Z. G., Zhang, K., Zhang, M., Guo, Z. G., Jiang, J. W., and Zhao, D. (2014) A

combinatorial approach towards water stable metal-organic frameworks forhighl efficient carbon dioxide separation. *ChemSusChem*, **7**, 2791-2795.

52. Yang, D.-A., Cho, H.-Y., Kim, J., Yang, S.-T., and Ahn, W.-S. (2012) CO_2 capture and conversion using Mg-MOF-74 prepared by a sonochemical method. *Energy & Environmental Science*, **5**, 6465-6473.

53. Achmann, S., Hagen, G., Hammerle, M., Malkowsky, I. M., Kiener, C., and Moos, R. (2010) Sulfur removal from low-sulfur gasoline and diesel fuel by metal-organic frameworks. *Chemical Engineering & Technology*, **33**, 275-280.

54. Chavan, S., Bonino, F., Valenzano, L., Civalleri, B., Lamberti, C., Acerbi, N., Cavka, J. H., Leistner, M., and Bordiga, S. (2013) Fundamental aspects of H_2S adsorption on CPO-27-Ni. *Journal of Physical Chemistry C*, **117**, 15615-15622.

55. van de Voorde, B., Hezinova, M., Lannoeye, J., Vandekerkhove, A., Marszalek, B., Gil, B., Beurroies, I., Nachtigall, P., and De Vos, D. (2015) Adsorptive desulfurization with CPO-27/MOF-74: an experimental and computational investigation. *Physical Chemistry and Chemical Physics*, **17**, 10759-10766.

56. Hamon, L., Serre, C., Devic, T., Loiseau, T., Millange, F., Frrey, G., and De Weireld, G. (2009) Comparative study of hydrogen sulfide adsorption in the MIL-53 (Al, Cr Fe), MIL-47 (V), MIL-100 (Cr), and MIL-101 (Cr) metal-organic frameworks at room temperature. *Journal of American Chemical Society*, **131**, 8775-8777.

57. Hamon, L., Leclerc, H., Ghoufi, A., Oliviero, L., Travert, A., Lavalley, J. C., Devic, T., Serre, C., Ferey, G., De Weireld, G., Vimont, A., and Maurin, G. (2011) Molecular in-sight into the adsorption of H_2S in the flexible MIL-53 (Cr) and rigid MIL-47 (V) MOFs: infrared spectroscopy combined to molecular simulations. *Journal of Physical Chemistry C*, **115**, 2047-2056.

58. Chen, Z., Ling, L., Wang, B., Fan, H., Shangguan, J., and Mi, J. (2016) Adsorptive desulfurization with metal-organic frameworks: A density functional theory investigation. *Applied Surface Science*, **387**, 483-490.

59. Fernandez, C. A., Thallapally, P. K., Motkuri, R.,K., Nune, S. K., Sumrak, J. C., Tian, J., and Liu, J. (2010) Gas-induced expansion and contraction of a fluorinated metal-organic framework. *Crystal Growth & Design*, **10**, 1037-1039.

60. Ahmed, I., and Jhung, S. H. (2014) Composites of metal-organic frameworks: preparation and application in adsorption. *Materials Today*, **17**, 136-146.

61. Glover, T. G., Peterson, G. W., Schindler, B. J., Britt, D., and Yaghi, O. M. (2011) MOF-74 building unit has a direct impact on toxic gas adsorption. *Chemical Engineering Science*, **66**, 163-170.

62. Petit, C., and Bandosz, T. J. (2012) Exploring the coordination chemistry of MOF-graphite oxide composites and their applications as adsorbents. *Dalton Transactions*, **41**, 4027-4035.

63. Britt, D., Tranchemontagne, D., and Yaghi, O. M. (2008) Metal-organic frameworks with high capacity and selectivity for harmful gases. *Proceedings of the National Academy of Sciences of the USA*, **105**, 11623–11627.

64. Karra, J. R., and Walton, K. S. (2008) Effect of open metal sites on adsorption of polar and nonpolar molecules in metal organic framework Cu-BTC. *Langmuir*, **24**, 8620-8626.

65. Xiao, B., Wheatley, P. S., Zhao, X., Fletcher, A. J., Fox, S., Rossi, A. G., Megson, I. L.,

Bordiga, S., Regli, L., Thomas, K. M., and Morris, R. E. (2007) High-capacity hydrogen and nitric oxide adsorption and storage in a metal organic framework. *Journal of American Chemical Society*, **129**, 1203-1209.

66. Jhung, S. H., Lee, J.-H., Yoon, J. W., Serre, C., Ferey, G., Chang, J.-S. (2007) Microwave synthesis of chromium terephthalate MIL-101 and its benzene sorption ability. *Advanced Materials*, **19**, 121-124.

67. Shi, J., Zhao, Z., Xia, Q., Li, Y., Li, Z. (2011) Adsorption and diffusion of ethyl acetate on the chromium-based metal-organic framework MIL-101. *Journal of Chemical Engineering Data*, **56**, 3419-3425.

68. Taylor, J. M., Vaidhyanathan, R., Iremonger, S. S., and Shimizu, G. K. H. (2012) Enhancing water stability of metal- organic frameworks via phosphonatemonoester linkers. *Journal of American Chemical Society*, **134**, 14338–14340.

69. Greathouse, J. A., and Allendorf, M. D. (2006) The interaction of water with MOF-5 simulated by molecular dynamics. *Journal of American Chemical Society*, **128**, 10678-10679.

70. Der Voort, P. V., Leus, K., Liu, Y.-Y., Vandichel, M., Speybroeck, V. V., Waroquier, M., and Biswas, S. (2014) Vanadium metal-organic frameworks: structures and applications. *New Journal of Chemistry*, **34**, 1853-1867.

71. Valenzano, L., Civalleri, B., Chavan, S., Bordiga, S., Nilsen, M. H., Jakobsen, S., Lillerud, K. P., and Lamberti, C. (2011) Disclosing the complex structure of UiO-66 metal organic framework: A synergic combination of experiment and theory. *Chemistry of Materials*, **23**(7), 1700-1718.

72. Park, K. S., Ni, Z., Cote, A. P., Choi, J. Y., Huang, R., Uribe-Romo, F. J., Chae, H. K., O'Keeffe, M., and Yaghi, O. M. (2006) Exceptional chemical and thermal stability of zeolitic imidazolate frameworks. *Proceedings of the National Academy of Sciences of the USA*, **103**, 10186-10191.

73. Ehrenmann, J., Henninger, S.,K., and Janiak, C. (2011) Water adsorption characteristics of mil-101 for heat-transformation applications of MOFs. European Journal of Inorganic Chemistry, 2011(4), DOI: doi:10.1002/ejic.201190006.

74. Jeremias, F., Khutia, A., Henninger, S. K., and Janiak, C. (2012) MIL-100(Al, Fe) as water adsorbents for heat transformation purposes - a promising application. *Journal of Materials Chemistry*, **22**, 10148-10151.

75. Cychosz, K. A., and Matzger, A. J. (2010) Water stability of microporous coordination polymers and the adsorption of pharmaceuticals from water. *Langmuir*, **26**, 17198-17202.

76. Wu, T., Shen, L., Luebbers, M., Hu, C., Chen, Q., Ni, Z., and Masel, R. I. (2010) Enhancing the stability of metal-organic frameworks in humid air by incorporating water repellent functional groups. *Chemical Communications*, **46**, 6120-6122.

77. Zu, D.-D., Lu, L., Liu, X.-Q., Zhang, D.-Y., and Sun, L.-B. (2014) Improving hydrothermal stability and catalytic activity of metal-organic frameworks by graphite oxide incorporation. *The Journal of Physical Chemistry C*, **118**, 19910-19917.

78. Yang, C., Kaipa, U., Mather, Q. Z., Wang, X., Nesterov, V., Venero, A. F., and Omary, M. A. (2011) Fluorous metalorganic frameworks with superior adsorption and hydrophobic properties toward oil spill cleanup and hydrocarbon storage. *Journal of the American Chemical Society*, **133**, 18094-18097.

79. Hasan, Z., and Jhung, S. H. (2015) Removal of hazardous organics from water using metal-organic frameworks (MOFs): Plausible mechanisms for selective adsorptions. *Journal of Hazardous Materials*, **283**, 329-339.

80. Haque, E., Lee, J. E., Jang, I. T., Hwang, Y. K., Chang, J.-S., Jegal, J., and Jhung, S. H. (2010) Adsorptive removal of methyl orange from aqueous solution with metal-organic frameworks, porous chromium-benzenedicarboxylates. *Journal of Hazardous Materials*, **181**, 535-542.

81. Haque, E., Lo, V., Minett, A. I., Harris, A. T., and Church, T. L. (2014) Dichotomous adsorption behaviour of dyes on an amino-functionalised metal-organic framework, amino-MIL-101(Al). *Journal of Materials Chemistry A*, **2**, 193-203.

82. Haque, E., Jun, J. W., and Jhung, S. H. (2011) Adsorptive removal of methyl orange and methylene blue from aqueous solution with a metal-organic frame- work material, iron terephthalate (MOF-235). *Journal of Hazardous Materials*, **185**, 507-511.

83. Leng, F., Wang, W., Zhao, X. J., Hu, X. L., and Li, Y. F. (2014) Adsorption interaction between a metal-organic framework of chromium–benzenedicarboxylates and uranine in aqueous solution. *Colloids & Surfaces A*, **441**, 164-169.

84. Chen, C., Zhang, M., Guan, Q., and Li, W. (2012) Kinetic and thermodynamic studies on the adsorption of xylenol orange onto, MIL-101(Cr). *Chemical Engineering Journal*, **183**, 60-67.

85. Liu, B., Yang, F., Zou, Y., and Peng, Y. (2014) Adsorption of phenol and p-nitrophenol from aqueous solutions on, metal-organic frameworks: effect of hydrogen bonding. *Journal of Chemical Engineering Data*, **59**, 1476-1482.

86. Xie, L., Liu, D., Huang, H., Yang, Q., and Zhong, C. (2014) Efficient capture of nitrobenzene from waste water using metal-organic frameworks. *Chemical Engineering Journal*, **246**, 142-149.

87. Khan, N. A., Jung, B. K., Hasan, Z., and Jhung, S. H. (2015) Adsorption and removal of phthalic acid and diethylphthalate from water with zeolitic imidazolate and metal-organic frameworks. *Journal of Hazardous Materials*, **282**, 194-200.

88. Tong, M., Liu, D., Yang, Q., Devautour-Vinot, S., Maurin, G., and Zhong, C. (2013) Influence of framework metal ions on the dye capture behavior of MIL-100 (Fe, Cr) MOF type solids. *Journal of Materials Chemistry A*, **1**, 8534-8537.

89. Wang, H.-N., Liu, F.-H., Wang, X.-L., Shao, K.-Z., and Su, Z.-M. (2013) Three neutral metal-organic frameworks with micro and meso-pores for adsorption and separation of dyes. *Journal of Materials Chemistry A*, **1**, 13060-13063

90. Han, R., Zou, W., Wang, Y., and Zhu, L. (2007) Removal of uranium (VI) from aqueous solutions by manganese oxide coated zeolite: discussion of adsorption isotherms and pH effect. *Journal of Environmental Radioactivity*, **93**, 127-143.

91. Feng, Y., Jiang, H., Li, S., Wang, J., Jing, X., Wang, Y., and Chen, M. (2013) Metal–organic frameworks HKUST-1 for liquid-phase adsorption of Uranium. *Colloids and Surfaces A*, **431**, 87-92.

92. Batzill, M. (2012) The surface science of graphene: Metal interfaces, CVD synthesis, nanoribbons, chemical modifications, and defects. *Surface Science Reports*, **67**, 83-115.

93. Furukawa, H., Cordova, K. E., O'Keeffe, M., and Yaghi, O. M. (2013) The chemistry and applications of metal-organic frameworks. *Science*, **341**, 1230444-1230456.

94. Tao, X., Ma, W., Li, J., Huang, Y., Zhao, J., and Yu, J. C. (2003) Efficient degradation of organic pollutants mediated by immobilized iron tetrasulfophthalocyanine under visible light irradiation. *Chemical Communications*, 80-81.

95. Kyzas, G. Z., and Matis, K. A. (2015) Nanoadsorbents for pollutants removal: a review. *Journal of Molecular Liquids*, **203**, 159-168.

96. House, K. Z., Harvey, C. F., Aziz, M. J., and Schrag, D. P. (2009) The energy penalty of post-combustion CO_2 capture & storage and its implications for retrofitting the U.S. installed base. *Energy & Environmental Science*, **2**, 193-205.

97. Cai, W., Yu, J., Anand, C., Vinu, A., and Jaroniec, M. (2011) Facile synthesis of ordered mesoporous alumina and alumina-supported metal oxides with tailored adsorption and framework properties. *Chemistry of Materials*, **23**, 1147-1157.

98. Leon, M., Diaz, E., Bennici, S., Vega, A., Ordonez, S., and Auroux, A. (2010) Adsorption of CO_2 on hydrotalcite-derived mixed oxides: Sorption mechanisms and consequences for adsorption irreversibility. *Industrial and Engineering Chemistry Research*, **49**, 3663-3671.

99. Yoshikawa, K., Sato, H., Kaneeda, M., and Kondo, J. N. (2014) Synthesis and analysis of CO_2 adsorbents based on cerium oxide. *Journal of CO_2 Utilization*, **8**, 34-38.

100. Wang, X. P., Yu, J. J., Cheng, J., Hao, Z. P. and Xu, Z. P. (2008) High-temperature adsorption of carbon dioxide on mixed oxides derived from hydrotalcite-like compounds. *Environmental Science and Technology*, **42**, 614-618.

101. Zhao, Z., Dai, H., Du, Y., Deng, J., Zhang, L., and Shi, F. (2011) Solvo- or hydrothermal fabrication and excellent carbon dioxide adsorption behaviors of magnesium oxides with multiple morphologies and porous structures. *Materials Chemistry and Physics*, **128**, 348-356.

102. Song, G., Zhu, X., Chen, R., Liao, Q., Ding, Y.-D., and Chen, L. (2016) An investigation of CO_2 adsorption kinetics on porous magnesium oxide. *Chemical Engineering Journal*, **283**, 175-183.

103. Kumar, S., Saxena, S. K., Drozd, V., and Durygin, A. (2015) An experimental investigation of mesoporous MgO as a potential pre-combustion CO_2 sorbent. *Materials for Renewable and Sustainable Energy*, **4**:8, DOI: 10.1007/s40243-015-0050-0.

104. Vu, A. T., Park, Y., Jeon, P. R., and Lee, C. H. (2014) Mesoporous MgO sorbent promoted with KNO_3 for CO_2 capture at intermediate temperatures. *Chemical Engineering Journal*, **258**, 254-264.

105. Ding, Y.-D., Song, G., Liao, Q., Zhu, X., and Chen, R. (2016) Bench scale study of CO_2 adsorption performance of MgO in the presence of water vapor. *Energy*, **112**, 101-110.

106. Abatzoglou, N., and Boivin, S. (2009) A review of biogas purification processes. *Biofuels, Bioproducts and Biorefining*, **3**, 42-71.

107. Carnes, C. L., Klabunde, K. J. (2002) Unique chemical reactivities of nanocrystalline metal oxides toward hydrogen sulfide. *Chemistry of Materials*, **14**, 1806-1811.

108. Rodriguez, J. A., and Maiti, A. (2000) Adsorption and decomposition of H_2S on MgO(100), NiMgO(100), and ZnO(0001) surfaces: A first-principles density functional study. *The Journal of Physical Chemistry B*, **104**, 3630-3638.

109. Davidson, J. M., Lawrie, C. H., and Sohail, K. (1995) Kinetics of the absorption of

hydrogen sulfide by high purity and doped surface area zinc oxide. *Industrial and Engineering Chemistry Research,* **34**, 2981-2989.

110. Ozdemir, S., and Bardakci, T. (1999) Hydrogen sulfide removal from coal gas by zinc titanate sorbent. *Separation and Purification Technology,* **16**, 225-234.

111. Ayala, R. E., and Marsh, D. W. (1991) Characterization and long-range reactivity of zinc ferrite in high-temperature desulfurization processes. *Industrial and Engineering Chemistry Research,* **30**, 55-60.

112. Lew, S., Jothimurugesan, K., and Stephanopoulos, M. F. (1992) The reaction of zinc titanate and zinc oxides solids. *Chemical Engineering Science,* **47**, 1421-1431.

113. Woods, M. C., Gangwal, S. K., Harrison, D. P., and Jothimurugesan, K. (1991) Kinetic of the reactions of a zinc ferrite sorbent in high-temperature coal gas desulfurization. *Industrial and Engineering Chemistry Research,* **30**, 100-107.

114. Akyurtlu, J. F., and Akyurtlu, A. (1995) Hot gas desulfurization with vanadium-promoted zinc ferrite sorbents. *Gas Separation and Purification,* **9**, 17-25.

115. Tamhankar, S. S., Bagajewicz, M., Gavalas, G. R., Sharma, P. K., and Flytzani-Stephanopoulos, M. (1986) Mixed-oxide sorbents for high- temperature removal of hydrogen sulfide. *Industrial and Engineering Chemistry Process Design and Development,* **25**, 429-437.

116. Flytzani-Stephanopoulos, M., Gavalas, G. R., and Tamhankar, S. S. (1998) High Temperature Regenerative H_2S Sorbents, US Patent 4729889.

117. Focht, G. D., Ranade, P. V., and Harrison, D. P. (1988) High-temperature desulfurization using zinc ferrite: Reduction and sulfidation kinetics. *Chemical Engineering Science,* **48**(11), 3005-3013.

118. Kobayashi, H., Shirai, M., and Nunokawa, M. (2002) High-temperature sulfidation behavior of reduced zinc ferrite in simulated coal gas revealed by in situ X-ray diffraction analysis and Mossbauer spectroscopy. *Energy & Fuels,* **16**, 601-607.

119. Huang, W.-F., Chen, H.-T., and Lin, M. C. (2009) Density functional theory study of the adsorption and reaction of H_2S on TiO_2 Rutile (110) and anatase (101) surfaces. *The Journal of Physical Chemistry C,* **113**, 20411-20420.

120. Novochinskii, I. I., Song, C., Ma, X., Liu, X., Shore, L., Lampert, J., and Farruato, R. J. (2004) Low-temperature H_2S removal from steam-containing gas mixtures with ZnO for fuel cell application. 1. ZnO particles and extrudates. *Energy & Fuels,* **18**, 576-583.

121. Bagreev, A., Rahman, H., and Bandosz, T. J. (2001) Thermal regeneration of activated carbon previously used as hydrogen sulfide adsorbent. *Carbon,* **39**, 1319-1326.

122. Polychronopoulou, K., Fierro, J. L. G., and Efstathiou, A. M. (2004) Novel Zn-Ti-based mixed metal oxides for low-temperature adsorption of H_2S from industrial gas streams. Applied Catalysis B: Environmental, 57, 125-137.

123. Baird, T., Campbell, K. C., Holliman, P. J., Hoyle, R., Noble, G., Stirling, D., and Williams, B. P. (2003) Mixed cobalt–iron oxide absorbents for low-temperature gas desulfurization. *Journal of Materials Chemistry,* **13**, 2341-2347.

124. Xue, M., Chitrakar, R., Sakane, K., and Ooi, K. (2003) Screening of adsorbents for removal of H_2S at room temperature. *Green Chemistry,* **5**, 529-534.

125. Bems, B., Schur, M., Dassenoy, A., Junkes, H., Herein, D., and Schlogl, R. (2003) Relations between synthesis and microstructural properties of copper/zinc hydroxycarbonates. *Chemistry - A European Journal,* **9**, 2039-2052.

126. Karvan, O., Sirkecioglu, A., and Atakul, H. (2009) Investigation of nano-CuO/mesoporous SiO2 materials as hot gas desulphurization sorbents. *Fuel Processing Technology*, **90**, 1452-1458.
127. Bae, J. W., Kang, S. H., Dhar, G. M., and Jun, K. W. (2009) Effect of Al_2O_3 content on the adsorptive properties of $Cu/ZnO/Al_2O_3$ for removal of odorant sulfur compounds. *International Journal of Hydrogen Energy*, **34**, 8733-8740.
128. Jiang, D., Su, L., Ma, L., Yao, N., Xu, X., Tang, H., and Li, X. (2010) Cu–Zn–Al mixed metal oxides derived from hydroxycarbonate precursors for H_2S removal at low temperature. *Applied Surface Science*, **256**, 3216-3223.
129. Polychronopoulou, K., Galisteo, F. C., Granados, M. L., Fierro, J. L. G., Bakas, T., and Efstathiou, A. M. (2005) Novel Fe–Mn–Zn–Ti–O mixed-metal oxides for the low-temperature removal of H2S from gas streams in the presence of H_2, CO_2, and H_2O. *Journal of Catalysis*, **236**, 205-220.
130. Lonkar, S. P., Pillai, V., Abdala, A., and Mittal, V. (2016) In situ formed graphene/ZnO nanostructured composites for low temperature hydrogen sulfide removal from natural gas. *RSC Advances*, **6**, 81142-81150.
131. Oliveira, L. C. A., Rios, R. V. R. A., Fabris, J. D., Garg, V., Sapag, K., and Lago, R. M. (2002) Activated carbon/ iron oxide magnetic composites for the adsorption of contaminants in water. *Carbon*, **40**, 2177-2183.
132. Khosravi, M., and Azizian, S. (2014) Adsorption of anionic dyes from aqueous solution by iron oxide nanospheres. *Journal of Industrial and Engineering Chemistry*, **20**, 2561-2567.
133. Tchounwou, P. B., Yedjou, C. G., Patlolla, A. K., and Sutton, D. J. (2012) Heavy metals toxicity and the environment. *EXS*, **101**, 133-164.
134. Agrawal, A., and Sahu, K. K. (2006) Kinetic and isotherm studies of cadmium adsorption on manganese nodule residue. *Journal of Hazardous Materials*, **137**, 915-924.
135. Pradeep, T., and Anshup (2009) Noble metal nanoparticles for water purification: A critical review. *Thin Solid Films*, **517**, 6441-6478.
136. Warner, C. L., Chouyyok, W., Mackie, K. E., Neiner, D., Saraf, L. V., Droubay, T. C., Warner, M. G., and Addleman, R. S. (2012) Manganese doping of magnetic iron oxide nanoparticles: tailoring surface reactivity for a regenerable heavy metal sorbent. *Langmuir*, **28**(8), 3931-3937.
137. Feng, L., Cao, M., Ma, X., Zhu, Y., and Hu, C. (2012) Superparamagnetic high-surface-area Fe_3O_4 nanoparticles as adsorbents for arsenic removal. *Journal of Hazardous Materials*, **217-218**, 439-446.
138. Hao, Y.-M., Man, C., and Hu, Z.-B. (2010) Effective removal of Cu (II) ions from aqueous solution by amino-functionalized magnetic nanoparticles. *Journal of Hazardous Materials*, **184**(1-3), 392-399.
139. Nassar, N. N. (2010) Rapid removal and recovery of Pb(II) from wastewater by magnetic nanoadsorbents. *Journal of Hazardous Materials*, **184**(1-3), 538-546.
140. Wang, H. Q., Yang, G. F., Li, Q. Y., Zhong, X. X., Wang, F. P., Li, Z. S., and Li, Y. H. (2011) Porous nano- MnO_2: large scale synthesis via a facile quick-redox procedure and application in a supercapacitor. *New Journal of Chemistry*, **35**, 469-475.

141.Li, J. D., Shi, Y. L., Cai, Y. Q., Mou, S. F., and Jiang, G. B. (2008) Adsorption of di-ethyl-phthalate from aqueous solutions with surfactant-coated nano/microsized alumina. *Chemical Engineering Journal*, **140**, 214-220.

142.Engates, K. E., and Shipley, H. J. (2011) Adsorption of Pb, Cd, Cu, Zn, and Ni to titanium diox-ide nanoparticles: effect of particle size, solid concentration, and exhaustion. *Environmental Science and Pollution Research*, **18**, 386-395.

143.Liang, P., Shi, T. Q., and Li, J. (2004) Nanometer-size titanium dioxide separation/preconcentration and FAAS determination of trace Zn and Cd in water sample. *International Journal of Environmental Analytical Chemistry*, **84**, 315-321.

144.Wang, X. B., Cai, W. P., Lin, Y. X., Wang, G. Z., and Liang, C. H. (2010) Mass production of micro/nanostructured porous ZnO plates and their strong structurally enhanced and selective adsorption performance for environmental remediation. *Journal of Materials Chemistry*, **20**, 8582-8590.

145.Lee, J. H., Kim, B. S., Lee, J. C., and Park, S. (2005) Removal of Cu++ ions from aqueous Cu-EDTA solution using ZnO nanopowder. In: *Eco-Materials Processing & Design VI*, Kim, H. S., Park, S.-Y., Hur, B. Y., and Lee, S. W. (eds.), Trans Tech Publications, Korea, pp. 510-513.

3

Potential of Ionic Liquids for Gas Separation

3.1 Introduction

Natural gas, which is largely hydrocarbon in nature, is one of the most important energy sources for several decades, and has seen its demand as a fuel to grow gradually worldwide [1]. However, besides the main component methane (CH_4), natural gas is generally contaminated with various impurities like hydrogen sulfide (H_2S) and carbon dioxide (CO_2), commonly known as acid gases. These acid gases need to be separated at the well for preventing the technological issues, which may occur during the transportation as well as the liquefaction of the gas [2]. The presence of acid gases in the natural gas can cause shrinkage in the natural gas heating value, and can also lead to corrosion in facilities [3]. Further, H_2S is immensely toxic gas and releases SO_2 into the atmosphere after the combustion process [4]. As per the regulations, the ultimate levels of CO_2 and H_2S in the sales grade natural gas are limited to 2 mol % and 4 ppm respectively [5]. In order to keep the acid gas levels within the limit for meeting the sales gas specifications, different processes have been developed for treating the natural gas streams. Gas sweetening process is the acid gas eradication process where the acid gases such as CO_2, H_2S and other sulfur containing components like mercaptans are removed from the natural gas stream. As the requirement for natural gas has enhanced in recent years, the need for developing highly efficient technologies for acid gas sweetening has emerged. Depending on the various operating conditions and different levels of CO_2 and H_2S present in the natural gas, several technologies such as absorption, adsorption, membranes and cryogenic condensation are used for the acid gas separation process.

Currently, the most widely employed techniques for the acid gas separation are the chemical and physical absorption processes [6]. The benefits of chemical absorption process, utilizing the aqueous amine solutions, include low hydrocarbon loss, high efficiency for acid gas separation at different conditions and low operational cost. However, this method suffers from various drawbacks like excessive corrosion, amine loss and intrinsically higher regeneration cost [7]. The physical absorption employing physical solvents such as

Haleema Saleem and Vikas Mittal*, The Petroleum Institute (part of Khalifa University of Science and Technology), Abu Dhabi, UAE
*Current address: Bletchington, Wellington County, Australia

Selexol has some advantages like simple regeneration and lower regeneration energy requirement when compared to the chemical absorption process. Nevertheless, physical absorption process has some drawbacks such as higher affinity of solvents towards the hydrocarbons which can lead to hydrocarbon losses as well as appropriateness for feed streams with adequately higher CO_2 pressure [8]. Hence, it is advantageous to develop new solvents with more favorable characteristics, therefore, substantial research effort has been devoted to develop suitable solvents for capturing the acid gases.

The substitution of conventional alkanolamine solutions with ionic liquids (ILs) for the removal of H_2S and CO_2 is a field of growing research [9]. The ILs are molten salts generated by the combination of various organic cations (e.g. pyrrolidinium, imidazolium, ammonium, pyridinium, phosphonium), polyatomic inorganic anions (e.g. hexafluorophosphate, tetrafluoroborate, chloride) or organic anions (e.g. tri-fluoro-methylsulfonate, bis[(trifluoromethyl)sulfonyl]imide,) [10-12]. For instance, Figure 3.1 depicts the reaction of an ionic liquid with CO_2.

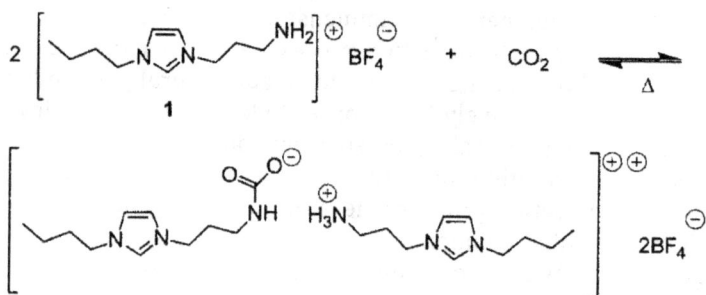

Figure 3.1 Proposed reaction between a task-specific ionic liquid (1) and CO_2. Reproduced from Reference 9 with permission from American Chemical Society.

During the past several years, ILs have acquired boundless attention in many fields due to their interesting features like excellent thermal stability, low volatility, lower regeneration energy, negligible vapor pressure, low corrosivity, fast absorption/desorption kinetics and tunable functionality [13,14]. Hence, these are regarded as environmentally benign solvents, when compared to other volatile organic solvents [15-17]. As these salts are liquids at room temperature, hence they are also known as room temperature ionic liquids (RTILs). The ILs are also called as the "designer" solvents because they can be tailored for different specific applications [18], by the suitable selection of cation, anion and substituents on the cations. Among the various applications of ILs, the two major areas of current research are (1) analyzing the possibility of utilizing ILs for the acid gas (CO_2 and H_2S) separation, and (2) the fixation as well as sequestration of CO_2, which is considered to be an important greenhouse gas [17,19]. Figure 3.2 demonstrates the use of equimolar solutions of primary alkanolamines (i.e., MEA) and secondary alkanolamines (i.e., DEA) with imidazolium based ionic liquids for CO_2 scrubbing [19]. Several theoretical as well as experimental studies have been performed to analyze the properties of various ILs for the removal of acid gases [20-22]. In a recent study, Jalili *et al.* [22] studied the solubility of hydrogen sulfide in three ionic liquids, 1-butyl-3-methylimidazolium hexafluorophosphate ([bmim][PF$_6$]), 1-butyl-3-methylimidazolium tetrafluoroborate ([bmim][BF$_4$]), and 1-butyl-3-methylimidazolium bis(trifluoromethylsulfonyl)imide ([bmim][Tf$_2$N]), at temperatures ranging from (303.15 to 343.15) K and pressures up to 1 MPa. As observed in Figure 3.3 depicting the comparison of the performance of N-methyldiethanolamine (MDEA) with ILs for the dissolution of H_2S, the low-pressure part with a low slope was attributed to absorption due to stoichiometric reaction between alkanolamine and the acid gas. However, as the overall molality of the sour gas reaches the overall molality of MDEA, the total pressure was observed to increase and additional acid gas must be dissolved physically. Comparing this phenomenon with the curves for ILs used for the study, it was observed that the solubility of H_2S in the ILs was typical of that of physical solvents. In summary, knowledge of different acid gases solubility as well as rate of solubility, i.e., diffusion coefficients of gases at different

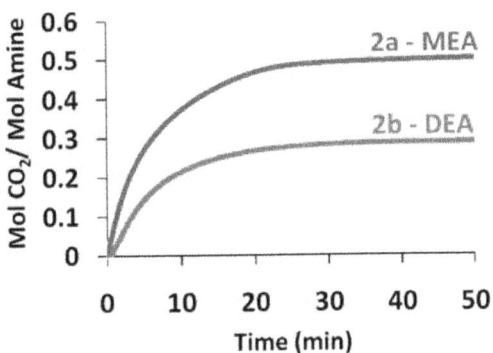

Figure 3.2 CO_2 uptake in equimolar solutions of 2a-MEA and 2b-DEA, where 2a and 2b are imidazolium based ionic liquids. Reproduced from reference 19 with permission from American Chemical Society.

pressures and temperatures, is critical in the assessment of ILs as a gas dissociation medium and also for the design as well as operation of gas sweetening
processes. Further, from the analysis of the solubility of the gases in ILs, beneficial information regarding the interaction of the gases with the ILs can also
be obtained [23].

In this chapter, the different types of ILs employed for the efficient removal
of acid gases during the natural gas sweetening process have been discussed.
Various IL techniques such as RTILs, TSILs (task specific ILs), PILs (polymerized ILs), or SILs (supported ILs) can be employed for the H_2S or CO_2 removal,

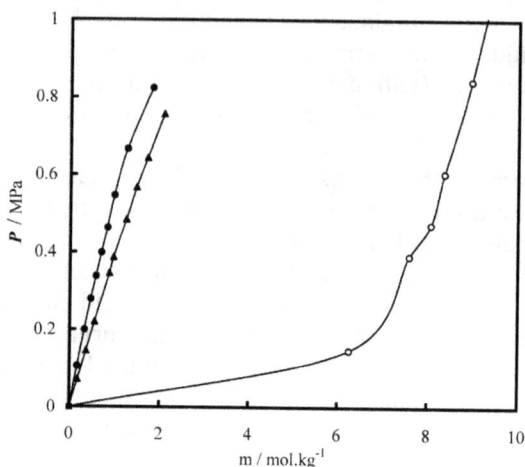

Figure 3.3 Comparison between the solubility of hydrogen sulfide as molality, m, in the ILs
and an aqueous solution of N-methyldiethanolamine (MDEA) at 313.15 K. ●, H_2S in
[bmim][Tf₂N]; ▲, H_2S in [bmim][BF₄]; O, H_2S in aqueous MDEA solution. Reproduced from
reference 22 with permission from American Chemical Society.

be employed for the H_2S or CO_2 removal, ranging from low temperature to
high temperature applications. In addition to this, the review deals with the
analysis of theoretical predictions, experimental results as well as emerging
ideas on the various ILs for the acid gas separation, with specific emphasis on
CO_2 and H_2S. Several observations in terms of advantages as well as disadvantages of employing various ILs are also detailed. Further, the review sheds
light on the influence of different anions, cations, alkyl substituents, polymer
backbones of anions and cations, and crosslinking on the acid gas sorption
ability and separating effectiveness.

3.2 Different IL Based Techniques

In this section, analysis of different ILs such as pure ILs, polymerized ILs and supported ILs, used for the acid gas separation, is carried out. Among the different supported ILs, ILs supported on membranes, MOFs as well as graphene have been provided more attention due to their potential for large scale applications.

3.2.1 Pure Ionic Liquids

Extensive reviews regarding gas solubility, exclusively CO_2, in ILs and utilization of ILs in H_2S/CO_2 removal from industrial gas streams have been reported in the literature [7,24,25]. In one such review, Bara *et al.* [25] reported an extensive analysis of imidazolium based room temperature ionic liquids for CO_2 separation. Figure 3.4 demonstrates the CO_2 solubility performance of oligo ethylene glycol, nitrile and analogous *n*-alkyl functionalized [Rmim][Tf$_2$N] RTILs. Overall, for enhancing the CO_2 solubility in ILs, many studies shave investigated the effect of various parameters like alkyl chain size of cations, different cation types, alkyl chains fluorination on the cation, various anion types and degree of fluorination of anions [23,26-28]. The major observation obtained from the various research studies is that the CO_2 solubility in the ILs with anions containing $-CF_3$ groups like [PF$_6$]$^-$, [Tf$_2$N]$^-$ and [BF$_4$]$^-$ is greater because of the CO_2 philic nature of the fluoroalkyl group, i.e., more number of fluorinated anions causes increased CO_2 solubility in ILs.

Imidazolium based ILs are of distinguished interest among the various ILs employed for the acid gas removal. Different studies on the solubility of H_2S and CO_2 in different imidazolium-based ILs, have been performed at varying pressure and temperature [25]. The first study on the H_2S solubility on imidazolium ILs was reported by Jou and Mather [20]. The authors reported the H_2S solubility in the IL 1-butyl 3-methyl imidazolium hexa-fluorophosphate ([C4mim][PF$_6$]) at pressure till 9.6 MPa and temperatures ranging from 298.15-403.15 K. As mentioned earlier, Jalili *et al.* [22] analyzed the H_2S solubility in three different imidazolium based ILs and concluded that the H_2S solubility in 1-butyl 3-methyl imidazolium based ILs followed the order [BF$_4$]$^-$ < [PF$_6$]$^-$ < [Tf$_2$N]$^-$. In another study by Jalili *et al.* [29], the solubility as well as diffusion coefficient of H_2S and CO_2 gases in the IL 1-ethyl 3-methyl imidazolium ethylsulfate ([emim][EtSO$_4$]) were determined, at pressures till 1.6 MPa along with temperatures 303.15-353.15 K [29]. The comparison confirmed that the solubility of H_2S was more than the CO_2 solubility in the IL, and the

Functional Nanomaterials & Nanotechnologies

diffusion coefficient of H_2S was almost twice the magnitude as that of CO_2 in the IL.

Nematpour *et al.* [30] studied the solubility of H_2S and CO_2 gases in the IL 1-ethyl 3-methyl imidazolium tri-fluoromethanesulfonate ([C_2mim][OTf]) at

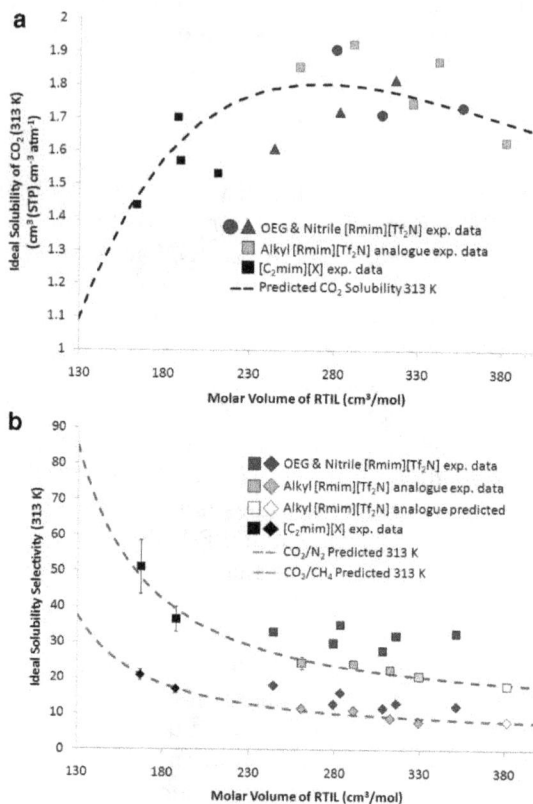

Figure 3.4 (a) Ideal CO_2 solubility as a function of molar volume, for OEG, nitrile and analogous n-alkyl functionalized [Rmim][Tf$_2$N] RTILs; (b) ideal solubility selectivity as a function of molar volume. Reproduced from reference 25 with permission from American Chemical Society.

pressures up to 3 MPa along the temperature range 303.15-353.15 K. The results indicated that at same pressure and temperature, the H_2S solubility in [C_2mim][OTf] was more than four times than the CO_2 solubility. The molality of H_2S at 2.0 MPa pressure and 303.15 K temperature was almost 1.3 times and 2.5 times that of $[C_2mim][Tf_2N]$ and $[C_2mim][eFAP]$, respectively. The order of CO_2 solubility in the ILs followed the order $[C_2mim][eFAP]$ > $[C_2mim][Tf_2N]$ > $[C_2mim][OTf]$. As the temperature was increased to 353.15 K, the H_2S solubility in $[C_2mim][OTf]$ reduced to 1.8 times the solubility in $[C_2mim][eFAP]$. Thus, as compared to other ILs, $[C_2mim][OTf]$ was confirmed to have better potential for the separation of H_2S and CO_2 gases from each other.

Soriano *et al.* [31] studied the CO_2 solubility in the IL 1-ethyl 3-methyl imidazolium tri-fluoromethanesulfonate [Emim][triflate] till 5.9 MPa pressure in the temperature range of 303.2-343.2 K with the help of a thermogravimetric microbalance. The CO_2 solubility was examined from absorption equilibrium data keeping fixed pressure and temperature. The maximal solubility was observed at 303.2 K. An enhancement in the solubility was observed with increase in pressure, while it reduced as the temperature increased. A two parameter extended Henry's law equation was employed to analyze the experimental results. In another study by Pomelli *et al.* [21], H_2S solubility in various $[C_4mim]^+$ based ILs with diverse anions and a series of $[Tf_2-N]$ ILs having distinct cations was analyzed at 1400 kPa and 298.15 K, with the help of a medium pressure NMR method. From the obtained results, it was concluded that the interaction between the anion of ILs and H_2S was the major factor contributing to the the higher H_2S solubility in the ILs. Safavi *et al.* [32] reported experimental data for the solubility of CO_2 and H_2S in 1-octyl-3methyl imidazolium hexa-fluorophosphate ($[C_8mim][PF_6]$) at pressures till 2 MPa in the temperature range from 303.15-353.15 K. It was observed that the solubility of CO_2 and H_2S enhanced with increasing pressure, however, it reduced with the increase in temperature. In this study, the H_2S solubility was about three times than that of CO_2 in the specific IL examined. The authors also examined the influence of cation alkyl chain length on H_2S and CO_2 solubility by comparing the experimental results with the data of previous reports. It was noted that the CO_2 as well as H_2S solubility increased as the cation alkyl chain length increased. In addition to this, the anion effect on the solubility was analyzed by comparing the H_2S as well as CO_2 solubility in $[C_8mim][PF_6]$ with that of $[C_8mim][Tf_2N]$. It was concluded that the H_2S and CO_2 solubility got increased in the IL with anion $[Tf_2N]^-$.

Muldoon *et al.* [28] studied the CO_2 solubility in 1-hexyl 3-methyl-1H-imidazol-3-ium tris(1,1,2,2,2-penta-fluoroethyl) tri-fluorophosphate. It was observed that the ILs with increased fluoroalkyl chains on either anion or cation enhanced the CO_2 solubility, when compared to the lesser fluorinated ILs. Zhang *et al.* [33] examined the CO_2 solubility using ILs having various anions and cations and it was confirmed that the ILs having the [eFAP]⁻ were better solvents for the removal of CO_2, when compared to other ILs. The authors also conducted molecular simulation analysis to gain further insights about the enhancement of CO_2 solubility in [eFAP]⁻ based ILs [34]. Shiflett *et al.* [35] ana-lyzed the results obtained by the vapour-liquid-liquid equilibrium (VLLE) measurements of single gases CO_2, H_2S, as well as their binary mixtures by us-ing [C₄mim][PF₆] at pressures till 6.5 MPa and temperatures from 273 K to 342 K. It was observed that the selectivity (CO_2/H_2S) was almost independent of the quantity of IL added, and ranged from 3.2 to 4.0. However, in the case of large CO_2/H_2S mole ratios (9/1), the incorporation of IL enhanced the CO_2/H_2S selectivity from 1.2 to 3.7, at 298.15 K temperature. In another study by Jalili *et al.* [36], the solubility of CO_2 and H_2S in 1-ethyl-3-methyl imidazoli-um tris(penta-fluoroethyl)tri-fluorophosphate ([C₂mim][eFAP]) was analyzed at pressures till 2 MPa and temperature range 303-353 K. The results indicat-ed that H_2S was more soluble in the specific IL than the CO_2 gas. The quantity of H_2S dissolved was greater than twice the CO_2 quantity, at fixed pressure and temperature. Thus, the aforementioned studies suggest that the pure IL sol-vent can be potentially employed for the separation of H_2S and CO_2 gases dur-ing the industrial gas sweetening processes.

3.2.2 Polymeric Ionic liquids

An emerging family of functional polymers, typically called as poly(ionic liq-uid)s or polymeric ionic liquids (PILs) have exclusive characteristics of ILs as well as macromolecular framework [37]. PILs are polyelectrolytes containing IL species in the form of anions or cations in monomers, which are linked to the polymer chain, thus, forming an extensive molecular design. The PILs can be generated by the monomer polymerization or by the co-polymerization of monomers with IL species being in repeating units. This specific class of mate-rials possess excellent CO_2 sorption, good thermal stability and faster adsorp-tion-desorption rates than the RTILs [38,39]. Due to the weak intermolecular forces, the monomeric IL molecules are liquid at the room temperature. How-ever, as the PILs are macro-molecules with strong intermolecular forces be-tween the chains, these materials are solids with greater CO_2 sorption ability than the corresponding monomeric ILs. The PILs also possess increased dura-bility, stability, excellent processability and improved control over their nano- and meso-structures.

Majority of the research studies dealing with PILs have concentrated on the addition of polymerizable species into IL cations and subsequently investigated the CO_2 and H_2S sorption as well as separation capability of PILs [37,40,41]. These studies indicated the capability of PILs as acid gas sorption material, however, more research effort is required to study different polycation-anion pairs that could govern the acid gas sorption capacity of PILs. Different functionalities can be incorporated on the interior or surface of polymeric framework, which could offer remarkable features to the porous polymers like the selectivity and reversibility. The key specifications for ideal CO_2 sorption are low cost, greater CO_2 solubility, lesser energy requirement for regeneration, long term reusability and environmental friendliness. The selection of proper cations is of great significance in the development of a PIL skeleton for the removal of acid gases. The nature of cation exhibits an essential role in describing the features of PIL, which is different than the RTIL, which are anion dependent.

Tang *et al.* [38] examined the potential of three different types of PILs as CO_2 sorption material, namely poly[2-(methacryloyloxy)ethyl-3-butyl imidazolium tetra-fluoroborate] (PIL-3), poly[1-(p-vinylbenzyl)-3-butyl-imidazolium hexa-fluorophosphate] (PIL-2b) and poly[1-(p-vinylbenzyl)3-butylimidazolium tetra-fluoroborate] (PIL-2a). At 22 °C and 592.3 mmHg CO_2, the CO_2 removal ability of RTIL [bmim][BF4] (1-butyl-3-methyl imidazolium tetra-fluoroborate) was found to be 1.30 mol%. However, in the case of PILs, the CO_2 sorption values of PIL-3, PIL-2a and PIL-2b were analyzed to be 1.78 mol% (1.4 moles), 2.27 mol% (1.7 moles) and 2.8 mol% (2.1 moles), respectively. In this study, the IL monomers exhibited no CO_2 sorption because of their crystalline structures. On the other hand, the PILs attained their 90% CO_2 sorption abilities within 3-4 minutes. In addition to this, the PILs attained entire CO_2 sorption within 30 minutes, however, the RTIL [bmim][BF$_4$] required more than 400 minutes to obtain the equilibrium [42].

Majority of PILs are based on the aliphatic backbone and are brittle, whereas the PILs based on condensation polymerization possess certain aromaticity in the backbone of main chain [43]. Films of such PILs have the capability to withstand pressure till 10 atm, during the permeation analysis. Tang *et al.* [44] proved that the P[VBTMA] based PILs with distinct anions [BF$_4$, Tf$_2$N and PF$_6$] exhibited significant CO_2 sorption capacity. The PILs can be generated either by IL monomer polymerization [45] or by anion exchange process of a precursor polymer [46]. Tang *et al.* [45,47] observed that the tetra-alkylammonium based PILs displayed reversible sorption/desorption at quicker rates as well as greater CO_2 sorption abilities, almost 6.0-7.6 times, when compared to RTILs. In addition to this, tetra-alkylammonium based PILs containing same anions offered higher CO_2 sorption, when compared to the imidazolium based PILs, which were studied earlier by the authors [42]. The steady interaction of the tetra-alkylammonium cation with the CO_2 is due to the greater positive charge density, when compared to the imidazolium cation where the positive charges are delocalized. Different types of PILs, mainly ammonium and imidazolium based polymers, have been widely prepared and compared to examine their CO_2 sorption abilities. Heintz *et al.* [41] also examined the solubility of CO_2 and N_2/H_2S mixture in a polymeric ammonium polyether-based IL with chloride anion at pressure range 0.23-3 MPa and temperature range 300 K to 500 K.

In a study by Bhavsar *et al.* [40], PILs were generated based on P[VBTMA][Cl] (poly(vinylbenzyltrimethylammonium chloride)) and P[DADMA][Cl] (poly(diallyldimethyl ammonium chloride)) as precursors, using anion exchange technique. Na or Li salts were employed to prepare water insoluble PILs, while the Ag salt of needed anion was employed to generate water soluble PILs with greater than 96% anion exchange in a single step. The PILs containing carboxylate anions exhibited great CO_2 sorption as well as higher selectivity over N_2 and H_2. With increasing order of molar masses, the CO_2 sorption in PILs containing inorganic anions was observed to be enhanced. P[DADMA][Bz] and P[DADMA][Ac] displayed significant CO_2 sorption along with higher CO_2 based selectivities S_{CO2}/S_{N2} of 41.5 and 114.3, respectively. The capability of PIL with [Ac] anion was confirmed further by detecting similar higher CO_2 sorption properties for the PIL based on the cation P[VBTMA]. The CO_2 sorption in a series of PILs with changing anion was found to be enhanced by reducing the PIL density and raising anion basicity. Xiong *et al.* [48] generated imidazolium based ILs with symmetrical ester as well as hydroxyl groups and the equivalent polymer was prepared by melt condensation polymerization. The properties and structure of the generated PILs were characterized by X-ray diffraction (XRD), differential scanning calorimetry (DSC) gel permeation chromatograph (GPC), scanning electron microscopy (SEM) and proton nuclear magnetic resonance. Further, the CO_2 sorption characteristics of the IL monomers and corresponding PILs were examined at a temperature of 25 °C and pressure of 648.4 mmHg CO_2 with the help of a thermogravimetric analyzer. Among the various PILs and IL monomers, the CO_2 sorption ability of 1,3-bis(2-hydroxylethyl)-imidazolium hexafluorophosphate ([HHIm]PF$_6$, 10 mol%) was observed to be highest. In addition to this, the sorption equilibrium of [HHIm]PF$_6$ was attained in a shorter time-period. Thus, many PILs have been developed as CO_2 sorbents, however, there is a requirement for additional exploration for understating the relationship between gas permeability and macromolecular/molecular composition [49]. The literature on PILs is generally dominated by PF_6, Tf_2N and BF_4 as anions. Several other evaluations have also been carried on anions such as OTf, dca and SbF_6 [50]. In addition to this, the CO_2 sorption abilities of imidazolium and ammonium based PILs displayed higher sorption, when compared to other polymers like polymethacrylates, polyethylenes and polystyrenes/polycarbonates [51]. Among the various cations, the ammonium cation present in simple/porous PILs exhibited greater CO_2 separation ability when compared to other cations.

In another study, Wilke *et al.* [52] reported the synthesis of a mesoporous poly(ionic liquid) network via a hard-templating pathway. The structure was stabilized by achieving a high degree of cross-linking. The synthesized mesoporous PIL exhibited faster CO_2 adsorption than the non-porous counterpart. Figure 3.5 demonstrates the synthetic pathway for the synthesis of mesoporous PIL along with template method for the network generation. The crosslinkable monomer was prepared by reaction of 1- vinylimidazole with 4-vinyl benzyl chloride. The authors observed favorable interactions of the PIL with CO_2 and the presence of large mesopores in the network led to enhanced mass transfer. The PIL also exhibited higher selectivity for CO_2 as compared to N_2.

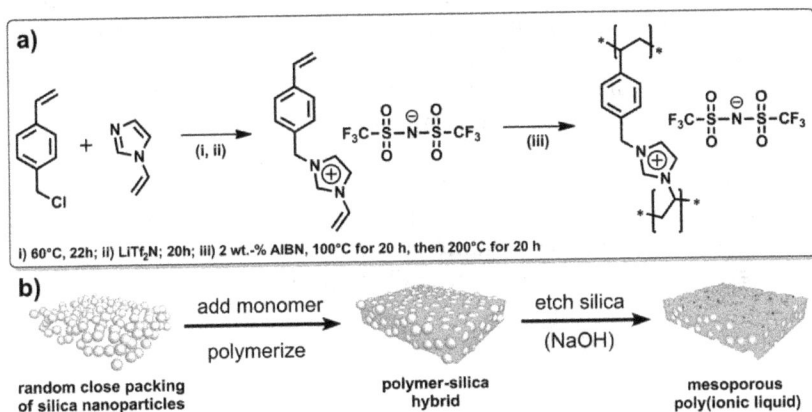

i) 60°C, 22h; ii) LiTf₂N; 20h; iii) 2 wt.-% AIBN, 100°C for 20 h, then 200°C for 20 h

random close packing of silica nanoparticles — add monomer / polymerize → polymer-silica hybrid — etch silica (NaOH) → mesoporous poly(ionic liquid)

Figure 3.5 (a) Synthetic pathway for the generation of PIL and (b) schematic of the template method used to generate the network. Reproduced from Reference 52 with permission from American Chemical Society.

In a study by Shahrom *et al.* [53], nitrile imidazolium based TSILs having different alkyl chain length and various anions ([SBA], [DDS], [DOSS], [TFMS], [BS]), PILs containing [VBTEA], [VBTMA] and [METMA] cations and [NO_3],

[BF$_4$], [Cl], [TFMS] and [PF$_6$] anions and amino acid-based polymerized ionic liquids [AAPILs] incorporating [Arg], [Pro] as anions and [VBTMA] as cation were prepared. It was observed that the CO$_2$ sorption enhanced with the enhancement in alkyl chain length, and the [C$_2$CNDim][DOSS] displayed the maximum CO$_2$ sorption (0.55 mol fraction at 298 K and 10 bar) among the different TSILs. In the case of PILs, the CO$_2$ sorption capacity enhanced in polymeric forms when compared to that of monomers. For further increasing the efficiency of PIL for the CO$_2$ adsorption, AAPILs were examined, and the results indicated that the AAPILs were able to adsorb more CO$_2$ than the PILs and TSILs, due to the presence of functionalized amine tethered at the anion in AAPIL. The results also indicated that at 298 K and lower pressure (0.7 bar), the CO$_2$ sorption of the AAPILs [VBTMA][Pro] and [VBTMA][Arg] were 0.38 and 0.530 mol fractions, respectively. On comparing the PILs and ILs, the CO$_2$ based selectivities of certain PILs were found to be higher, which clearly indicated that PILs are next generation promising materials for CO$_2$ sorption.

3.2.3 Supported Ionic Liquids

Despite the fact that ILs are promising candidates for the acid gas capture, the viscosity, energy necessity for recycling and high cost of preparation are the three main drawbacks which limit their practical applications [54]. To overcome these limitations, a new approach known as supported ILs (SILs) was proposed [55], where the ILs are encapsulated into porous solids for bettering the removal efficiency [56]. SILs are, thus, derivatives of ILs, and the IL immobilization processes combine their properties with the advantageous features of the substrates. As a result, the SILs have been explored in nearly all the fields involving ILs and have exhibited features like efficient performance, low cost operation and environmental friendly nature.

Ionic liquids Supported on Membranes

Supported liquid membranes (SLMs), the porous supports where the pores are impregnated using a solvent, have exhibited remarkable potential in various applications [57]. In the recent years, the concept of combining RTILs with membranes has been developed for the CO$_2$ separation [58,59]. By utilizing the ionic liquid membrane (ILM) technology, the drawbacks of ILs can be overcome. In this system, IL could be stabilized by impregnation inside the support membrane's pores. The ILM consists of feed as well as permeate phases, which are separated by the membrane holding the IL, allowing the concurrent extraction and stripping at each side of ILM. The use of ILM technique offers advantages like the requirement of lesser quantity of IL as carrier and no need of additional steps for IL recycling. Further, this technique also offers some other benefits like simple fabrication of compact and flexible devices, lower capital as well as operating costs, lesser energy requirements, etc. Due to its higher viscosity as well as negligible vapor pressure, the stability of ILM is higher when compared to the traditional SLM based on different organic solvents [60,61]. Thus, ILMs exhibit encouraging application potential for the removal of CO_2 and other gases. Various types of membrane processes accommodating ILs have been studied, which include supported IL membranes (SILMs), polymer/IL gel membranes, polymerized ionic liquid (PIL) membranes and membrane absorption utilizing membrane contactors (MCs) employing ILs as absorbent.

In a study by Scovazzo *et al.* [62], the selective separation of the gas pairs CO_2/N_2 and CO_2/CH_4 employing continued flows of mixed gases, at different CO_2 concentrations (till 2 bars of CO_2 partial pressure) was carried out. In this system, the ILs [emim][BF$_4$], [emim+][TfO$^-$] as well as [emim$^+$][dca$^-$] were supported in polyvinylidene fluoride (PVDF) membrane, while the ILs [emim$^+$][BETI$^-$], [hmim+][Tf$_2$N$^-$] and [emim$^+$][Tf$_2$N$^-$] were supported in a polyethersulfone (PES) membrane. The maximum CO_2/N_2 and CO_2/CH_4 selectivities were 21.2 and 27, by employing the ILs [emim+][Tf$_2$N$^-$] and [emim$^+$][BF$_4$], respectively. The segregation of CO_2, from CH_4 and N_2, was also favorably performed with IL polymer films, which were prepared from IL monomers along with polymerizable groups like acrylate and styrene [63]. The separation of CO_2 from CH_4 was also studied by utilizing task specific ILs like [NH$_2$pmim$^+$][CF$_3$SO$_3^-$] and [NH$_2$pmim$^+$][Tf$_2$N–], where the functional groups present in the ILs are able to chemically complex with CO_2 [64]. This type of SILMs, having amine terminated ILs, promoted the transportation of CO_2 through the membrane, exhibiting greater selectivities when compared to the IL [bmim$^+$][NTf$_2^-$] for the removal of CO_2 from the CO_2/CH_4 gas mixture. Effective gas separations by employing SILM have also been reported in other studies such as H_2S/CH_4 [65] and CO_2/He [66]. Park *et al.* [65] studied the multiphase separation method, using RTIL and polymer to develop SILMs, for the separation of acidic gases from the crude natural gas. PVDF material was used for the membrane, while the BMImBF$_4$ was employed as the RTIL. SEM analysis was carried out for the structural analysis of generated SILMs. For the examination of the permeation behavior, the SILMs were tested with H_2S, CO_2 and CH_4 at different operating conditions. Due to the fact that the gases like H_2S and CO_2 have greater affinity towards the RTILs than the CH_4, the permeability coefficients of H_2S and CO_2 were found to be significantly higher at 160-1100 and 30-180 Barrer, respectively. In addition to this, the selectivity of H_2S/CH_4 and CO_2/CH_4 were observed to be 130-260 and 25-45, respectively. The influence of various operating conditions in gas separation has also been examined. Neves *et al.* [67] analyzed the potential of utilizing the SILMs based on 1-n-alkyl 3 methyl imidazolium cation for the CO_2/CH_4 as well as CO_2/N_2 separations. Here, the effect of existence of water vapor in the gas stream on the gas selectivity and permeability was studied. It was found that the water vapor presence in the gas stream enhanced the gas permeability of SILM, however, reduced the CO_2/CH_4 and CO_2/N_2 selectivity remarkably, when compared to the dry gas stream. This reduction in the selectivity was due to the water cluster development within the membrane, and the influence was more evident for the less hydrophobic ILs. In another study by Cserjesi *et al.*, [68] the analysis of permeability for CO_2, N_2, H_2 and CH_4 gases and the selectivity of SILM based on broad variety of ILs at various trans-membrane pressures and temperatures was performed. It was observed that the enhancement in temperature increased the permeability, while the enhancement in trans-membrane pressure reduced the permeability. Also, the SILMs exhibited relatively higher long-term stability, as their permeability did not exhibit any remarkable variation in the course of the experiments.

IL based membranes have also been utilized for gas permeation at higher temperatures [66]. In these type of membranes, the diffusivity selectivity benefits the smaller molecules like H_2 or He, while CO_2 is highly benefitted by solubility selectivity. Myers *et al.* [69] synthesized task specific ILs containing functional groups which are able to form complexes with CO_2. This was employed as fiber supported liquid membrane (FSLM) and the CO_2/H_2 selectivity of the FSLM reached till 10-20 at about 85 °C temperature. Albo *et al.* [70] analyzed the CO_2 removal in the IL [emim][EtSO$_4$] (1-ethyl-3-methyl-imidazolium ethylsulfate) using a polypropylene (PP) hollow fiber membrane contactor for simulated flue gas at ambient temperatures (Figure 3.6). Here, a mathematical model considering a parallel flow configuration was employed for a cross flow system, for explaining the mass transfer rate.

In another study by Gomez-Coma *et al.* [71], the analysis of the temperature effect on the CO_2 removal efficiency using two ILs, [emim][EtSO$_4$] (1-Ethyl-3-methyl-imidazolium ethylsulfate) and [emim][Ac] (1-Ethyl-3-methyl-imidazolium acetate) was carried out. In this system, a PP hollow fiber module served as the membrane device where the absorption of CO_2 took place and the temperature ranged from room temperature to 333 K. The CO_2 separation efficiency was attained from the experimental data, and it was observed that only in the case of [emim][Ac], the temperature dependence was observed. Here, the CO_2 removal efficiency doubled from temperature 291 K to 318 K.

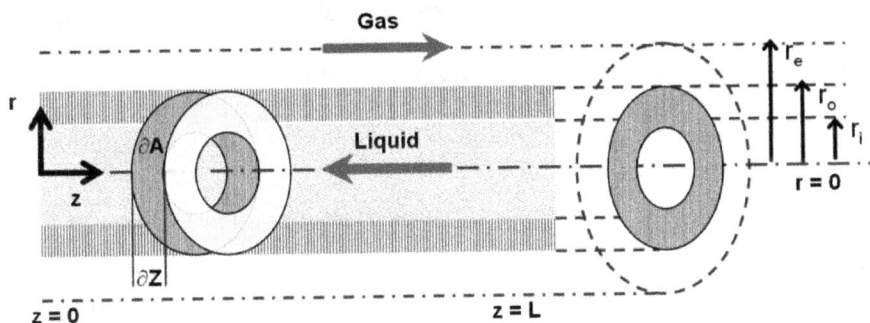

Figure 3.6 Axial and radial coordinates of the fiber. Reproduced from Reference 70 with permission from American Chemical Society.

Mulikutla *et al.* [72] analyzed the CO_2 reactive absorption in the novel non-volatile absorbent containing IL [bmim][DCA] (1-butyl 3-methyl-imidazolium-dicyanamide) which consisted of 20 wt% polyamidoamine (PAMAM) dendrimer Gen 0. Here, a humidified simulated flue gas which contained almost 14% CO_2 was employed, and favorable separation of bulk of CO_2 was observed. The absorption and regeneration was attained by using rectangular cross flow hollow fiber membrane modules which were made of porous PP fibers with porous polymerized fluorosiloxane coating on the outside diameter. Chau *et al.* [73] studied the novel cyclic 5 valve pressure swing membrane absorption process (PSMAB) for the separation of CO_2 from low temperature shifted syngas in hollow fiber membrane contactor using IL absorbent ([bmim][DCA]) at about 100 °C. At higher temperature, the CO_2 removal efficiency was remarkably increased by the addition of PAMAM dendrimer Generation 0 to the IL for a 60% He, 40% CO_2 feed. The absorption behavior in this process depends on the selective sorption, absorbent viscosity and hollow fiber module design.

The IL based supported membranes possess several limitations like poor stability which leads to poor long term performance and operation being achievable only at low pressures [63,74,75]. For overcoming these limitations, the inclusion of IL nature in the polymer backbone is developing as a promising technology. Due to the limitations of PILs based on aliphatic backbone, such as inability to form film and the limitation of obtained membranes to be useful only at lower pressure about 2 atm [63], efforts have been taken for transforming these into membranes by utilizing techniques such as copolymerization or crosslinking for efficient flat film formation [76]. The casting of IL monomer on a porous polymer support, and followed by UV polymerization has also been studied [76]. The gas removal behavior of the produced membranes could be analyzed at lower pressure, i.e., till 2 atm upstream pressure, due to its fragile nature. By utilizing the crosslinking methodology, Li et al. [74] generated composite membrane of IL and PIL, as shown in Figure 3.7. In

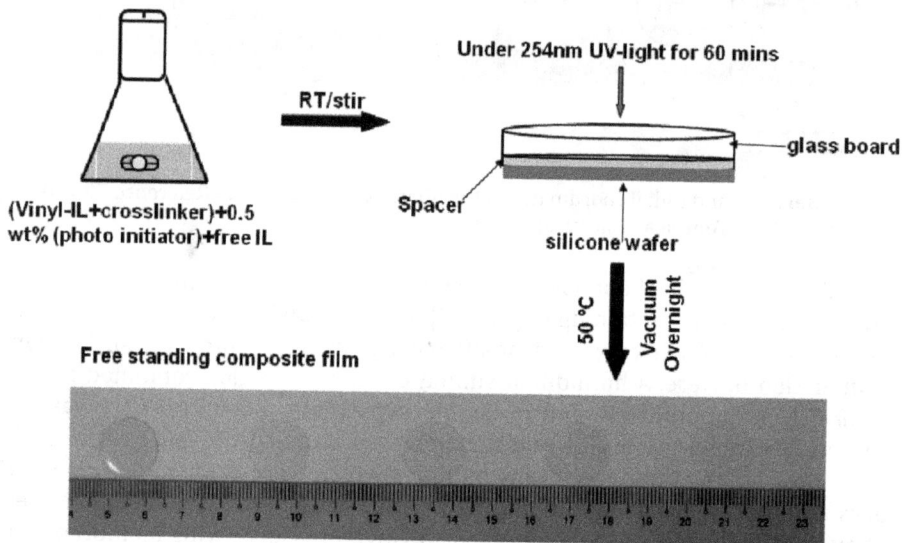

Figure 3.7 Synthesis procedure of poly(RTIL)-RTIL composite films Reproduced from Reference 74 with permission from American Chemical Society.

the study by Hu et al. [77], PEG grafted onto PIL, such as P[MATMA][BF$_4$] and P[VBTMA][BF$_4$] generated chemically, mechanically as well as thermally stable CO_2 selective membranes. PEG475 was observed to be less effective than the PEG2000. The membranes generated using grafted polymers were found to be less brittle when compared to the membrane made of pure P[MATMA][BF$_4$] and (P[VBTMA][BF$_4$]). The P[VBTMA][BF$_4$]-g-PEG2000 and P[MATMA][BF$_4$]-g-PEG2000 exhibited similar CO_2/CH_4 separation, when compared to the polymeric membranes. The selectivity of P[MATMA][BF$_4$]-g-PEG2000 and P[VBTMA][BF$_4$]-g-PEG2000 for the CO_2/N_2 and CO_2/CH_4 separation was mainly because of the solubility differences, and not due to the differences in their diffusivities.

Ionic Liquids Supported on Metal Organic Frameworks

In the past several decades, metal organic frameworks (MOFs) have obtained significant attention because of their exclusive properties like large pore volume, higher surface area and tunable structure [17]. Many research studies have demonstrated that the MOFs can be used as a promising porous support for the ILs [78-81]. Jiang *et al.* [78] studied the composite of [BMIM][PF₆] which was supported on a MOF IRMOF-1, for the CO_2/N_2 separation by means of molecular simulations (Figure 3.8). It was confirmed that the selectivity could be increased when the ILs are added into the pores of the IRMOF-1. The IL present in the composite displayed an ordered structure, as seen from the radial distribution functions, because of the confinement effect. In the open pores of IRMOF-1, the large [BMIM]⁺ cation resided, while the small anion [PF₆]⁻ was located in the metal cluster corner. In comparison with [PF₆]⁻, [BMIM]⁺ illustrated higher mobility. By enhancing the IL ratio in the composite, the confinement effect could be increased, thereby reducing the motility of [PF₆]⁻ and [BMIM]⁺. Ions present in the composite offered stronger interaction with CO_2, especially the [PF₆]⁻ anion was the most favorable site for the adsorption of CO_2. The composite offered selective CO_2 adsorption from the CO_2/N_2 mixture, with selectivity remarkably high in comparison with several

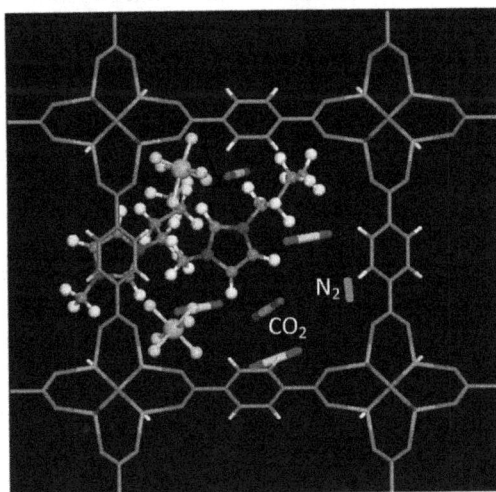

Figure 3.8 Simulation snapshot of CO_2/N_2 mixture in IL/IRMOF-1 at $W_{IL/IRMOF-1} = 0.4$. Reproduced from Reference 78 with permission from American Chemical Society.

other supported ILs. In addition, by enhancing the IL ratio in the composite, the selectivity was observed to increase. Subsequently, the authors computationally investigated the efficiency of MOF supported IL membranes for the CO_2/N_2 separation [81]. The IRMOF-1 supported IL membranes were analyzed for the capture of CO_2 by atomistic simulations. The role of hydrophilic/hydrophobic framework and the effect of anions of the ILs was analyzed. In this system, the IL consisted of same cation 1-n-butyl-3-methylimidazolium [BMIM]$^{(+)}$, however, four different types of anions namely tetrafluoroborate [BF$_{(4)}$]$^{(-)}$, thiocyanate [SCN]$^{(-)}$, hexafluorophosphate [PF$_{(6)}$]$^{(-)}$ and bis(trifluoromethylsulfonyl)imide [Tf$_{(2)}$N]$^{(-)}$ were used. The anions possessed a stronger interaction with IRMOF-1 and highly ordered structure in the IRMOF-1, when compared to the cation. The bulky as well as chain like [Tf$_{(2)}$N]$^{(-)}$ and [BMIM]$^{(+)}$ occupied place near the phenyl ring, whereas the small anions [SCN]$^{(-)}$, [BF$_{(4)}$]$^{(-)}$ and [PF$_{(6)}$]$^{(-)}$ were observed to be located near the metal cluster, specifically the quasi-spherical[BF$_{(4)}$]$^{(-)}$ and [PF$_{(6)}$]$^{(-)}$. Among the four different anions, [BMIM]$^{(+)}$ exhibited strongest interaction with IRMOF-1, while [Tf$_{(2)}$N]$^{(-)}$ had the weakest interaction. By enhancing the weight ratio of IL to IRMOF-1, the selectivity of CO_2/N_2 at infinite dilution could be increased. At a specified weight ratio $W_{IL/IRMOF-1}$, the selectivity decreased in the order [SCN]$^{(-)}$> [BF$_{(4)}$]$^{(-)}$> [PF$_{(6)}$]$^{(-)}$> [Tf(2)N]$^{(-)}$. The simulation analysis clearly acknowledged that the anion had substantial influence on the microscopic properties of ILs and also proposed that the MOF supported ILs are capable material for the CO_2 capture.

Vicent-Luna *et al.* [79] investigated different IL/Cu-BTC composites for the CO_2 removal, with special focus on the effect of various anions and the quantity of ILs. A molecular simulation study was carried out to analyze the effect on gas adsorption (CO_2, CH_4, N_2 and their mixtures), when the RTILs are incorporated into the pore of Cu-BTC MOF. It was found that the existence of RTILs in the pores of MOF increases remarkably the adsorption of CO_2 at lower pressure, whereas the N_2 and CH_4 adsorption was unaffected (Figure 3.9). In another study conducted by Tzialla *et al.* [80], it was observed that the permeability and CO_2 selectivity of a [omim][TCM]/ZIF 69 composite membrane were greater when compared to the bulk IL and pure Zeolitic imidazolate framework ZIF-69 membrane (Figure 3.10). The ZIF-69 membrane was developed on porous α-alumina substrates by seeded secondary growth and was subsequently functionalized using a CO_2-selective alkylmethyl imidazolium cation/tri-cyanomethanide anion based IL. In this system, the ZIF intergrain boundaries as well as the defects were fixed by a medium offering higher CO_2 selectivity. Thus, the selectivity of the hybrid membrane was remarkably greater than that of as grown ZIF membranes, due to the presence of ZIF channels, and the permeability was greater when compared to bulk IL. It was

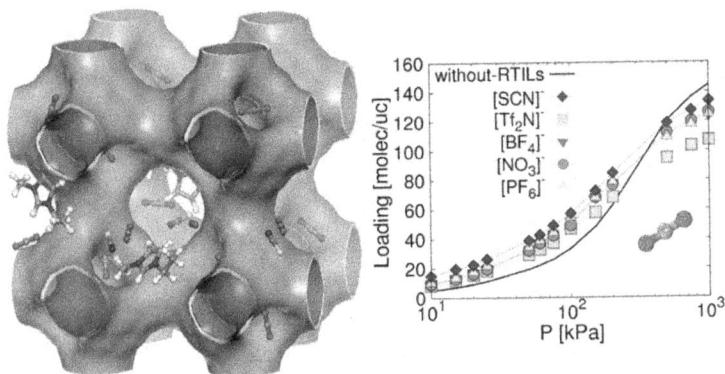

Figure 3.9 Comparison of various anions for CO_2 removal. Reproduced from Reference 79 with permission from American Chemical Society.

observed that the CO_2 permeation was 20 times faster than the N_2 through the undamaged ZIF pores, and 65 times faster than through the IL phase. The developed membranes under a 2 bar transmembrane pressure and at room temperature displayed CO_2 permeance of 3.7×10^{-11} mol m^{-2}s^{-1}Pa^{-1} and 5.6×10^{-11} molm^{-2}s^{-1}Pa^{-1}, and real CO_2/N_2 selectivities of 64 and 44 for CO_2/N_2 mixtures containing 75% and 44% (v/v) CO_2, respectively. Thus, the study presented a promising solution for the issues related to the defect formation during the growth of ZIFs, zeolite and inorganic membranes used for the removal of CO_2.

Xue *et al.* [82] conducted a computational study to analyze the dispersion characteristics of ILs in MOFs and covalent organic frameworks (COFs). In addition, the separation behavior of the developed composites for CO_2/N_2 and CO_2/CH_4 mixtures was also examined. For the IL 1-n-butyl-3-methyl imidazolium thiocyanate [BMIM][SCN], eight COFs and five MOFs, with various pore structure as well as chemical properties, were picked as the supports. The results indicated that the stronger Coulombic interactions provided by the MOF framework caused the better dispersion of IL molecules in their pores, when compared to COFs. The gas removal ability was remarkably increased by the addition of [BMIM][SCN] into COFs and MOFs. The superior dispersion of the IL in the support provided improved removal efficiency of the composite, and

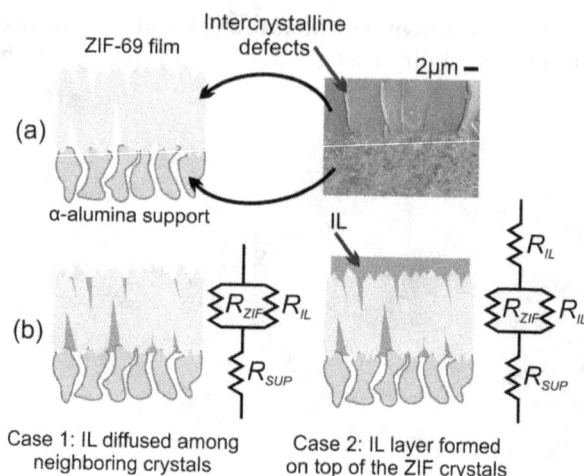

Figure 3.10 Schematic illustration of ZIF-69 membrane morphologies. Reproduced from Reference 80 with permission from American Chemical Society.

this phenomenon was more visible for the CO_2/CH_4 mixture, when compared to the CO_2/N_2 mixture. The study also clearly mentioned that employing IL supports with strong adsorption sites such as co-ordinatively unsaturated-metal sites cannot accomplish remarkable increase in the gas removal ability of the composites. In another study, Li et al. [83] performed molecular simulations to analyze the performance of IL/Cu-TDPAT composites for the separation of the H_2S/CH_4 mixture. In this study, the ILs consisted of four types of anions ([PF_6]-, [Cl]-, [BF_4]-, [Tf_2N]-]) and   identical cation [BMIM]+. The results indicated that the H_2S adsorption ability could be remarkably increased by the addition of IL into the pores of Cu-TDPAT, and the composite with [Cl]- anion exhibited the highest heat of adsorption. Within the pressure range analyzed, the H_2S/CH_4 adsorption selectivities of each composite were remarkably greater than that of the pure Cu-TDPAT, and the selectivity mainly exhibited an enhancement with increasing the IL loading. The study also confirmed that the [BMIM][Cl]/Cu-TDPAT composite attained the best separation performance in both PSA and VSA processes. Thus, the observations made in the study provided beneficial information for the development of new assuring IL/MOF composites for H_2S capture from the natural gas stream. Overall, it was also confirmed that the application of MOFs as the IL support is an effective approach to prepare new promising adsorbents for the removal of H_2S and CO_2 during the natural gas purification.

Ionic Liquids Supported on Graphene

Graphene, a material with two-dimensional (2D) lattice of carbon atoms having one atom thickness, has attained significance in adsorption related applications because of its exclusive properties among the various carbon nanomaterials. The performance of graphene oxide (GO) in the CO_2 adsorption behavior of chitosan based aerogel was studied and the adsorption capacity was observed to become almost double by the addition of 20% GO, due to the enhanced specific surface area [84]. Further, it was also reported that the hydrogen facilitated exfoliation-co-reduction of graphite oxide, which resulted in the formation of highly wrinkled graphene, offered greater affinity sites for holding the CO_2 molecules [85]. In a study by Ghosh *et al.* [86], the CO_2 adsorption on graphene at 195 K temperature and 100 kPa was examined. The analysis of CO_2 adsorption behavior of graphene confirmed that the adsorption energy of chemisorption was in the range of 297.9-301.7 kJ.mol^{-1}, while that of physisorption was in the range 8.8-13.8 kJ.mol^{-1}, depending on the CO_2 molecule orientation [87]. As the adsorption behavior depends on the surface anchoring sites as well as the lateral surface properties (porosity and specific surface area), the task specific functionalization of graphene with an appropriate moiety present the potential to tune the adsorption behavior of graphene [88]. It was also observed that the CO_2 solubility in ILs can be increased through supporting ILs using graphene based substrate.

Tamilarasan *et al.* [89] studied the integration of IL or PIL with graphene to examine the increase in CO_2 adsorption properties. Here, graphene was subjected to non-covalent functionalization by IL or PIL, followed by determination of CO_2 adsorption as well as desorption behavior at lower pressures (<100 kPa). The IL uniformly enveloped the surface of graphene, whereas the PIL formed greatly distributed porous nanoparticles. The IL functionalization enhanced the CO_2 adsorption capacity by 2% than graphene, whereas the PIL functionalized graphene exhibited 22% greater adsorption capacity. Thus, the study demonstrated the benefit of polymerizing the IL for the CO_2 adsorption process. Further, it was also noted that the adsorption behavior of the integrated system was greater, when compared to the single constituents (either PIL or IL or graphene). The authors also observed that the PIL functionalization provided more adsorption favorability, with a greater adsorption energy. Due to the ease of regeneration of adsorbent, the isosteric heat of adsorption was determined to be in the range of 18-28 kJ.mol^{-1}. These observations underline the potential of PIL integration with greater surface area nanostructures for the enhancement in the adsorption ability. In another study by the authors, the preparation and application of adsorptivity based amine rich IL (ARIL) i.e., 3,5-diamino-1-methyl-1,2,4-triazolium tetra-fluoroborate grafted graphene (HEG/ARIL) for CO_2 adsorption was studied [90]. At standard pressure and temperature, the IL appeared in solid state, and formed solid like short range ordering on the graphitic substrate. It was noted from the molecular vibrational spectroscopy that the ARIL molecules were physically adsorbed on the graphene surface, with no identifiable chemical bonding. By grafting graphene with ARIL, the adsorbate confinement of graphene got significantly increased. Further, on the HEG/ARIL, the entropy and isosteric heat of CO_2 adsorption was observed to be lower than that of pure HEG. Thus, it was concluded that the ARIL functionalization of graphene enhanced the CO_2 adsorption behavior and reduced the interaction between graphene and CO_2, due to the fact that it increased the density of lesser affinity adsorption sites.

Bian *et al.* [91] proposed a new method of IL- assisted growth of $Cu_3(BTC)_2$ on GO sheets and used it for increasing the CO_2 adsorption ability as well as adsorption rate. To analyze the influence of anions and cations of ILs on the GO-IL/MOF composite structure, three ILs, namely, 1-butyl-3methyl-imidazolium tetra-fluoroborate ([Bmim]BF_4), triethylene tetramine tetrafluoroborate (TETA-BF_4) and triethylene tetramine acetate (TETA-Ac) were examined. With imidazole or amine cations adsorbed at the GO surface and the closely attached anions, the GO-ILs can contribute a number of active sites for the Cu^{2+} cations absorption through coordination. Among the different GO-IL/MOF composites, GO-TAc/MOF-60 exhibited a superimposed structure, which could result in more adsorption activity sites and lessen the transfer distance. Further, the GO sheets present in the GO-IL/MOF composites contributed channels for the quicker transfer and exhibited a greater CO_2 adsorption ability of 5.62 mmol/g at 100 kPa and 25 °C. The material also exhibited greater CO_2 kinetic separation behavior. Thus, the composite was observed to have a better cyclic adsorption/desorption stability. The relation between the CO_2 adsorption ability and the composite specific structures was experimentally determined to obtain a suitable method for designing and preparing hierarchical MOF composites.

3.3 Summary and Outlook

The increasing level of CO_2 in the atmosphere plays a significant role towards the global warming as well as climate change and can cause substantial damage to the environment. In addition, CO_2 and H_2S are present in the natural gas streams and are required to be removed for enhancing the usefulness of natural gas as well as to safeguard the facilities agonist corrosion and other harmful effects. H_2S is specifically highly toxic and releases SO_2 into the atmosphere following combustion. To reduce the levels of these gases, ionic liquids offer an exceptionally versatile as well as tunable platform to generate an extensive variety of sorbents for the H_2S and CO_2 capture. This review has highlighted the current developments in the area of ILs as sorbents for H_2S and/or CO_2 separation, and has also enlisted the challenges impeding the increased efficiency of these materials. In this category, various IL bases techniques such as RTIL, PIL and SILs have been extensively reviewed. This brief scrutiny on the present trends on the IL mediated CO_2 and H_2S capture suggests that these acid gas capture by ILs is a feasible practice. From the analysis of different literature studies, it was observed that the parameters like anions types, cation types, alkyl chain size of cations, alkyl chains fluorination on the cation, pressure, degree of fluorination of anions, temperature and the anion interaction with acid gases have a major role in determining the gas solubility of ILs. Further, it was observed that the conversion of ILs into the macromolecules (PILs) can remarkably increase the CO_2 or H_2S sorption behavior. The proper selection of cations is of considerable significance in the development of a PIL skeleton for the removal of CO_2. Further, the benefits of using supported ILs such as ILs supported on membranes, MOFs and graphene were discussed. Supported ionic liquid membranes combine the benefits of IL with the solid supports and it can be regarded as excellent materials for gas separation because of the higher selectivity and satisfactory permeability. MOFs have been proved to be promising porous supports for the ILs. It was also seen that the integration of IL or PIL with graphene enhanced its CO_2 adsorption properties. Even though the unavailability of low cost and diverse ILs is the major concern hindering the large scale application of IL systems for CO_2 and H_2S separation, however, developments in the synthesis methods would result in large scale production to become feasible in near future. Thus, with optimized choice of the IL system components, low coat and efficient performance, the IL based technologies represent strong potential to replace other absorption and adsorption based methods for gas separation.

References

1. Berg, S. V. (1998) Lessons in electricity market reform: regulatory processes and performance. *The Electricity Journal*, **11**(5), 13-20.
2. Kohl, A. L., and Nielsen, R. B. (1997) *Gas Purification*, 5th edition, Gulf Publishing Company, USA.

3. Rufford, T. E., Smart, S., Watson, G. C. Y., Graham, B. F., Boxall, J., Diniz da Costa, J. C., and May, E. F. (2012) The removal of CO_2 and N_2 from natural gas: a review of conventional and emerging process technologies. *Journal of Petroleum Science and Engineering*, **94**, 123-154.

4. Koech, P. K., Rainbolt, J. E., Bearden, M. D., Zheng, F., and Heldebrant, D. J. (2011) Chemically selective gas sweetening without thermal-swing regeneration. *Energy and Environmental Science*, **4**, 1385-1390.

5. Abdulrahman, R. K., and Sebastine, I. M. (2013) Natural gas sweetening process simulation and optimization: a case study of Khurmala field in Iraqi Kurdistan region. *Journal of Natural Gas Science and Engineering*, **14**, 116-120.

6. Bhide, B. D., Voskericyan, A., and Stern, S. A. (1998) Hybrid processes for the removal of acid gases from natural gas. *Journal of Membrane Science*, **140**, 27-49.

7. Karadas, F., Atilhan, M., and Aparicio, S. (2010) Review on the use of ionic liquids (ILs) as alternative fluids for CO_2 capture and natural gas sweetening. *Energy & Fuels*, **24**, 5817-5828.

8. Revelli, A.-L., Mutelet, F., Jaubert, J.-N. (2010) High carbon dioxide solubilities in imidazolium-based ionic liquids and in poly(ethylene glycol) dimethyl ether. *The Journal of Physical Chemistry B*, **114**, 12908-12913.

9. Bates, E. D., Mayton, R. D., Ntai, I., and Davis, J. H. (2002) CO_2 capture by a task-specific ionic liquid. *Journal of the American Chemical Society*, **124**, 926-927.

10. Rahmati-Rostami, M., Ghotbi, C., Hosseini-Jenab, M., Ahmadi, A. N., and Jalili, A. H. (2009) Solubility of H_2S in ionic liquids [hmim][PF6], [hmim][BF4], and [hmim][Tf2N]. *The Journal of Chemical Thermodynamics*, **41**, 1052-1055.

11. Plechkova, N. V., and Seddon, K. R. (2008) Applications of ionic liquids in the chemical industry. *Chemical Society Reviews*, **37**, 123-150.

12. Wang, J., Luo, J., Feng, S., Li, H., Wan, Y., and Zhang, X. (2016) Recent development of ionic liquid membranes. *Green Energy & Environment*, **1**(1), 43-61.

13. Zhang, X., Zhang, X., Dong, H., Zhao, Z., Zhang, S., and Huang, Y. (2012) Carbon capture with ionic liquids: Overview and progress. *Energy and Environmental Science*, **5**, 6668-6681.

14. Ju, Y.-J., Lien, C.-H., Chang, K.-H., Hu, C.-C., and Wong, D. S.-H. (2012) Deep eutectic solvent-based ionic liquid electrolytes for electrical double-layer capacitors. *Journal of the Chinese Chemical Society*, **59**, 1280-1287.

15. Seddon, K. (2002) Ionic liquids: designer solvents for green synthesis. *Chemical Engineer*, **730**, 33-35.

16. Duchet, L., Legeay, J. C., Carrie, D., Paquin, L., Vanden Eynde, J. J., and Bazureau, J. P. (2010) Synthesis of 3,5-disubstituted 1,2,4-oxadiazoles using ionic liquid-phase organic synthesis (IoLiPOS) methodology. *Tetrahedron*, **66**, 986-994.

17. Gurdal, Y., and Keskin, S. (2013) Predicting noble gas separation performance of metal organic frameworks using theoretical correlations. *The Journal of Physical Chemistry C*, **117**, 5229-5241.

18. Rogers, R. K., and Seddon, K. R. (2002) *Ionic Liquids: Industrial Applications to Green Chemistry*, Oxford University Press, USA.

19. Camper, D., Bara, J. E., Gin, D. L., and Noble, R. D. (2008) Room-temperature ionic liquid– amine solutions: Tunable solvents for efficient and reversible capture of CO_2. *Industrial and Engineering Chemistry Research*, **47**, 8496-8498.

20. Jou, F.-Y., and Mather, A. E. (2007) Solubility of hydrogen sulfide in [bmim][PF6]. *International Journal of Thermophysics*, **28**, 490-495.

21. Pomelli, C. S., Chiappe, C., Vidis, A., Laurenczy, G., and Dyson, P. J. (2007) Influence of the interaction between hydrogen sulfide and ionic liquids on solubility: Experimental and theoretical investigation. *The Journal of Physical Chemistry B*, **111**, 13014-13019.

22. Jalili, A. H., Rahmati-Rostami, M., Ghotbi, C., Hosseini-Jenab, M., and Ahmadi, A. N. (2009) Solubility of H_2S in ionic liquids [bmim][PF6], [bmim][BF4], and [bmim][Tf2N]. *Journal of Chemical & Engineering Data*, **54**, 1844-1849.

23. Anthony, J. L., Maginn, E. J., and Brennecke, J. F. (2002) Solubilities and thermodynamic properties of gases in the ionic liquid 1-n-butyl-3-methylimidazolium hexafluorophosphate. *The Journal of Physical Chemistry B*, **106**, 7315-7320.

24. Lei, Z., Dai, C., and Chen, B. (2014) Gas solubility in ionic liquids. Chemical Reviews, 114, 1289-1326.

25. Bara, J. E., Carlisle, T. K., Gabriel, C.,J., Camper, D., Finotello, A., Gin, D. L., and Noble, R. D. (2009) Guide to CO_2 separations in imidazolium-based room-temperature ionic liquids. *Industrial & Engineering Chemistry Research*, **48**, 2739-2751.

26. Aki, S. N. V. K., Mellein, B. R., Saurer, E. M., and Brennecke, J. F. (2004) High-pressure phase behavior of carbon dioxide with imidazolium-based ionic liquids. *The Journal of Physical Chemistry B*, **108**, 20355-20365.

27. Aziz, N., Yusoff, R., and Aroua, M. K. (2012) Absorption of CO_2 in aqueous mixtures of N-methyldiethanolamine and guanidinium tris (pentafluoroethyl) trifluorophosphate ionic liquid at high-pressure. *Fluid Phase Equilibria*, **323**, 120-125.

28. Muldoon, M. J., Aki, S. N. V. K., Anderson, J. L., Dixon, J. K., and Brennecke, J. F. (2007) Improving carbon dioxide solubility in ionic liquids. *The Journal of Physical Chemistry B*, **111**, 9001-9009.

29. Jalili, A. H., Mehdizadeh, A., Shokouhi, M., Ahmadi, A. N., Hosseini-Jenab, M., and Fateminassab, F. (2010) Solubility and diffusion of CO_2 and H_2S in the ionic liquid 1-ethyl-3-methylimidazolium ethylsulfate. *The Journal of Chemical Thermodynamics*, **42**, 1298-1303.

30. Nematpour, M., Jalili, A. H., Ghotbi, C., and Rashtchian, D. (2016) Solubility of CO_2 and H_2S in the ionic liquid 1-ethyl-3-methylimidazolium trifluoromethanesulfonate. *Journal of Natural Gas Science and Engineering*, **30**, 583-591.

31. Soriano, A. N., Doma, Jr., B. T., and Li, M.-H. (2009) Carbon dioxide solubility in 1-ethyl-3-methylimidazolium trifluoromethanesulfonate. *The Journal of Chemical Thermodynamics*, **41**(4), 525-529.

32. Safavi, M., Ghotbi, C., Taghikhani, V., Jalili, A. H., and Mehdizadeh, A. (2013) Study of the solubility of CO_2, H_2S and their mixture in the ionic liquid 1-octyl-3-methylimidazolium hexafluorophosphate: Experimental and modelling. *The Journal of Chemical Thermodynamics*, **65**, 220-232.

33. Zhang, X., Liu, Z., and Wang, W. (2008) Screening of ionic liquids to capture CO_2 by COSMO-RS and experiments. *AIChE Journal*, **54**, 2717-2728.

34. Zhang, X., Huo, F., Lio, Z., Wang, W., Shi, W., and Maginn, E. J. (2009) Absorption of CO_2 in the ionic Liquid 1-n-Hexyl-3-methylimidazolium tris (pentafluoroethyl) tri-

fluorophosphate ([hmim][FEP]): A molecular view by computer simulations. *The Journal of Physical Chemistry B*, **113**, 7591-7598.

35. Shiflett, M. B., and Yokozeki, A. (2010) Separation of CO_2 and H_2S using room-temperature ionic liquid [bmim][PF_6]. *Fluid Phase Equilibria*, **294**, 105-113.

36. Jalili, A. H., Shokouhi, M., Maurer, G., and Hosseini-Jenab, M. (2013) Solubility of CO_2 and H_2S in the ionic liquid 1-ethyl-3-methylimidazolium tris(pentafluoroethyl)trifluorophosphate. *The Journal of Chemical Thermodynamics*, **67**, 55-62.

37. Mecerreyes, D. (2011) Polymeric ionic liquids: Broadening the properties and applications of polyelectrolytes. *Progress in Polymer Science*, **36**, 1629-1648.

38. Tang, J., Sun, W., Tang, H., Radosz, M., and Shen, Y. (2005) Enhanced CO_2 absorption of poly(ionic liquid)s. *Macromolecules*, **38**, 2037-2039.

39. Yu, G., Man, Z., Li, Q., Li, N., Wu, X., Asumana, C., and Chen, X. (2013) New crosslinked-porous poly-ammonium microparticles as CO_2 adsorbents, *Reactive and Functional Polymers*, **73**, 1058-1064.

40. Bhavsar, R. S., Kumbharkar, S. C., and Kharul, U. K. (2012) Polymeric ionic liquids (PILs): Effect of anion variation on their CO_2 sorption. *Journal of Membrane Science*, **389**, 305-315.

41. Heintz, Y. J., Sehabiague, L., Morsi, B. I., Jones, K. L., Luebke, D. R., and Pennline, H. W. (2009) Hydrogen sulfide and carbon dioxide removal from dry fuel gas streams using an ionic liquid as a physical solvent. *Energy & Fuels*, **23**, 4822-4830.

42. Tang, J., Tang, H., Sun, W., Radosz, M., and Shen, Y. (2005) Poly (ionic liquid)s as new materials for CO_2 absorption. *Journal of Polymer Science, Part A: Polymer Chemistry*, **43**, 5477-5489.

43. Li, P., Zhao, Q., Anderson, J. L., Varanasi, S., and Coleman, M. R. (2010) Synthesis of copolyimides based on room temperature ionic liquid diamines. *Journal of Polymer Science, Part A: Polymer Chemistry*, **48**, 4036-4046.

44. Tang, J., Shen, Y., Radosz, M., and Sun, W. (2009) Isothermal carbon dioxide sorption in poly(ionic liquid)s. *Industrial & Engineering Chemistry Research*, **48**, 9113-9118.

45. Tang, J., Tang, H., Sun, W., Plancher, H., Radosz, M., and Shen, Y. (2005) Poly(ionic liquid)s: a new material with enhanced and fast CO_2 absorption. *Chemical Communications*, 3325-3327.

46. Marcilla, R., Blazquez, J. A., Rodriguez, J., Pomposo, J. A., and Mecerreyes, D. (2004) Tuning the solubility of polymerized ionic liquids by simple anion-exchange reactions. *Journal of Polymer Science, Part A: Polymer Chemistry*, 42, 208-212.

47. Tang, J., Tang, H., Sun, W., Radosz, M., and Shen, Y. (2005) Low-pressure CO_2 sorption in ammonium-based poly(ionic liquid)s. *Polymer*, **46**, 12460-12467.

48. Xiong, Y. B., Wang, H., Wang, Y. J., and Wang, R. M. (2011) Novel imidazolium-based poly(ionic liquid)s: preparation, characterization, and absorption of CO_2. *Polymers for Advanced Technologies*, **23**(5), 835-840.

49. Green, O., Grubjesic, S., Lee, S., and Firestone, M. A. (2009) The design of polymeric ionic liquids for the preparation of functional materials. *Journal of Macromolecular Science C: Polymer Reviews*, **49**, 339-360.

50. Bara, J. E., Gin, D. L., and Noble, R. D. (2008) Effect of anion on gas separation performance of polymer-room-temperature ionic liquid composite membranes. *Industrial & Engineering Chemistry Research*, **47**, 9919-9924.

51. Kato, S., Tsujita, Y., Yoshimizu, H., Kinoshita, T., and Higgins, J. S. (1997) Characterization and CO_2 sorption behaviour of polystyrene/polycarbonate blend system. *Polymer*, **38**, 2807-2811.

52. Wilke, A., Yuan, J., Antonietti, M., and Weber, J. (2012) Enhanced carbon dioxide adsorption by a mesoporous poly(ionic liquid). *ACS Macro Letters*, **1**, 1028-1031.

53. Shahrom, M. S. R., Wilfred, C. R., and Taha, A. B. K. Z. (2016) CO_2 capture by task specific ionic liquids (TSILs) and polymerized ionic liquids (PILs and AAPILs). *Journal of Molecular Liquids*, **219**, 306-312.

54. Zheng, J., Li, S., Wang, Y., Li, L., Su, C., Liu, H., Zhu, F., Jiang, R., and Ouyang, G. (2014) In situ growth of IRMOF-3 combined with ionic liquids to prepare solid-phase microextraction fibers. *Analytica Chimica Acta*, **829**, 22-27.

55. Selvam, T., Machoke, A., and Schwieger, W. (2012) Supported ionic liquids on non-porous and porous inorganic materials - A topical review. *Applied Catalysis A*, **445-446**, 92-101.

56. Yu, Y., Mai, J., Wang, L., Li, X., Jiang, Z., and Wang, F. (2014) Ship-in-a-bottle synthesis of amine-functionalized ionic liquids in NaY zeolite for CO_2 capture. *Scientific Reports*, **4**, 5997-6005.

57. Muthuramam, G., and Palanivelu, K. (2006) Transport of textile dye in vegetable oils based supported liquid membrane. *Dyes and Pigments*, **70**, 99-104.

58. Scovazzo, P., Visser, A. E., Davis, J. H., Rogers, R. D., Koval, C. A., DuBois, D. L., and Noble, R. D. (2002) Supported ionic liquid membranes and facilitated ionic liquid membranes. *Ionic Liquids*, **818**, 69-87.

59. Scovazzo, P., Kieft, J., Finan, D. A., Koval, C., DuBois, D., and Noble, R. (2004) Gas separations using non-hexafluorophosphate [PF_6]$^-$ anion supported ionic liquid membranes. *Journal of Membrane Science*, **238**(1-2), 57-63.

60. Luis, P., Van Gerven, T., and van der Bruggen, B. (2012) Recent developments in membrane-based technologies for CO_2 capture. *Progress in Energy and Combustion Science*, **38**(3), 419-448.

61. Al Marzouqi, M. H., Abdulkarim, M. A., Marzouk, S. A., El-Naas, M. H., and Hasanain, H. M. (2005) Facilitated transport of CO_2 through immobilized liquid membrane. *Industrial & Engineering Chemistry Research*, **44**, 9273-9278.

62. Scovazzo, P., Havard, D., McShea, M., Mixon, S., and Morgan, D. (2009) Long-term, continuous mixed-gas dry fed CO_2/CH_4 and CO_2/N_2 separation performance and selectivities for room temperature ionic liquid membranes. *Journal of Membrane Science*, **327**, 41-48.

63. Bara, J. E., Lessmann, S., Gabriel, C. J., Hatakeyama, E. S., Noble, R. D., and Gin, D. L. (2007) Synthesis and performance of polymerizable room-temperature ionic liquids as gas separation membranes. *Industrial & Engineering Chemistry Research*, **46**, 5397-5404.

64. Hanioka, S., Maruyama, T., Sotani, T., Teramoto, M., Matsuyama, H., Nakashima, K., Hanaki, M., Kubota, F., and Goto, M. (2008) CO_2 separation facilitated by taskspecific ionic liquids using a supported liquid membrane. *Journal of Membrane Science*, **314**, 1-4.

65. Park, Y.-I., Kim, B.-S., Byun, Y.-H., Lee, S.-H., Lee, E.-W., and Lee, J.-M. (2009) Preparation of supported ionic liquid membranes (SILMs) for the removal of acidic gases from crude natural gas. *Desalination*, **236**, 342-348.

66. Ilconich, J., Myers, C., Pennline, H., and Luebke, D. (2007) Experimental investigation of the permeability and selectivity of supported ionic liquid membranes for CO_2/He separation at temperatures up to 125 °C. *Journal of Membrane Science*, **298**, 41-47.

67. Neves, L. A., Crespo, J. G., and Coelhoso, I. M. (2010) Gas permeation studies in supported ionic liquid membranes. *Journal of Membrane Science*, **357**, 160-170.

68. Cserjesi, P., Nemestothy, N., and Belafi-Bako, K. (2010) Gas separation properties of supported liquid membranes prepared with unconventional ionic liquids. *Journal of Membrane Science*, **349**, 6-11.

69. Myers, C., Pennline, H., Luebke, D., Ilconich, J., Dixon, J. K., Maginn, E. J., and Brennecke, J. F. (2008) High temperature separation of carbon dioxide/hydrogen mixtures using facilitated supported ionic liquid membrane. *Journal of Membrane Science*, **322**, 28-31.

70. Albo, J., Luis, P., and Irabien, A. (2010) Carbon dioxide capture from flue gases using a cross-flow membrane contactor and the ionic liquid 1-ethyl-3-methylimidazolium ethyl sulfate. *Industrial & Engineering Chemistry Research*, **49**, 11045-11051.

71. Gomez-Coma, L., Garea, A., and Irabien, A. (2014) Non-dispersive absorption of CO_2 in [emim][EtSO4] and [emim][Ac]: temperature influence. *Separation and Purification Technology*, **132**, 120-125.

72. Mulukutla, T., Obuskovic, G., and Sirkar, K. K. (2014) Novel scrubbing system for post-combustion CO_2 capture and recovery: experimental studies. *Journal of Membrane Science*, **471**, 16-26.

73. Chau, J., Jie, X., and Sirkar, K. K. (2016) Polyamidoamine-facilitated poly(ethylene glycol)/ionic liquid based pressure swing membrane absorption process for CO_2 removal from shifted syngas. *Chemical Engineering Journal*, **305**, 212-220.

74. Li, P., Pramoda, K. P., and Chung, T. S. (2011) CO_2 separation from flue gas using polyvinyl-(room temperature ionic liquid)-room temperature ionic liquid composite membranes. *Industrial & Engineering Chemistry Research*, **50**, 9344-9353.

75. Yuan, J., Mecerreyes, D., and Antonietti, M. (2013) Poly(ionic liquid)s: an update. *Progress in Polymer Science*, **38**, 1009-1036.

76. Bara, J. E., Hatakeyama, E. S., Gabriel, C. J., Zeng, X., Lessmann, S., Gin, D. L., and Noble, R. D. (2008) Synthesis and light gas separations in cross-linked gemini room temperature ionic liquid polymer membranes. *Journal of Membrane Science*, *316*, 186-191.

77. Hu, X., Tang, J., Blasig, A., Shen, Y., and Radosz, M. (2006) CO_2 permeability, diffusivity and solubility in polyethylene glycol-grafted polyionic membranes and their CO_2 selectivity relative to methane and nitrogen. *Journal of Membrane Science*, **281**, 130-138.

78. Chen, Y., Hu, Z., Gupta, K. M., and Jiang, J. (2011) Ionic liquid/metal-organic frameworks composite for CO_2 capture: A computational investigation. *The Journal of Physical Chemistry C*, **115**, 21736-21742.

79. Vicent-Luna, J. M., Gutierrez-Sevillano, J. J., Anta, J. A., and Calero, S. (2013) Effect of room-temperature ionic liquids on CO_2 separation by a Cu-BTC metal-organic framework. *The Journal of Physical Chemistry C*, **117**, 20762-20768.

80. Tzialla, O., Veziri, C., Papatryfon, X., Beltsios, K. G., Labropoulos, A., Iliev, B., Adamova, G., Schubert, T. J. S., Kroon, M. C., Francisco, M., Zubeir, L. F., Romanos, G. E., and Karanikolos, G. N. (2013) Zeolite imidazolate framework-ionic liquid hybrid

membranes for highly selective CO_2 separation. *The Journal of Physical Chemistry C,* **117**, 18434-18440.

81. Gupta, K. M., Chen, Y., Hu, Z., and Jiang, J. (2012) Metal-organic frameworks supported ionic liquid membranes for CO_2 capture: Anion effects. *Physical Chemistry and Chemical Physics,* **14**, 5785-5794.

82. Xue, W.-Li, Z., Huang, H., Yang, Q., Liu, D., Xu, Q., and Zhong, C. (2016) Effects of ionic liquid dispersion in metal-organic frameworks and covalent organic frameworks on CO_2 capture: A computational study. *Chemical Engineering Science,* **140**, 1-9.

83. Li, Z., Xiao, Y., Xue, W., Yang, Q., and Zhong, C. (2015) Ionic liquid/metal-organic framework composites for H_2S removal from natural gas: A computational exploration. *The Journal of Physical Chemistry C,* **119**, 3674-3683.

84. Alhwaige, A. A., Agag, T., Ishida, H., and Qutubuddin, S. (2013) Biobased chitosan hybrid aerogels with superior adsorption: Role of graphene oxide in CO_2 capture. *RSC Advances,* **3**, 16011-16020.

85. Bienfait, M., Zeppenfeld, P., Dupont-Pavlovsky, N., Muris, M., Johnson, M. R., Wilson, T., DePies, M., and Vilches, O. E. (2004) Thermodynamics and structure of hydrogen, methane, argon, oxygen, and carbon dioxide adsorbed on single-wall carbon nanotube bundles. *Physical Reviews B,* **70**, 035410.

86. Ghosh, A., Subrahmanyam, K. S., Krishna, K. S., Datta, S., Govindaraj, A., Pati, S. K., and Rao, C. N. R. (2008) Uptake of H_2 and CO_2 by graphene. *The Journal of Physical Chemistry C,* **112**, 15704-15707.

87. Lee, K.-J., and Kim, S.-J. (2013) Theoretical investigation of CO_2 adsorption on graphene. *Bulletin of Korean Chemical Society,* **34**, 3022-3026.

88. Mishra, A. K., and Ramaprabhu, S. (2012) Nanostructured polyaniline decorated graphene sheets for reversible CO_2 capture. *Journal of Materials Chemistry,* **22**, 3708-3712.

89. Tamilarasan, P., and Ramaprabhu, S. (2015) Integration of polymerized ionic liquid with graphene for enhanced CO_2 adsorption. *Journal of Materials Chemistry A,* **3**, 101-108.

90. Tamilarasan, P., and Ramaprabhu, S. (2016) Amine-rich ionic liquid grafted graphene for sub-ambient carbon dioxide adsorption. *RSC Advances,* **6**, 3032-3040.

91. Bian, Z., Zhu, X., Jin, T., Gao, J., Hu, J., and Liu, H. (2014) Ionic liquid-assisted growth of $Cu_3(BTC)_2$ nanocrystals on graphene oxide sheets: Towards both high capacity and high rate for CO_2 adsorption. *Microporous and Mesoporous Materials,* **200**, 159-164.

4

Recent Developments in Self-healing Coatings for Corrosion Protection

4.1 Introduction

Generally, corrosion involves the degradation of metals due to oxidation and reduction processes occurring during interaction of metallic surfaces with aggressive environments. These electrochemical processes result in the impairment of materials' physical and mechanical properties such as strength and ductility. In general, the corrosion processes can be divided into following types depending on the environment [1]:

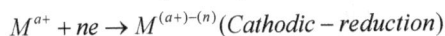

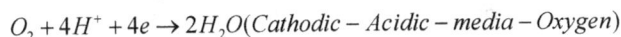

$$M \rightarrow M^{a+} + ne \, (Anodic)$$

$$M^{a+} + ne \rightarrow M^{(a+)-(n)} \, (Cathodic - reduction)$$

$$M^{a+} + ne \rightarrow M \, (Cathodic - deposition)$$

$$2H^+ + 2e \rightarrow H_2 \uparrow (Cathodic - Acidic - media)$$

$$O_2 + 4H^+ + 4e \rightarrow 2H_2O \, (Cathodic - Acidic - media - Oxygen)$$

$$2H_2O + O_2 + 4e \rightarrow 4OH^- (Cathodic - Neutral - Basic - media - Oxygen - pH \uparrow)$$

where M represents the metal or metallic cation. It can be seen from above mentioned corrosion reactions that there is only one anodic reaction whereas the corresponding cathodic reactions depend on the environment, reduction ability of metal or presence of more noble metal.

Figure 4.1 shows the reactions occurring on thermodynamically unstable metallic surface exposed to oxygenated neutral (pH 7) corrosive solution. The nature of non-metallic products on the metal surface can be sometimes protective to avoid further corrosion reactions. The economic losses due to corrosion reach billions of dollars per year globally [2]. To slow down the corrosion to a manageable rate, many methods are employed such as alloying, materials selection, design, cathodic protection, sacrificial anodes, environmental alteration (if possible), inhibitors for solutions and coatings (metallic and nonmetal-

Ali U. Chaudhry and Vikas Mittal**, The Petroleum Institute (part of Khalifa University of Science and Technology), Abu Dhabi, UAE
*Current address: Texas A & M University, Qatar; **Current address: Bletchington, Wellington County, Australia

Functional Nanomaterials & Nanotechnologies

lic) [3]. Among non-metallic coatings, polymers are widely used on metallic surfaces as passive physical barrier (coatings or films) to the corrosive species. Along with the barrier nature of the polymeric coatings, the insulating nature of polymers also prevents external flow between anodic and cathodic areas [4]. Most commonly employed resins for polymer coatings, paints, or primers are the epoxies, phenolics, vinyls, polyolefin, rubbers, alkyds, polyurethanes, fluorocarbons, polyesters etc. [5]. In a recent report, the global demand for anti-corrosion coating has been predicted to reach to ~$27 billion by 2019 [6].

Application of polymeric coatings on substrates is a complex process and usually defects appear in the coatings which allow faster deterioration upon ingression of corrosive solution. The passive nature of the polymeric coatings can be changed into active by introducing anti-corrosion inhibitor/fillers in the matrices. Along with the anti-corrosion properties, fillers also improve mechanical, electrical, optical, rheological, adhesion, and weathering properties of the polymer coatings [7]. During corrosion, the active inhibitors/fillers may interfere with the corrosion process when the main barrier is damaged or becomes permeable, thus, leading to the corrosion reaction to be stopped or slowed to a manageable rate. Generally, the protection of metallic surface in the presence of corrosion inhibitor/pigments can be attributed to the interactions between the free electrons of inhibitor and empty orbital of metals, thus, resulting in the development of surface complexes, passive or barrier films [7].

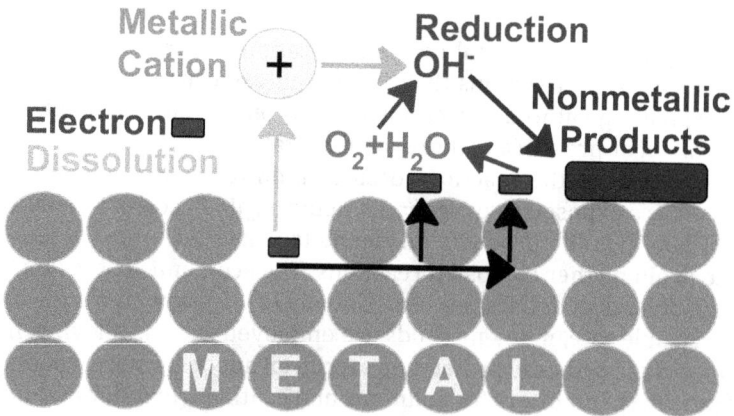

Figure 4.1 Fundamental processes of corrosion on metal surface in oxygenated neutral environment.

Anti-corrosion fillers/inhibitors for polymeric coatings can be categorized depending on the protection mechanisms offered by them when incorporated in the coatings. Barrier fillers provide tortuous/longer diffusional paths for the corrosive solution due to the high aspect ratio, plate-like or layered structure e.g. carbon nanotubes, graphene, clay, hexagonal boron nitride, etc., thus, leading to the slowdown of the corrosion process to a manageable rate. The protection action provided by the electrochemical fillers is due to their redox ability e.g. chromates based anti-corrosion fillers/inhibitors. These fillers usually produce cations by leaching out in corrosive solution. These cations interfere with the corrosion processes and produce complex corrosion products or barrier/passive layer on the metal surface which further slows down the corrosion. The other protection mechanism exhibited by the electrochemical fillers could be sacrificial and cathodic protective effects e.g. zinc metal. These fillers consume themselves instead of metal due to the lower position in galvanic series and provide protection to the metal [4,7]. These fillers also provide barrier effect to the coating by forming corrosion products which further fill the voids of polymeric coatings. The later protection mechanism is generally called self-healing. Certain kinds of fillers also have mixed protection behavior e.g. the electrochemical fillers can act as barrier due to their shape and aspect ratio e.g. ferrites and cerium based fillers [8,9]. Similarly, to obtain synergistic effect of anti-corrosion fillers, electrochemical/redox properties can be added to the barrier fillers through surface modifications such as composites of conducting polymers with graphene [10], carbon nanotubes [11], clay [12], and titanium dioxide (TiO_2) [13].

There are many disadvantages associated with the use of conventional electrochemical anti-corrosion inhibitors/fillers in polymer coatings such as deactivation or inert behavior due to filler-interaction with the coating material, loss of coating barrier properties due to solubility of the inhibitors in corrosive solution, etc. [14]. Along with this, the inhibition action of the anti-corrosion fillers/inhibitors mainly depends on many factors like solubility of inhibitor in a corrosive solution, formation of passive/barrier layer on the metal surface which separates it from the corrosive solution, nature of the passive layer, and nature of electrolyte [7]. Controlled release of the loaded corrosion inhibitors during corrosion process or due to externally applied signal/trigger is a promising way to overcome drawbacks associated with the conventional electrochemical anti-corrosion inhibitors/fillers. The development of self-healing fillers enables the passive coatings to provide smart response to the changes in the local environment. Depending on the kind of changes in the local environments i.e. pH, redox reactions, etc., many techniques were adopted to introduce self-healing functionality to the polymeric coatings. The incorporated active fillers can respond intelligently and immediately to the changes in the coating or at the coating-metal interface. The main approach to add self-healing ability to the polymer coating is the use of an active carrier loaded with corrosion inhibitor. The active carrier could be nano-containers, nanotube capsules, polymer backbone [15], clay, metallic oxides, porous and hollow particles, etc. [16]. The carrier is incorporated in the host material or polymer coating and reacts to the changes occurring due to corrosion. This approach avoids the unnecessary release or interaction of corrosion inhibitor with the polymer coating, maintains the coating integrity and leads to better compatibility between the fillers and host polymer.

Before discussing about the self-healing fillers, it is necessary to discuss briefly about the techniques used to determine the self-healing ability of incorporated filler or coating. The electrochemical testing techniques used to evaluate the efficiency of self-healing ability of incorporated fillers are mainly divided based on bulk/global and local electrochemical corrosion measurements. The techniques like open circuit potential (OCP), electrochemical impedance spectroscopy (EIS), potentiodynamic polarization (PP) and linear polarization resistance (LPR) come under the category of bulk/global measurements. On the other hand, for local electrochemical corrosion measurements, techniques like scanning vibrating electrode technique (SVET), local electrochemical impedance spectroscopy (LEIS) and scanning Kelvin probe (SKP) are widely used [14]. Bulk techniques are usually unable to distinguish the intensity of corrosion processes occurring on the surface. On the contrary, localized corrosion measurements are performed by positioning the detector close to the corroding surface (≤100 microns). This method also helps to create map of the corrosion data which explains the non-uniformity of the corrosion on the metal surface. In literature, the use of local corrosion measurements techniques has been reported to be superior for determining the self-healing abilities of the incorporated fillers.

This chapter highlights recent developments achieved to add functionality in passive coatings by synthesizing self-healing fillers and incorporating into polymers or hybrid coating matrices. The classification of self-healing fillers is based on the loading techniques of the inhibitor. Thus, this article also reviews the synthesis and potential applications of self-healing fillers. Specifically, four techniques including self-healing nano-containers, layer by layer assembly of polyelectrolyte self-healing coatings, ceramic materials as self-healing fillers, conducting polymers as self-healing fillers have been reviewed owing to their advanced performance.

4.2 Self-healing Nano-containers

Incorporation of nano-containers impregnated with corrosion inhibitor is a common method to add self-healing ability to the polymeric coatings. This method gained considerable attention and provides stability to the coating by avoiding the excessive leaching out of inhibitor which further avoids the blistering and formation of micro-holes in the coating. Further, these nano-containers have been reported to provide strength, extended shelf-life, and exceptional bonding to the host materials. During the destruction of dense physical barrier, these nano-containers slowly release the inhibitor as triggered by changes in pH or potential. The first reported approach in this category was the use of urea-formaldehyde microcapsules loaded with inhibitors [16]. Kumar *et al.* [17] used urea formaldehyde (UF) microcapsules filled with different types of inhibitors/film formers such as camphor, tung oil, spar varnish, isodecyl diphenyl phosphate, alkyl-ammonium salt in xylene. These filled UF capsules were incorporated in commercial paints and after accelerated corrosion testing of 2016 hours, the filled capsules were observed to reduce the under-film corrosion. The protection behavior was attributed to the controlled release of the corrosion inhibitor from the capsules during production of cracks in coating. Similarly, Liao *et al.* [18] used UF microcapsules filled with a mixture of epoxy resin and healing agents and confirmed that the cracks produced in polymer coatings during salt spray testing were healed satisfactorily. Jadhav *et al.* [19] also reported the UF microcapsules filled with linseed oil as self-healing fillers. Figure 4.2 exhibits the mechanism of controlled release of the inhibitor from nano-containers and the corrosion inhibition process [20].

Attempts have also been made to use inexpensive and economically viable nano-containers for the encapsulation of corrosion inhibitors. Shchukin *et al.* [21] used low cost cylindrical shaped halloysite nano-containers filled with 2-mercaptobenzothiazole as corrosion inhibitor (Figure 4.3). The nano-containers were covered with polyelectrolyte multilayers to avoid the undesirable leakage and interaction of inhibitor with the coating. The release of the inhibitor was triggered by changes in local pH due to the sensitivity of polyelectrolyte layers [21]. Similarly, Lvov *et al.* [22] also used the benzotriazole filled halloysite alumino-silicate nanotubes and observed effective results in terms of self-healing.

Further, several research studies have exhibited the self-healing abilities of filled containers/capsules as self-healing anti-corrosion fillers for polymer coatings such as polyurethane microcapsule with hexamethylene diisocyanate

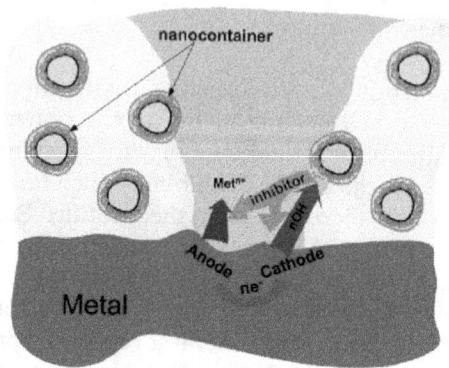

Figure 4.2 Mechanism of controlled release of the inhibitor from nano-containers and the corrosion inhibition process. Reprinted from Reference 20 with permission from American Chemical Society.

[23], layered double hydroxides and cerium molybdate hollow nano-spheres with mercaptobenzothiazole [24,25], cerium molybdate with 8-hydroxyquinoline (8-HQ) or 1-H-benzotriazole-4-sulfonic acid (1-BSA) [26], silica-polystyrene with 8-hydroxyquinoline [27], mesoporous silica nanoparticles with 1*H*-benzotriazole (BTA) [28], calcium carbonate microbeads with cerium nitrate, salicylaldoxime and 2,5-dimercapto-1,3,4-thiadiazolate [29], poly(urea–formaldehyde) with perfluorooctyl triethoxysilane [30], pH responsive polymeric nanocapsules with diethanolamine, ethanolamine, propylamine, triethanolamine, dipopylamine and 5-mino 1-pentanol [31], pH and sulfide ion sensitive mesoporous silica nanoparticles with 1H-benzotriazole [32], polyurea microcapsules with 4,4'-bis-methylene cyclohexane diisocyanate [33], poly(urea-formaldehyde) with hexamethylene diisocyanate and 1H,1H,2H,2H-perfluorooctyltriethoxysilane [34], etc.

Fu *et al.* [35] reported acid and alkaline dual stimuli-responsive mechanized hollow mesoporous silica nanoparticles as smart nano-containers for anti-corrosion coatings. As shown in Figure 4.4, the smart nano-containers delayed the penetration of the corrosive species in the coating. In addition, the nano-containers were also observed to repair the damaged aluminum oxide layer, thus, exhibiting self-healing behavior to provide the long term anti-corrosion behavior. The nano-containers were observed to encapsulate the corrosion inhibitor at neutral pH, whereas the release was observed at both acidic and alkaline conditions, thus, confirming the functional nature of the nano-containers.

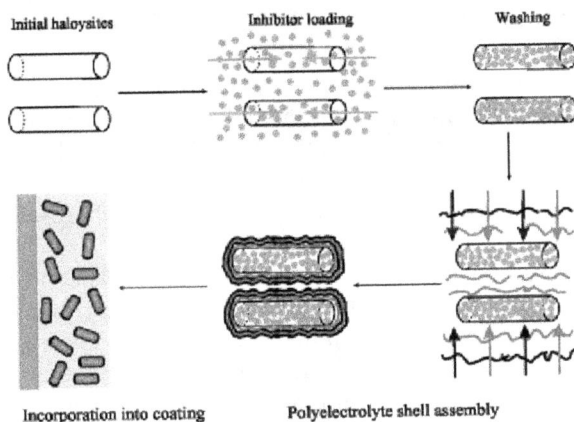

Figure 4.3 Fabrication of halloysite nano-containers as self-healing filler. Reprinted from Reference 21 with permission from American Chemical Society.

Contrary to the polymer micro-capsules, ceramic based containers are limited to 20-30 wt% loading of the corrosion inhibitors [36]. In addition to the earlier mentioned studies on nano-container synthesis, metallic oxide like TiO_2 nano-containers loaded with the corrosion inhibitor 8-hydroxyquinoline were also synthesized [37]. The continuous increase of the total impedance value with the time of exposure to the corrosive solution indicated the release of the inhibitors from the nano-container which further improved the anti-corrosion ability of the epoxy coating [37].

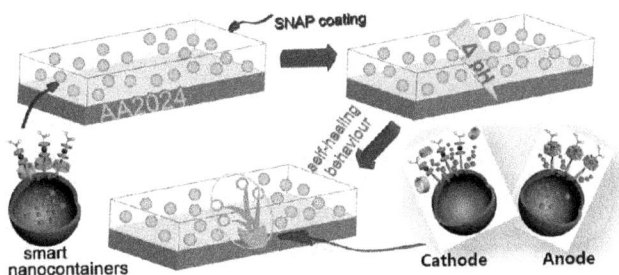

Figure 4.4 Schematic of the self-healing behavior of the nano-containers generated using mechanized hollow mesoporous silica nanoparticles. Reproduced from Reference 35 with permission from American Chemical Society.

Mesoporous nano-capsules usually consist of empty core and shell which results in the significant loading of the corrosion inhibitor and permits extended and stimuli-triggered liberation of the self-healing agent [36]. Maia *et al.* synthesized silica nano-capsules loaded with 2-mercaptobenzothiazole and incorporated into water-based coating [38]. Release of the inhibitor at different pH indicated that the self-healing agent released at low pH and high concentration of NaCl and could be used in systems where slow release is required.

In another study, Yang *et al.* [39] reported microcapsules containing diisocyanate for generating self-healing polymers. The capsuled were fabricated via interfacial polymerization of polyurethane (PU). Figure 4.5 also shows the surface and shell morphology of microcapsules obtained at various agitation rates. The microcapsules were observed to encapsulate efficient amounts of the healing agent and were stable with time.

4.3 Layer by Layer (lbl) Assembly of Polyelectrolyte Self-healing Coatings

Layer by layer assembly involves the development of alternate multiple layers of polyelectrolyte on metal surface containing corrosion inhibitors (sandwich like structure). It is a versatile and quite simple method for the fabrication of various kinds of self-healing coatings for corrosion protection.

Andreeva *et al.* [40] developed layer by layer assembly of pH-sensitive polyelectrolyte and corrosion inhibitor combination directly on the metal surface. Similar results were also demonstrated by Wang *et al.* [41]. Using the similar approach of lbl assembly, Yabuki *et al.* [42] developed multi-layer coatings consisting of superabsorbent particles and vinyl ester based self-healing coatings on carbon steel. The protection mechanism offered by these systems was attributed to the migration of spherical particles towards scratch area and prevention of the diffusion of oxygen towards corrosion sites. Syed *et al.* [43] developed multilayer coating consisting of polyaniline-polyacrylic acid/polyethyleneimine with self-healing and redox catalytic ability. The synergistic corrosion protection was achieved due to the improvement in the anti-corrosion ability of the coating by polyaniline-polyacrylic acid, whereas polyethyleneimine layer acted as self-healing medium. Andreeva *et al.* [44] also described the protection mechanism as: i) pH changes due to the corrosion process and stimulates the polyelectrolyte (PE) coating, ii) pH buffering causes the rearrangement of polymer chains and as a result releases the corrosion inhibitor.

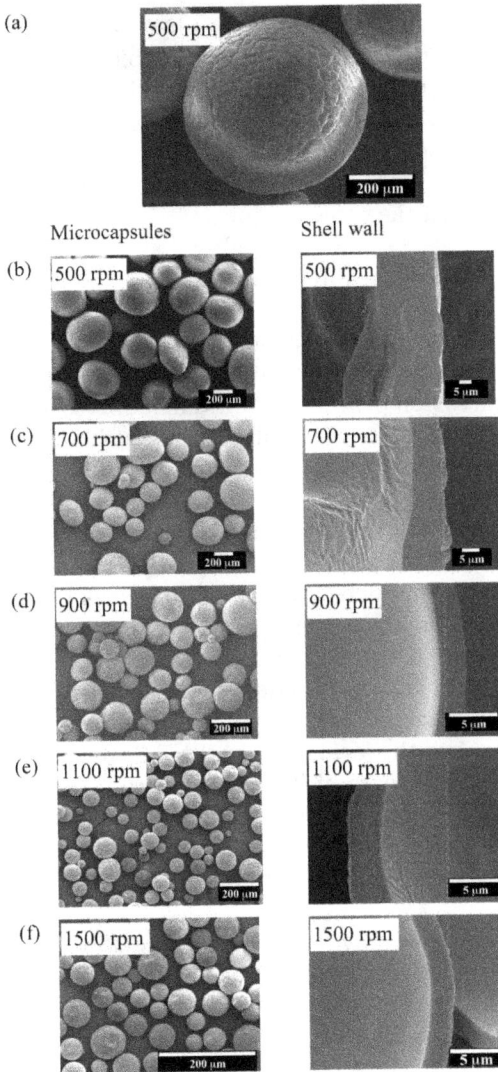

Figure 4.5 Surface and shell morphology of microcapsules obtained at various agitation rates. Reproduced from Reference 39 with permission from American Chemical Society.

The quick response to the corrosion process is a key parameter in designing the self-healing coatings. The short response can be generated by optimizing the electrostatic interactions and swelling characteristics of polymers. In order to improve the corrosion resistance with rapid self-healing ability, Fan *et al.* [45] prepared lbl coatings consisting of cerium based conversion layer, graphene oxide layer as corrosion inhibitor, and poly(ethylene imine)/poly(acrylic acid), (PEI/PAA) multilayers for self-healing ability. During corrosion testing, graphene was observed to act as barrier filler and provided resistance against the corrosive ions.

4.4 Ceramic Materials as Self-healing Filler

The use of ceramic based nano-particulate fillers in polymer coatings have been repeatedly reported in the literature for many purposes, e.g. to enhance barrier and anti-corrosion properties or as an additive. Several research studies have also reported the use of metallic oxides such as TiO_2 and CeO_2 as self-healing fillers for anti-corrosive polymer coatings. Yabuki *et al.* [46] reported that TiO_2 particles acted as containers of the inhibiting agent bisphenol A (BPA) which was the precursor of the polymer coating. The TiO_2/polymer composite coatings were applied on aluminum alloy and examined in artificial seawater with an artificial defect. The inhibiting agent was produced due to the dissolution of the host polymer by the action of hydroxyl ions (OH^-). The self-healing mechanism occurred due to the formation of defect-free film contained BPA. During the anodic reaction of aluminum, the corresponding cathodic reaction was observed to produce OH^- ions. The OH^- ions diffused in to the coating and changed the local pH near the TiO_2 particles. In response, TiO_2 particles promoted the release of BPA from the surface. Energy dispersive X-ray fluorescence spectrometry (EDX) analysis also confirmed the presence of enriched carbon film at the defect site. These results confirmed that carbon contained by the barrier film was originated from the vinyl ester polymer [46].

Intrinsically conducting polymers (ICPs) can also cause adverse effect on the physical and mechanical properties of coatings, when used as anti-corrosion fillers. Rohwerder *et al.* [47] reported that during corrosion reactions, the oxidation or reduction of ICPs changes their density and volume, which results in the deterioration of coatings' properties. The dispersion of ICPs in coatings is also a difficult process. These challenges may be solved by using inorganic fillers modified with ICPs [47,48]. Mittal *et al.* [49] modified the surface of inorganic pigments by in-situ polymerization of polyaniline and incorporated in the primer of vinyl acrylate. The corrosion rate was the least in the case of coating containing polyaniline modified inorganic. Radhakrishnan *et al.* [50] also used conducting polymer modified nano-TiO_2 and exhibited the self-healing ability of polyvinyl butyral coatings. The protection mechanism was described in terms of fillers' ability to behave as n-type and p-type barrier at the same time which hindered the transfer of charge during the corrosion process. The coating containing higher concentration of fillers exhibited enhanced self-healing behavior.

Cerium based compounds such as cerium oxide, cerium nitrate, cerium molybedate, etc., have been reported in many applications such as catalysts, additives, anti-oxidant, UV absorber and as an anti-corrosion pigments [51]. As an alternative to the hazardous anti-corrosion pigments, Ce^{+3} fulfils the essential requirements as functional and effective corrosion inhibitors [52]. Aramaki *et al.* [53] prepared self-healing polymer coating on zinc electrode containing sodium silicate and cerium nitrate. After immersion in the corrosive solution, X-ray photoelectron spectroscopy confirmed the presence of passive film containing $Zn(OH)_2$, $ZnSi_2O_5$ and Ce^{3+}–$Si_2O_5^{-2}$ salt or complex formed on the scratched surface.

The use of environmentally friendly and naturally derived materials like chitin, cellulose and starch for coating purposes is getting significant importance due to environmental concerns [54]. Similarly, environmental restrictions by Environmental Protection Agency (EPA) on using strong corrosion inhibitors like chromates based anti-corrosion pigments have promoted the development of non-toxic pigments [4]. The complete green approach to produce self-healing coatings i.e. composite coatings comprising of environmentally friendly polymers and bio-fillers has also the potential of alleviating the issues due to the use of environmentally polluting materials. Carneiro *et al.* [55] produced the "green" coating by doping chitosan (and its derivative) with cerium cations (Ce^{+3}), where the chitosan layer also acted as reservoir for the metallic cations. The formation of a complex between the cations and chitosan functional groups also helped for the controlled release/leaching of the cations during corrosion process. The release of metallic cations from chitosan coating was also studied in corrosive solution containing metal samples. The relation between oxide resistance and time indicated that the coating containing 1 wt% cations was better at the beginning owing to improved barrier properties, whereas the coating loaded with 10 wt % exhibited active protection for longer period of immersion. The coating without inhibitor exhibited no self-healing effect. In another study, Hassannejad *et al.* [56] also studied the self-healing ability of CeO_2-chitosan nanocomposite coatings. The coating exhibited increased impedance value on increasing exposure time. The corrosion protection behavior was attributed to the diffusion of cerium ions in the chitosan coating towards the corrosion sites forming thin layer of cerium oxide.

Mesoporous nanoparticles have also been reported in the literature as carriers of corrosion inhibitor for self-healing coatings. Shchukin *et al.* [20,57] developed self-healing fillers by entrapping the corrosion inhibitor (benzotriazole) between polyelectrolyte layers coated on SiO$_2$. The release of inhibitor was initiated by the pH changes during corrosion process. Ion exchange nanoclay particles were also introduced as carriers for storage and delayed release of self-healing agents due to their rich intercalation chemistry and good compatibility with polymer materials. Hang *et al.* [58] modified the montmorillonite clay with indole-3 butyric acid (IBA) and observed improved self-healing properties of epoxy coating after prolong immersion in NaCl solution. The release of inhibitor was favored at higher pH due to low interaction between inhibitor and clay. The reduced interaction at higher pH between inhibitor and clay was due to acid dissociation constant, pKa = 4.9 for IBA. Similarly, composites of Ce(III) and montmorillonite clay were prepared where the triggering mechanism of cerium ion was due to sodium or zinc ions [59].

4.5 Conducting Polymers as Self-healing Fillers

ICPs have gained popularity in the numerous applications such as drug delivery, batteries, actuators, adhesives, EMI shielding, memory devices, light-emitting diodes, antistatic films, chemicals sensors or anti-corrosion fillers. These copious applications of ICPs are due to their inherent properties such as stimuli to electro- or chemical reactions, high electrical conductivity, fusibility, low density, good thermal stability and optical properties, low cost, doping primacy, and environment friendliness [15,60]. ICPs have also provoked a great deal of interest as self-healing agents owing to their electrochemical properties, redox-responsive liberation of inhibitor and compatibility as a filler with polymer coatings [61]. ICPs such as polyaniline (PA), polypyrrole (Ppy), polythiophene, poly ortho-anisidine (PoA) and their derivatives have redox abilities which can be used to inhibit the corrosion process. For instance, as shown in Figure 4.6, self-healing polyvinyl butyral (PVB) based organic coating formulations were prepared by incorporating polypyrrole-carbon black (PPyCB) composite as an inhibiting pigment [62].

The mechanism of corrosion protection offered by ICPs is still not established, but various theories have been put forward, such as barrier protection, electrochemical inhibition, cathodic or anodic protection [48,63-64]. The factors like morphological structure of ICPs and compatibility with the host resins are responsible to enhance the barrier properties of coatings. In case of electrochemical corrosion inhibition, ICPs slow down the electrochemical reaction by forming monomolecular layer on the metallic surface due to the adsorption of ICPs on the surface. ICPs also help to passivate the metal surface

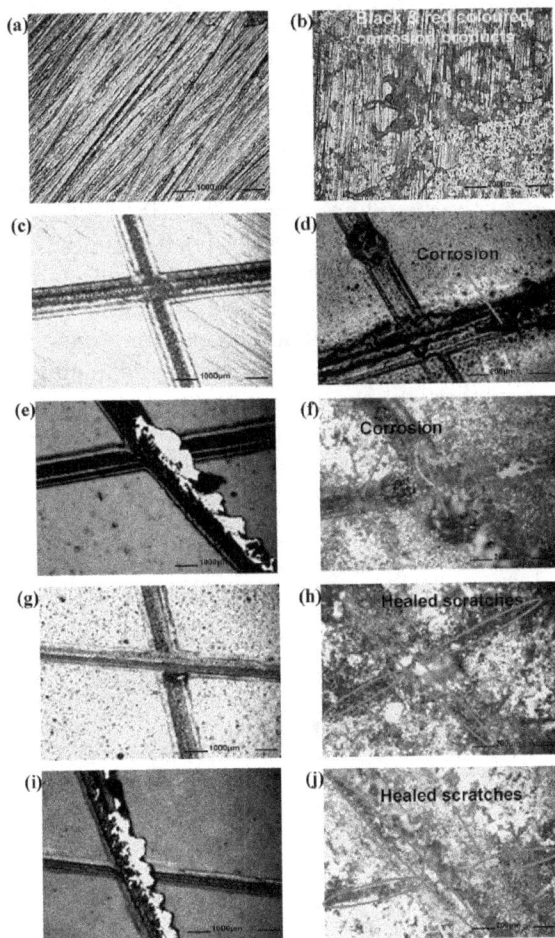

Figure 4.6 Optical images of scratches during immersion in 4 wt% NaCl solution. Bare steel (a) before and (b) after 2d; PVB (c) before and (d) after 7 d; PVB/PPyCB5 (e) before and (f) after 4 d; PVB/PPyCB10 (g) before and (h) after 3 d; PVB/PPyCB20 (i) before and (j) after 3 d. Reproduced from Reference 62 with permission from Royal Society of Chemistry.

by enhancing anodic reaction which results in the formation of oxide layer and keeps electrode potential in passive region. The inhibition of cathodic reaction is due to the transfer of the cathodic reaction from the metal/electrolyte interface towards the electrolyte/polymer interface [48]. ICP doped with higher molecular weight molecule can release its dopant during anodic reaction due to galvanic coupling between corroded iron and ICP. The released dopant may react with the metal cations and form a complex and slow down the corrosion process. This process can be referred as self-healing process [65]. In case of polyaniline, different oxidation states of polyaniline are stable at different pH e.g. at lower pH, emeraldine Salt (conducting), whereas at higher pH, emeraldine base (insulating/reduced) are stable [66]. In this manner, if a corresponding cathodic reaction takes place at neutral or high pH, polyaniline gets reduced and liberates its counter ion. Further, polyaniline also exists in other insulating and deprotonating states like leuco base, penigraniline, etc. (Figure 4.7). The other possible protection mechanism offered by ICPs could be anodic protection mechanism (ennobling mechanism) and oxygen reduction reaction (ORR) [65]. During ennobling mechanism, ICPs maintains the metal passivity by supporting the formation of oxide layer on the metal surface due to their oxidizing nature. During corrosion process, the stored charge in ICPs is transferred to the defect due to reduction of ICPs and promotes the formation of oxide layer on the metal surface.

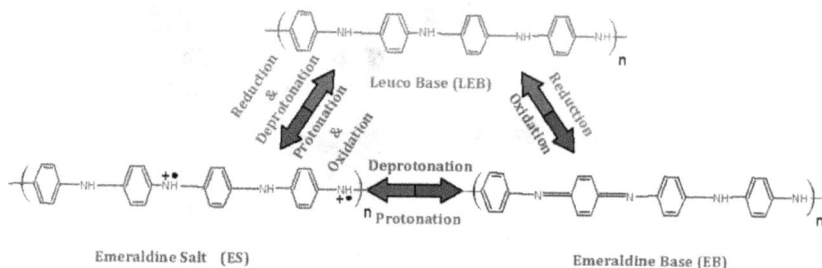

Figure 4.7 Different oxidation and protonating states of polyaniline.

In literature, the reduction potential (vs. SHE at 7 pH) of some of ICPs is -0.8-0.3 Volts for polypyrrole, -0.4-1.0 Volts for polyaniline and 0.8-1.2 Volts for polythiophene [65,67]. These reduction potential are below the reduction potential of oxygen which can be useful for ICPs reduction. The use of dopant may affect the performance depending on the dopant's molecular size, nature, etc. Hetropolyacids have general formula $[XM_{12}O_{40}]^{n-}$ and have high electro-activity as compared with conventional acidic dopants [68]. Molybdenum and tungsten hetropolyanions have been used in many catalysis systems. The use of these hetropolyanions can be beneficial along with ICP during the corrosion process, as these may act as corrosion inhibitor upon release during cathodic process. Paliwoda-Porebska *et al.* [69] studied the self-healing behavior of coatings containing polypyyrole doped with $[MoO_4]^{-2}$ or $[PMo12O40]^{-3}$. The stability of the $[PMo12O40]^{-3}$ anion was found to be pH dependent as at low pH, the anion was more stable, while beyond pH 4, it decomposed to $[MoO_4]^{-2}$ and $[HPO_4]^{-2}$. The main inhibition effect was attributed to the presence of decomposed anions from $[PMo_{12}O_{40}]^{-3}$. Scanning Kelvin probe experiments on polypyyrole coating exhibited significant inhibition of the delamination owing to the intelligent release of the anions during reduction of polypyyrole [69].

The use of ICPs as carriers such as capsules loaded with self-healing agents can be advantageous and promising to the anti-corrosion coatings. In this case, the triggering process of ICPs would be redox-stimulus which is the same approach as the release of high molecular weight doped anion. Lv *et al.* [15] synthesized polyaniline or polypyrrole capsules using one-step miniemulsion polymerization technique (Figure 4.8). The capsules were also loaded with hydrophobic self-healing agent such as diglycidyl ether or dicarboxylic acid terminated polydimethylsiloxane during the polymerization of polyaniline or polypyrrole. The controlled release of self-healing agent was studied in different solution where the capsules exhibited delayed and quick release under oxidation and reduction respectively.

One of the biggest advantages of ICPs is their ability to electro-polymerize as films on different metal substrates using various electrochemical techniques such as potentiodynamic, potentiostatic and galvanostatic. During electro-deposition of ICP monomer, desired filler or dopant are added to suitable solvent in an electrochemical cell containing working electrode for ICP coating. Electro-polymerization technique produces films with improved adhesion to metal surface and uniformity/homogeneity along with good anti-corrosion properties [60]. Kowalski *et al.* [70] demonstrated the ability of films to repair defects and restore the passive state of a metal surface by using bi-layered polypyrrole films on the metal surface. The permselective nature of the bi-layer film worked by restricting permeation of the cations and also controlling the release of healing ions. The bi-layered film consisted of inner layer doped with heteropolyanions whereas the outer layer contained dodecylsulfate [70]. The outer layer provided electrochemical barrier to the ingression of chlorine ions and barrier for the self-healing anions towards the solution. The material was observed to provide 20 h of protection after artificial defect was formed in the coating as shown by rise in the profile of OCP.

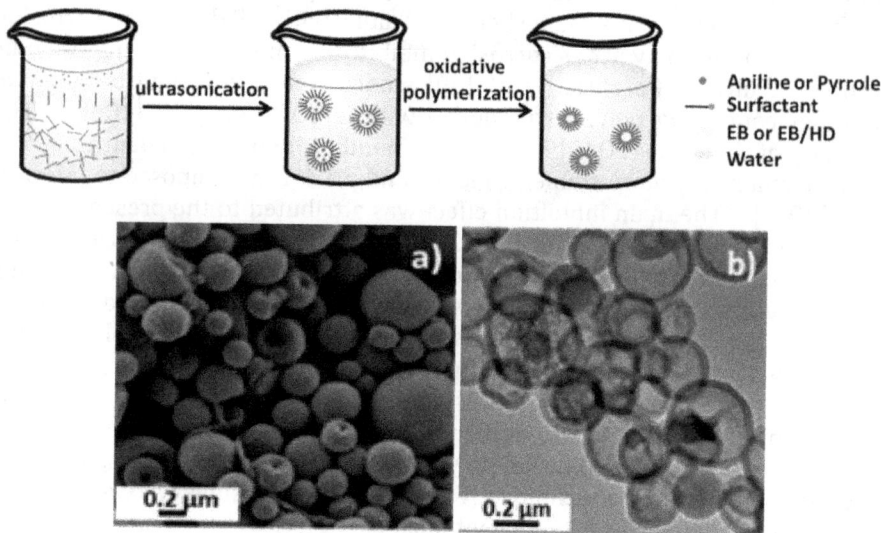

Figure 4.8 One step in-situ polymerization of aniline or pyrrole monomers using miniemulsion technique to produce conducting polymer nano-capsules in the presence of ethylbenzene (EB) or ethylbenzene/hexadecane (EB/HD) and a) SEM and b) TEM micrographs of the PA capsules. Reprinted from Reference 15 with permission from American Chemical Society.

Similarly, Karpakam *et al.* [71] electro-synthesized the polyaniline-molybdate (polyaniline-$[MoO_4]^{-2}$) on mild steel using cyclic voltammetry technique. The self-healing ability of the films was due to intelligent release of anion from reduced polyaniline which further reacted with the iron cations to form iron-molybdate complex on the metal surface. Similar results were also reported in case of tungstate doped polyaniline coating on steel [72].

4.6 Miscellaneous Systems

Zhu *et al.* [73] reported self-healing electrically conductive superhydrophobic poly(3,4-ethylenedioxythiophene) (PEDOT) coatings. The coatings were generated by chemical vapor deposition of a fluoroalkylsilane (POTS) onto the PEDOT film. After etching the coating with O_2 plasma, the coating exhibited ef-

fective self-healing ability as it regained spontaneously the superhydrophobicity under ambient conditions for 20 h. The coating also recovered its superhydrophobicity after being corroded by strong acid solution or strong base solution. Figure 4.9 demonstrates the principle of self-healing ability of electronically conductive superhydrophobic PEDOT coatings. The authors analyzed
the distribution of POTS in the coating using EDS and XPS. It was observed
that the coating had significant amount of POTS, migration of which to the surface is necessary for exhibiting the self-healing ability.

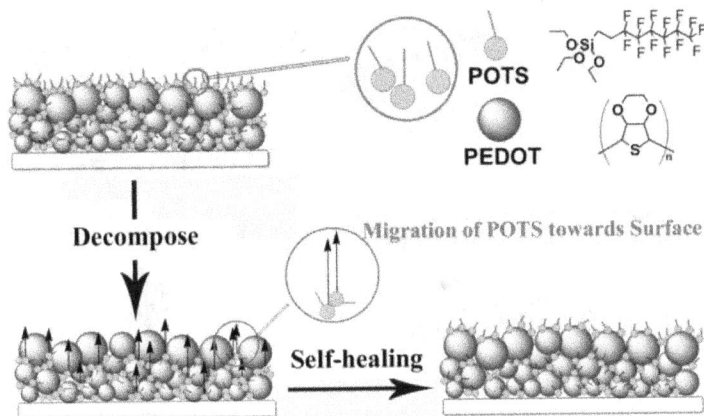

Figure 4.9 Illustration of the principle of self-healing ability of electronically conductive
superhydrophobic PEDOT coatings. Reproduced from Reference 73 with permission from
American Chemical Society.

Chen *et al.* [74] reported self-healing superoleophobic and anti-biofouling
coatings generated by the self-assembly of hydrophilic polymeric chain modified hierarchical microgel spheres (MHMS). Figure 4.10 demonstrates the
schematic for the preparation of MHMS-based coating. Though the coatings
were developed for underwater applications, however, these coatings also
represent the potential for use as self-healing anti-corrosion coatings.

4.7 Conclusion

Large number of self-healing coatings systems have been developed, especially for achieving effective anti-corrosion performance. Self-healing characteristics lead to efficient performance through longer service life of the coatings

which leads to the reduction in maintenance costs. In addition, these systems also pave way for the development of environmental friendly coating

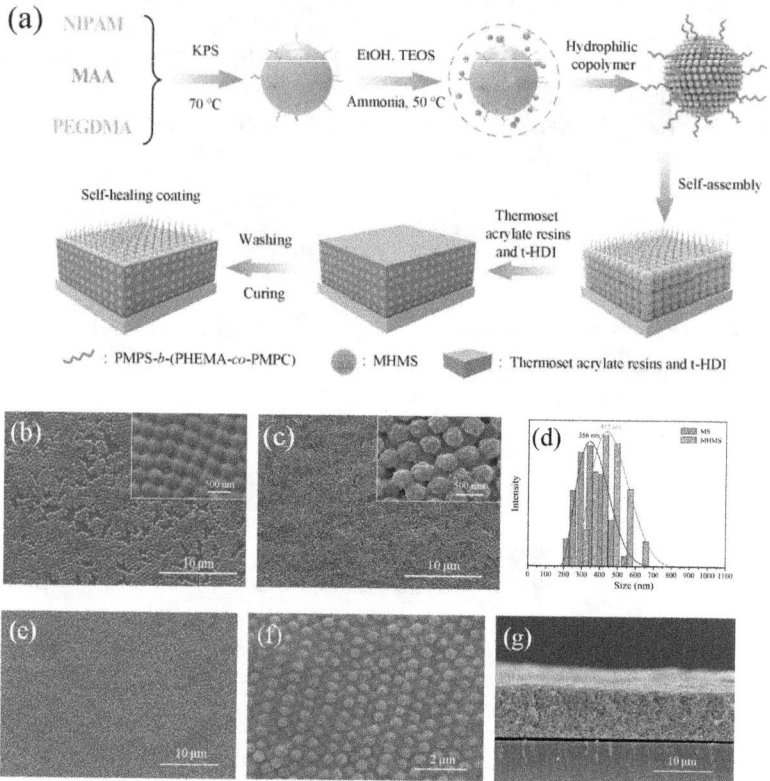

Figure 4.10 (a) Illustration for the preparation of MHMS-based coating; (b) SEM images of microgel spheres (MS); (c) SEM images of MHMS; (d) size distribution of MS and MHMS; (e), (f) SEM image of the MHMS-based coating and (g) cross-sectional SEM image of the MHMS-based coating. Reproduced from Reference 74 with permission from American Chemical Society.

formulations by effectively replacing the conventional anti-corrosion fillers/inhibitors which are occasionally environmental unfriendly. Various anti-corrosion systems reported in the literature include self-healing nano-containers, layer by layer assembly of polyelectrolyte self-healing coatings, ceramic materials as self-healing fillers, conducting polymers as self-healing fillers, etc. Though challenges exist for the large scale application of the self-healing coatings systems by replacing the conventional anti-corrosion inhibitors, however, the high potential of these systems would ensure overcoming these challenges in near future.

References

1. Jones, D. A. (1992) *Principles and Prevention of Corrosion*, Macmillan Publishing Company, USA.
2. Riaz, U., Nwaoha, C., and Ashraf, S. M. (2014) Recent advances in corrosion protective composite coatings based on conducting polymers and natural resource derived polymers. *Progress in Organic Coatings*, **77**(4), 743-756.
3. Callister, W. D., Jr., and Rethwisch, D. G. (2014) *Materials Science and Engineering: An Introduction*, 7th edition, John Wiley & Sons, USA.
4. Chaudhry, A. U., Mittal, V., and Mishra, B. (2015) Inhibition and promotion of electrochemical reactions by graphene in organic coatings. *RSC Advances*, **5**(98), 80365-80368.
5. *NACE Basic Corrosion Course*. Onlne: http://corrosion-doctors.org/Corrosion-History/Course.htm (assessed 19th April 2017).
6. *Anti-Corrosion Coating Market*. Online: http://www.prnewswire.com/news-releases/anti-corrosion-coating-market-worth-26583-million-by-2019-293825341.html (assessed 16th April 2017).
7. Chaudhry, A. U., Mittal, V., and Mishra, B. (2015) Nano nickel ferrite ($NiFe_2O_4$) as anti-corrosion pigment for API 5L X-80 steel: An electrochemical study in acidic and saline media. *Dyes and Pigments*, **118**, 18-26.
8. Brodinova, J., Stejskal, J., and Kalendova, A. (2007) Investigation of ferrites properties with polyaniline layer in anticorrosive coatings. *Journal of Physics and Chemistry of Solids*, **68**(5-6), 1091-1095.
9. Eduok, U., Faye, O., Tiamiyu, A., and Szpunar, J. (2017) Fabricating protective epoxy-silica/CeO2 films for steel: Correlating physical barrier properties with material content. *Materials and Design*, **124**, 58-68.
10. Chang, C.-H., Huang, T.-C., Peng, C.-W., Yeh, T.-C., Lu, H.-I., Hung, W.-I., Weng, C.-J., Yang, T.-I., and Yeh, J.-M. (2012) Novel anticorrosion coatings prepared from polyaniline/graphene composites. *Carbon*, **50**(14), 5044-5051.
11. Ionita, M., and Pruna, A. (2011) Polypyrrole/carbon nanotube composites: Molecular modeling and experimental investigation as anti-corrosive coating. *Progress in Organic Coatings*, **72**(4), 647-652.
12. Yeh, J.-M., Chin, C.-P., and Chang, S. (2003) Enhanced corrosion protection coatings prepared from soluble electronically conductive polypyrrole-clay nanocomposite materials. *Journal of Applied Polymer Science*, **88**(14), 3264-3272.
13. Mahulikar, P. P., Jadhav, R. S., and Hundiwale, D. G. (2011) Performance of polyaniline/TiO2 nanocomposites in epoxy for corrosion resistant coatings. *Iranian Polymer Journal*, **20**(5), 367-376.

14. Fayyad, E. M., Almaadeed, M. A., Jones, A., and Abdullah, A. M. (2014) Evaluation techniques for the corrosion resistance of self-healing coatings. *International Journal of Electrochemical Science*, **9**, 4989-5011.
15. Lv, L.-P., Zhao, Y., Vilbrandt, N., Gallei, M., Vimalanandan, A., Rohwerder, M., Landfester, K., and Crespy, D. (2013) Redox responsive release of hydrophobic self-healing agents from polyaniline capsules. *Journal of the American Chemical Society*, **135**(38), 14198-14205.
16. Montemor, M. F. (2014) Functional and smart coatings for corrosion protection: A review of recent advances. *Surface and Coatings Technology*, **258**, 17-37.
17. Kumar, A., Stephenson, L. D., and Murray, J. N. (2006) Self-healing coatings for steel. *Progress in Organic Coatings*, **55**(3), 244-253.
18. Liao, L.-P., Zhang, W., Xin, Y., Wang, H.-M., Zhao, Y., Li, W.-J. (2011) Preparation and characterization of microcapsule containing epoxy resin and its self-healing performance of anticorrosion covering material. *Chinese Science Bulletin*, **56**(4), 439-443.
19. Jadhav, R. S., Hundiwale, D. G., and Mahulikar, P. P. (2011) Synthesis and characterization of phenol–formaldehyde microcapsules containing linseed oil and its use in epoxy for self-healing and anticorrosive coating. *Journal of Applied Polymer Science*, **119**(5), 2911-2916.
20. Zheludkevich, M. L., Shchukin, D. G., Yasakau, K. A., Mohwald, H., and Ferreira, M. G. S. (2007) Anticorrosion coatings with self-healing effect based on nanocontainers impregnated with corrosion inhibitor. *Chemistry of Materials*, **19**(3), 402-411.
21. Shchukin, D. G., Lamaka, S. V., Yasakau, K. A., Zheludkevich, M. L., Ferreira, M. G. S., and Mohwald, H. (2008) Active anticorrosion coatings with halloysite nanocontainers. *The Journal of Physical Chemistry C*, **112**(4), 958-964.
22. Lvov, Y. M., Shchukin, D. G., Mohwald, H., and Price, R. R. (2008) Halloysite clay nanotubes for controlled release of protective agents. *ACS Nano*, **2**(5), 814-820.
23. Huang, M., and Yang, J. (2011) Facile microencapsulation of HDI for self-healing anticorrosion coatings. *Journal of Materials Chemistry*, **21**(30), 11123-11130.
24. Montemor, M. F., Snihirova, D. V., Taryba, M. G., Lamaka, S. V., Kartsonakis, I. A., Balaskas, A. C., Kordas, G. C., Tedim, J., Kuznetsova, A., Zheludkevich, M. L., and Ferreira, M. G. S. (2012) Evaluation of self-healing ability in protective coatings modified with combinations of layered double hydroxides and cerium molibdate nanocontainers filled with corrosion inhibitors. *Electrochimica Acta*, **60**, 31-40.
25. Kartsonakis, I. A., Athanasopoulou, E., Snihirova, D., Martins, B., Koklioti, M. A., Montemor, M. F., Kordas, G., and Charitidis, C. A. (2014) Multifunctional epoxy coatings combining a mixture of traps and inhibitor loaded nanocontainers for corrosion protection of AA2024-T3. *Corrosion Science*, **85**, 147-159.
26. Kartsonakis, I. A., and Kordas, G. (2010) Synthesis and characterization of cerium molybdate nanocontainers and their inhibitor complexes. *Journal of the American Ceramic Society*, **93**(1), 65-73.
27. Haase, M. F., Grigoriev, D. O., Mohwald, H., and Shchukin, D. G. (2012) Development of nanoparticle stabilized polymer nanocontainers with high content of the encapsulated active agent and their application in water-borne anticorrosive coatings. *Advanced Materials*, **24**(18), 2429-2435.

28. Borisova, D., Mohwald, H., and Shchukin, D. G. (2011) Mesoporous silica nanoparticles for active corrosion protection. *ACS Nano*, **5**(3), 1939-1946.

29. Snihirova, D., Lamaka, S. V., and Montemor, M. F. (2012) "SMART" protective ability of water based epoxy coatings loaded with CaCO₃ microbeads impregnated with corrosion inhibitors applied on AA2024 substrates. *Electrochimica Acta*, **83**, 439-447.

30. Huang, M., Zhang, H., and Yang, J. (2012) Synthesis of organic silane microcapsules for self-healing corrosion resistant polymer coatings. *Corrosion Science*, **65**, 561-566.

31. Choi, H., Kim, K. Y., and Park, J. M. (2013) Encapsulation of aliphatic amines into nanoparticles for self-healing corrosion protection of steel sheets. *Progress in Organic Coatings*, **76**(10), 1316-1324.

32. Zheng, Z., Huang, X., Schenderlein, M., Borisova, D., Cao, R., Mohwald, H., and Shchukin, D. (2013) Self-healing and antifouling multifunctional coatings based on pH and sulfide ion sensitive nanocontainers. *Advanced Functional Materials*, **23**(26), 3307-3314.

33. Sun, D., Zhang, H., Tang, X.-Z., and Yang, J. (2016) Water resistant reactive microcapsules for self-healing coatings in harsh environments. *Polymer*, **91**, 33-40.

34. Wu, G., An, J., Tang, X., Xiang, Y., and Yang, J. (2014) A versatile approach towards multifunctional robust microcapsules with tunable, restorable, and solvent-proof superhydrophobicity for self-healing and self-cleaning coatings. *Advanced Functional Materials*, **24**(43), 6751-6761.

35. Fu, J.-J., Chen, T., Wang, M.-D., Yang, N.-W., Li, S.-N., and Liu, X.-D. (2013) Acid and alkaline dual stimuli-responsive mechanized hollow mesoporous silica nanoparticles as smart nanocontainers for intelligent anticorrosion coatings. *ACS Nano*, **7**(12), 11397-11408.

36. Shchukin, D. G. (2013) Container-based multifunctional self-healing polymer coatings. *Polymer Chemistry*, **4**(18), 4871-4877.

37. Balaskas, A. C., Kartsonakis, I. A., Tziveleka, L.-A., and Kordas, G. C. (2012) Improvement of anti-corrosive properties of epoxy-coated AA 2024-T3 with TiO₂ nanocontainers loaded with 8-hydroxyquinoline. *Progress in Organic Coatings*, **74**(3), 418-426.

38. Maia, F., Tedim, J., Lisenkov, A. D., Salak, A. N., Zheludkevich, M. L., and Ferreira, M. G. S. (2012) Silica nanocontainers for active corrosion protection. *Nanoscale*, **4**(4), 1287-1298.

39. Yang, J., Keller, M. W., Moore, J. S., White, S. R., and Sottos, N. R. (2008) Microencapsulation of isocyanates for self-healing polymers. *Macromolecules*, **41**, 9650-9655.

40. Andreeva, D. V., Fix, D., Mohwald, H., and Shchukin, D. G. (2008) Self-healing anticorrosion coatings based on pH-sensitive polyelectrolyte/inhibitor sandwichlike nanostructures. *Advanced Materials*, **20**(14), 2789-2794.

41. Wang, X., Liu, F., Zheng, X., and Sun, J. (2011) Water-enabled self-healing of polyelectrolyte multilayer coatings. *Angewandte Chemie International Edition*, **50**(48), 11378-11381.

42. Yabuki, A., and Okumura, K. (2012) Self-healing coatings using superabsorbent polymers for corrosion inhibition in carbon steel. *Corrosion Science*, **59**, 258-262.

43. Syed, J. A., Tang, S., and Meng, X. (2016) Intelligent saline enabled self-healing of multilayer coatings and its optimization to achieve redox catalytically provoked anti-corrosion ability. *Applied Surface Science*, **383**, 177-190.

44. Andreeva, D. V., Skorb, E. V., and Shchukin, D. G. (2010) Layer-by-layer polyelectrolyte/inhibitor nanostructures for metal corrosion protection. *ACS Applied Materials & Interfaces*, **2**(7), 1954-1962.

45. Fan, F., Zhou, C., Wang, X., and Szpunar, J. (2015) Layer-by-layer assembly of a self-healing anticorrosion coating on magnesium alloys. *ACS Applied Materials & Interfaces*, **7**(49), 27271-27278.

46. Yabuki, A., Urushihara, W., Kinugasa, J., and Sugano, K. (2011) Self-healing properties of TiO_2 particle–polymer composite coatings for protection of aluminum alloys against corrosion in seawater. *Materials and Corrosion*, **62**(10), 907-912.

47. Rohwerder, M., and Michalik, A. (2007) Conducting polymers for corrosion protection: What makes the difference between failure and success? *Electrochimica Acta*, **53**(3), 1300-1313.

48. Khan, M. I., Chaudhry, A. U., Hashim, S., and Iqbal, M. Z. (2010) Investigation of corrosion-protective performance of polyaniline covered inorganic pigments. *Nucleus*, **47**(4), 287-293.

49. Mittal, V., Chaudhry, A. U., and Khan, M. I. (2011) Comparison of anti-corrosion performance of polyaniline modified ferrites. *Journal of Dispersion Science and Technology*, **33**(10), 1452-1457.

50. Radhakrishnan, S., Siju, C. R., Mahanta, D., Patil, S., and Madras, G. (2009) Conducting polyaniline–nano-TiO_2 composites for smart corrosion resistant coatings. *Electrochimica Acta*, **54**(4), 1249-1254.

51. Ecco, L. G. (2014) *Waterborne Paint System Based on CeO2 and Polyaniline Nanoparticles for Anticorrosion Protection of Steel*. Ph.D. Thesis, University of Trento. Online: http://eprints-phd.biblio.unitn.it/1334/ (assessed 17th April 2017).

52. Zand, R. Z., Verbeken, K., and Adriaens, A. (2013) Influence of the cerium concentration on the corrosion performance of Ce-doped silica hybrid coatings on hot dip galvanized steel substrates. *International Journal of Electrochemical Science*, **8**(1), 548-563.

53. Aramaki, K. (2002) Preparation of chromate-free, self-healing polymer films containing sodium silicate on zinc pretreated in a cerium(III) nitrate solution for preventing zinc corrosion at scratches in 0.5 M NaCl. *Corrosion Science*, **44**, 1375-1389.

54. Luckachan, G. E., and Mittal, V. (2015) Anti-corrosion behavior of layer by layer coatings of cross-linked chitosan and poly(vinyl butyral) on carbon steel. *Cellulose*, **22**(5), 3275-3290.

55. Carneiro, J., Tedim, J., Fernandes, S. C. M., Freire, C. S. R., Silvestre, A. J. D., Gandini, A., Ferreira, M. G. S., and Zheludkevich, M. L. (2012) Chitosan-based self-healing protective coatings doped with cerium nitrate for corrosion protection of aluminum alloy 2024. *Progress in Organic Coatings*, **75**(1-2), 8-13.

56. Hassannejad, H., and Nouri, A. (2016) Synthesis and evaluation of self-healing cerium-doped chitosan nanocomposite coatings on AA5083-H321. *International Journal of Electrochemical Science*, **11**(3), 2106-2118.

57. Shchukin, D. G., Zheludkevich, M., Yasakau, K., Lamaka, S., Ferreira, M. G. S., and Mohwald, H. (2006) Layer-by-layer assembled nanocontainers for self-healing corrosion protection. *Advanced Materials*, **18**(13), 1672-1678.
58. Hang, T. T. X., Truc, T. A., Olivier, M.-G., Vandermiers, C., Guerit, N., and Pebere, N. (2010) Corrosion protection mechanisms of carbon steel by an epoxy resin containing indole-3 butyric acid modified clay. *Progress in Organic Coatings*, **69**(4), 410-416.
59. Motte, C., Poelman, M., Roobroeck, A., Fedel, M., Deflorian, F., and Olivier, M.-G. (2012) Improvement of corrosion protection offered to galvanized steel by incorporation of lanthanide modified nanoclays in silane layer. *Progress in Organic Coatings*, **74**(2), 326-333.
60. Khan, M. I., Chaudhry, A. U., Hashim, S., Zahoor, M. K., and Iqbal, M. Z. (2010) Recent developments in intrinsically conductive polymer coatings for corrosion protection. *Chemical Engineering Research Bulletin*, **14**(2), 73-86.
61. Saji, V. S., and Thomas, J. (2007) Nanomaterials for corrosion control. *Current Science*, **92**(1), 51-55.
62. Niratiwongkorn, T., Luckachan, G. E., and Mittal, V. (2016) Self-healing protective coatings of polyvinyl butyral/polypyrrole-carbon black composite on carbon steel. *RSC Advances*, **6**(49), 43237-43249.
63. Spinks, G. M., Dominis, A. J., Wallace, G. G., and Tallman, D. E. (2002) Electroactive conducting polymers for corrosion control. Part 2. Ferrous metals. *Journal of Solid State Electrochemistry*, **6**(2), 85-100.
64. Tallman, D. E., Spinks, G., Dominis, A., and Wallace, G. G. (2002) Electroactive conducting polymers for corrosion control. Part 1. General introduction and a review of non-ferrous metals. *Journal of Solid State Electrochemistry*, **6**(2), 73-84.
65. Abu-Thabit, N. Y., and Makhlouf, A. S. H. (2014) Recent advances in polyaniline (PANI)-based organic coatings for corrosion protection. In: *Handbook of Smart Coatings for Materials Protection*, Makhlouf, A. S. H. (ed.), Woodhead Publishing, UK, pp. 459-486.
66. Dominis, A. J., Spinks, G. M., and Wallace, G. G. (2003) Comparison of polyaniline primers prepared with different dopants for corrosion protection of steel. *Progress in Organic Coatings*, **48**(1), 43-49.
67. Mishra, B., Chaudhry, A., and Mittal, V. (2017) Development of polymer-based composite coatings for the gas exploration industry: Polyoxometalate doped conducting polymer based self-healing pigment for polymer coatings. *Materials Science Forum*, **879**, 60-65.
68. Pielichowski, K., and Hasik, M. (1997) Thermal properties of new catalysts based on heteropolyanion-doped polyaniline. *Synthetic Metals*, **89**(3), 199-202.
69. Paliwoda-Porebska, G., Stratmann, M., Rohwerder, M., Potje-Kamloth, K., Lu, Y., Pich, A. Z., and Adler, H.-J. (2005) On the development of polypyrrole coatings with self-healing properties for iron corrosion protection. *Corrosion Science*, **47**(12), 3216-3233.
70. Kowalski, D., Ueda, M., and Ohtsuka, T. (2010) Self-healing ion-permselective conducting polymer coating. *Journal of Materials Chemistry*, **20**(36), 7630-7633.

71. Karpakam, V., Kamaraj, K., Sathiyanarayanan, S., Venkatachari, G., and Ramu, S. (2011) Electrosynthesis of polyaniline–molybdate coating on steel and its corrosion protection performance. *Electrochimica Acta*, **56**(5), 2165-2173.

72. Kamaraj, K., Karpakam, V., Sathiyanarayanan, S., Azim, S. S., and Venkatachari, G. (2011) Synthesis of tungstate doped polyaniline and its usefulness in corrosion protective coatings. *Electrochimica Acta*, **56**(25), 9262-9268.

73. Zhu, D., Lu, X., and Lu, Q. (2014) Electrically conductive PEDOT coating with self-healing superhydrophobicity. *Langmuir*, **30**(16), 4671-4677.

74. Chen, K., Zhou, S., and Wu, L. (2016) Self-healing underwater superoleophobic and antibiofouling coatings based on the assembly of hierarchical microgel spheres. *ACS Nano*, **10**, 1386-1394.

5

Recent Advances in Nanocarbon-based Polymer Composite Materials as Membranes in Fuel Cell Applications

5.1 Introduction

The most significant environmental hazards in the world today is the air pollution from the internal combustion engines. The environmental need for clean and efficient energy has paved the way to the development of fuel cells. Fuel cells currently attract significant research attention because of their advantages compared to other energy producing devices such as saving in conventional fuels such as oil or natural gas, higher energy conversion efficiency than diesel or gas engines, no moving parts, lack of sonic pollution compared to internal combustion engines, no atmospheric polluting emissions of NO_x, SO_x, CO, CO_2, etc. Fuel cells generate electrical energy by an electrochemical reaction in which the fuel (e.g. H_2, ethanol, or methanol) combines with oxygen (from air) to produce water, heat and electricity [1].

Fuel cells can be categorized based on different parameters which include the electrolyte type employed in them, fuel cell structure, type of reactants, operating conditions, etc. Fuel cells are of five types, based on the nature of electrolyte used in them; i) alkaline fuel cells (AFC) using alkaline electrolyte, ii) phosphoric acid fuel cells (PAFC) using phosphoric acid electrolyte, iii) proton exchange membrane fuel cells (PEMFCs) using proton exchange membrane electrolyte, iv) molten carbonate fuel cells (MOFC) using molten carbonate salt electrolyte, and v) solid oxide fuel cells (SOFC) using solid oxide electrolyte [2]. Among these, PEMFCs have merits like higher energy efficiency, quiet operation and environmental friendliness for stationary, mobile and portable applications [3]. Direct methanol fuel cells (DMFC) are the most suited energy sources for transportation purposes because of the use of inexpensive and portable fuel, simple structural design, higher energy density per unit volume as compared to PEMFCs at low operating temperatures, and simple operation [4-6].

Seba Sara Varghese[a,b], Saino Hanna Varghese[c], Sundaram Swaminathan[d], Krishna Kumar Singh[b] and Vikas Mittal[a,]*
[a]The Petroleum Institute (part of Khalifa University of Science and Technology), Abu Dhabi, UAE; [b]Birla Institute of Technology and Science, Dubai, UAE; [c]Catholicate College, India; [d]DIT University, India
Current address: Bletchington, Wellington County, Australia

The proton exchange membrane (PEM) is the indispensable part of the PEMFCs and DMFCs, which acts as the electrolyte that transports protons from anode to the cathode and effectively divides the anode and cathode reactants [7]. The materials used as PEMs should have high proton conduction, low fuel or water crossover, low electronic conduction, moderate cost, durability, high thermal and mechanical stability above 120 °C, good mechanical strength, and high hydrolytic and oxidative stability [8].

The currently used polymer electrolyte membranes (PEMs) are perfluorosulfonated (PFSA) membranes, as they possess excellent proton conducting property, poor electrical conduction, good mechanical, physical, chemical and thermal stabilities at moderate temperatures [9]. Among various PFSA polymers, Nafion membrane is the most commonly used PEM for both PEMFCs and DMFCs. The hydrophilic side chains of the Nafion membrane with sulfonic acid functional groups are responsible for its remarkable proton conduction property, as the proton conductive clusters formed by the sulfonic acid groups contribute to the proton transport through the Nafion membrane [10]. For instance, Figure 5.1 demonstrates 2-D SAXS scattering patterns of as-received Nafion, showing a slight morphological anisotropy in the ionic domains from the membrane calendaring process in comparison with uniaxially oriented Nafion [10]. However, high cost, high fuel gas/liquid permeability, water dependence for proton conduction, low proton conductivity

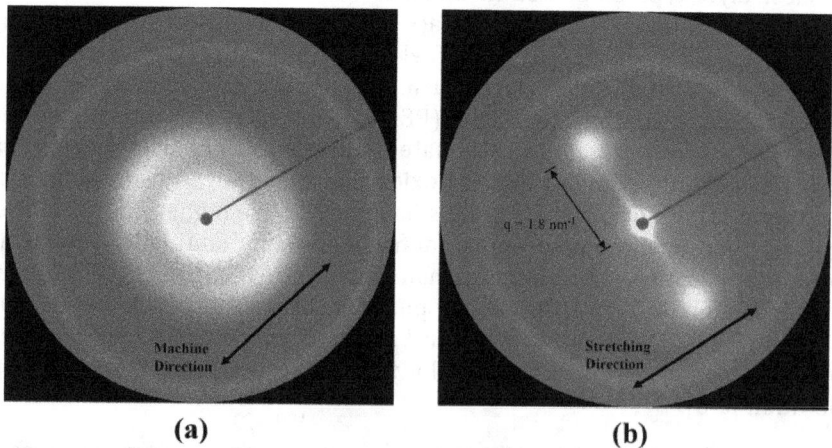

(a)   (b)

Figure 5.1 2-D SAXS scattering patterns of (a) as-received Nafion, and (b) uniaxially oriented Nafion (λ_b = 5.4). Reproduced from Reference 10 with permission from American Chemical Society.

above 100 °C and under low humidity, low thermal and mechanical stability above 120 °C, limit the application of Nafion [11]. These problems need to be resolved for enabling practical applications of the fuel cells, as the properties of PEMs determine the performance of the PEMFCs and DMFCs. Alternative low cost membranes such as modified PFSA polymers, sulfonated aromatic hydrocarbon based polymers, acid-functionalized aromatic hydrocarbon polymers, etc., were also proposed as potential PEMs in PEMFCs and DMFCs.

PEMFCs operating at high temperature (>100 °C) have shown significantly enhanced performance as high temperature operation improves the CO tolerance of the electrocatalysts, enhances electrode reaction rates, reduces the need for excess precious Pt catalysts and simplifies water management systems of PEMFCs. So polymer electrolytes having desirable transport properties and good thermo-mechanical stability at elevated temperatures for PEMFCs have received great scientific attention. The limited capability of Nafion membrane at high temperature makes it unsuited for high temperature applications because of the dramatic reduction of water in these membranes at temperatures above 100 °C. Hence research has also centered on the search for polymer membranes suitable for usage at elevated temperature and under reduced humidity.

The high methanol crossover of Nafion membrane from the anode to the cathode is a serious problem in DMFCs, as it causes fuel wastage, catalyst poisoning and reduces the efficiency and the cell performance. Several studies have also been performed to lower the methanol permeability through Nafion while preserving the good proton conductivity. The slow oxidation kinetics of methanol below 100 °C and significant methanol permeation through PEM in DMFCs results in low energy efficiencies of ~20-25% [12-17]. Hence, many research studies have been directed on boosting the energy efficiency of DMFCs for enabling portable power applications.

The development of desirable PEMs for fuel cells has been one of the hottest topics for more than 40 years. The high thermal stability, low fuel crossover, high CO tolerance, zero water-osmotic drag and excellent mechanical properties of polybenzimidazole (PBI)-based membranes make them attractive as promising electrolyte materials for high temperature applications [18]. PBI-based membranes can achieve desired proton conductivities only by the use of dopants such as sulfuric or phosphoric acid (PA). Previous experimental researches have shown that the characteristics of PBI-based PEMs could be improved by suitable modifications such as sulfonation, acid-base blend, and incorporation of fillers or additives [19]. PEMFCs and DMFCs with Nafion-based membranes also need some modifications to lower the methanol crossover, as well as to make these highly proton conductive and mechanically strong. The modification of polymer membranes with nano-fillers has demonstrated to be one of the promising approaches for improving the membrane's properties and for achieving desired fuel cell performance.

To satisfy the needs of good mechanical strength, proton transport property and reduced methanol crossover for PEMs under fuel cell operating conditions, extensive research efforts were carried out to modify Nafion and other polymer membranes with nanocarbon based materials such as carbon nanotubes (CNTs) and graphene oxide (GO). Various polymer composite membranes based on CNTs and GO have been prepared for PEMs in fuel cells. The nanocarbon-modified polymer membranes have exhibited superior features like excellent mechanical properties and resistance to methanol crossover, while preserving the proton conductivity as high as possible, compared to unmodified polymer membranes and, thus, remarkably improved fuel cell performance. This chapter reviews the state-of-the-art research on composite membranes based on polymers and nanocarbon materials such as CNTs, graphene oxide (GO) used as PEMs for fuel cells. Firstly, polymer composite membranes based on CNTs and functionalized CNTs are discussed. The next section focuses on recent reports that describe the modification of widely used polymer membranes with GO and functionalized GO. The chapter concludes with future prospects of nanocarbon-based polymer composite PEMs for applications in fuel cells.

5.2 Types of Polymer/Nanocarbon Composite Membranes

5.2.1 Polymer/CNT Composite Membranes

CNTs, the one-dimensional carbon allotrope with cylindrical nanostructure [20], have demonstrated great utility in applications such as electronics, sensors, hydrogen storage, gas adsorption, biomedicine, spintronics, energy storage, photonics, etc. [21-26], because of advanced characteristics like high aspect ratio, large specific surface area, excellent thermal and electrical conductivity, good mechanical strength and nanometer scale size. CNTs have also attracted particular scientific attention in fuel cell applications as materials for electrode supports and as catalyst supports [27-30]. Figure 5.2 demonstrates the schematic of the use of multi-walled carbon nanotubes as a platinum support for proton exchange membrane fuel cells [27]. Various research studies have revealed that the incorporation of CNTs could significantly improve the mechanical performance and the electrical conductivity of polymer matrices, hence, CNTs have also been used as nanofillers for reinforcement of polymers [31-34]. The use of CNTs as filler in polymer membranes is limited by the short-circuiting problem in fuel cells arising from the high electrical conductivity of CNTs [35, 36].

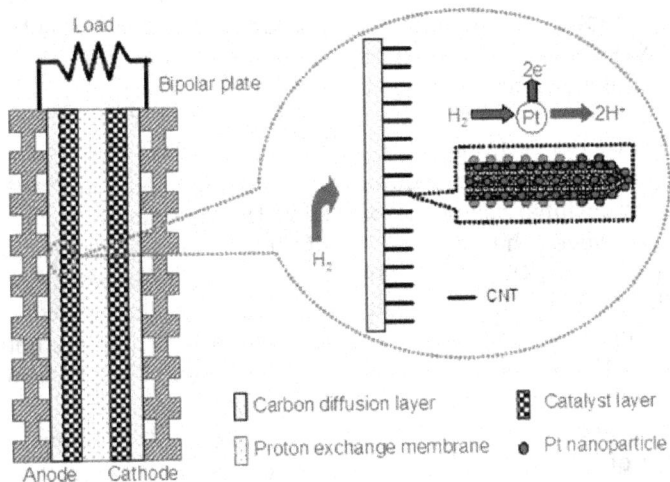

Figure 5.2 Schematic of the use of multi-walled carbon nanotubes as a platinum support for proton exchange membrane fuel cells. Reproduced from Reference 27 with permission from American Chemical Society.

Jana and Bhunia [37] prepared silane based nanostructured composite membranes having different proportion of aqueous orthophosphoric acid, nano-clay (Cloisite® 30B) and multi-walled carbon nanotubes (MWCNTs) by sol–gel method and investigated the dependence of thermal stability and proton conductivity of the composite membrane on the content of orthophosphoric acid, nano-clay and MWCNT in the composites. The C–O–C and Si–O–Si bonds between the silanes resulted in improved thermal stability of the composite membranes up to 200 °C. The thermal stability was observed to decrease with the increase in phosphoric acid content, whereas an increase in thermal stability was observed in the case of increased quantity of nano-clay and nanotubes in the membranes. At 80 °C, 100 °C, 120 °C and 140 °C and at 30% relative humidity (RH), the measured proton conductivity of the membranes increased with increase in weight proportion of nano-clay and MWCNTs in the composite membranes, which could be attributed to the presence of chemisorbed water in the composites. The uniform distribution of the nanoparticles throughout the polymer matrix was evident from the phase morphology study [37].

The use of Nafion/CNT as proton-conducting membrane for FCs has been investigated by several research groups. Liu *et al.* [35] demonstrated that the risk of short-circuit in Nafion/CNTs composites could be avoided by keeping the CNTs content lower than the percolation threshold. The authors developed MWCNTs reinforced Nafion composite membrane for the H_2O_2 fuel cell using ball-milling and solution-casting method. The as-prepared Nafion/CNTs composite membrane had uniform dispersion of MWCNTs within the Nafion polymer matrix. The addition of 1 wt% of CNTs resulted in excellent mechanical strength of the composite membrane. The composite membrane exhibited decreased dimensional change compared to the commercial Nafion NRE-212 membrane. However, the performance of the CNTs reinforced Nafion and the commercial Nafion NRE-212 membrane in fuel cell tests was almost similar [35], which revealed that the addition of CNTs did not improve the proton conductivity of the Nafion membrane. Also, the ball-milling method followed by solution-casting used for the preparation of Nafion/CNTs composite membrane is expensive and not suitable for mass production [36,38].

Thomassin and coworkers [36] used melt mixing technique for MWCNT modification of Nafion membranes. Since no electrical conductivity was observed in the Nafion membrane for MWCNT content lower than 2 wt%, the short circuit risk was minimized. The improved dispersion of MWCNTs within the Nafion membranes resulted in drastically reduction in methanol permeability of about 60%, with only a slight reduction in the ionic conductivity of the Nafion membrane. The ionic conductivity of Nafion membranes was later enhanced by the functionalization of MWCNTs by carboxylic acid groups. The Nafion membrane with 2 wt% of modified-MWCNT content exhibited a 160% increase in Young's modulus as compared to pure Nafion membrane [36].

Wang *et al.* [38] used a more convenient and effective solution-cast method for the preparation of MWCNTs reinforced Nafion membrane. MWCNTs were oxidized by H_2O_2 and subsequently an improved solution-casting method by sodium hydroxide (NaOH) addition in the MWCNTs/Nafion/*N,N*-dimethylacetamide (DMAC) solution was used to prepare MWCNTs reinforced Nafion membranes. The addition of NaOH and the oxidation of MWCNTs greatly improved the dispersion quality of MWCNTs in Nafion/DMAC solution. The observed better dispersion of oxidized MWCNTs in the Nafion membrane could be attributed to the interactions of –OH with –COOH groups of the oxidized MWCNTs and the sodium sulfonate groups of the Nafion. The incorporation of MWCNTs improved the tensile strength of the reinforced membrane by

54%. The polarization curves of the fuel cells with NaOH-MWCNTs/Nafion re-inforced membranes with different MWCNTs contents (from 1 to 4 wt%) measured at 80 °C and under fully humidified conditions indicated slightly better fuel cell performance of reinforced membranes than commercial NRE-212 membrane. The increase in the MWCNTs content resulted in reduced fuel cell performance owing to the minor reduction in the proton conductivity of the membrane by substituting polymer with MWCNTs. The authors concluded from the analysis of the comparison of the mechanical properties, proton conductivities and cell performance of the composite membranes that the content of MWCNTs in NaOH-MWCNTs/Nafion reinforced membrane should not exceed 3 wt% [38].

Several studies have focused on the modification of aromatic hydrocarbon based polymers using CNTs for membranes in fuel cells. Kang *et al.* [39] synthesized high molecular-weight poly(2,5-benzimidazole) (ABPBI)/CNT composite membranes via *in situ* polymerization of the AB-monomer in the presence of SWCNT or MWCNT in a mildly acidic polyphosphoric acid. The authors observed uniform dispersion of SWCNTs and MWCNTs throughout the PBI polymer matrix. The incorporation of CNTs into the ABPBI polymer matrix significantly improved the tensile properties of the ABPBI polymer. The composite membranes exhibited toughness of $\sim$200 MPa which was close to that of the nature's toughest spider silk, $\sim$215 MPa. The electrical conductivities of ABPBI films improved by $\sim$19 and 52,700 times by the modification with SWCNT and MWCNT, respectively. The composite membranes of ABPBI with SWCNT and MWCNT without acid impregnation exhibited proton conducting nature with maximum conductivities of 0.018 and 0.017 Scm^{-1} respectively at 130 °C and 50% RH [39].

Guerrero Moreno *et al.* [40] evaluated the composite membranes of PBI and non-functionalized MWCNTs (PBI-CNT) as membrane electrolytes in PEMFC. At 180 °C, the proton conductivities of the PBI doped with H_3PO_4 (PBIPA) and the doped PBI-CNT (PBICNTPA) membranes were measured to be 6.3×10^{-2} and 7.4×10^{-2} Scm^{-1} respectively. The better phosphoric acid retention in the PBICNTPA composite membrane due to the incorporation of CNTs in PBI resulted in increased proton conductivity of the PBIPA membrane. At 180 °C, the single fuel cell having Pt catalyzed hydrogen fed anode and a similar oxygen cathode using PBICNTPA as membrane electrolyte without feed gas humidification showed higher open circuit voltage (OCV) (0.96 versus 0.8 V) and higher maximum power density (174.5 versus 153 $mWcm^{-2}$) than that using a PBIPA membrane, respectively. It was also observed that the OCV of the composite membrane did not change in a second cell test, whereas the OCV of the PBIPA membrane dropped. The polarization curves of fuel cells using PBIPA and PBICNTPA membranes at varying temperatures suggested enhanced fuel cell performance with rise in temperature for both PBIPA and PBICNTPA membranes. The comparison of the fuel cell performance using PBICNTPA membrane and PBI only membrane indicates that the use of CNTs reduces gas permeation through the composite membrane and minimizes power losses in fuel cells. On comparing the observed cell performance with that from previous works up to 300 mA cm^{-2}, it was observed that PBICNT membrane exhibited superior cell performance [41-43]. The thermal oxidative stability and the spatial thermal distribution of the PBIPA membrane also enhanced with the incorporation of CNTs owing to the enhanced thermal properties of the composite membrane. PBICNTPA membrane with 1 wt% CNT exhibited 32% greater tensile strength and 147% greater Young's modulus compared to the corresponding values for a pristine PBI membrane. The mechanical properties of PBIPA and PBICNTPA membranes compared to earlier reports revealed that the values of tensile strength and Young's Modulus for these membranes were comparable or greater than the reported values [41-46]. The enhancement of the mechanical stability and the proton conductivity of the PBI membrane by the incorporation of CNTs indicate the potential of PBI-CNT membrane as electrolyte in PEMFCs [40].

5.2.2 Polymer/Functionalized CNTs Composite Membranes

The dispersion of CNTs in polymer membranes is crucial for obtaining improved membrane properties. The functionalization of CNTs has proved to be an effective approach in the preparation of polymer/CNT composites for improving the dispersion ability of CNTs within the polymer matrix. The presence of CNTs may result in the disruption of the separation of hydrophilic and hydrophobic polymer phases, thus, leading to very little effect on proton conductivity, however, this could be resolved by introducing the sulfonic acid groups on CNTs.

Kannan *et al.* [47] observed improved conductivity of protons in Nafion-based membranes through the incorporation of SWCNTs pre-functionalized with sulfonic acid groups (s-SWCNTs) into the Nafion matrix. The sulfonic acid groups in the SWCNT connect the hydrophobic regions of the membrane and, thus, form a channel-like network for proton transport through the membrane, which thereby increases its proton conductivity. The s-SWCNT-functionalized Nafion membrane exhibited superior fuel cell performance and mechanical stability to Nafion 1135 in H_2/O_2 fuel cells [47]. The same group

[48] modified Nafion membranes using s-MWCNTs so as to manipulate the hydrophilic domain size of Nafion membranes for improving the proton conductivity. The hydrophilic domain size showed remarkable increase with increase in s-MWCNT loading from 0.01% to 0.05% and decreased further at CNT loadings higher than 0.05%. The authors reported that the increased ionic cluster size in s-MWCNT-modified Nafion membrane increased the proton conductivity. The single cell measurements of the Nafion S-MWCNT composite 0.05% membranes with maximum ionic cluster size of ~70 Å showed maximum power density of 380 mW cm^{-2}, which was greater than those of recast Nafion (230 mW cm^{-2}) and Nafion 115 with ionic cluster size of ~50 Å (250 mW cm^{-2}) membranes. Figure 5.3 also shows the single-cell polarization plots obtained with membrane electrode assemble (MEAs) using commercial Nafion, Nafion recast, and composite membranes at 70 °C along with that of the 0.05% Nafion composite with unfunctionalized pristine MWCNT (NapM 0.05%) [48]. These measurements indicated that nanotube-tailored Nafion membranes enhanced the performance of the H$_2$/O$_2$ fuel cell.

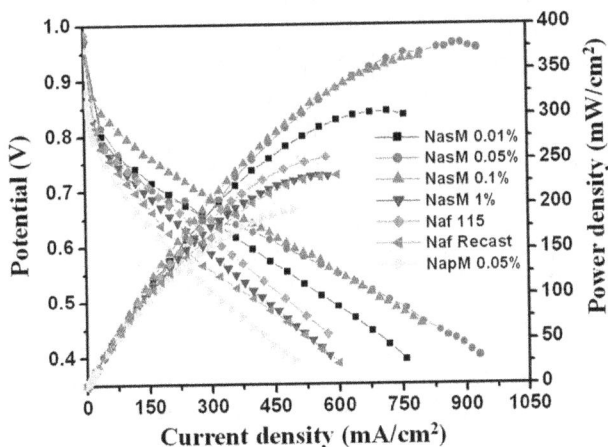

Figure 5.3 Fuel cell single-cell polarization plots of MEAs with Nafion and composite membranes at 70 °C. Reproduced from Reference 48 with permission from American Chemical Society.

Cele *et al.* [49] investigated the proton conductivity, thermal and mechanical stability of CNT based Nafion composite membranes using pure CNTs (pCNTs), oxidized CNTs (oCNTs) and amine functionalized CNTs (fCNTs) prepared by melt-mixing-compression-moulding process to analyze the influence of surface oxidation and functionalization of CNTs in Nafion membranes. The addition of 1 wt% of CNTs improved the thermal stability of all composite membranes. Composite membranes containing 1 wt% of oCNTs showed drastic improvements in thermo-mechanical stability than that of pure Nafion and other CNT-containing Nafion composite membranes, due to high compatibility of oCNT outer surfaces with Nafion matrix. The DC electrical conductivity of the CNT-containing Nafion composite membrane improved with the use of pCNT filler in Nafion, whereas the proton conductivity decreased with the addition of CNTs. Therefore, the authors suggested the need for reduction of filler amount to maintain low electrical conductivity and to maximize the proton conductivity of Nafion membrane. Out of the various CNT-based Nafion composite membranes, oCNTs-containing Nafion composite membrane was claimed as the best polymer electrolyte membrane for fuel cells, as the surface oxidation of CNTs in Nafion membranes improved the properties of pure Nafion, compared to that of other Nafion composite membranes with pure CNTs and amine functionalized CNTs [49].

Functionalized CNTs/Nafion nanocomposite membranes synthesized by surfactant-assisted method exhibited highly enhanced dispersion of CNTs in the nanocomposite membranes [50]. The reinforcement of 1 wt% sulfonated CNTs (s-CNTs) and oxidized CNTs (ox-CNTs) into Nafion membrane by solution casting method lowered the methanol uptake and enhanced the thermal property of cast Nafion, while maintaining the ion exchange capacity. These results indicate that s-CNTs or ox-CNTs/Nafion nanocomposite membranes prepared by surfactant-assisted method could be used as PEMs for fuel cells. Wu *et al.* [46] observed improved ionic conductivity and fuel cell performance in KOH-doped PBI-CNT composite membranes by adding <1% of functionalized MWCNTs via solution casting method, revealing the potential of PBI-CNT composite membranes in high temperature alkaline direct methanol fuel cell (ADMFC) applications. Single cell evaluation using 6 M KOH as anode fuel and humidified oxygen as cathode oxidant generated the maximum power density of 104.7 mW cm^{-2} at 90 °C under 2 M methanol feed concentration.

A self-humidifying composite membrane with a layer of Pt/CNTs dispersed Nafion sandwiched between two outer layers of plain Nafion was developed by Liu *et al.* [51] using solution casting method. The Pt/CNTs were observed to be uniformly dispersed in the composite membrane. The N-Pt/CNTs-N membrane-based PEMFC using dry and humidified H$_2$ and O$_2$ reactants exhibited better performance compared to N-CNTs-N and Nafion NRE-212 membrane. With dry reactants, PEMFC containing N-Pt/CNTs-N membrane exhibited nearly 90% of the performance achieved with humidified reactants whereas the Nafion NRE-212 membrane exhibited only 72% of the performance displayed with humidified reactants. The superior performance of the composite membrane could be attributed to the catalytic recombination of the reactants permeating from the anode to the cathode which resulted in the production of water. In addition to this, the Pt/CNTs in the composite membrane could improve the mechanical strength of the Nafion membrane. The short circuiting issue of the electrically conductive Pt/CNTs was avoided by the two plain Nafion layers on the outer surface. The authors claimed that the reinforced composite membrane could overcome the limitations associated with Nafion/CNTs membrane by incorporating the self-humidifying layer in the membrane [51].

The polymer-functionalized CNTs are also reported to have good compatibility with the polymer matrix so as to avoid any possible micro-phase separation in the nanocomposites [52-54]. Several research studies have reported the use of polymer-functionalized CNTs as membrane electrolytes for FCs. For instance, Chen *et al.* [55] synthesized proton-conducting membrane by blending polysiloxane-functionalized MWCNTs with Nafion. The polysiloxane-functionalized MWCNTs were synthesized by the authors by covalently grafting hydrophilic layers composed of poly(oxyalkylene)diamines (EO) and tetraethyl orthosilicate-reinforced polysiloxane in a layer-by-layer manner onto the nanotube walls. The electron conduction between the CNTs was effectively prohibited by the polysiloxane layer on the CNT sidewalls which acted as an electrical resistance layer for MWCNTs. The addition of polysiloxane layer onto the CNT sidewalls allowed uninterrupted pathways for fast proton conduction (2.8×10^{-2} S cm^{-1} at 30 °C) through Nafion membrane *via* the ionic interactions between the amino groups of the CNTs and the sulfonated groups of Nafion. The fast proton conduction was maintained even at high temperatures (6.3×10^{-2} S cm^{-1} at 130 °C). These results suggested the potential use of functionalized CNT/Nafion nanocomposites as proton exchange electrolyte materials for high temperature fuel cell applications [55]. Figure 5.4 shows the TEM images of the pristine CNTs and CNT-EO. The polymer shell around the individual nanotubes was observed to be thin and uniform, thus, confirming uniform surface modification.

Liu *et al.* [56] reported ozone-mediated synthesis of PEMs using Nafion-functionalized MWCNTs (MWCNT-Nafion), in an effort to impart the high sulfonic acid content of Nafion to CNTs. The authors observed high compatibility of MWCNT-Nafion nanohybrid with the Nafion matrix through the Nafion chains attached to the MWNTs. At 60 °C temperature, the resulting composite membrane with 0.05 wt% MWCNT-Nafion loading (N/MN-0.05) exhibited an

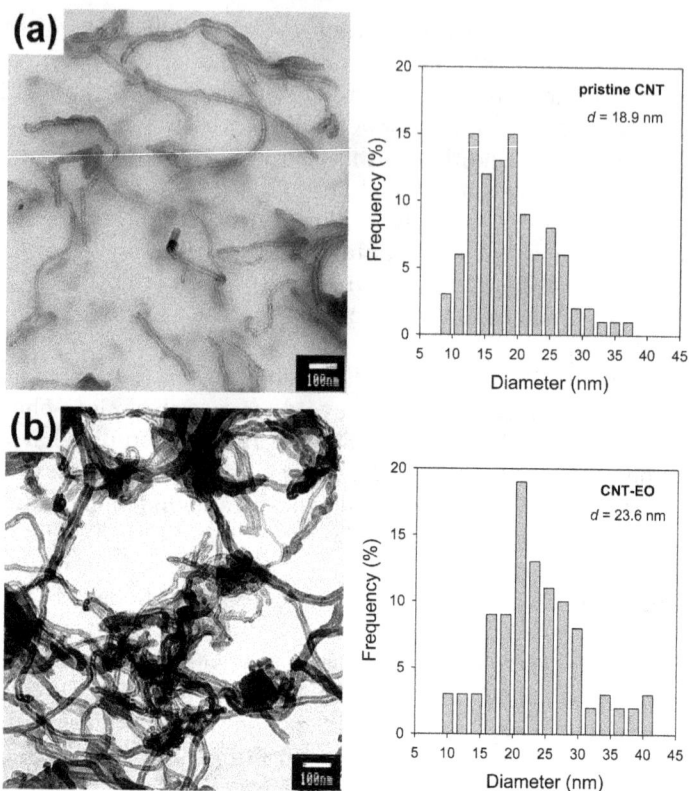

Figure 5.4 TEM images and average size distribution of (a) pristine CNTs and (b) modified CNTs CNT-EO. Reproduced from Reference 55 with permission from American Chemical Society.

increase in mechanical strength and proton conductivity of about 1.5 times and 5 times as compared to pristine Nafion. The current density at 0.6 V and the maximum power density were observed to be 1556 mA cm^{-2} and 650 mW cm^{-2}, respectively, for N/MN-0.05 composite membrane, which are 1.5 times greater than those observed using pristine Nafion membrane. The high proton conductivity of Nafion/MWCNT-Nafion composite membrane consequently yielded high performance for PEMFC application. The superior thermal stability, improved mechanical strength and good single cell performance of the N/MN-0.05 composite membrane indicated that Nafion/MWCNT-Nafion composite membranes are promising candidates as PEMs in fuel cell applications

[56]. In another study, Hasani-Sadrabadi *et al.* [57] fabricated nanocomposite membranes by incorporating polybenzimidazole-decorated CNT (PBI-CNT) doped with phosphotungstic acid (PWA) into the Nafion matrix. The PWA-doped PBI-decorated CNT/Nafion composite membrane exhibited good proton conductivity at elevated temperatures and low humidity. Moreover, the nanocomposite membrane had improved power generation capacity of 386 mW cm^{-2}, at low humidity and high temperatures than the commercial Nafion commercial Nafion 117 membrane of 73 mW cm^{-2} [57].

Composite membranes of functionalized CNTs with other polymer electrolytes have also been reported. Kannan *et al.* [58] described the ability of phosphonated MWCNTs (P-MWCNTs) in tuning the properties of PBI-based membrane electrolytes for FC applications. Single-cell evaluation of composite membrane exhibited a 50% and 40% enhancement in proton conductivity and power density compared to a PA-doped PBI membrane. The fuel cell maximum power densities of 780 mW cm^{-2} and 600 mW cm^{-2} were observed for the composite and pristine PBI membrane under similar conditions. The PA-doped PBI/P-MWCNT membrane exhibited greatly improved ultimate strength compared to PA-doped PBI membrane because of the added MWCNTs in the polymer matrix. The high proton conductivity and good mechanical stability indicate the potential of phosphonic acid-functionalized MWCNT based hybrid membranes as polymer electrolytes for fuel cells [58]. In an earlier study, Kannan *et al.* [42] also employed composite membrane based on PA-doped PBI and P-MWCNTs for high temperature PEMFCs. The improvement in the proton conductivity was mainly because of the formation of proton conducting network in the composite membrane arising from the increase in the net phosphonic acid by the phosphonated groups on MWCNTs. Thus, the incorporation of P-MWCNTs enhanced the fuel cell performance of P-MWCNT/PA-doped PBI composite membranes. The composite membrane reduced the activation energy of the oxygen reduction reaction (ORR). The proton conductivity, stability and overall performance of composite membranes were observed to be greatly enhanced compared with PBI membranes. The optimum P-MWCNT doping level was reported to be 1 wt%. The acid leaching from the polymer matrix was found to be reduced by the use of composite membranes [42].

Composite membranes consisting of CNTs modified with sulfonic acid or PtRu nanoparticles and sulfonated poly(arylene sulfone) (sPAS) were synthesized by Joo *et al.* [59] for DMFC by solution casting. The functionalized CNTs were uniformly dispersed within the sPAS matrix. The obtained SO$_3$CNT-sPAS or PtRu/CNT-sPAS composite membranes exhibited improved tensile strength and toughness compared to the pure sPAS membrane. The addition of functionalized CNTs to sPAS matrix resulted in improved ionic conductivity and reduced methanol permeability compared with a sPAS membrane. This was due to the fact that sPAS/CNT composite membranes contained smaller and uniform ion clusters compared to the sPAS membrane. The single cell DMFC tests demonstrated that PtRu/CNT-sPAS membrane resulted in low ohmic resistance and high OCV compared with the neat sPAS membrane, which ultimately leads to high power density [59]. In another study, Tripathi *et al.* [60] developed nanocomposites based on sulfonated poly(ether ether ketone) (SPEEK) polymer and covalently functionalized MWCNTs obtained by grafting MWCNTs with sulfonic acid groups. The authors investigated the influence of f-MWCNT content on membrane properties for enabling hydrogen or alcohol fuel cell applications. The introduction of f-MWCNTs in the SPEEK membrane with 5 to 10 wt% content resulted improvement of proton conductivity, a jump in electrical conductivity, an increase in mechanical stability, a decrease in swelling, a decrease in water permeability down to $1/3^{rd}$ of that of plain SPEEK membrane and also a decrease in methanol permeability by $1/20^{th}$ of that of plain SPEEK membrane. The increase in the bound water with increasing content of f-MWCNTs in membrane favored the operation of hydrogen fuel cells above 100 °C and under low external humidity conditions [60].

Suryani *et al.* [41] reported ozone mediated synthesis of PBI- and Nafion-functionalized MWCNTs (MWCNT- PBI and MWCNT-Nafion) and investigated the effect of different functionalization of MWCNTs on membrane properties. Both MWCNT-PBI and MWCNT-Nafion were used as additives in the preparation of nanocomposite membranes for PEMFC. The high compatibility between PBI matrix and functionalized-MWCNTs was evident from the morphological analysis of the PBI/MWCNT nanocomposite membranes. The thermal and mechanical properties of the nanocomposite membranes were observed to be enhanced by the addition of the functionalized MWCNTs. The authors observed the formation of proton conducting pathways in composite membranes, and significant increase in the proton conductivities of the membranes by PBI functionalization of MWCNT. The PA-doped PBI/MWCNT nanocomposite membranes containing 0.2 wt% of MWCNT–Nafion (N-0.2) and 0.2 wt% MWCNT–PBI (P-0.2) exhibited proton conductivities of 0.05 and 0.08 S cm^{-1} at 160 °C under anhydrous conditions, as compared to the proton conductivity of the PA-doped PBI membrane (0.03 S cm^{-1}). At 150 °C, maximum power densities of 700 and 600 mW cm^{-2} were observed for N-0.2 and P-0.2, respectively, which were greater than that measured with pristine PBI (530 mW cm^{-2}). The composite membranes exhibited improved proton conductivities and superior fuel cell performance compared to pristine PBI membrane. Nafion functionalization of MWCNTs produced ionically cross-linked structures in the nanocomposite membranes, which enhanced the compatibility between PEM and electrocatalyst, and ultimately led to efficient fuel cell performance due to the improved interfacial contact of membrane electrode assembly [41].

Composite membranes consisting of PBI and functionalized MWNTs have been investigated as PEMs [43]. The functionalization of MWNTs with sodium poly(4-styrene sulfonate) and imidazole provided remarkably improved mechanical reinforcement of the composite membranes due to much better compatibility with PBI. The proton conductivities measured at 160 °C under anhydrous condition for the MWNT-poly(NaSS)/PBI and MWNT-imidazole/PBI composite membranes were up to 5.1×10^{-2} and 4.3×10^{-2} S cm^{-1} respectively, which were higher than the pristine PBI membrane (2.8×10^{-2} S cm^{-1}). At 170 °C, the fuel cell performance of the MWNT-poly(NaSS)/PBI and MWNT-imidazole/PBI composite membranes was also observed to be greatly improved compared to the pristine PBI. In an study, Gong *et al.* [61] demonstrated the use of a super proton conductor, boron phosphate (BPO$_4$) to functionalize CNTs. The BPO$_4$ coated CNTs prepared by polydopamine-assisted sol-gel method were used as additive to modify SPEEK. The use of polydopamine enabled uniform adhesion of BPO$_4$ nanoparticles on to CNTs and the BPO$_4$ coating layer on CNTs reduced the short-circuiting risks of CNTs. The SPEEK/BPO$_4$@CNT composite membranes exhibited significantly improved tensile properties, dimensional and thermal stability compared to pure SPEEK membrane due to the homogeneous dispersion of BPO$_4$@CNTs in the SPEEK polymer matrix. The newly formed proton conducting channels in the SPEEK/BPO$_4$@CNT composite membranes yielded improved proton conductivity compared with pure SPEEK membrane (45% and 150% at 20 °C and at 80 °C, respectively). The peak power density released by the H$_2$/O$_2$ fuel cell using SPEEK/ BPO$_4$@CNT composite membrane was much higher than achieved by using pure SPEEK membrane. These results suggested that composites of SPEEK and proton conductor-functionalized CNTs possessed great potential as high performance PEM materials [61]

Jana *et al.* [62] investigated the effect of varying proportions of functionalized multi-walled carbon nanotubes (f-MWNTs) on the DMFC performance properties of polyethylene co-acrylic acid-based ionomer (Lotek 4200) based membranes. The authors observed that the proton conductivity and the methanol permeability of the membranes were strongly influenced by the introduction of f-MWNTs, as the incorporation of f-MWNTs in polymer matrix modified the polymer structure. At 1.0 M methanol feed concentration, it was observed that the DMFC performance of the composite membranes was comparable to that of a Nafion117 and the methanol crossover flux was 2.5-fold lower than that of Nafion117. In single-cell tests, the DMFC performance of composite membrane was superior to Nafion117 up to 10.0 M methanol concentration [62]. These results indicated that the functionalized CNTs can be used as nanofillers for the modification of polymer electrolyte membranes for achieving desired performance in fuel cells.

5.2.3 Polymer/Graphene Oxide Composite Membranes

GO has been used as nanofiller in polymer composite membranes due to its unique properties such as presence of oxygen rich functional groups, high surface area, high mechanical strength, poor electrical conductivity and exceptional structural resilience [63,64]. The addition of GO as additive to polymer membranes greatly enhances their proton conductivity and methanol-blocking efficiency, thus, GO/polymer composite membranes have the potential to be employed as one of the best PEMs for fuel cells. Kumar *et al.* [65] observed high ionic conductivities of 4.1×10^{-2} S cm^{-1} to 8.2×10^{-2} S cm^{-1} at temperatures of 25–90 °C, for a free standing GO paper membrane (with thickness of 100 μm), and underlined the potential of GO for use as PEM for DMFC application. The acidic functional groups, hydrogen bonds, and connecting oxygen atoms in the GO structure resulted in abundant proton conducting paths, which ultimately led to high proton conductivity. However, the methanol permeability of the GO paper membrane was observed to be significantly higher than that of Nafion 115 membrane, which resulted in poor DMFC performance despite its improved proton conductivity. By underlining the strong potential of GO as a filler in polymer matrix, the study suggested that the future work on GO based composite membranes with binder could provide great mechanical strength, enhanced proton conductivity and good flexibility with reduced methanol permeability for DMFCs.

Ionic conducting composite membranes of GO and poly (ethylene oxide) (PEO) were reported to be desired membrane electrolyte for low-temperature PEMFC application, without any modification [66]. The reason behind the ionic conductivity of the membrane is the release of protons from the COOH groups on GO sheets in the PEO polymer matrix. It was observed that the ionic conductivity of the PEO/GO membrane increased from 0.086 to 0.134 S cm^{-1} with increase in temperature from 25 to 60 °C, at 100% RH. Above 60 °C, the mechanical properties of the composite membrane got deteriorated. The

PEO/GO membrane released a maximum power density of 53 mW cm^{-2} in a single cell operated at around 60 °C.

Composite membranes based on Nafion and GO as PEMs for fuel cell applications have also been discussed. Specifically, the Nafion/GO composite has been exploited as polymer electrolyte membrane for DMFC applications [67]. The interaction of the non-polar Nafion backbone with the hydrophobic GO structure and that of the polar ionic clusters of Nafion with the hydrophilic GO functional groups leads to the modification of the microstructures of the Nafion domains. The incorporation of GO into Nafion matrix favorably altered the physical and transport properties of Nafion for fuel cells. The incorporation of GO enhanced the mechanical properties of the Nafion due to strong interfacial interactions. Due to the same reason, the composite membrane also exhibited higher thermal stability than pure Nafion. The Nafion/GO composite membranes had a 14% increase in tensile strength and a 26% increase in Young's modulus as compared the Nafion membrane. The prepared Nafion/GO composite membranes with 0.5 wt% loading of GO sheets exhibited a 40% reduction in methanol crossover compared to the value for pristine Nafion membrane. The proton conductivity of the composite membrane reduced to ~98.1% of that of pure Nafion. The addition of 0.5 wt% GO in Nafion 112 membrane increased the selectivity (ratio of proton conductivity to methanol permeability) by ~150%, due to the reduced size of the ionic channels and the presence of hydrophilic functional groups of GO. The DMFC with Nafion/GO composite membrane exhibited remarkable improvement in performance with power densities much higher than earlier observed for Nafion-based DMFC. Also, the composite membrane with 5M methanol had ~3 times higher power density than Nafion and ~14 times higher than that of the composite membrane at 1M methanol feed concentration, which indicated that the modification of ionic clusters of Nafion with GO nanofiller suppressed the methanol crossover even at high methanol concentrations, thus, resulting in the observed behavior [67].

Wang *et al.* [68] prepared GO/Nafion composite membranes to provide mechanical reinforcement to Nafion and optimized the GO loading based on fuel cell performance and mechanical properties of membrane. The addition of GO improved the tensile strength and dimensional stability of Nafion significantly. The mechanical reinforcement and dimensional stability provided by the individually dispersed GO sheets in Nafion resin enhanced the integrity of the membrane electrode assembly within the fuel cell and provided better fuel cell durability. GO/Nafion composite membrane with 3 wt% exhibited superior mechanical properties as compared to the recast Nafion, but the fuel cell performance was similar to that of recast Nafion. Apart from providing mechanical reinforcement, the use of GO as additive in Nafion preserved the low electrical conductivity of Nafion due to its electrically insulating nature [68]. GO/Nafion composites fabricated by LbL assembly were also applied as membrane electrolytes in DMFCs [69]. GO/Nafion composite membranes had lower uptake of water than recast Nafion 117 membrane. The ion exchange capacity (IEC) of composite membrane was only slightly higher than that of the recast Nafion 117 membrane. The enhanced IEC was due to the presence of carboxyl and hydroxyl functional groups at the GO sheet edges which released H$^+$ upon humidification. The oxidative stability of Nafion 117 membrane was improved remarkably by the GO coating on the Nafion surface. GO/Nafion composite membranes exhibited proton conductivities lower than that of Nafion 117 membrane owing to their decreased water uptake arising from the shrinkage in the proton diffusion channels. The methanol permeability of the composite membranes was much lower than that measured for pristine Nafion membrane under similar conditions. This observation was due to the reduced swelling ratio of the composite membranes which blocked the passage of methanol molecules through the proton conducting channels. The lowest methanol permeability (6.7×10^{-8} cm^2 s^{-1}), achieved for the composite membrane with 80-layered GO nanosheets, was about two orders of magnitude lower than that of pristine Nafion membrane (1.82×10^{-6} cm^2 s^{-1}) at 30 °C. The composite membranes exhibited decreasing methanol permeability with increase in the number of GO layers, which indicated the advantage of modifying Nafion with GO. The comparison of the performance of all composite membranes with that of the recast Nafion 117 membrane was also performed by defining relative selectivity (RS) as the ratio of the selectivity of the composite membranes to that of the Nafion 117. The RS of the GO/Nafion composite membrane with 80-layered GO nanosheets was 5-fold greater than that of the Nafion 117 membrane and also RS was found to be increasing with the increase in the number of GO layers. The observed dramatic suppression of methanol crossover by the modification with GO nanosheets, thus, guarantee the potential of composite membranes based on GO and polymer for future DMFCs. The single cell using GO/Nafion composite membrane with 50-layered GO achieved a maximum power density of 64.38 mW cm^{-2} at 60 °C and ambient pressure with RH of 100% under high methanol concentration of 5 M, whereas for the cell utilizing recast Nafion 117 membrane, a maximum power density of 41.60 Mw cm^{-2} was obtained, which indicated the superior fuel cell performance of GO/Nafion composite membranes compared to recast Nafion 117 membrane [69].

Several reports have revealed the application of GO as an excellent methanol-blocking film on Nafion surface for DMFCs. GO-laminated Nafion membrane was observed to be beneficial as PEM for DMFC operating under high methanol feed concentrations, compared to GO-dispersed poly(vinyl alcohol) (PVA) composite membrane [70]. Lin *et al.* [70] fabricated a dual layer membrane by laminating the Nafion 115 with a highly ordered and parallel oriented GO paper by vacuum filtration, through transfer printing followed by hot-pressing. The 1 μm thick, 2D GO paper adhered to the Nafion base membrane acted as efficient methanol barrier under high methanol feed concentrations. The methanol permeability of the GO-Nafion nanocomposite membrane was observed to be approximately 70% lower than that of Nafion 115, whereas the proton conductivity decreased by 22% by GO-lamination of Nafion 115. The ratio of proton conductivity to methanol crossover of GO-laminated Nafion 115 was observed to be 40% higher than Nafion 115. The GO-Nafion nanocomposite membrane was observed to be superior to neat Nafion membrane for DMFC during operation under 8 M methanol feed concentration [70].

Sulfonated polyimide (SPI)/GO composite membranes as proton exchange membranes were fabricated by He *et al.* [71] by incorporating GO with different sheet sizes into SPI. The SPI/GO composite membranes exhibited an increase in mechanical properties and proton conductivities, along with a decrease in methanol permeabilities by the addition of GO into SPI. It was observed that the size of GO sheets affected the properties of the composite membranes. SPI/GO composite membrane with the smallest size of GO achieved the best proton conductivity, lowest methanol permeability and significantly enhanced fuel cell performance compared to that of pure SPI and other SPI/GO composite membranes, at 0.5 wt% loading of GO in SPI. The observed outstanding properties of composites by the incorporation of smaller size GO could be ascribed to the formation of highly uniform microstructure and continuous proton conducting pathways due to strong hydrogen bonding between SPI and the smaller sized GO. The DMFC with SPI/GO composite membrane containing 0.5% GO exhibited a 1.4-fold higher power density compared with that of the SPI membrane at 25 °C [71].

PA-doped PBI/GO nanocomposite membranes were prepared by Uregen *et al.* [72] by solution blending method for high temperature fuel cell applications. The proton conductivity and acid retention properties of PBI/GO nanocomposite membrane were improved by the dispersion of GO in PBI polymer matrix. The PBI/GO composite membrane exhibited the highest proton conductivity for 2 wt% GO loading, which was 79% higher than that of a neat PBI membrane at 180 °C. In high temperature PEMFC tests, the PBI/GO composite membrane with 2 wt% GO content exhibited better performance than pure PBI membrane under 165 °C and non-humidifying conditions, with peak power densities of 0.31 and 0.38 W cm^{-2} obtained for composite and neat PBI membranes respectively. The enhanced proton conductivity of the nanocomposite membrane was mainly attributed to the facile proton hopping induced by the acidic functional groups (epoxy oxygen and carboxylic acid groups) of GO and also the interconnection of proton conducting channels in PBI membrane by GO. However, at temperatures above 165 °C, it was observed that the conductivity decreased due to the degradation of acidic functional groups in GO. These results demonstrate the strong potential of utilizing PBI/GO composites as electrolyte materials for high temperature PEMFCs [72].

5.2.4 Polymer/Functionalized Graphene Oxide Composite Membranes

Polymer electrolyte/functionalized GO composite membranes have also been reported for use in fuel cell applications. Choi *et al.* [73] reported the incorporation of sulfonated GOs (SGOs) into Nafion matrix (Figure 5.5) and observed

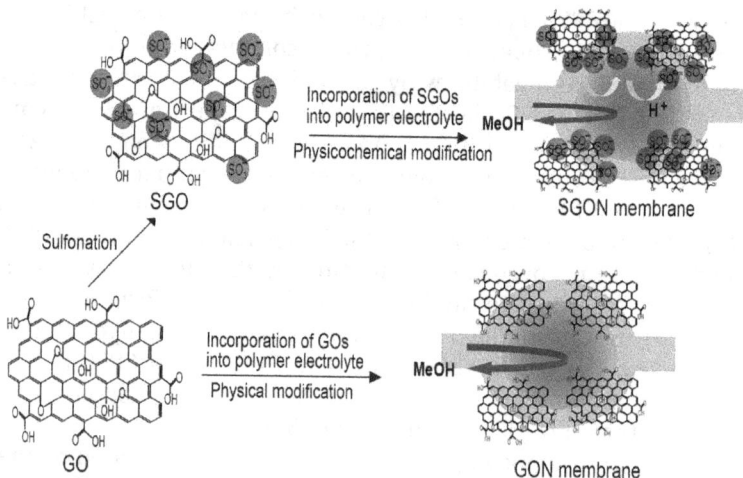

Figure 5.5 Schematic of generating tunable transport properties using functionalized graphene. Reproduced from Reference 73 with permission from American Chemical Society.

that the transport properties of SGO/Nafion composite membranes were superior to those of pristine Nafion and GO-incorporated nanocomposites. The physical and chemical modification of ion-conducting channels through chemical tuning of the graphene functionality controlled the confinement of bound water within the ionic channels and, thus, resulted in selectively facilitated transport behavior of SGO/Nafion composite membranes, as compared to pristine Nafion and GO-incorporated nanocomposites. The composite membrane exhibited enhanced proton conductivity (~66%) with low activation energy, and reduced methanol permeability (~35%) at high temperature for high performance DMFCs. In another study, composites of Nafion with sulfonic acid-functionalized GO were also presented as novel PEMs for high temperature PEM fuel cell applications due to the exhibited high proton conductivity and fuel cell performance [74]. At 120 °C with 25% humidity, the proton conductivity and fuel cell performance was observed to be 4 times greater than that of recast Nafion. Figure 5.6 demonstrates the polarization and power density curves for 10 wt% functionalized GO/Nafion and recast Nafion membranes.

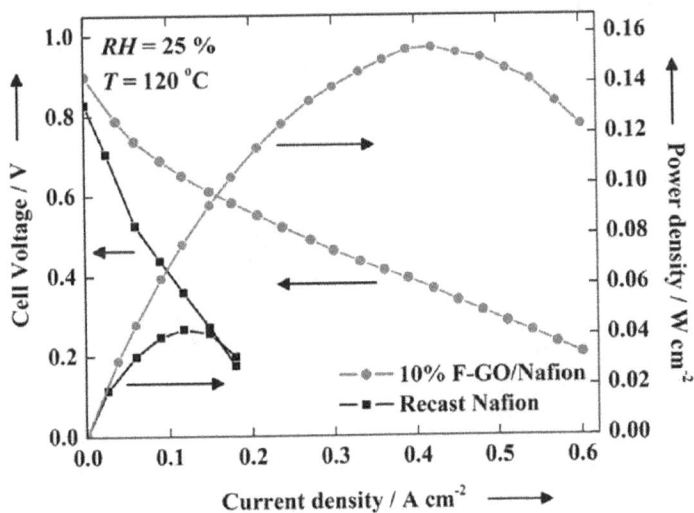

Figure 5.6 Polarization and power density curves for 10 wt% functionalized GO/Nafion and recast Nafion membranes. Reproduced from Reference 74 with permission from American Chemical Society.

In a recent study, Sahu *et al.* [75] employed sulfonic acid-functionalized graphene (S-graphene) as filler in Nafion for the preparation of composite membrane for PEMFCs under low RH. The incorporation of S-graphene in

Nafion increased the sulfonate moieties per unit volume of each graphene domain and, thus, reconstructed the hydrophilic domains of Nafion, thereby providing fast transport of protons through the composite membrane at low RH. The high density of sulfonic acid groups on S-graphene dispersed in Nafion and the high surface area of S-graphene resulted in significantly enhanced IEC and proton conductivity of Nafion-S-graphene, which ultimately led to better PEMFC performance of composite membrane at low RH than that attained by employing pure Nafion. It was observed that the proton conductivity of Nafion-S-graphene (1%) composite membrane was five times higher that of the pristine Nafion membrane, at 20 % RH. The comparison of mechanical properties of Nafion-S-graphene composite membrane with 1% S-graphene content and pristine Nafion membrane exhibited higher mechanical stability of composite membrane. Nafion-S-graphene-based PEMFC delivered a peak power density of 300 mW cm^{-2} at 70 °C under 20% RH and ambient pressure, whereas a pristine recast Nafion-based PEMFC released a peak power density of only 220 mW cm^{-2} (Figure 5.7). The high surface roughness of the Nafion-S-graphene resulting from the high surface area and the crumpling nature of S-graphene increased the amount of water uptake and improved its compatibility with electrodes during the fabrication of membrane electrode assemblies. Therefore, Nafion–S-graphene hybrid membranes are promising for applications at low RH and elevated temperature PEMFCs [75].

Enotiadis *et al.* [76] also prepared GO/Nafion composite membranes with GO carrying hydrophilic functional groups such as -NH$_2$, -OH, -SO$_3$H and observed that water retention and proton conduction of Nafion was favorably improved by the addition of organo-functionalized GO materials. The nano-composite membranes were observed to be much stronger and could withstand higher temperatures compared to that of Nafion. In another study, Lee *et al.* [77] also explored GO as potential filler in low humidifying PEMFCs. Higher water uptake was observed in Nafion/GO composite membrane as compared to casting Nafion due to the hydrophilic nature of GO and the uptake increased with GO content. Nafion/Pt-G composite membrane obtained via *in-situ* deposition of Pt nanoparticles onto GO using microwave method exhibited lower water uptake than that of Nafion/GO membrane as GO reduced to hydrophobic graphene during Pt deposition. The water uptake of Nafion/Pt-G composite membrane also decreased with Pt-G content. However, the proton conductivities of the composite membranes followed the opposite trend. The proton conductivities of Nafion/GO composite membrane were lower than those of the Nafion/Pt-G, irrespective of the content of GO or Pt-G. The reason behind the higher proton conductivity of Nafion/Pt-G composite membrane was the presence of electronically conductive Pt nanoparticles.

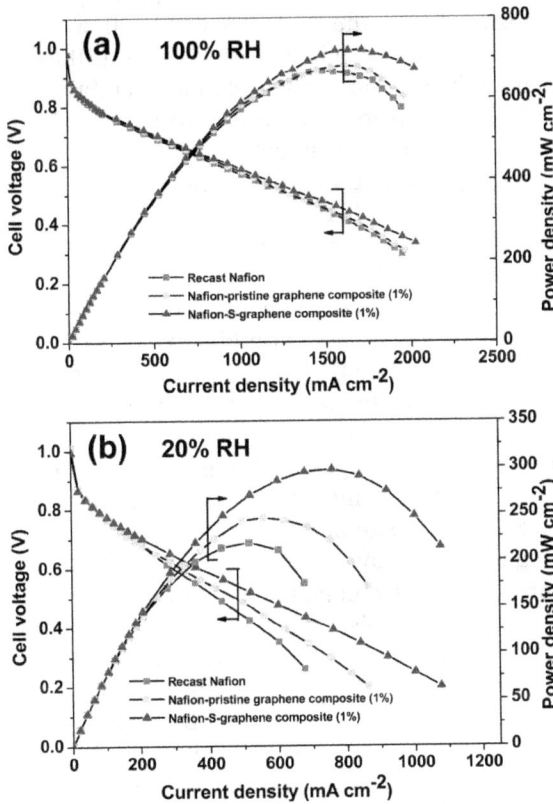

Figure 5.7 Performance of H_2/O_2 PEFC with recast Nafion, Nafion-graphene, and Nafion-S-graphene composite membranes at (a) 100% RH and (b) 20% RH at 70 °C under atmospheric pressure. Reproduced from Reference 75 with permission from American Chemical Society.

Nafion/GO composite membranes with 0.5 wt% and 3.0 wt% filler also exhibited reduced proton conductivities compared with those of pristine Nafion membrane due to blocking of the ionic clusters in Nafion by GO. It is noteworthy that incorporating 4.5 wt% of GO in Nafion led to higher proton conductivity than Nafion membrane owing to the fact that the abundant oxygen rich functional groups facilitated proton transfer channels in Nafion/GO membrane and, thus, overcame the blockage of Nafion's ionic clusters by GO. Signif-

icant increase in tensile strength was observed for both Nafion based composite membranes, thus, allowing their use at high temperature. DMFCs with
Nafion/GO composite membrane with 0.5 wt%, 3.0 wt% and 4.5 wt% GO
showed superior performance under various RHs (100%, 60% and 40%)
compared to that with pristine casting Nafion membrane. The cell performance of Nafion/Pt-G composite membrane was found to be inferior to that of
Nafion/GO composite membrane under similar conditions, due to the poor
water retention of graphene and the loss of electrons caused by the electrical
network formation by Pt nanoparticles on GO. The measured OCVs of Nafi-
on/Pt-G and Nafion/GO composite membranes experienced no significant variation with RH down to 40% [77].

Xu *et al.* [78] prepared PBI/GO and PBI/SGO composite membranes by
adding graphite oxide and sulfonated graphite oxide into PBI membranes with
low phosphoric loadings (1.8-2.0 PRU) for high temperature PEMFCs. The
PBI/GO and PBI/SGO composite membranes with low content of H_3PO_4 exhibited higher proton conductivity than that of the PBI/PA membrane with high
phosphoric loadings. The hydrogen bonding in GO resulted in high proton
conductivity of the composite membranes. The authors reported that both
PBI/GO and PBI/SGO composite membranes generated superior fuel cell performance and higher peak power densities than that of the PBI/PA membranes. Therefore, PBI/SGO composite membranes with low PA loadings
could be used as potential PEMs for high temperature operation [78]. In another study, Liu *et al.* [79] reported nanohybrid membranes based on SGO
nanosheets and chitosan (CS). The uniform dispersion of SGO nanosheets in
the chitosan membrane resulted in improved thermal and mechanical stabilities of nanohybrid membrane due to the inhibition of mobility of CS chains
through the strong electrostatic interactions between the acid group of SGO
and –NH2 group of CS matrix. The addition of SGO generated acid-base pairs at
the CS–SGO interface, which served as proton-hopping locations. The high surface area, nanosheet structure and better dispersion of SGO in CS matrix allowed the formation of continuous channels for efficient proton migration to
the nanohybrid membrane, thus, resulting in significantly enhanced proton
conductivity under both hydrated (100%) and anhydrous (0% RH) conditions. It was observed that the incorporation of 2.0% SGO could provide a
122.5% increase of hydrated conductivity and 90.7% increase of anhydrous
conductivity when compared with CS control membrane. The lower water uptake of SGO-filled membranes compared to GO-filled membranes reduced the
area swelling, which resulted in reinforced structural stabilities of nanohybrid
membranes and thus, making them suitable for practical fuel cell application.

The maximum current density and the maximum power density of a H_2/O_2 fuel cell based on SGO-filled CS membrane increased by 58% and 64% with 2 wt% of SGO at 120 °C and under anhydrous conditions [79].

Gahlot *et al.* [80] prepared SGO/SPES nanocomposite membranes by incorporating various concentrations of SGO into sulfonated poly(ether sulfone) (SPES). The SGO sheets were observed to uniformly distribute throughout the SPES matrix. The IEC, proton conductivity, water retention, selectivity, methanol crossover resistance and mechanical strength of SPES were improved by the incorporation of SGO in SPES (Figure 5.8). The strong interfacial interactions between SGO and SPES matrix prevented the formation of hydrophilic channels in the nanocomposite membrane, which led to reduced methanol permeability and higher proton conductivity. The SGO/SPES nanocomposite with 5% SGO possessed larger methanol crossover resistance and maximum proton conductivity. The observed good proton conductivity of the nanocomposite membrane at temperatures varying from 30 to 90 °C with low activation energy suggested the applicability of SGO/SPES membrane as PEM for DMFCs at high temperature [80].

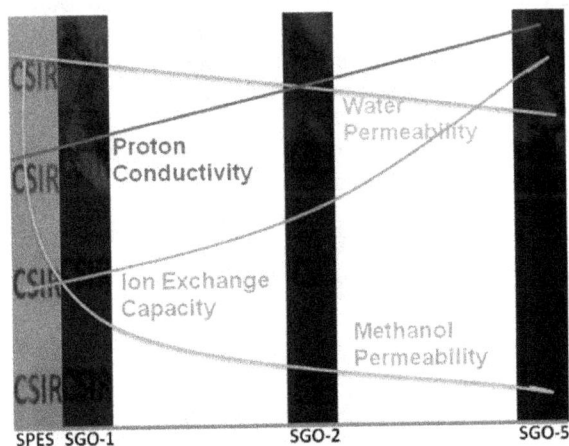

Figure 5.8 Effect of addition of SGO on various properties of SPES membrane. Reproduced from Reference 80 with permission from American Chemical Society.

Kumar *et al.* [81] also observed that the composite membrane of SPEEK with SGO exhibited improved proton conductivity as compared to the neat SPEEK membrane due to the interconnection of proton transport networks within the SPEEK membrane by the incorporated SGO and creation of more proton conducting pathways enabled by presence of water between GO layers. The morphological analysis confirmed the uniform distribution of SGO sheets in the SPEEK structure. At 30% RH, the composite membrane exhibited a proton conductivity of 0.055 S cm^{-1} which was higher than that of a neat SPEEK membrane (0.015 S cm^{-1}). The excellent fuel cell performance of SGO/SPEEK composite membrane compared to neat SPEEK membrane indicates the suitability of using SGO/SPEEK as alternative low-cost PEM for fuel cells [81].

In another study, Pandey *et al.* [82] reported that the SPI/sulfonated propylsilane graphene oxide (SPSGO) composite membrane containing 8 wt% of SPSGO (SPI/SPSGO-8) exhibited significant improvement in proton conductivity (nearly equal to that of Nafion 117)as well as thermal and mechanical stability, along with 10-fold enhanced bound water content. The incorporation of the multi-functionalized SPSGO in the SPI matrix also promoted internal self-humidification and reduced the methanol crossover as compared to the pristine SPI membrane. The membrane properties desirable for DMFC application were observed to be greatly influenced by the loading of acid functionalized GO in the SPI matrix (Figure 5.9). A maximum power density of 75.06 mW cm^{-2} was exhibited by SPI/SPSGO-8 in single-cell DMFC tests, as compared to the commercial Nafion 117 membrane (62.40 mW cm^{-2}) under 2M ethanol fuel at 70 °C. The strong H-bonding between the multifunctional groups, the presence of hydrophobic graphene sheets and polymer chains

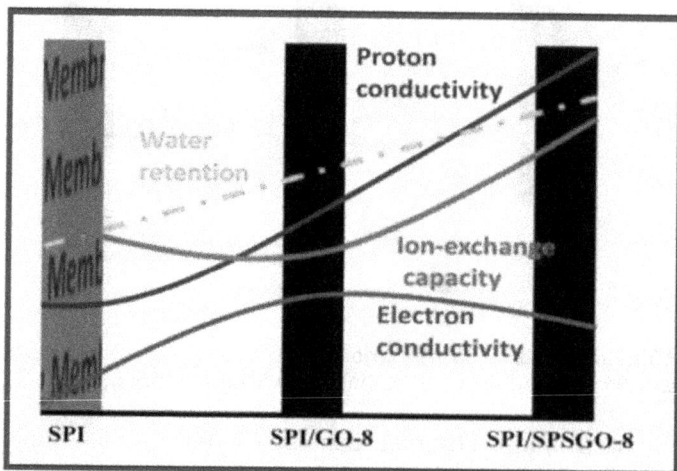

Figure 5.9 Effect of GO and SPSGO loading on SPI membrane properties. Reproduced from Reference 82 with permission from American Chemical Society.

contributed to the proton conducting property by providing a suitable design of proton conducting channels in the membrane matrix.

Graphite oxide/ polybenzimidazole containing tertiary butyl group (GO/BuIPBI) and isocyanate modified graphite oxide/BuIPBI (iGO/ BuIPBI) composites were explored as PEMs for high temperature fuel cell applications [83]. It was observed that the incorporation of 1% isocyanate functionalized GO could enhance the chemical stability of the BuIPBI matrix due to hydrogen bonding between the iGO filler and BuIPBI. The GO/BuIPBI and iGO/BuIPBI composite membranes doped with PA had higher proton conductivities than neat BuIPBI membrane and exhibited increasing trend with increase in temperature. The enhanced proton conductivities of the composite membranes could be ascribed to the proton transport through the membrane induced by the well-connected proton conduction pathways provided by the GO. However, as the temperature was increased as high as 140 °C, it was observed that the conductivities of the composite membranes decreased due to the destruction of carboxyl and epoxy functional groups in GO. The authors also observed that the proton conductivity was influenced by the PA doping level in the membrane. The iGO/BuIPBI membrane with 10 wt% of iGO in BuIPBI and with 76.6% PA doping had the highest proton conductivity of 0.027 S cm^{-1} at 140 °C without humidity, compared to that of BuIPBI and GO/BuIPBI [83].

Recently, Shukla *et al.* [84] prepared sulfonated graphene nanoribbons (sGNR) by unzipping and sulfonation of MWCNTs by a hydrothermal synthetic route which had better dispersion and functional compatibility with SPEEK polymer matrix. Nanocomposite PEMs of SPEEK with sGNR exhibited improvement in water uptake, ion exchange capacity and proton conductivity. The SPEEK/sGNR (0.1 wt%) composite membrane displayed improved mechanical stability than the pristine SPEEK membrane. The SPEEK/sGNR composite membrane had current density of 840 mA cm^{-2} at 0.6 V and maximum power density of 660 mW cm^{-2} compared to the current density of 480 mA cm^{-2} at 0.6 V and maximum power density of 331 mW cm^{-2} for pristine SPEEK membrane in fuel cell tests. The SPEEK/sGNR nanocomposite membranes were also observed to be highly durable. The observed results indicated better fuel cell performance and durability of SPEEK/sGNR composite membranes. The authors suggested that sGNR could be used as effective additives in different polymer matrices so as to achieve durable and cheaper nanocomposites as PEMs for fuel cells [84]. In another study, SPI membrane, incorporating 0.5 wt% GO/poly(sodium-4-styrenesulfonate) modified graphene (PSS-G) was observed to exhibit an improvement in proton conductivity of 490.3% and a drastically decreased methanol permeability of 508.9%, with ~25 times increase in selectivity (proton conductivity/methanol permeability) at 80 °C. The SPI/PSS-G composite membrane also had enhanced tensile strength of 134.4% at 30 °C [85].

Ye *et al.* [86] prepared PEM based on protic ionic liquids (PILs) composite membranes and ionic liquid polymer modified graphene sheets (denoted as PIL(NTFSI)-G) using solution blending method. The sulfonated polyimide (SPI)/PIL(NTFSI)-G/PIL composite membranes with 0.5 wt% graphene loading exhibited enhanced ionic conductivity of 7.5×10^{-3} S cm^{-1}, nearly four times higher than that of a pure SPI/PIL membrane at 160 °C. GO incorporation also resulted in PIL cost-saving of 20%. The Young's modulus and tensile strength of the composite membrane exhibited an increase of 127% and 345% respectively with 0.9 wt% addition of graphene, in comparison with the plain SPI/PIL membrane. The 3D network of homogeneously distributed graphene sheets throughout the polymer matrix in the SPI/PIL(NTFSI)-G/PIL composite membrane facilitated high-level mechanical reinforcement through enhanced nanofiller-polymer adhesion. The homogeneously dispersed graphene nanosheets in SPI matrix created 3D ion transport channels throughout the membrane, which resulted in high ionic conductivity [86].

In a similar study, He *et al.* [87] demonstrated the preparation of highly conductive nanocomposite membrane by incorporating polydopamine-modified graphene oxide (DGO) sheets into SPEEK matrix [87]. The strong electrostatic interaction between the $-NH_2$ and $-NH-$ groups of DGO sheets and the $-SO_3H$ groups of SPEEK chains generated acid-base pairs with closely linked proton donor and acceptor and allowed facile proton transport between the donor and acceptor in the nanocomposite membrane. The proton conductivity of SPEEK/DGO nanocomposite membrane was observed to be much higher than SPEEK membrane under both hydrated and anhydrous conditions. The enhanced proton conductivity of the composite membrane led to much higher H_2/O_2 cell performance, with an increase in maximum current density of 47% and increase in maximum power density of 38 % under 120 °C and anhydrous conditions by the addition of 5 wt% of DGO to SPEEK. SPEEK/DGO composite membrane exhibited enhanced thermal stability due to the tuning of the nanophase-separated structure and inhibition of the chain motion and thermal decomposition of the composite membrane due to strong electrostatic interactions of DGO with the SPEEK polymer matrix. The strong mutual interactions between the DGO and SPEEK also resulted in an increase in mechanical stability of SPEEK/DGO composites. The tensile strength and the Young's modulus of the composite reached up to 57.5 MPa and 896.1 MPa by 2.5% DGO loading. These findings indicated that SPEEK/DGO nanocomposite membranes could be employed as promising PEMs for practical fuel cell applications under high temperature and anhydrous condition [87].

Mishra *et al.* [88] reported that the presence of both SPEEK and GO in Nafion membrane enhanced the membrane's proton conductivity. Nafion containing 1wt% SPEEK and 0.75wt% of highly oxidized GO (HGO) exhibited maximum proton conductivity of 322.2 mS cm^{-1} at 100% RH and 90 $^{\circ}$C, as compared to that of pure Nafion (198 mS cm^{-1}). Nafion-HGO (NHG) composite membranes had the highest storage modulus among virgin Nafion and other composite membranes such as Nafion-SPEEK (NS), Nafion-GO (NG), Nafion-SPEEK-GO (NSG) and Nafion-SPEEK-HGO (NSHG) in the temperature range of 50-150 $^{\circ}$C, which could be attributed to the increase in the secondary interactions of HGO with Nafion arising from higher degree of oxidation and smaller size of GO. The chemical stability of Nafion membrane was observed to be enhanced by the addition of GO, SPEEK or SPEEK-GO. The storage modulus decreased with the addition of highly sulfonated PEEK to NG and NHG composite membranes owing to the high stiffness of GO and HGO as compared to SPEEK. Benefiting from the enhanced proton conductivities and water uptake, NSHG composite membrane with 0.75 wt% of GO and HGO exhibited higher power density and current density than virgin Nafion and other composite membranes, which provided superior fuel cell performance [88]. In another study, Yuan *et al.* [89] reported the LbL assembly of polyelectrolyte poly(diallyldimethylammonium chloride) (PDDA) and GO multilayer film on Nafion membranes and demonstrated the capability of Nafion-PDDA-GO as an efficient methanol-blocking membrane for DMFC. The composite membrane suppressed the methanol crossover with 67% decrease in methanol diffusion coefficient as compared to the pristine Nafion membrane, and also enhanced membrane strength. The methanol crossover limiting current density across the Nafion membrane with PDDA-GO bilayer structure was about 1/3rd of pristine Nafion membrane, which demonstrated that the dense and uniform bilayer structure on Nafion reduced the methanol crossover by about 63%. The passive DMFC using Nafion-PDDA-GO composite membrane exhibited higher energy efficiency and enhanced power density as compared to the pristine Nafion membrane [89].

Nanocomposite anion-exchange membranes (AEMs) were prepared by dispersing GO functionalized with PDDA into polyvinyl alcohol/PDDA semi-interpenetrating polymer networks (PVA/PDDA SIPNs) via solvent casting followed by thermal cross-linking [90]. The OH- conductivity and the mechanical stability of the membranes prepared by physical cross-linking was higher than that of the chemically cross-linked ones. The optimal conditions for fabricating nanocomposite AEMs were determined by studying the effects of PVA/PDDA ratio, FGO content, cross-linking method and temperature on the material properties of nanocomposite membranes. The OH- conductivity reached the highest value (12.1 mS cm^{-1} at 30 °C and 21 mS cm^{-1} at 80 °C) for the PVA/PDDA membrane having a PVA/PDDA weight ratio of 70/30 by the inclusion of 20 wt% FGO. The high OH- conductivity of the nanocomposite membrane at high FGO content (20 wt%) was due to the transfer of OH- through the membrane facilitated by the high-water uptake of the membrane having hydrophilic FGO with high surface area. The hydrogen fuel cell performance of the nanocomposite membrane was observed to be at the maximum with 10 wt% FGO. The use of FGO as filler in the PVA/PDDA membrane also improved the thermo-mechanical stability of the nanocomposite membrane at 100% RH. The FGO based SIPNs synthesized in an environmentally friendly approach utilizing deionized water as solvent are promising membranes for AEM fuel cells.

Zhao *et al.* [91] reported the use of sulfonated polymer brush functionalized GO (SP-GO) as filler in SPEEK polymer matrix. The effect of the SP-GO integrated into SPEEK membrane on membrane properties and single PEMFC performance was studied. SP-GO fillers were observed to be more uniformly distributed within SPEEK than GO due to the hydrogen bonding between the sulfonic acid groups of SP-GO and polar groups of SPEEK, which resulted in improved interfacial compatibility between GO and SPEEK. The uniformly dispersed SP-GO connected the ionic clusters in SPEEK and formed wide and well-connected proton-transfer pathways along the SPEEK/SP-GO interface through interfacial interactions. Hence, SP-GO-filled composite membranes achieved enhanced proton conductivities and reduced activation energy for proton conduction as compared to the SPEEK control membrane and the GO-filled membrane. The proton conductivity of the composite membrane exhibited further increase with increase in the polymer brush length and the content of SP-GO filler. The composite membrane with 10 wt% SSGO had a 95.5% increase in proton conductivity at 65 °C under hydrated condition and a 178% increase in conductivity at 150 °C under anhydrous conditions, compared to SPEEK control membrane. The addition of SP-GO fillers in SPEEK polymer also lowered the H$_2$ permeability of the composite membranes. The improved proton conductivity and the low H$_2$ permeability of the composite membranes led to higher PEMFC performance than Nafion under anhydrous conditions and high temperature [91].

In another study, Chu *et al.* [92] explored the potential of PBI/zwitterion-coated GO (ZC-GO) membranes for DMFC applications, with ZC-GO containing ammonium and sulfonic acid groups as filler in PBI. The authors investigated the influence of ZC-GO content on the properties of composite membranes for DMFCs by varying the amount of ZC-GO incorporated into PBI. The membrane properties such as water uptake, proton conductivity, methanol permeability, and swelling ratio of the PBI/ZC-GO increased with increase in the amount of ZC-GO. PBI/ZC-GO composite membrane with 25 wt% ZC-GO exhibited a higher water uptake and a lower swelling ratio compared with Nafion 117 membrane, due to the restricted hydrophilic domains owing to the hydrogen bonding between ZC-GO and PBI. The methanol permeability of PBI/ZC-GO with 25 wt% ZC-GO was observed to be higher than that of pure PBI membrane due to the formation of proton transport channels through the membrane by the adsorbed water, but was much lower than that of Nafion 117 membrane or other hybrid membranes under similar conditions. The increase in ZC-GO content facilitated transfer of more protons due to the formation of hydrogen bonds between PBI and ZC-GO. The increase in proton mobility in the composite membrane with increasing ZC-GO content resulted from the extension of the number of available proton exchange sites per cluster. The conductivities of PBI/ZC-GO membranes with different ZC-GO percentage exhibited increase in conductivities with increase in temperature. The reason behind this phenomenon was the faster migration of ions and increased diffusivity at higher temperatures. The PBI/ZC-GO composite membrane with 25 wt% ZC-GO exhibited the highest proton conductivity of 4.12×10^{-2} S cm^{-1} at 90 °C at 100% RH. Recently, Ye *et al.* [93] used simple blending to incorporate graphene nanosheets into poly(vinyl alcohol) (PVA) and investigated the transport properties of the prepared graphene/PVA nanocomposites. The continuous and well-connected ionic channels introduced in the graphene/PVA composite membrane by the uniformly dispersed graphene nanosheets in the PVA matrix improved the transport properties of the composite membrane. As a result, the graphene/PVA composite membranes exhibited an increase in ionic conductivity of ~126% and a decrease in methanol permeability of ~55% with a 0.7 wt% graphene loading. The significant improvement of the transport properties observed for the graphene/PVA composite membranes resulted in better cell performance than the PVA membrane, with ~148% improved maximum power density at 60 °C. The addition of graphene into PVA also resulted in enhanced mechanical properties (~73% improvement in tensile strength) at graphene loading of 1.4 wt%, owing to interfacial interactions between graphene and PVA matrix through H-bonding.

Kim *et al.* [94] prepared phosphotungstic acid (PW) coupled GO-Nafion (Nafion/PW-mGO) membrane and investigated PEFCs utilizing (Nafion/PW-mGO) composite membrane at elevated temperature and low RH operating conditions. The composite membrane exhibited improved proton conductivity compared to pristine and recast Nafion membranes. Nafion/PW-mGO composite membrane provided maximum power density, greater than those of the Nafion-212 and recast Nafion membranes at 80 °C and under low RH. Maximum power density delivered by a Nafion/PW-mGO membrane operated at 80 °C and under low RH was found to be about 4-fold higher than the Nafion-212 membrane. The fuel cell performance of the Nafion/PW-mGO composite membrane was remarkably better than those of the Nafion-212 and recast Nafion membranes under low RH. The higher PEFC performance at elevated temperature was ascribed to the hygroscopic nature and strong acid strength of PW which allowed water retention in the composite membrane through hydrogen bonding with protons and facile hopping of protons, thus, resulting in increased proton transfer. Yang *et al.* [95] also prepared novel composite membranes of PBI with triazole modified GO (MGO) for high temperature PEMFCs. The composite membrane exhibited simultaneous improvement in both tensile strength and proton conductivity with the incorporation of well-dispersed triazole functionalized GO. PA-doped PBI/MGO membrane had high proton conductivity of 0.135 S cm^{-1} at 180 °C. The presence of triazole groups in MGO provided facile proton hopping pathways through hydrogen bonds with PA, which led to superior proton conduction through the hybrid membrane at high temperature. The fuel cell performance of PBI/MGO membrane was remarkably better than that of the PBI membrane with the same acid doping level. For example, at 180 °C, the maximum power densities of 537 mW cm^{-2} and 506 mW cm^{-2} were obtained with fuel cells based on composite and pristine membranes, respectively [95].

Bai *et al.* [96] described highly conductive PEMs based on chitosan and phosphorylated graphene oxide (PGO). The authors synthesized GO with phosphorylated polymeric layer by distillation-precipitation polymerization and subsequently PGO nanosheets were embedded into the CS matrix. The thermal and mechanical stabilities of PGO-filled membranes were observed to be higher than that of CS and GO-filled membranes due to the strong electrostatic interactions between the PA groups of PGO and the amine groups of CS in PGO-filled membranes. The attractive interactions between the PA groups on PGO surface and the CS chains resulted in better dispersion of PGO nanosheets in the CS membrane and, thus, facilitated efficient proton migration pathways along PGO surface. The nanohybrid membrane exhibited improved proton conductivities under both hydrated and anhydrous conditions. At 160 °C and under anhydrous conditions, the conductivity achieved for the nanohybrid membrane with 2.5% PGO was 22.2 and 95.5 times higher than those of CS control and Nafion membranes respectively. This superior proton conductivity of the nanohybrid membrane led to remarkably higher H_2/O_2 cell performance compared to CS control and GO-filled membranes at 0% RH.

Pandey and Shahi [97] prepared SPI/sulphonated imidized GO (SIGO) composite membranes for PEFCs. SPI/SIGO composite membranes exhibited enhanced mechanical, thermal and oxidative stabilities. The bound water content in PEM matrix increased with SIGO content and, thus, composite membranes had improved water-retention properties (with bound water 5.12% higher than that of Nafion 117 membrane) and high proton conduction ability. The proton conductivity of pristine PBI membrane increased after the incorporation of SIGO in SPI, due to the inter-connected proton conducting channels formed by the acidic groups and was observed to be comparatively higher than Nafion 117 membrane at 30 °C. The proton-transfer activation energy was improved by the presence of SIGO. The integration of covalently bonded SIGO in polymer matrix resulted in narrow hydrophilic-hydrophobic phase separation in SPI/SIGO composite and SPI/SIGO composite membrane exhibited lower methanol permeability than both pristine SPI and Nafion 117 membrane at 30 °C. The composite membrane exhibited high maximum power density in comparison with pristine PBI and Nafion 117 membrane in single-cell DMFC performance at 70 °C. These results demonstrated that SPI/SIGO composite membrane could be utilized as suitable materials for high performance DMFC application.

Dai *et al.* [98] introduced octadecylamine-functionalized GO (GOA) and partially quaternized GOA (GOAN) into cross-linked quaternized poly(styrene-*b*-isobutylene-*b*-styrene) (QSIBS) to prepare cross-linked QSIBS/GOA and QSIBS/GOAN composite membranes. The authors evaluated these hybrid composite membranes as anion exchange polymer electrolytes in DMFCs. The introduction of GOA into QSIBS improved the ionic conductivity of QSIBS, the ionic conductivity of which was much lower than that of Nafion 115. The QSIBS/GOA composite membranes exhibited an increasing ionic conductivity with temperature due to the increase in the mobility of both ionic and polymeric chains. The QSIBS/GOA membrane exhibited 138% increase in ionic conductivity compared to that of the pure QSIBS with the increase in GOA content up to 0.50 wt%. However, further increase in GOA amount reduced the ionic conductivity to 37% with 1 wt% GOA. The decreased conductivity beyond 0.5 wt% GOA loading could be ascribed to the "barrier effect" arising from the aggregation of the GOA in the hybrid membrane. The GOA integrated QSIBS membrane exhibited a 32.1% decrease in methanol crossover compared to that for a pure QSIBS at 0.5 wt% GOA loading. The QSIBS/GOAN composite membrane containing 0.5 wt% GOAN loading exhibited higher ionic conductivity than that of QSIBS/GOA, but was comparable to that of Nafion 115. Methanol permeabilities of hybrid composite membrane were much lower compared to Nafion 115. QSIBS/GOAN with 0.5 wt% filler loading possessed the highest selectivity (ratio of ionic conductivity to methanol permeability), which was about 12 times greater than that of the Nafion 115 membrane. The excellent performance was due to the barrier effect of GOAN, which favored the methanol crossover, and also due to the interconnected ion transport channels between GOAN and QSIBS by the hydrophilic groups in the GOAN, which favored the ionic transport. These results demonstrated that QSIBS/GOAN could be utilized as promising AEMs for DMFCs [98].

5.3 Conclusion and Future Prospects

Fuel cells are preferred over other energy producing devices for stationary, transportation and marine applications due to their advantages such as high energy density, low polluting emissions, silent operation, simple maintenance and longer operating durations than batteries. Perfluorosulfonic acid based polymers are the most promising candidates as proton exchange membrane electrolytes because of their excellent proton conducting property, good thermal, mechanical, physical and chemical stabilities at moderate temperatures and in a wide range of relative humidities. The high cost, high methanol permeability, low proton conductivity and loss of mechanical stability at high temperature and under low humidity practically limited the applications of PFSA polymers as proton conduction membranes for fuel cells. The modification of Nafion membrane with various materials has proved to be a promising method in attaining desired properties such as high proton conductivities, low methanol permeability and good mechanical properties. Specifically, the incorporation of nanocarbon materials into polymer matrices can remarkably improve their applicability as PEMs.

CNTs and GO are advanced nanomaterials that could be used as additives in the fabrication of electrolyte membranes for potential fuel cell applications. CNT/polymer composites were employed to improve the mechanical properties and reduce the methanol crossover with negligible effect on the proton transport behavior. The integration of GO into polymer matrices has shown to enhance the mechanical properties, proton transport, methanol blocking properties of PEMs and hence can be used as remarkable membranes for fuel cells. The homogenous dispersion of the filler in host polymeric matrices and strong interfacial interactions of the filler with the polymer matrix are important factors determining the performance of polymer/nanocarbon composite membranes. The modification of nanocarbon materials with different functional groups leads to a better compatibility between the nanocarbon materials and polymer matrix. It was revealed that the use of functionalized nanocarbon materials as filler in polymeric membranes could greatly improve the properties of state-of-the-art polymer membranes. Various kinds of functionalized CNTs and GO were explored and employed in the fabrication of hybrid membranes. The addition of sulfonic or phosphonic acid groups to CNTs and GO resulted in an increase in the proton conductivity of membrane electrolytes. CNTs and GO can be more effectively solubilized through functionalization with matrix polymer and polymer/nanocarbon composite membranes based on matrix polymer-functionalized nanocarbon materials have shown supreme properties such as improved proton conductivity, alleviated methanol crossover effect and good mechanical strength even at high temperature and low humidity conditions, which could ultimately lead to a better fuel cell performance compared to the currently used PEM electrolytes. Still further research is needed to improve the membrane properties so as to obtain desirable electrolyte membranes.

Future research efforts in this area will focus on the preparation of high performance hybrid membranes by the incorporation of modified nanocarbon materials into polymer electrolytes and the performance evaluation of the fuel cells based on the newly developed PEMs. The mass production of fuel cells and potential utilization of newly developed hybrid composite membranes based on modified nanocarbon materials can revolutionize the field of electricity in the near future.

References

1. Li, X., Fields, L., and Way, G. (2006) Principles of fuel cells. *Platinum Metals Review,* **50**, 200-201.
2. Peighambardoust, S. J., Rowshanzamir, S., and Amjadi, M. (2010) Review of the proton exchange membranes for fuel cell applications. *International Journal of Hydrogen Energy,* **35**, 9349-9384.
3. Barbir, F., and Gomez, T. (1997) Efficiency and economics of proton exchange membrane (PEM) fuel cells. *International Journal of Hydrogen Energy,* **22**, 1027-1037.
4. Kamarudin, S. K., Daud, W. R. W., Ho, S. L., and Hasran, U. A. (2007) Overview on the challenges and developments of micro-direct methanol fuel cells (DMFC). *Journal of Power Sources,* **163**, 743-754.

5. Kamarudin, S. K., Achmad, F., and Daud, W. R. W. (2009) Overview on the application of direct methanol fuel cell (DMFC) for portable electronic devices. *International Journal of Hydrogen Energy*, **34**, 6902-6916.

6. Ismail, A., Kamarudin, S. K., Daud, W. R. W., Masdar, S., and Yosfiah, M. R. (2011) Mass and heat transport in direct methanol fuel cells. *Journal of Power Sources*, **196**, 9847-9855.

7. Rikukawa, M., and Sanui, K. (2000) Proton-conducting polymer electrolyte membranes based on hydrocarbon polymers. *Progress in Polymer Science*, **25**, 1463-1502.

8. Ye, Y.-S., Rick, J., and Hwang, B.-J. (2012) Water soluble polymers as proton exchange membranes for fuel cells. *Polymers*, **4**(2), 913-963.

9. Kerres, J. A. (2001) Development of ionomer membranes for fuel cells. *Journal of Membrane Science*, **185**, 3-27.

10. Mauritz, K. A., and Moore, R. B. (2004) State of understanding of Nafion. *Chemical Reviews*, **104**, 4535-4586.

11. Appleby, A. J., and Foulkes, F. R. (1989) *Fuel Cell Handbook*, Van Nostrand Reinhold, USA.

12. Surampudi, S., Narayanan, S. R., Vamos, E., Frank, H., Halpert, G., LaConti, A., Kosek, J., Prakash, G. K. S., and Olah, G. A. (1994) Advances in direct oxidation methanol fuel cells. *Journal of Power Sources*, **47**, 377-385.

13. Ren, X., Wilson, M. S., and Gottesfeld, S. (1996) High performance direct methanol polymer electrolyte fuel cells. *Journal of The Electrochemical Society*, **143**, L12-L15.

14. Antonucci, P. L., Arico, A. S., Creti, P., Ramunni, E., and Antonucci, V. (1999) Investigation of a direct methanol fuel cell based on a composite Nafion®-silica electrolyte for high temperature operation. *Solid State Ionics*, **125**, 431-437.

15. Arico, A. S., Srinivasan, S., and Antonucci, V. (2001) DMFCs: From fundamental aspects to technology development. *Fuel Cells*, **1**, 133-161.

16. Song, S., Zhou, W., Liang, Z., Cai, R., Sun, G., Xin, Q., Stergiopoulos, V., and Tsiakaras, P. (2005) The effect of methanol and ethanol cross-over on the performance of PtRu/C-based anode DAFCs. *Applied Catalysis B: Environmental*, **55**, 65-72.

17. Kang, K., Park, S., Gwak, G., Jo, A., Kim, M., Lim, Y.-D., Kim, W.-G., Hong, T., Kim, D., and Ju, H. (2014) Effect of variation of hydrophobicity of anode diffusion media along the through-plane direction in direct methanol fuel cells. *International Journal of Hydrogen Energy*, **39**, 1564-1570.

18. Li, Q., He, R., Jensen, J. O., and Bjerrum, N. J. (2004) PBI-based polymer membranes for high temperature fuel cells - Preparation, characterization and fuel cell demonstration. *Fuel Cells*, **4**, 147-159.

19. Qingfeng, L., Hjuler, H. A., and Bjerrum, N. J. (2001) Phosphoric acid doped polybenzimidazole membranes: Physiochemical characterization and fuel cell applications. *Journal of Applied Electrochemistry*, **31**, 773-779.

20. Iijima, S. (1991) Helical microtubules of graphitic carbon. *Nature*, **354**, 56-58.

21. Baughman, R. H., Zakhidov, A. A., and de Heer, W. A. (2002) Carbon nanotubes - the route toward applications. *Science*, **297**, 787-792.

22. Tsukagoshi, K., Yoneya, N., Uryu, S., Aoyagi, Y., Kanda, A., Ootuka, Y., and Alphenaar, B. W. (2002) Carbon nanotube devices for nanoelectronics. *Physica B: Condensed Matter*, **323**, 107-114.

23. Saino Hanna, V., Remya, N., Baiju, G. N., Hanajiri, T., Maekawa, T., Yoshida, Y., and Kumar, D. S. (2010) Sensors based on carbon nanotubes and their applications: A review. *Current Nanoscience*, **6**, 331-346.

24. De Volder, M. F. L., Tawfick, S. H., Baughman, R. H., and Hart, A. J. (2013) Carbon nanotubes: Present and future commercial applications. *Science*, **339**, 535-539.

25. Lee, Y. H., An, K. H., Lim, S. C., Kim, W. S., Jeong, H. J., Doh, C.-H., and Moon, S.-I. (2002) Applications of carbon nanotubes to energy storage devices. *New Diamond & Frontier Carbon Technology*, **12**, 209-228.

26. Ajayan, P. M., and Zhou, O. Z. (2001) Applications of carbon nanotubes. In: *Carbon Nanotubes: Synthesis, Structure, Properties, and Applications*, Dresselhaus, M. S., Dresselhaus, G., and Avouris, P. (eds.), Springer, Germany, pp. 391-425.

27. Wang, C., Waje, M., Wang, X., Tang, J. M., Haddon, R. C., and Yushan (2004) Proton exchange membrane fuel cells with carbon nanotube based electrodes. *Nano Letters*, **4**, 345-348.

28. Danilov, M. O., Kolbasov, G. Y., and Melezhyk, A. V. (2008) Electrodes for fuel cells based on carbon nanotubes and catalysts. In: *Carbon Nanomaterials in Clean Energy Hydrogen Systems*, Baranowski, B., Zaginaichenko, S. Y., Schur, D. V., Skorokhod, V. V., and Veziroglu, A. (eds.), Springer, Netherlands, pp. 279-284.

29. Khantimerov, S. M., Kukovitsky, E. F., Sainov, N. A., and Suleimanov, N. M. (2013) Fuel cell electrodes based on carbon nanotube/metallic nanoparticles hybrids formed on porous stainless steel pellets. *International Journal of Chemical Engineering*, 2013, ID 157098.

30. Luo, C., Xie, H., Wang, Q., Luo, G., and Liu, C. (2015) A review of the application and performance of carbon nanotubes in fuel cells. *Journal of Nanomaterials*, 2015, ID 560392.

31. Wang, W., Ciselli, P., Kuznetsov, E., Peijs, T., and Barber, A. H. (2008) Effective reinforcement in carbon nanotube–polymer composites. *Philosophical Transactions of the Royal Society A: Mathematical, Physical and Engineering Sciences*, **366**, 1613-1626.

32. Coleman, J. N., Khan, U., Blau, W. J., and Gun'ko, Y. K. (2006) Small but strong: A review of the mechanical properties of carbon nanotube-polymer composites. *Carbon*, **44**, 1624-1652.

33. Coleman, J. N., Khan, U., and Gun'ko, Y. K. (2006) Mechanical reinforcement of polymers using carbon nanotubes. *Advanced Materials*, **18**, 689-706.

34. Arash, B., Wang, Q., and Varadan, V. K. (2014) Mechanical properties of carbon nanotube/polymer composites. *Scientific Reports*, 4, ID 6479.

35. Liu, Y.-H., Yi, B., Shao, Z.-G., Xing, D., and Zhang, H. (2006) Carbon nanotubes reinforced Nafion composite membrane for fuel cell applications. *Electrochemical and Solid-State Letters*, **9**, A356-A359.

36. Thomassin, J.-M., Kollar, J., Caldarella, G., Germain, A., Jerome, R., and Detrembleur, C. (2007) Beneficial effect of carbon nanotubes on the performances of Nafion membranes in fuel cell applications. *Journal of Membrane Science*, **303**, 252-257.

37. Jana, R. N., and Bhunia, H. (2008) Thermal stability and proton conductivity of silane based nanostructured composite membranes. *Solid State Ionics*, **178**, 1872-1878.

38. Wang, L., Xing, D. M., Zhang, H. M., Yu, H. M., Liu, Y. H., and Yi, B. L. (2008) MWCNTs reinforced Nafion® membrane prepared by a novel solution-cast method for PEMFC. *Journal of Power Sources,* **176,** 270-275.

39. Kang, J.-Y., Eo, S.-M., Jeon, I.-Y., Choi, Y. S., Tan, L.-S., and Baek, J.-B. (2010) Multifunctional poly(2,5-benzimidazole)/carbon nanotube composite films. *Journal of Polymer Science Part A: Polymer Chemistry,* **48,** 1067-1078.

40. Guerrero Moreno, N., Gervasio, D., Godinez Garcia, A., and Perez Robles, J. F. (2015) Polybenzimidazole-multiwall carbon nanotubes composite membranes for polymer electrolyte membrane fuel cells. *Journal of Power Sources,* **300,** 229-237.

41. Suryani, Chang, C.-M., Liu, Y.-L., and Lee, Y. M. (2011) Polybenzimidazole membranes modified with polyelectrolyte-functionalized multiwalled carbon nanotubes for proton exchange membrane fuel cells. *Journal of Materials Chemistry,* **21,** 7480-7486.

42. Kannan, R., Kagalwala, H. N., Chaudhari, H. D., Kharul, U. K., Kurungot, S., and Pillai, V. K. (2011) Improved performance of phosphonated carbon nanotube-polybenzimidazole composite membranes in proton exchange membrane fuel cells. *Journal of Materials Chemistry,* **21,** 7223-7231.

43. Jheng, L.-c., Huang, C.-y., and Hsu, S. L.-c. (2013) Sulfonated MWNT and imidazole functionalized MWNT/polybenzimidazole composite membranes for high-temperature proton exchange membrane fuel cells. *International Journal of Hydrogen Energy,* **38,** 1524-1534.

44. Zhang, L., Ni, Q.-Q., Shiga, A., Natsuki, T., and Fu, Y. (2011) Preparation of polybenzimidazole/functionalized carbon nanotube nanocomposite films for use as protective coatings. *Polymer Engineering & Science,* **51,** 1525-1532.

45. Shao, H., Shi, Z., Fang, J., and Yin, J. (2009) One pot synthesis of multiwalled carbon nanotubes reinforced polybenzimidazole hybrids: Preparation, characterization and properties. *Polymer,* **50,** 5987-5995.

46. Wu, J.-F., Lo, C.-F., Li, L.-Y., Li, H.-Y., Chang, C.-M., Liao, K.-S., Hu, C.-C., Liu, Y.-L., and Lue, S. J. (2014) Thermally stable polybenzimidazole/carbon nano-tube composites for alkaline direct methanol fuel cell applications. *Journal of Power Sources,* **246,** 39-48.

47. Kannan, R., Kakade, B. A., and Pillai, V. K. (2008) Polymer electrolyte fuel cells using Nafion-based composite membranes with functionalized carbon nanotubes. *Angewandte Chemie International Edition,* **47,** 2653-2656.

48. Kannan, R., Parthasarathy, M., Maraveedu, S. U., Kurungot, S., and Pillai, V. K. (2009) Domain size manipulation of perflouorinated polymer electrolytes by sulfonic acid-functionalized MWCNTs to enhance fuel cell performance. *Langmuir,* **25,** 8299-8305.

49. Cele, N. P., Sinha Ray, S., Pillai, S. K., Ndwandwe, M., Nonjola, S., Sikhwivhilu, L., and Mathe, " M. K. (2010) Carbon nanotubes based Nafion composite membranes for fuel cell applications. *Fuel Cells,* **10,** 64-71.

50. Kim, Y. H., Sayeed, M. A., Lee, H. K., Park, Y., Gopalan, A. I., Lee, K.-P., and Choi, S.-J. (2012) Functionalized Carbon Nanotubes Reinforced Polymer Electrolyte Membranes Prepared by a Surfactant-assisted Method for Fuel Cell Applications. *2012 12th IEEE International Conference on Nanotechnology (IEEE-NANO),* UK, pp. 1-2.

51. Liu, Y.-H., Yi, B., Shao, Z.-G., Wang, L., Xing, D., and Zhang, H. (2007) Pt/CNTs-Nafion reinforced and self-humidifying composite membrane for PEMFC applications. *Journal of Power Sources,* **163,** 807-813.

52. Lin, Y., Zhou, B., Shiral Fernando, K. A., Liu, P., Allard, L. F., and Sun, Y.-P. (2003) Polymeric carbon nanocomposites from carbon nanotubes functionalized with matrix polymer. *Macromolecules,* **36,** 7199-7204.

53. Lin, Y., Meziani, M. J., and Sun, Y.-P. (2007) Functionalized carbon nanotubes for polymeric nanocomposites. *Journal of Materials Chemistry,* **17,** 1143-1148.

54. Chang, C.-M., and Liu, Y.-L. (2010) Functionalization of multi-walled carbon nanotubes with non-reactive polymers through an ozone-mediated process for the preparation of a wide range of high performance polymer/carbon nanotube composites. *Carbon,* **48,** 1289-1297.

55. Chen, W.-F., Wu, J.-S., and Kuo, P.-L. (2008) Poly(oxyalkylene)diamine-functionalized carbon nanotube/perfluorosulfonated polymer composites: Synthesis, water state, and conductivity. *Chemistry of Materials,* **20,** 5756-5767.

56. Liu, Y.-L., Su, Y.-H., Chang, C.-M., Suryani, Wang, D.-M., and Lai, J.-Y. (2010) Preparation and applications of Nafion-functionalized multiwalled carbon nanotubes for proton exchange membrane fuel cells. *Journal of Materials Chemistry,* **20,** 4409-4416.

57. Hasani-Sadrabadi, M. M., Dashtimoghadam, E., Majedi, F. S., Moaddel, H., Bertsch, A., and Renaud, P. (2013) Superacid-doped polybenzimidazole-decorated carbon nanotubes: a novel high-performance proton exchange nanocomposite membrane. *Nanoscale,* **5,** 11710-11717.

58. Kannan, R., Aher, P. P., Palaniselvam, T., Kurungot, S., Kharul, U. K., and Pillai, V. K. (2010) Artificially designed membranes using phosphonated multiwall carbon nanotube-polybenzimidazole composites for polymer electrolyte fuel cells. *The Journal of Physical Chemistry Letters,* **1,** 2109-2113.

59. Joo, S. H., Pak, C., Kim, E. A., Lee, Y. H., Chang, H., Seung, D., Choi, Y. S., Park, J.-B., and Kim, T. K. (2008) Functionalized carbon nanotube-poly(arylene sulfone) composite membranes for direct methanol fuel cells with enhanced performance. *Journal of Power Sources,* **180,** 63-70.

60. Tripathi, B. P., Schieda, M., Shahi, V. K., and Nunes, S. P. (2011) Nanostructured membranes and electrodes with sulfonic acid functionalized carbon nanotubes. *Journal of Power Sources,* **196,** 911-919.

61. Gong, C., Zheng, X., Liu, H., Wang, G., Cheng, F., Zheng, G., Wen, S., Law, W.-C., Tsui, C.-P., and Tang, C.-Y. (2016) A new strategy for designing high-performance sulfonated poly(ether ether ketone) polymer electrolyte membranes using inorganic proton conductor-functionalized carbon nanotubes. *Journal of Power Sources,* **325,** 453-464.

62. Jana, R. N., Maity, B., Mallick, S., Majumdar, A., and Singh, P. (2015) Nanostructured ionomeric membranes for direct methanol fuel cell. *Indian Chemical Engineer,* **57,** 103-114.

63. Dreyer, D. R., Park, S., Bielawski, C. W., and Ruoff, R. S. (2010) The chemistry of graphene oxide. *Chemical Society Reviews,* **39,** 228-240.

64. Zhu, J., Liu, F., Mahmood, N., and Hou, Y. (2015) Graphene polymer nanocomposites for fuel cells. In: *Graphene-Based Polymer Nanocomposites in Electronics,* Sadasivuni, K. K., Ponnamma, D., Kim, J., and Thomas, S. (eds.), Springer International Publishing, Switzerland, pp. 91-130.

65. Kumar, R., Mamlouk, M., and Scott, K. (2011) A graphite oxide paper polymer electrolyte for direct methanol fuel cells. *International Journal of Electrochemistry,* **2011,** ID 434186.

66. Cao, Y.-C., Xu, C., Wu, X., Wang, X., Xing, L., and Scott, K. (2011) A poly (ethylene oxide)/graphene oxide electrolyte membrane for low temperature polymer fuel cells. *Journal of Power Sources,* **196,** 8377-8382.

67. Choi, B. G., Huh, Y. S., Park, Y. C., Jung, D. H., Hong, W. H., and Park, H. (2012) Enhanced transport properties in polymer electrolyte composite membranes with graphene oxide sheets. *Carbon,* **50,** 5395-5402.

68. Wang, L., Kang, J., Nam, J.-D., Suhr, J., Prasad, A. K., and Advani, S. G. (2015) Composite membrane based on graphene oxide sheets and Nafion for polymer electrolyte membrane fuel cells. *ECS Electrochemistry Letters,* **4**(1), F1-F4.

69. Sha Wang, L., Nan Lai, A., Xiao Lin, C., Gen Zhang, Q., Mei Zhu, A., and Lin Liu, Q. (2015) Orderly sandwich-shaped graphene oxide/Nafion composite membranes for direct methanol fuel cells. *Journal of Membrane Science,* **492,** 58-66.

70. Lin, C. W., and Lu, Y. S. (2013) Highly ordered graphene oxide paper laminated with a Nafion membrane for direct methanol fuel cells. *Journal of Power Sources,* **237,** 187-194.

71. He, Y., Tong, C., Geng, L., Liu, L., and Lu, C. (2014) Enhanced performance of the sulfonated polyimide proton exchange membranes by graphene oxide: Size effect of graphene oxide. *Journal of Membrane Science,* **458,** 36-46.

72. Uregen, N., Pehlivanoglu, K., Ozdemir, Y., and Devrim, Y. (2017) Development of polybenzimidazole/graphene oxide composite membranes for high temperature PEM fuel cells. *International Journal of Hydrogen Energy,* **42**(4), 2636-2647.

73. Choi, B. G., Hong, J., Park, Y. C., Jung, D. H., Hong, W. H., Hammond, P. T., and Park, H. (2011) Innovative polymer nanocomposite electrolytes: Nanoscale manipulation of ion channels by functionalized graphenes. *ACS Nano,* **5,** 5167-5174.

74. Zarrin, H., Higgins, D., Jun, Y., Chen, Z., and Fowler, M. (2011) Functionalized graphene oxide nanocomposite membrane for low humidity and high temperature proton exchange membrane fuel cells. *The Journal of Physical Chemistry C,* **115,** 20774-20781.

75. Sahu, A. K., Ketpang, K., Shanmugam, S., Kwon, O., Lee, S., and Kim, H. (2016) Sulfonated graphene–Nafion composite membranes for polymer electrolyte fuel cells operating under reduced relative humidity. *The Journal of Physical Chemistry C,* **120,** 15855-15866.

76. Enotiadis, A., Angjeli, K., Baldino, N., Nicotera, I., and Gournis, D. (2012) Graphene-based Nafion nanocomposite membranes: Enhanced proton transport and water retention by novel organo-functionalized graphene oxide nanosheets. *Small,* **8,** 3338-3349.

77. Lee, D. C., Yang, H. N., Park, S. H., and Kim, W. J. (2014) Nafion/graphene oxide composite membranes for low humidifying polymer electrolyte membrane fuel cell. *Journal of Membrane Science,* **452,** 20-28.

78. Xu, C., Cao, Y., Kumar, R., Wu, X., Wang, X., and Scott, K. (2011) A polybenzimidazole/sulfonated graphite oxide composite membrane for high temperature polymer electrolyte membrane fuel cells. *Journal of Materials Chemistry,* **21,** 11359-11364.

79. Liu, Y., Wang, J., Zhang, H., Ma, C., Liu, J., Cao, S., and Zhang, X. (2014) Enhancement of proton conductivity of chitosan membrane enabled by sulfonated graphene oxide under both hydrated and anhydrous conditions. *Journal of Power Sources,* **269**, 898-911.

80. Gahlot, S., Sharma, P. P., Kulshrestha, V., and Jha, P. K. (2014) SGO/SPES-based highly conducting polymer electrolyte membranes for fuel cell application. *ACS Applied Materials & Interfaces,* **6**, 5595-5601.

81. Kumar, R., Mamlouk, M., and Scott, K. (2014) Sulfonated polyether ether ketone - sulfonated graphene oxide composite membranes for polymer electrolyte fuel cells. *RSC Advances,* **4**, 617-623.

82. Pandey, R. P., Thakur, A. K., and Shahi, V. K. (2014) Sulfonated polyimide/acid-functionalized graphene oxide composite polymer electrolyte membranes with improved proton conductivity and water-retention properties. *ACS Applied Materials & Interfaces,* **6**, 16993-17002.

83. Xue, C., Zou, J., Sun, Z., Wang, F., Han, K., and Zhu, H. (2014) Graphite oxide/functionalized graphene oxide and polybenzimidazole composite membranes for high temperature proton exchange membrane fuel cells. *International Journal of Hydrogen Energy,* **39**, 7931-7939.

84. Shukla, A., Bhat, S. D., and Pillai, V. K. (2016) Simultaneous unzipping and sulfonation of multi-walled carbon nanotubes to sulfonated graphene nanoribbons for nanocomposite membranes in polymer electrolyte fuel cells. *Journal of Membrane Science,* **520**, 657-670.

85. Tseng, C.-Y., Ye, Y.-S., Cheng, M.-Y., Kao, K.-Y., Shen, W.-C., Rick, J., Chen, J.-C., and Hwang, B.-J. (2011) Sulfonated polyimide proton exchange membranes with graphene oxide show improved proton conductivity, methanol crossover impedance, and mechanical properties. *Advanced Energy Materials,* **1**, 1220-1224.

86. Ye, Y.-S., Tseng, C.-Y., Shen, W.-C., Wang, J.-S., Chen, K.-J., Cheng, M.-Y., Rick, J., Huang, Y.-J., Chang, F.-C., and Hwang, B.-J. (2011) A new graphene-modified protic ionic liquid-based composite membrane for solid polymer electrolytes. *Journal of Materials Chemistry,* **21**, 10448-10453.

87. He, Y., Wang, J., Zhang, H., Zhang, T., Zhang, B., Cao, S., and Liu, J. (2014) Polydopamine-modified graphene oxide nanocomposite membrane for proton exchange membrane fuel cell under anhydrous conditions. *Journal of Materials Chemistry A,* **2**, 9548-9558.

88. Mishra, A. K., Kim, N. H., Jung, D., and Lee, J. H. (2014) Enhanced mechanical properties and proton conductivity of Nafion–SPEEK–GO composite membranes for fuel cell applications. *Journal of Membrane Science,* **458**, 128-135.

89. Yuan, T., Pu, L., Huang, Q., Zhang, H., Li, X., and Yang, H. (2014) An effective methanol-blocking membrane modified with graphene oxide nanosheets for passive direct methanol fuel cells. *Electrochimica Acta,* **117**, 393-397.

90. Movil, O., Frank, L., and Staser, J. A. (2015) Graphene oxide-polymer nanocomposite anion-exchange membranes. *Journal of The Electrochemical Society,* **162**, F419-F426.

91. Zhao, L., Li, Y., Zhang, H., Wu, W., Liu, J., and Wang, J. (2015) Constructing proton-conductive highways within an ionomer membrane by embedding sulfonated polymer brush modified graphene oxide. *Journal of Power Sources,* **286**, 445-457.

92. Chu, F., Lin, B., Feng, T., Wang, C., Zhang, S., Yuan, N., Liu, Z., and Ding, J. (2015) Zwitterion-coated graphene-oxide-doped composite membranes for proton exchange membrane applications. *Journal of Membrane Science,* **496**, 31-38.

93. Ye, Y.-S., Cheng, M.-Y., Xie, X.-L., Rick, J., Huang, Y.-J., Chang, F.-C., and Hwang, B.-J. (2013) Alkali doped polyvinyl alcohol/graphene electrolyte for direct methanol alkaline fuel cells. *Journal of Power Sources,* **239**, 424-432.

94. Kim, Y., Ketpang, K., Jaritphun, S., Park, J. S., and Shanmugam, S. (2015) A polyoxometalate coupled graphene oxide-Nafion composite membrane for fuel cells operating at low relative humidity. *Journal of Materials Chemistry A,* **3**, 8148-8155.

95. Yang, J., Liu, C., Gao, L., Wang, J., Xu, Y., and He, R. (2015) Novel composite membranes of triazole modified graphene oxide and polybenzimidazole for high temperature polymer electrolyte membrane fuel cell applications. *RSC Advances,* **5**, 101049-101054.

96. Bai, H., Li, Y., Zhang, H., Chen, H., Wu, W., Wang, J., and Liu, J. (2015) Anhydrous proton exchange membranes comprising of chitosan and phosphorylated graphene oxide for elevated temperature fuel cells. *Journal of Membrane Science,* **495**, 48-60.

97. Pandey, R. P., and Shahi, V. K. (2015) Sulphonated imidized graphene oxide (SIGO) based polymer electrolyte membrane for improved water retention, stability and proton conductivity. *Journal of Power Sources,* **299**, 104-113.

98. Dai, P., Mo, Z.-H., Xu, R.-W., Zhang, S., Lin, X., Lin, W.-F., and Wu, Y.-X. (2016) Development of a cross-linked quaternized poly(styrene-b-isobutylene-b-styrene)/graphene oxide composite anion exchange membrane for direct alkaline methanol fuel cell application. *RSC Advances,* **6**, 52122-52130.

6

Deep Eutectic Solvents for Gas Separation, EOR and Other Applications

6.1 Introduction

In chemical processes, the selection of appropriate solvent is very important as it may correspond to almost 80% of the entire volume of chemicals employed in the process [1]. The application of solvents presents various health, safety as well as environmental challenges inclusive of process safety hazards, eco-toxicity, human impact and waste management issues [2]. Following the fundamentals of green chemistry, majority of the organic solvents do not achieve the pre-requisite for their application in the green technology because of the fact that these solvents have greater volatility and inherent toxicity [1]. Thus, the usage of solvents is required to be optimized as well as minimized for generating the processes with minimal operational and environmental concerns.

Over the past several decades, many efforts have been made for replacing the organic solvents. The approaches include employing no solvent at all, usage of easily recyclable systems like supercritical carbon dioxide (scCO$_2$) or fluorous solvents, utilization of involatile systems like ionic liquids (ILs), low melting mixtures (LMMs) and deep eutectic solvents (DESs) [3]. The ILs are salts that are mostly liquid at temperature lesser than 100 °C. The first IL, ethyl-ammonium nitrate ([EtN$^+$H$_3$][NO$_3^-$]), was reported in 1914 [4]. Ever since, exponential growth in the studies concerning ILs has been observed [5]. The greatest benefit of using ILs is that these can be tuned by mixing various anions (Cl$^-$, BF$_4^-$, PF$_6^-$, NTf$_2^-$, etc.) and cations (generally imidazolium based cations) [6]. Nevertheless, their disadvantage is the questionable environmental acceptability with regards to their synthesis and application [5,7]. Many research reports have revealed the poor biodegradability and toxicity of most ILs [8]. Further, their preparation demands enormous quantity of solvents and salts for completing the anion exchange. These disadvantages along with their higher price restrict the industrial evolution of ILs. In order to overcome these drawbacks, a new class of IL counterpart, i.e., the DESs were discovered and developed recently.

Haleema Saleem and Vikas Mittal*, The Petroleum Institute (part of Khalifa University of Science and Technology), Abu Dhabi, UAE
*Current address: Bletchington, Wellington County, Australia

DESs contain a combination of two or more components, which have the ability to correlate with each other [9]. Despite the fact that these low transition temperature mixtures (LTTMs) share several properties and characteristics with ILs, they portray a different class of solvents [10]. Contrary to ILs, which consist of one kind of distinct cation and anion, the DESs are generated from a eutectic mixture of Bronsted and Lewis acids as well as bases. DESs consist of different cationic and/or anionic species and possess a great depression in the melting point, when compared to pure materials [10]. The reduction in DESs' freezing point is linked to the co-operation between mixture components, mainly of hydrogen bonding character [11]. Among the various DES components, choline chloride (ChCl) is regarded as the most common component (Figure 6.1). ChCl is a non-toxic and biodegradable salt and is very similar to the B vitamins. The DESs have numerous benefits when compared to the traditional ILs, since they are: (1) less expensive, (2) easy to synthesize

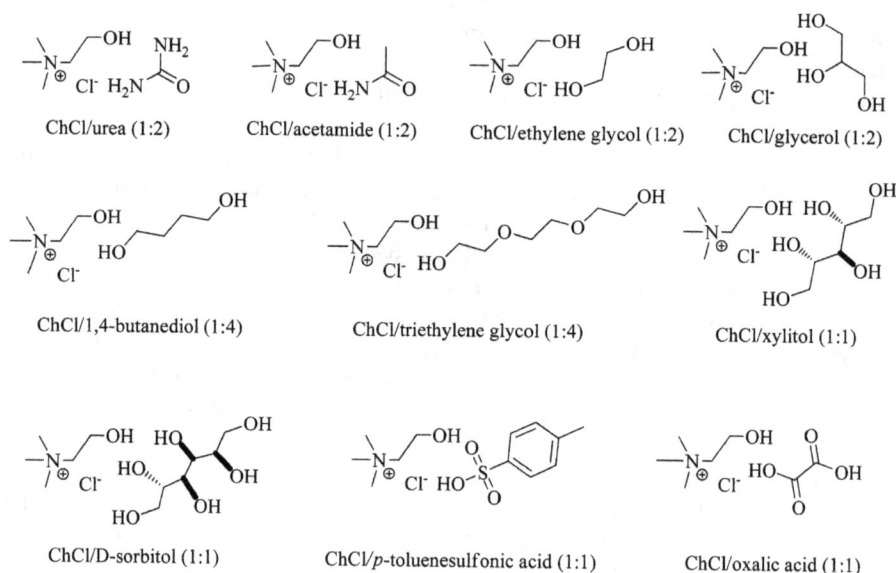

Figure 6.1 Structure of some common DESs based on choline chloride. Reproduced from Reference 13 with permission from American Chemical Society.

as materials purification is not required, (3) non- volatile and (4) mostly nontoxic and biodegradable [12]. The term DES indicates the liquids which are close to the mixtures' eutectic content, i.e. molar ratio of ingredients that con-tributes the least melting point. The DESs are formed as a result of a complex-ation between a hydrogen bond acceptor (HBA) or a halide salt and a hydro-gen bond donor (HBD). Figure 6.1 also depicts the structures of common DESs based on choline chloride [13]. DESs can be also be generated from natural sources (known as natural DESs), specifically through primary metabolites like sugars, amino acids and organic acids. The diversification of attainable consolidations of the starting materials contributes an effective tool for gov-erning the physical characteristics of DESs, which also makes the DESs more suitable than conventional ionic liquids in terms of performance and material properties.

The idea of DES was initially proposed by Abbott *et al.* [14] as a combina-tion of two or more constituents, which generates a eutectic mixture. As men-tioned above, the "deep eutectic solvent" term indicates a combination of HBD and a halide salt to generate the liquid [15]. The aforementioned liquid was designated as deep eutectic solvent for discriminating the DES from IL that consists of particularly distinct anion. The eutectic occurrence was initially established by means of a combination of ChCl and urea (molar ratio 1:2) with melting points of 302 °C and 133 °C, respectively. As a result, a eutectic mix with a melting point of 12 °C was generated. Lithium chloride and copper (II) oxide were favorably dissolved in the DES. The utilization of DESs attracted other researchers as their physico-chemical properties were same as ILs [16]. As a result, a range of liquids generated from eutectic mix of HBD and salts have now been developed. The Figure 6.2 also illustrates the schematic of schematic diagram of the solid–liquid boundaries of a mixture of two solids forming the DES depending on the composition of the mixture [17]. DESs are expressed by the generic formula Cat^+X^-zY, where Cat^+ denotes any sulfonium, phosphonium or ammonium cation, X^- represents a Lewis base (mostly halide anion), Y is a Bronsted or Lewis acid, and z denotes the number of Y mole-cules. Complex anion species is generated between X^- and either a Bronsted or Lewis acid Y [10].

During the past decade, a quick advancement in the DESs as designer sol-vents for different applications has been observed [18]. As mentioned earlier, the DESs, as advanced kind of green solvents, have several distinguished char-acteristics like excellent thermal stability, high viscosity and less vapor pres-sure [19]. Due to these significant benefits of DESs, there were used in applica-tions in various fields such as gas absorption, enhanced oil recovery (EOR), preparation of materials, solvent development/reaction medium, electro-chemistry, catalytic processes, hydro-metallurgy, etc. [20]. In this review, modern developments made in the field of DESs have been focused along with the different applications of DESs. In addition to this, an overview of the properties as well as the different types of DESs employed has also ben presented. Based on the available literature, it can be suggested that the DESs are potential candidates for gas separation, EOR, generation of new materials as well as novel structures and also in the field of metal extraction and metal oxides processing.

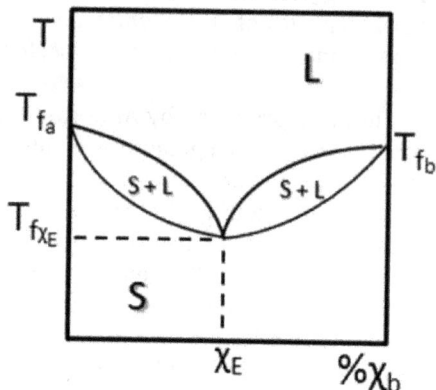

Figure 6.2 Schematic diagram of the solid-liquid phase of a mixture of two solids determined by the mixture composition. Reproduced from Reference 17 with permission from American Chemical Society.

6.2 Properties

DESs are considered to be similar to ILs in reference to the ease of being adoptable in different operations [21], which can be accomplished by the appropriate combination of HBDs and salts. Subsequently, a work specific DES can be designed, which is able to meet the physico-chemical properties necessary for specific application areas [6]. Due to the fact that the DESs have the ability to dissolve certain metal oxides (MOs), these solvents present modern appealing avenues in chemistry exclusively [22]. When compared to other solvents, the DESs possess high surface tensions, viscosities, excellent conductivities, and lower vapour pressures [23]. Because of such advantageous properties of DESs, these solvents have found numerous applications in the oil and gas industries. Particular physico-chemical properties of the DESs, which can be tailored according to the requirement are freezing point, viscosity, surface tension, density and conductivity. These properties are reviewed in the below sections.

6.2.1 Freezing Point

DES's freezing point (T_f) is controlled by the proper selection of HBD, the composition as well as the organic salt. Overall, the strength of the interaction between the anion and HBD plays a vital role. As an illustration, together with urea, the T_f of a choline salt based DES reduced in the order $F^- > NO3^- > Cl^- > BF4^-$, revealing an interrelationship with hydrogen bond strength. The NMR spectroscopy confirmed the presence of hydrogen bond network inside the eutectic mixtures. In addition, a deep cross-interrelationship between the fluoride anion (F^-) of choline fluoride and the NH_2 groups existing in urea molecule in a DES was observed by two-dimensional hetero-nuclear NOE (HOESY) measurements [14]. Abbott *et al.* [24] suggested that the T_f of HBD-salt eutectic mixtures is determined by (1) the mechanism how the couple HBD-anion interacts, (2) lattice energies of DESs, and (3) variation in entropy originating from the liquid phase generation. Kareem *et al.* [25] prepared novel eutectic mixtures (Figure 6.3) based on benzyl-triphenyl-phosphonium chloride and methyl-triphenyl-phosphonium bromide as salts and 2,2,2-triflouracetamide, ethylene glycol and glycerin as HBDs for different molar ratios of salt/HBD. As seen in Figure 6.3, it was observed that the salt/HBD molar ratio that contributed the eutectic mixture with least melting temperatures depended on both the HBD and the salt.

6.2.2 Viscosity

For the practical applications, the viscosity of a mixture or a substance is of great significance. Majority of the DESs display comparatively greater viscosity at room temperature (greater than 100 cP), when compared to the molecular solvents [26]. As the viscosity is linked to the free volume as well as the probability of discovering holes of appropriate dimensions for the solvent molecules or ions to progress into [26], hence, it can be confirmed that the ionic size also influences the viscosity. Low viscosity liquids are generated employing quaternary ammonium cations like fluorinated HBD and ethylammonium. Further, the presence of a comprehensive hydrogen bond network causes lesser mobility of free species inside the mixture, which provides higher viscosity to the DESs. In addition to this, other forces like van der

Waals or electrostatic interactions may generate higher viscosities [6]. The water content, composition and temperature also have a substantial significance [15]. As the viscosity of DESs reduce with increasing temperature, it was confirmed that these follow an Arrhenius-like behavior [27]. Compatible with the recognized density characteristics of various ChCl-glycerol mixtures,

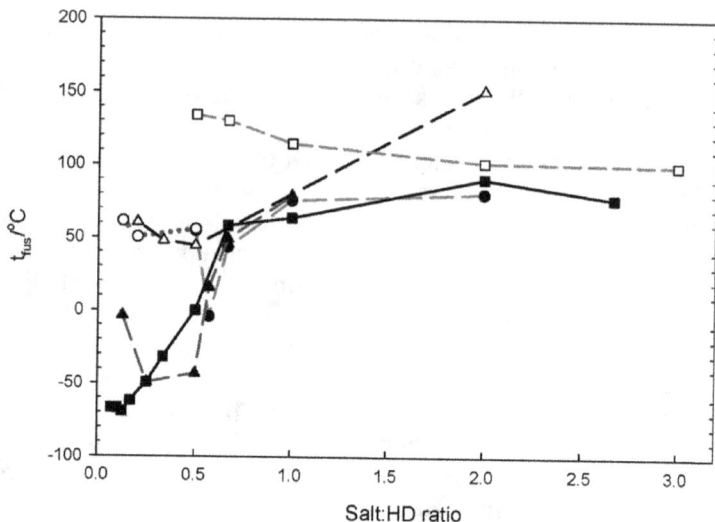

Figure 6.3 Freezing temperatures of different phosphonium-based eutectic mixtures as a function of salt/HBD molar ratio. ●, methyl salt/glycerine; ▲, methyl salt/ethylene glycol; ■, methyl salt/2,2,2-triflouracetamide; O, benzyl salt/glycerine; △, benzyl salt/ethylene glycol; and □, benzyl salt/2,2,2-triflouracetamide. Reproduced from Reference 25 with permission from American Chemical Society.

the viscosity reduces with rising ChCl quantity though the mixtures with ethylene glycol exhibit the opposite effect [28]. From these examples, it was noted that the viscosity is dependent on the composition as well as resulting interactions. Considering the fact that water has lesser viscosity than the DESs, the viscosity value reduces with rising the content of water [29]. Yadav and Pandey [29] also reported the dynamic viscosities of ChCl/urea [molar ratio 1:2, known as reline] and its aqueous mixtures in the temperature range 293.15-363.15 K, as shown is Figure 6.4. The authors observed that the variation of dynamic viscosity of aqueous mixtures of reline with temperature was better described by a Vogel–Fulcher–Tamman (VFT) model, as opposed to an Arrhenius-like behavior.

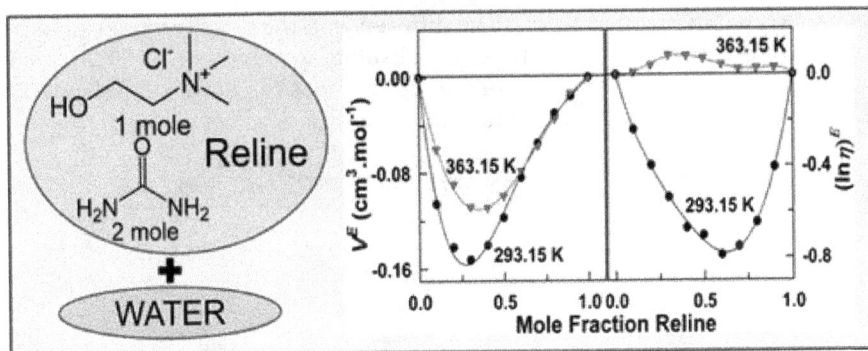

Figure 6.4 Dynamic viscosities of ChCl/urea [molar ratio 1:2, known as reline] and its aqueous mixtures, at a range of temperature 293.15-363.15 K. Reproduced from Reference 29 with permission from American Chemical Society.

6.2.3 Density

It is obvious that the density has a significant importance for the process design applications [30]. Similarly, the pressure and temperature impact on density are necessary for the advancement of appropriate equations of state which contribute immensely in the determination of thermo-dynamic characteristics needed for the industrial processes [31]. Mostly, the DESs display greater density values, when compared to water. The density of the DES is determined by the molecular organization as well as packing of DES [20]. Same as ILs, it was observed that the DESs consists of holes as well as empty vacancies that control the density. Generally, the density values of DESs reduce with temperature increase [32]. Moreover, the water content also affects the density, where the density reduces with increasing the water percentage [33]. In addition to this, the ratio of organic salt and HBD also influences the density. As an illustration, the incorporation of ChCl to glycerol causes a reduction in density, which can also be illustrated in terms of hole theory and free volume [32,15]. Figure 6.5 illustrates the change in density with the amount of HBD for ChCl:LA (lactic acid) based and ChCl:G (glycerol) based DESs. It was noted that for both kinds of DESs, absolute deviation enhanced with increasing the HBD mole fraction in the DES [36]. For analyzing the effect of generation of ester on the properties of DESs, the density of DES ChCl:glutaric acid, prepared by the heating and grinding methods was analyzed, and the results are demonstrated in Figure 6.6 [37]. The difference in the density values obtained using two methods was small (average absolute deviation of 0.15%). The authors also concluded that the grinding method was the more suitable method for the generation of DESs as the formation of ester was not favored.

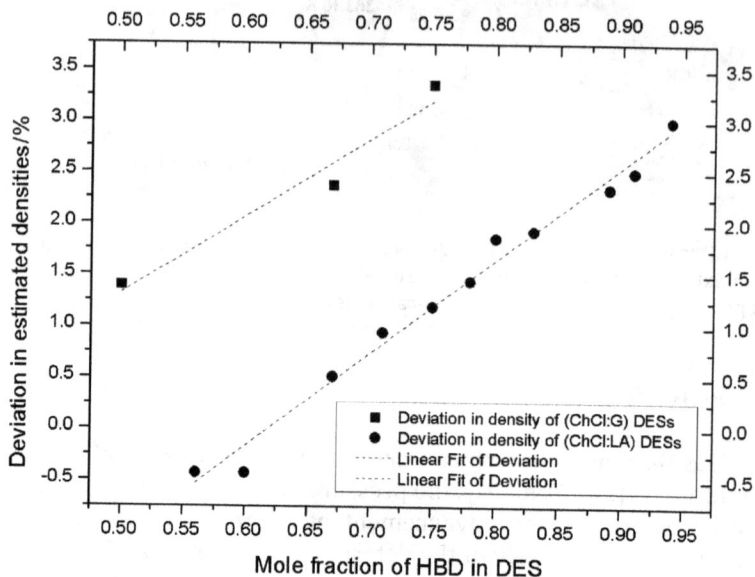

Figure 6.5 Change in density with the amount of HBD for ChCl:LA (lactic acid) based and ChCl:G (glycerol) based DESs. Reproduced from Reference 36 with permission from American Chemical Society.

6.2.4 Electrical Conductivity

Electrical conduction ability of DESs is associated with the viscosity; as the viscosity increases; the conductivity reduces. Hence, majority of the usable DESs possess low conductivities (at ambient temperature, <1 mS.cm^{-1}) [6], which might cause challenges for the specific electro-chemical applications. The low viscosity DESs like those accommodating ethylene glycol or imidazole with ChCl display higher conductivities. Hou *et al.* [38] developed a new series of the binary eutectic mixtures with organic salts (tetrabutylammonium bromide (Bu$_4$NBr), choline chloride (ChCl) and 1-ethyl-3-butylbenzotriazolium

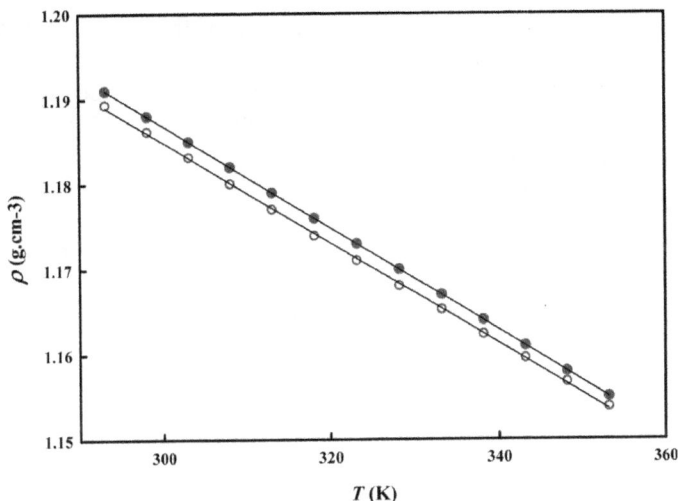

Figure 6.6 Experimental densities (ρ) vs. temperature for the DESs [Ch]Cl:Glu, generated using two distinct techniques: ($\bullet$) grinding and ($\circ$) heating. Reproduced from Reference 37 with permission from American Chemical Society.

hexafluorophosphate (C_2BtPF_6) and organic neutral molecule (imidazole (IM) solid) (denoted as Bu4NBr-IM, ChCl-IM and C2BtPF6-IM). The conductivities of the binary eutectic mixtures decreased in the order ChCl-IM > C_2BtPF_6-IM > Bu4NBr-IM below 80 °C. In addition, the conductivity of the eutectic mixtures was also observed to increase with temperature. Abbott *et al.* [28] applied the hole theory for analyzing the conductivity of DES, and despite the fact that significant results were attained in a few cases (e.g. ethylene glycol + ChCl 2:1 DES), feeble predictions were recorded for others (e.g. glycerol + ChCl 2:1 DES). This proved that the ion mobility and conductivity are controlled by appropriate holes' availability as well as the strength and type of HBD-ion interactions. Further, it was observed that the conductivity of DES enhanced with increase in the salt concentration [6]. However, it is not true for all DESs, and the change in the conductivity with the concentration of salt is dependent on the type of both HBD and salt.

6.2.5 Surface Tension

Until now, the studies pertaining to the surface tension of DESs have been very limited. Abbott *et al.* [24] reported the surface tension of $ZnCl_2$ as well as

ChCl derived DESs. The surface tension values of the DESs phenylacetic acid/ChCl [2:1 mole ratio] and malonic acid/ChCl [1:1 mole ratio] were almost 41.86 and 65.68 mN.m^{-1}, respectively. In addition, the surface tension values of 1,4-butanediol/ChCl [3:1 mole ratio], glycerol/ChCl [3:1 mole ratio] and ethylene glycol/ChCl [3:1 mole ratio] eutectic mixtures were observed to be 47.6, 50.8 and 45.4, and mN.m^{-1} at a temperature 20°C, respectively [28]. The surface tension is anticipated to act in an identical mode as viscosity as it is determined by the stability of inter-molecular synergy which controls the DES generation. In another study, the surface tension values of different glycerol/ChCl DESs were also observed to display a direct inter-relationship with the temperature [15].

Thus, a number of properties of DESs define their tenability and performance under various conditions and a control on these properties is necessary in order to generate DESs of required characteristics suitable for different applications.

6.3 Types of DESs

Depending on the types of complexing agent employed, the DESs can be classified into four types, namely type I, type II, type III and type IV [10]. Among the four types of DESs, type III is the most reported in the literature studies due to the simple preparation, biodegradability, cheaper raw materials, etc. [44]. Also, as mentioned above, many of the developed DESs are based on choline chloride due to non-toxicity as well as low cost. The structural details and other basic characteristics of these types of DESs are explained in the sections below.

6.3.1 Type I DESs

The type I DESs are formed from quaternary ammonium salts and MCl$_x$. This type could be regarded as a resembling kind to the adequately-researched imidazolium/metal halide salt systems. The different examples of type I DESs comprise the imidazolium/chloro-aluminate salt melts, along with solvents generated with imidazolium salts and different metal halides inclusive of FeCl$_2$, and the consecutive halides of metals such as LiCl, CuCl, AgCl, CuCl$_2$, ZnCl$_2$, SnCl$_2$, CdCl$_2$, YCl$_3$, LaCl$_3$ and SnCl$_4$. The type I DESs did not gain as much attention as the type III DESs (generated from HBD and ChCl) due to the inherent hygroscopic nature of the metal salts (e.g., AlCl$_3$) utilized to generate them [44].

6.3.2 Type II DESs

The range of non-hydrated metal halides that have an appropriate melting temperature to generate the type I DES is restricted, nevertheless, the capacity of DESs can be enhanced by employing ChCl and hydrated metal halides. Such DESs are termed as type II DESs. The second type has found superior applications in different large scale industrial operations because of inherent moisture/air insensitivity along with relatively lower cost of several hydrated metal salts [45]. For instance, the type II DESs have been utilized for depositing non-micro-cracked chromium (Cr), as the aqueous electro-deposition of the same is restricted by the legislation [45]. From a 2: 1 molar ratio mixture of $CrCl_3.6H_2O$ and ChCl, a liquid was generated. On the other hand, the formation of liquid from mixture of 6 molar equivalents of water and anhydrous $CrCl_3$ was impossible, indicating that the Cr coordination sphere is vital in the liquid formation. This also implies that the indicated system is IL like in nature and is not commonly a concentrated aqueous solution. More studies revealed the significance of cationic species as grain modifiers. By the incorporation of LiCl, the nano-crystalline black Cr was formed [46].

6.3.3 Type III DESs

The type III DESs, which are generated from HBD and ChCl, are of great interest because of their capability to solvate an extensive spectrum of transition metal species, including oxides and chlorides [14]. A wide range of HBDs have been analyzed till date, with the DESs generated utilizing alcohols, carboxylic acids and amides. These type of DESs are easy to synthesize, relatively cheap and comparatively non-reactive with water, along with majority of them being highly biodegradable. An example for type III DESs is [ChCl]-[amide/carboxylic acids] engineered by Abbott *et al.* [24]. The broad range of accessible HBDs indicates that this class of DESs is specifically adaptable. Physical characteristics of the liquid are greatly dependent on the HBD and it can be readily tailored for particular applications. In spite of the fact that the capabilities of these DESs are remarkably lesser than some imidazolium salt-distinct anionic ILs, these are still adequately broad to permit the metal deposition with greater effectiveness. The type III DESs have been observed to be specifically adaptable, with a broad variety of attainable utilizations inclusive of the metal oxides processing [22], separation of glycerol from biodiesel [47] and the preparation of cellulose derivatives [48]. In addition, as mentioned above, these solvate transition metal species extensively.

6.3.4 Type IV DESs

Most of the ILs that are in a flowing state at room temperature are generated employing an organic cation, mainly using sulfonium, ammonium and phosphonium moieties. Due to the higher charge density of inorganic cations, these are not able to generate low melting point eutectics. Nevertheless, the earlier research studies have presented that the mixtures of urea and metal halides can establish eutectics having melting points <150 °C [49,50]. Abbott *et al.* [20] generated type IV DESs by the incorporation of a variety of transition metals into the room temperature eutectics. It was anticipated that the aforementioned metal salts may not ionize in non-aqueous medium, however, $ZnCl_2$ was observed to generate eutectics with acetamide, 1,6- hexanediol, urea and ethylene glycol [20]. Many studies have described the physical characteristics of the eutectics and it was observed that the properties of these eutectics were similar to other classes of ILs and could be efficiently modelled using the hole theory.

6.4 Applications of DESs

6.4.1 Gas Separation

In this section, the application of DESs for the gas separation has been reviewed. In specific, the removal of CO_2 and SO_2 has been examined in detail. The prevailing knowledge on the gas separation employing the DESs, correlation of the capturing capacities as well as properties of DESs with those of ILs, identification of the defects as well as strengths of DESs, and suggestions related to prospective research directions on this topic have been explored.

CO_2 Capture

Greenhouse gas emissions, specifically the emission of CO_2, result in the serious problem of global warming. Reducing the CO_2 emissions by carbon capture and storage (CCS) has been a focus of intense research in the recent years. The separation of CO_2 is a critical step in CCS and is considered to be an energy intensive process. In addition, efforts have been concentrated on establishing energy sources which do not discharge CO_2, nevertheless, these are not yet at the stage where these could be enforced on a large scale. Also, it is very important to design a feasible synthetic process to generate recyclable solid sorbent which has an enhanced sorption capacity. ILs as the green solvents have been considered for application as liquid absorbents for the separation of CO_2. On the other hand, the combustible character [52], toxicity [53], poor biodegradability [54] and high cost limit the usage of ILs. In recent past, DESs based on ChCl were recommended as a new type of ILs, however, with added benefits in preparation, environmental impact as well as cost.

Gutierrez *et al.* [55,56] utilized resorcinol:ChCl:3-hydroxypyridine accommodating DES as the structure directing agent, liquid medium as well as the source of nitrogen and carbon for generating nitrogen doped hierarchical carbons. Zhu *et al.* [57] stated the greater catalytic effectiveness of a DES urea:ChCl supported on the molecular sieves in order to chemically fix CO_2 to the cyclic carbonates. The green as well as biodegradable catalyst was observed to be active as well as selective, and had also the capability for preparing cyclic carbonates from epoxides and CO_2. Following the chemical process, the products as well as solid catalyst could be effortlessly isolated due to the fact that DES was not soluble in products. The porous carbon nano-sheets prepared in DESs with manageable thicknesses, which could detach the CO_2 from N_2 have also reported [58].

DESs can also be employed themselves as the solvents for capturing CO_2 [59-63]. Leron and Li [60] examined the solubility of CO_2 in a DES consisting of 1 mole ChCl and 2 mole ethylene glycol at pressure till 6 MPa and the temperature from 303.15 to 343.15 K. CO_2 solubility in the DES was determined utilizing a thermo-gravimetric pressure balance that was integrated with a control system, which enabled real-time information monitoring as well as recording. The CO_2 solubility in the solvent was observed to enhance with pressure and reduce with rising temperature. Further, the values were not beyond the range of the CO_2 solubility in 1-butyl-3-methylimidazolium [Bmim]- as well as 1-ethyl-3-methylimidazolium [Emim]- based RTILs. The solubility information was favorably depicted by an extended Henry's law equation as a function of pressure and temperature with an average absolute deviation of 1.6%.

One of the most famous DESs is the mixture generated from 1 mole ChCl and 2 moles urea ($CO(NH_2)_2$; commercial name: Reline). Its melting temperature is 12 °C, which is significantly lesser than the melting temperature of its distinctive components [14]. The DES was observed to be very beneficial as a solvent in a variety of applications like catalytic reactions and preparation of micro-porous crystalline zeolites, metal-organic frameworks and coordination polymers [64,65]. Identical to several room temperature ILs (RTILs), the capability of this green solvent for the absorption of CO_2 has also been examined. Li *et al.* [66] stated the CO_2 solubilities in a range of ChCl–urea mixtures, and analyzed the influence of composition on the gas solubility in the DES. It was confirmed that at the same pressure and temperature, solubilities of CO_2 were very identical in mixtures with 1:2.5 and 1:1.5 mole ratios but remarkably greater in the combination with 1:2 mole ratio. In a study by Leron *et al.* [62], experimental analysis on the CO_2 solubility in ChCl–urea (1:2 mole ratio) DES at pressures till 6.0 MPa and temperature range 303.15 - 343.15 K was carried out. It was noted that the CO_2 solubility in the DES reduced with enhancing the temperature. Further, it was also observed that the CO_2 solubility enhanced with increase in pressure at constant temperature. The impact of buoyancy (Δm) on the CO_2 solubility in the DES was remarkable at greater pressures, however, essentially insignificant at lesser pressures. It was confirmed that the effect of buoyancy reduced with increase in the temperature. By utilizing an extended Henry's law equation with an average absolute deviation of 1.0%, the CO_2 solubility information was precisely characterized as a function of pressure and temperature.

Ali *et al.* [63] conducted research on the ammonium- and phosphonium-based DESs having various HBDs. The solubility of CO_2 in the generated DESs at a fixed temperature and pressure was experimentally analyzed. In addition, a mathematical model based on the Peng Robinson equation of state (PR EoS) was generated for correlating the solubility of CO_2 in these categories of DESs. This model was justified with the attained experimental information and then analyzed with additional precise DESs stated in the literature over a broad range of pressure and temperature values. The experimental data as well as the calculated data using PR EoS had a very good agreement. Hence, the generated model could be employed for studying the efficiency in utilizing DES in CO_2 removal processes or any other separation processes.

Sze *et al.* [67] examined the CO_2 capturing as well as releasing ability of ternary DES systems consisting of glycerol, ChCl and one of the three various superbases. Out of the systems examined, the superior performance system was observed to remove CO_2 at a capacity of approximately 10% by weight. Analysis was performed on ternary mixtures ChCl:glycerol:1,5-diazabicyclo[4.3.0]-non-5-ene (DBN) with molar ratios 1:2:x, where x = 3, 6, 7 or 8, respectively. As illustrated is Figure 6.7, the DES system with 1:2:7 molar ratio offered the highest CO_2 capturing ability (105 mg/g of DES). In a study by Francisco *et al.* [68], a new liquid solvent was synthesized by combining two naturally as well as effortlessly accessible solid starting materials (ChCl and lactic acid). ChCl is a HBA, whereas lactic acid is a strong HBD. The combination generated a stable liquid in a broad spectrum of compositions, displaying extremely low transition temperatures. The lactic acid:ChCl combinations were analyzed as the solvents for gas separations. Analysis of the physical characteristics confirmed the tunability of physical characteristics as well as phase behavior by varying the composition. Glass transition temperature, surface tension, density and viscosity were examined as a function of lactic acid:ChCl ratio. The physical characteristics of the new DESs were very identical to that of ILs, however, these did not possess the main drawbacks of typical ILs, for instance, the DESs were renewable, cheap and environmental friendly solvents. Due to the potential characteristics such as tunability of physical as well as thermodynamic properties by varying the composition, excellent solvation properties generated from hydrogen bonding interactions, broad liquid range, recoverability and biodegradability, the lactic acid:ChCl DESs are considered to be potential candidates as solvents for CO_2 separation.

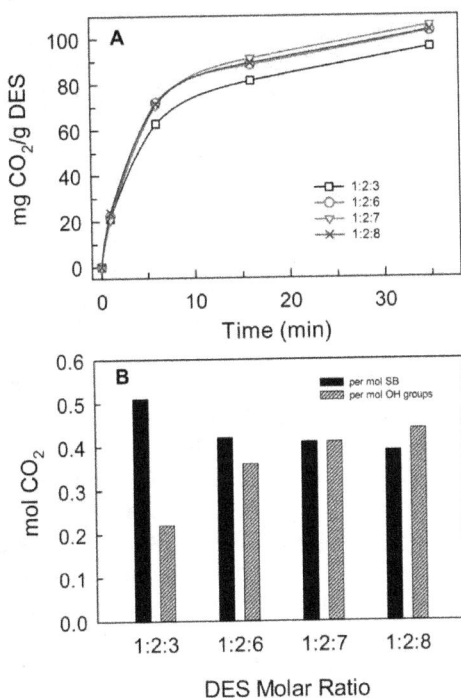

Figure 6.7 (A) Quantity of CO_2 separated per gram of DES as a function of time; (B) total CO_2 captured per mole of DBN and per mole of alcohol (OH) group for the same DES mixtures with increasing relative DBN amounts. Reproduced from Reference 67 with permission from American Chemical Society.

Considering the literature findings, it can be confirmed that the choline-based DESs are promising for the application as liquid absorbents for CO_2 separation. For instance, the DES generated from ChCl and urea (molar ratio 1:2) displayed a better performance with a calculated CO_2 absorption of 3.559 mmol/g at 60 bar pressure and 303.15 K temperature [62]. The above discussion clearly revealed that the DESs have the ability to absorb more CO_2 than the toxic ILs because of their tunable chemical as well as physical properties. However, uncertainties and obstacles still prevail, and further examination on the micro-structure as well as properties has to be performed using experimental measurements and model developments. Further, detailed analysis is needed for determining the CO_2 absorption mechanism.

Sulfur Dioxide (SO₂) Capture

The generation of SO_2, principally from the fossil fuels' burning, leads to severe environmental issues like smog and acid rain. Till date, the technologies of flue gas desulfurization (FGD) perform a vital role in the separation of SO_2 [69]. The advancement of efficient and renewable absorbents for the removal as well as recovery of SO_2 is very critical to from health, safety and environment perspectives. Similar to the CO_2 removal studies, majority of the prevailing SO_2 removal technologies that employ DES systems are concentrated on the application of ChCl as the IL, due to the fact that the material is non-toxic, cheap, biodegradable and can be either derived from biomass or easily accessible as bulk commodity material.

Yang *et al.* [70] analyzed the SO_2 absorption using ChCl-glycerol DESs at different temperatures and SO_2 partial pressures. Molar ratios of ChCl and the glycerol varied from 1:4 to 1:1. The solubility of SO_2 was observed to enhance with increase in the ChCl concentration. At 1 atm pressure and 20 °C temperature, the SO_2 absorption ability of the DESs with a glycerol/ChCl mole ratio of 1:1 was as large as 0.678 g SO_2 per g DES. In addition, the absorbed SO_2 could be effortlessly discharged, and the superior characteristics of greater absorption ability as well as accelerated absorption/desorption rates continued for five consecutive absorption/desorption cycles. However, ChCl-glycerol DES did not reach greater absorption abilities of pure SO_2, the absorption process was physical as well as the solubility data adapted to Henry's law, which overall led ChCl-glycerol DES inadequate for absorbing low-partial pressure SO_2 competently. Later, Sun *et al.* [71] generated four types of ChCl-based DESs using four organic compounds namely malonic acid (MA), ethylene glycol (EG), thiourea and urea as the HBDs. All DESs exhibited excellent thermal stability and were observed to be stable at 363 K, which is very advantageous for the usage in FGD. The SO_2 absorption abilities of the DESs were analyzed at varying temperatures as well as SO_2 partial pressures. Results obtained from the absorption analysis confirmed that ChCl-thiourea (1:1) and ChCl-EG (1:2) DESs exhibited superior absorption efficiency, and the absorption abilities were 2.96 and 2.88 mol SO_2 per mol DES at 1 atm and 293 K, respectively, as demonstrated in Figure 6.8. Further, the SO_2 regeneration experiments were performed and it was noted that all the solvents were regenerated at 343 K with N_2 bubbling. Also, the absorption abilities of DESs persisted without any remarkable loss after six absorption/desorption cycles.

In another study by Liu *et al.* [72], DESs based on caprolactam (CPL) and low molecular weight organic compounds, namely benzoic acid, imidazole, acetamide, o-toluic acid and furoic acid were synthesized and designated as CPL-benzoic acid (molar ratio 1:1), CPL-imidazole (molar ratio 1:1), CPL-acetamide (molar ratio 1:1), CPL-o-toluic acid (molar ratio 2:1) and CPL-furoic acid (molar ratio 1:1), respectively. The DESs were illustrated as promising absorbents for the capture of SO_2 in the temperature range 30-70 °C and 1 atmospheric pressure of pure gas. The characteristics of the DESs were identical to the traditional ILs and entirely distinct from the molecular solvents.

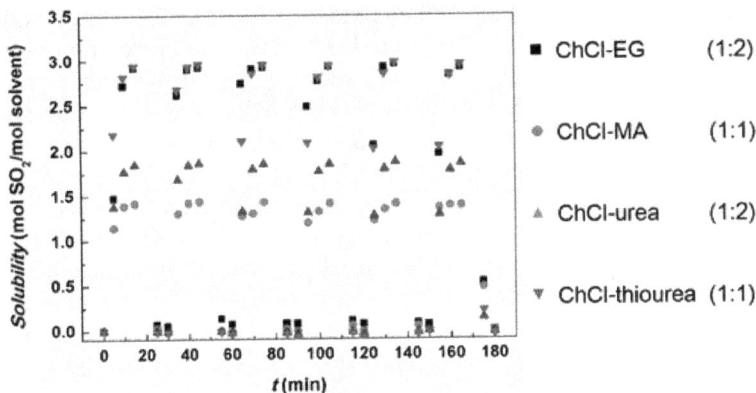

Figure 6.8 SO_2 absorption abilities of different DES systems. Reproduced from Reference 71 with permission from American Chemical Society.

From the examined temperature range, it was noted that the density enhanced linearly with increasing temperature. Further, Vogel–Tammann–Fulcher (VTF) equation was employed for describing the temperature dependence of conductivity or viscosity. At 30 °C, the CPL-imidazole (1:1) DES exhibited highest conductivity (32.3 μs/cm) and lowest viscosity (48.6 mPa·s). It was noticed that SO_2 absorption abilities of CPL-organic amine based DESs were greater, when compared to the CPL-organic acid based DESs. The SO_2 solubility in CPL-acetamide (1:1) DES was greater than the BMImBF4 IL and lower than dimethylformamide (DMF) solvent. The SO_2 solubility in CPL-acetamide (1:1) solvent was 0.497 g/g of mass fraction at temperature 30 °C and the absorption was essentially reversible. Similar to the ChCl based DESs, the SO_2 absorption of CPL-acetamide DES was observed to be a physical process, which induced CPL-acetamide to not absorb low-partial pressure SO_2 conveniently. The SO_2 absorption/desorption in CPL-acetamide (1:1) was carried out for 5 times and proportionate absorption capacity was observed.

As noted in the earlier literature studies, the DESs for the absorption of SO_2 mainly displayed a physical process and the solubility data in accordance with Henry's law make it challenging for the application of DESs to separate low partial pressure SO_2 in the flue gas. Hence, it is essential to design environmentally benign functional DESs, which have the ability to absorb lesser partial pressure SO_2 in the flue gas. In a study by Zhang *et al.* [73], DESs were designed for absorbing lesser partial pressure SO_2 from the simulated flue gas. Two types of biodegradable functional DESs based on ethylene glycol (EG) as HBD and l-carnitine (L-car) as well as betaine (Bet) as HBA were synthesized with mole ratios of HBA to HBD from 1:3 to 1:5. The aforementioned DESs were analyzed for absorbing SO_2 with varying partial pressures at distinct temperatures. The obtained results confirmed that the two DESs could be used to absorb the low-partial pressure SO_2 effectively. At SO_2 partial pressure of 0.02 atm and 40 °C temperature, the SO_2 absorption abilities of the DESs with HBA/HBD molar ratio of 1:3 were found to be 0.820 mol SO_2/mol HBA for the DES [L-car+EG] and 0.332 mol SO_2/mol HBA for DES [Bet+EG]. In addition, there was no apparent loss of SO_2 absorption abilities in the course of the regeneration cycles. The SO_2 absorption of two DESs at a temperature of 40 °C and partial pressure of 0.0037 atm was performed and it was observed that the two DESs had very quick absorption rates for SO_2 with the low-partial pressure because of the practically linear absorption curve at the introductory absorption period. Further, the absorption mechanism was suggested demonstrating that there were acid–base reactions between the acidic SO_2 and the Lewis base COO− present in HBA. DESs based on L-car and betaine as HBA are potential absorbents for SO_2 removal in the FGD application.

Guo *et al.* [74] synthesized economically as well as environmental friendly DES systems having distinct molar ratios of tetra-butylammonium bromide

(TBAB) and caprolactam (CPL) and examined the system for the separation of SO_2 at atmospheric pressure and different temperatures (298.2 to 403.2 K). From Figure 6.9, it can be observed that the equilibrium solubility of SO_2 in distinct mixtures exhibited a firm dependence on the temperature as well

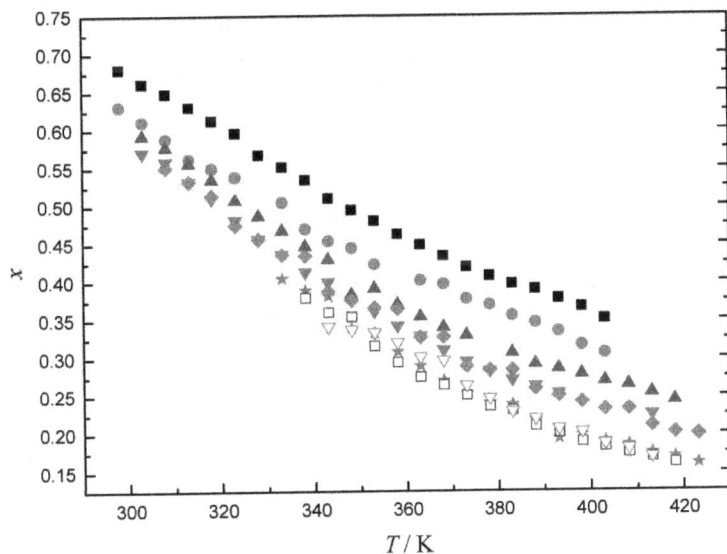

Figure 6.9 Solubility of SO_2 as a function of temperature in [CPL][TBAB] DES system with varying molar mixing ratios of CPL and TBAB. Reproduced from Reference 74 with permission from American Chemical Society.

as the mole ratio of parent chemicals at the atmospheric pressure. Mole fraction solubility of SO_2 reduced remarkably both with enhancing temperature (at specified mole ratio) and rising mole fraction of CPL in the DES [CPL][TBAB] at a particular temperature. Utmost solubility noticed for the DES with a molar mixing ratio of 1:1 was 0.680 g/g at temperature 298.2 K, however, it decreased to 0.351 g/g at temperature 373.2 K and constant atmospheric pressure. Duan *et al.* [75] made efforts to improve the viscosity of [CPL][TBAB] by the mixing water to enhance the affinity as well as solubility of SO_2 in the binary mixture. Nevertheless, the addition of water reduced the viscosity, and also minimized the SO_2 removing ability of [CPL][TBAB]:H_2O. By heating the densest phase, the removal of SO_2 could be performed effortlessly, and the materials could be reused for absorption.

6.4.2 Enhanced Oil Recovery

The demand for global energy has been developing firmly because of the fast growth in world's population. The growing energy requirement develops into the major justification for applying various technologies for improving the effectiveness of petroleum production. Field as well as laboratory studies reported that just 30-50% of oil is generally recovered succeeding the water injection from original oil in place (OOP) [76]. For extracting the remaining oil, it is essential to employ effective methods and the EOR or tertiary oil recovery is the technique achieving greater importance in this respect [77]. Currently, the three main methods employed for EOR are (1) thermal recovery, (2) gas and (3) chemical injections. Out of the aforementioned techniques, chemical flooding is considered as the superior EOR technique [78]. It can be classified into three major processes namely (i) alkaline flooding, (ii) surfactant flooding and (iii) polymer flooding [76]. Various studies have been carried out on laboratory scale and as field applications on various thermal methods like steam injection and *in-situ* combustion, analyzing their efficiency in increasing the heavy oil recovery. Thermal methods have the benefit of transferring the thermal energy to the reservoir as well as heating the heavy oil. Thus, the viscosity of heavy oil reduces remarkably, thereby increasing the mobility ratio [79]. In order to design advanced steam flooding with efficient heavy oil production, different methodologies have been further developed. One of the methodologies is to co-inject the steam using additives like solvents, caustic steam, hydrocarbon and non-hydrocarbon gases, foam and surfactant [80]. A substantial concern in the usage of chemical additives in steam flooding is the thermal decomposition of chemicals occurring at higher temperature which remarkably decreases the efficiency of these chemicals. The chemicals employed in conventional EOR have been observed to have lesser thermal stability. In contemplation of addressing the challenges related to polymers as well as the surfactants in the flooding operations, the exploration for fresh thermally-stable chemicals or advancement in the current ones has become critical. In the recent past, some research studies have employed ILs in upgrading as well as enhancing the heavy crude oil recovery. Results attained from the upgrading experiments revealed that ILs have a better viscosity lowering property and, thus, can cause higher oil recovery [81]. The capability of ILs was also examined as alternate surfactants in chemical flooding. As mentioned earlier, the analogs of ILs, i.e., DESs have many benefits like low price, biodegradability, non-toxicity, chemical compatibility with water, biocompatibility as green solvents and easy preparation. Besides, the aforestated characteristics, existing polar components, greater viscosity, as well as surface active agents in DESs provide them capacity for the usage in heavy oil recovery.

In recent years, the DESs from group IV have been analyzed as new EOR agents for the heavy oil recovery improvement from sandstone cores at the reservoir environment. These DESs include ChCl:glycerol, ChCl:urea [82,83,84], ChCl:malonic acid [85] and ChCl:ethylene glycol [86]. In all the aforementioned events, the DES flooding exhibited assuring outcomes in terms of recovery improvement through enhancement in viscosity ratio as well as benign wettability amendment of rock surface as the major mechanisms. In a study by Mohsenzadeh *et al.* [82], ChCl-based DESs were analyzed experimentally for the first time for studying the impact on various oil recovery mechanisms in heavy oil/formation brine/Berea sandstone rock system. The authors compared the efficiency of two distinct types of DESs, namely ChCl:glycerol (DES1 1:2 molar ratio) and ChCl:urea (DES2 1:2 molar ratio), in improving the heavy oil recovery. The impact of the two DESs diluted with formation brine on the surface and interfacial tensions, altering wettability, impulsive imbibitions, emulsification, as well as tertiary residual heavy oil saturation minimization were analyzed experimentally at varying temperatures. The heavy oil having 16 °API and formation brine from an Omani heavy oil field were employed. Core flood analysis was carried out at the reservoir conditions utilizing the Berea sandstone core plugs. The results confirmed that in the course of brine flooding, the enhancement in the heavy oil recovery occurred by raising the temperature. Cumulative oil recovery by brine injection was 42%, 40% and 36.7% at temperatures 80 °C, 60 °C and 45 °C, respectively. The viscosity ratios of the brine to heavy oil enhanced with increase in temperature. Further, the results from the Amott tests, contact angle analysis and oil-brine interfacial tension (IFT) examinations confirmed that as the temperature enhanced, the IFT reduced and the wettability alteration increased. It was concluded that the DES2 consistently outperformed the DES1 in increasing the oil recovery. This was due to the fact that DES1 was less alkaline, when compared to the DES2.

In another study by Mohsenzadeh *et al.* [87], sequential DES and steam flooding was suggested and experimentally analyzed for the enhancement in the heavy oil recovery and achieving capability for *in-situ* heavy oil upgrading. ChCl:glycerol (DES1) and ChCl :urea (DES2) DESs were employed in the study. By utilizing the 16.5 °API heavy oil as well as Berea sandstone core plugs, the primary DES as well as secondary DES flooding at varying concentration followed by high-temperature steam flooding were performed. From the TGA analysis of the undiluted DESs, it was noted that the two DESs initiated to decompose at approximately 210 °C. For DES1 and DES2, the ultimate decomposition temperatures were recorded at 320°C and 370 °C, respectively. Thus, the thermal stability analysis confirmed that the selected DESs exhibited greater thermal stability when compared to other EOR chemicals, thus, confirming high potentials to be utilized as the additives in the thermal EOR processes. The steam flooding occurring after the undiluted DES injection as well as 2-fold diluted DES injection led to an extra heavy oil recovery of 12%, when compared to the primary or secondary steam flooding alone. Nevertheless, by employing greater dilution of DES, i.e., about 20 fold diluted solution, same or even lesser total recovery factors in certain cases were obtained, when compared to the secondary steam flooding. In terms of the recovery increase in the period of subsequent steam flooding, DES1 was found to be less effective when compared to DES2 at the same concentrations. The examination of physico-chemical properties of produced oil for various cases indicated the supportive role of DES in improving the *in-situ* heavy oil. Relatively, the DES1 exhibited superior comprehensive performance than the DES2, in terms of the *in-situ* heavy oil upgrading.

Studies were also performed to investigate the critical parameters influencing the heavy oil recovery performance by DESs injection [83]. Mohsenzadeh *et al.* [83] carried out a study for examining the impacts of DESs concentration, brine salinity, secondary as well as tertiary DES injections on the recovery performance. These were investigated through core flooding experiments, contact angle measurements and IFT measurements. The same DESs namely ChCl:glycerol (DES1 with molar ratio 1:2) and ChCl:urea (DES2 with molar ratio 1:2) were used for the analyses. The IFT between the heavy oil and various solutions were determined by the spinning drop tensiometer at 45 °C. It was observed that both DES1 and DES2 solutions enhanced the IFT of the formation brine due to the powerful synergy between the functional groups present in DES2 [(–NH₂ and C=O) and DES1 (OH–)] and the functional groups existing in the oil because of the van der Waals forces. Nevertheless, greater IFT was seen during the usage of DES1 mainly due to the existence of stronger force (HO– vs –NH₂) as well as larger population (3HO– vs 2 –NH₂) in DES1, when compared to DES2 [82]. In addition, the results also confirmed that the IFT enhanced with increasing DES concentration in brine for both DES1 and DES2 solutions. The contact angle measurements were performed between aged rocks and oil droplet in distinct DES solutions and expressed as a qualitative index of rock wettability. DES2 was observed to have more impact to enhance the contact angles, when compared to DES1, at the identical concentration of DES in brine. In addition, the contact angle values revealed that reducing the concentration of both DESs from 50 vol% to 5 vol% led to minor variations in wettability alterations. The authors concluded that the two DES solutions had good potential for changing the wettability from the oil-wet state towards neutral-wet condition. It was confirmed that the wettability alteration in the period of DES injection had remarkable effect on the oil recovery and it could be the governing as well as superior mechanism for the heavy oil recovery increase. For both DESs, the total recovery at the tertiary mode was almost 3-6% greater at 5 vol% and 16% greater at 50 vol% concentration, when compared to secondary mode. Further, it was seen that the DESs could allow broad range of reservoir salinities and no remarkable variation in oil recovery was observed at distinct brine salinity in the period of injection of tertiary DESs.

Despite the fact that the chemical injection into oil reservoir is regarded as an efficient EOR method, there are certain negative effects on reservoir like formation damage because of fluids incompatibility or the chemical interactions occurring with reservoir rocks as well as fluids. In the oil and gas reservoirs, the formation damage is considered to be one of the most critical justifications for dropping well injectivity and productivity. This damage might occur by drilling fluid invasion close to the wellbore or during the EOR tasks in the field like chemical flooding, gas injection, water flooding as well as thermal methods. Mohsenzadeh *et al.* [84] also analyzed the possible impact of the DESs on sandstone formation damage. The authors employed the same DESs i.e., DES1 (ChCl:glycerol with molar ratio 1:2) and DES2 (ChCl:urea with molar ratio 1:2) for the analysis. The core flooding analysis was carried out at varying temperatures and reservoir pressure to determine possible permeability variation by DESs injection. In order to gain insights about the formation damage mechanism, various analyses such as critical injection rate after which fines transfer occurs, scanning electron microscopy (SEM), water shock impact on the formation damage, X-ray diffraction (XRD), dry core weight analysis after and before each core flooding examination and computed tomography (CT) scanning investigation of fresh as well as treated core samples were carried out. The permeability analysis by the core flooding experiments confirmed that water shock event led to uncompromising damage (65% decline in absolute permeability) when the distilled water was infused succeeding brine flooding. The DES1 and DES2 displayed positive effect on lowering the water shock loss at distinct temperatures. DES1 was observed to exhibit superior performance when compared to DES2. The core weight analysis proved that when the permeability losses were greater, the extent of core weight enhancement was also observed to be higher. In addition, results of quantitative XRD and SEM confirmed that the precipitation as well as deposition within the pores were the major sources of formation damage. Moreover, the CT-scanning images exhibited better comparisons between damaged and fresh cores. It was also noted that in spite of the DES solutions' function in prohibiting serious water shock deterioration as well as clay stabilization, there was still certain formation damage brought about by the re-crystallization and precipitation processes lowering the core samples' permeability.

For enhancing the heavy oil recovery using ChCl-malonic acid DES, several studies have also been performed. Al-Weheibi *et al.* [85] examined the capability of DES for recovering the residual heavy oil left after the waterflooding. For this study, two DESs with ChCl-malonic acid of molar ratios 1:1 and 1:0.5 were examined. The two solvents were preliminarily analyzed by calculating the conductivity, viscosity, density and pH at distinct temperatures (20-80 °C). The core flood test was performed at reservoir condition (temperature = 45-80 °C, pressure = 1200 psi,) utilizing Berea sandstone core samples as well as fluids from the field of interest (formation brine and crude oil). The results obtained from the core flood test revealed that the solvents generated 7-14% of the residual heavy oil after the brine flooding as tertiary recovery stage. Further, it was also confirmed that both the DESs exhibited superior performance in increasing the oil recovery at elevated temperatures. The analysis of absolute permeabilities after and before the injection of DESs/brine solutions displayed no deterioration to the formation. For the tertiary oil recovery enhancement, the wettability alteration was the dominant mechanism.

In addition to the above-mentioned DESs for EOR, the DES based on ChCl-ethylene glycol has also be utilized for this purpose. Shuwa *et al.* [86] analyzed the ability of the ethylene glycol-ChCl DES solution on increasing the heavy oil recovery by conducting core flood tests. The test was carried out at reservoir conditions (temperature = 45-80 °C and pressure = 1200 psi) with Berea sandstone core specimens as well as fluids from the field (crude oil and formation brine). Enhancement in IFT between the oil and DES/brine solution was observed. Further, from the spontaneous water imbibition analysis, it was confirmed that a reduction in oil wetness brought about an enhancement in the oil production. Succeeding the water flooding process, almost 52% of residual oil was recovered employing the DES in the core flooding process. The obtained results confirmed an advancement in oil recovery with an increase in the temperature of the reservoir. Formation damage did not occur due to the interaction between the DES and core materials, confirmed from the permeability analysis of the DES/brine solution after and before the injection process.

The dominant mechanisms for the tertiary oil recovery advancement were observed to be wettability alteration and the viscous forces. Overall, on the basis of the analyses performed and in comparison with the conventional methods of EOR, the DESs have been proved to be effective and promising.

6.4.3 Preparation of Materials

The traditional synthesis of inorganic materials is generally performed in organic solvents or polar solvents like water. For instance, solvo-thermal synthesis is the method of developing single crystals by thermal treatment under pressure, from a non-aqueous solution [6]. In recent past, the iono-thermal preparation has been developed, where ILs or DESs are employed as both solvent as well as template. Iono-thermal preparation can be easily performed in low pressure conditions because of the lesser volatility of the DESs or ILs. Mostly, the capability for tuning the ionic characteristics of DESs offers modulating reaction surroundings under which the novel materials having beneficial frameworks might be generated. A recently developed method in this field is the substitution of ILs by safe as well as cheap DESs. As an illustration, in the iono-thermal strategy, the DESs based on renewable carboxylic acids/ChCl or urea/ChCl were employed as an alternate medium to the ILs. As stated in the previous sections, the DESs have numerous marked benefits, when compared to ILs. In addition, the existence of neutral components in DESs having greater boiling points such as urea, carboxylic acids or carbohydrates contribute new as well as complementary features, when compared to the ILs which are entirely ionic. The DESs also have the capability to perform multiple functions during materials preparation, inclusive of the classic role of solvent, reactant, templating agent, etc. Due to the aforementioned characteristics, the DESs are currently attaining remarkable recognition for the iono-thermal preparation of materials as well as the design of novel structures. Till date, several inorganic materials with substantial structures and properties, inclusive of metal oxides, metal-organic frameworks, nanomaterials, carbon and zeolite-type materials have been generated utilizing iono-thermal process. In the following sections, the materials generated in DESs for different usages like gas separation, catalysis and hydrogen storage have been illustrated.

Metal-organic Frameworks

The preparation of metal-organic frameworks (MOFs) acquired importance because of their promising utilization in fields like gas separation, hydrogen storage, etc. [88,89]. Taking benefits of the ecological, physico-chemical and economic benefits of DESs in iono-thermal preparation of materials, the eutectic mixtures are currently considered as potential interesting media for the huge-scale preparation of advanced MOF frameworks. Unlike the ILs (the binary system containing cations and anions), which were utilized for the preparation of MOF, the DESs consist of neutral ligands like urea which can bring about structure-directing impact, thereby, tuning the DESs to be the appropriate solvents for the MOF synthesis. Till date, several efforts have been made for the iono-thermal preparation of advanced porous MOFs in DESs, because of the diversified roles which DESs can perform in such preparation. Zhang *et al.* [90] studied the multifaceted function of the DESs in the creation of open metal sites as well as porosity in the MOF structures for the gas storage application. The authors initially described the preparation of a range of MOFs in the DES, consisting of 1,4-benzenedicarboxylate (bdc), trivalent metal (Sm, Ln, Nd, Y) and a single or two elements from the DES. From the analysis, it was noted that about six various topologies were attained, and the DES performed as structure-directing agent as well as solvent. With reference to the Ln-accommodating structure, the extra-framework choline ion was maintained with Cl$^-$ confined to the layers of polymers. Urea molecules were chemically linked to the structure, along with external choline ion. Further, choline could additionally be held by a chemical bond to MOF framework by way of its OH groups. The urea molecules, which are neutral, could perform as an external structure-directing media. In addition, the authors reported the role of DES in the straight chemical linkage of ethylene urea hemihydrate (e-urea) and m-urea to the polymer structure. Entire novel structures have revealed that the DES can perform as a solvent, and can also act as a ligand supplier and/or template. In another study by Himeur *et al.* [91], the iono-thermal preparation of three advanced lanthanide (Ln)/organic based MOFs present in a DES di-methyl-urea/ChCl were reported. Usage of hydrated mineral Ln precursors inevitably induced the contamination of DES with water. Durable attachment of water molecules, which were isolated, in DES by way of the hydrogen bonding system of connections extremely restricted its reactivity.

Recently, Zhan *et al.* [92] synthesized eight exclusive three dimensional (3-D) Ln-thiophene-2,5-di-carboxylate materials in the DES e-urea/ChCl. In the group of the prepared materials, six compounds were observed to be iso-structural, crystallizing in the $P2_1/c$ space group of monoclinic conformity [Ln(TDC)$_2$](choline) where Ln = Nd, Gd, Er, Eu, Dy, Tb]. In the mentioned frameworks, the choline cation compensated the anionic frameworks' charge. An identical material was achieved employing the IL 1-methyl-3-ethyl-imidazolium bromide, confirming that the guest cation did not perform the template's role. The structure of [Nd$_2$(TDC)$_3$(ethylene-urea)$_4$]·3(ethylene-urea) exhibited three types of guest ethylene-urea molecules as indicated in Figure 6.10. Ethylene-urea A generated durable hydrogen bonds with host structure, ethylene-urea B displayed hydrogen bonding with host structure, however, ethylene-urea C had only poor interactions with host structure. Chen *et al.* [93] stated the preparation of Zn(II)-B(III)-imidazole structure with exceptional pentagonal channels by preparation in DES m-urea/ChCl. In Zn$_2$(im)Cl$_2$[B(im)$_4$], the neutral ligand and the cation did not incorporate the material framework, however , the DES assisted for stabilizing the halogen-metal bonds. Later, the DES solvent permitted the preparation of material which integrated two different imidazolate structures, with Zn–im connection as well as B(im)$_4^-$ complex.

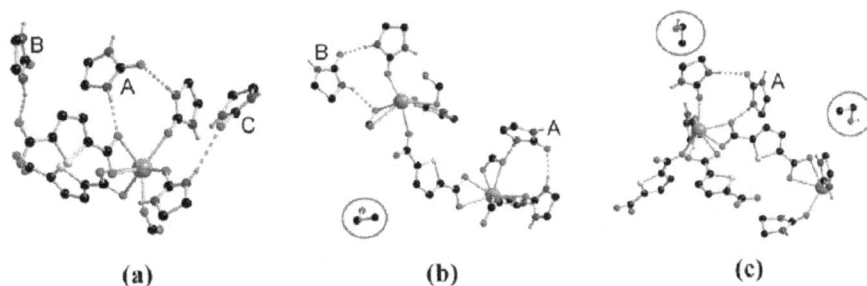

(a)   **(b)**   **(c)**

Figure 6.10 (a) Three types of ethylene-ureas (A, B and C) in [Nd$_2$(TDC)$_3$(ethylene-urea)$_4$]·3(ethylene-urea); (b) e-urea C exchanged by ethanol and (c) e-urea B and C exchanged by ethanol. Reproduced from Reference 92 with permission from American Chemical Society.

Metal Phosphates as well as Phosphites with Open-framework

Zeolites or zeolite-like materials are of considerable significance because of their utilization in oil and gas industries. Currently, most of the solid acid catalysts employed in refinery as well as petro-chemical industries are based on zeolites. The zeolite-analogous structure materials are generally prepared in organic solvents or water by a conventional solvo-thermal method by employing the structure directing agents. Cooper *et al.* [64] firstly introduced the DES-mediated iono-thermal method for the generation of micro-porous crystalline zeolite analogues. For the preparation of zeo-types, the DESs based on IL ([EMIm]Br) or urea/ChCl were employed as alternate solvents. Four various alumino-phosphate zeotypes, namely SIZ-5, SIZ-4, SIZ-3, and SIZ-1, were generated in the presence of [EMIm]Br, under various experimental surroundings, where the imidazolium-based IL acted both as template as well as solvent. Unlike IL, the application of DES ChCl/urea brought about a different kind of framework, i.e. SIZ-2. Al–O–P interchange was retained and the structural composition $(NH_4)_3[Al_2(PO_4)_3]$ confirmed that the SIZ-2 too exhibited a disrupted framework. Comparison with the preparation in the IL, it was observed that the incorporation of fluoride to the DES urea/ChCl mix induced the generation of a non-zeolitic alumino-phosphate termed as AlPO-CJ2. Overall, the application of DESs or ILs could adjust the way how the material precursors are put together. As stated by Copper *et al*. [64], this is a critical particularity of these preparations. Depending on the favorable synthesis of SIZ-n type materials, Parnham *et al*. [94] stated the iono-thermal preparation of materials employing different DESs as the template delivery agents. 2-imidazolindone(e-urea), 1,3-Dimethyl urea (DMU) as well as tetrahydro-2-pyrimidione derivatives were examined as the eutectic combinations in consolidation with tetra-ethylammonium bromide or ChCl. The synthesis of materials based on aluminum-phosphate with advanced structures has been achieved utilizing this approach.

Even though the iono-thermal preparation was developed in a comparatively 'dry' and especially 'ionic' surroundings, existence of water may not be eliminated. If existing in very small quantity, water did not exhibit a disadvantageous influence on the DES-mediated iono-thermal preparation of certain materials. Lesser reactivity of water (in very small quantity) was due to the powerful interaction with the anion present in DES that remarkably restricted its reactivity [95]. Particularly, it signifies that advanced inorganic compounds which are reactive to water could be hypothetically synthesized by the iono-thermal method in DESs. In this respect, three distinct DESs based on ChCl/carboxylic acid were employed for the generation of different alumino-phosphate layered materials having chemical configurations which were improbable to be available employing conventional hydro-thermal methods. It was confirmed that the structure of material attained by iono-thermal preparation was affected by the DES chemical composition (e.g. charge balance, template species, mineralizers etc.).

Zinc phosphates are also a critical subset of the large phosphate family. Influenced by the methods stated for the synthesis of alumino-phosphate-based materials, Liao *et al*. [65] illustrated the iono-thermal generation of a $[Zn(O_3PCH_2CO_2)].NH_4$ framework utilizing a urea/ChCl DES as the medium of reaction. Thorough analysis confirmed that NH^{4+}, produced *in situ* by the urea partial decomposition at greater temperature, performed an important task in the generation of the targeted material.

As already stated, the application of DESs with distinct ionic compositions contribute noticeable benefits as green solvents, and also provides rational design of fresh materials with open frameworks. Similar to the zinc-phosphate and alumino-phosphate, zirconium-phosphates having one-dimensional (1 D), two-dimensional (2 D) and three-dimensional (3 D) connectivities are generally synthesized utilizing a solvo-thermal method in organic solvents or water. Liu *et al.* [96] reported the preparation of three zirconium phosphate structures employing H_3PO_4, HF and zirconium (IV) oxychloride in DESs consisting of tetra-methylammonium chloride (TMA)/oxalic acid dehydrate as well as urea/TMA. Recently, Liu *et al.* [97] also prepared a unique open-structure Zr phosphate using small 7- as well as 8-ring channels (ZrPOF-EA, $[(C_2H_7NH)_8(H_2O)_8]-[Zr_{32}P_{48}O_{176}F_8(OH)_{16}])$ by the iono-thermal preparation in ethylammonium chloride-oxalic acid DES. In recent studies, the ChCl-1,3-dimethyl urea DES has also been stated as an effective solvent for the synthesis of cobalt phosphate-borate which includes the $[B_5O_6(OH)_4]^-$ anion among the layers of cobalt phosphate [98].

Carbon Materials

The carbon materials are widely employed for different applications such as gas storage and separation, electrochemical devices, catalyst supports, etc. The iono-thermal preparation process is considered as an effective method for the synthesis of crystalline porous materials. In this respect, the ILs have been established to be flexible precursors for the synthesis of porous carbon materials [99]. As reported by Gutierrez *et al.* [100], the generation of carbon materials can also be attained in an identical way in the DESs. As an illustration, the polycondensation of resorcinol-formaldehyde (RF) in the DESs for the synthesis of carbon as well as carbon–carbon nanotube (CNT) composites was performed [100]. During the polycondensation of formaldehyde and resorcinol, the hydroxyl-methyl compounds were generated. The hydroxyl-methyl species reacted to generate methylene as well as methylene-ether joined compounds. The authors systematically examined the water impact, catalyst impact and the function of the DES. During the preparation of the material, the RF polycondensation initially took place in a DES ethylene glycol/ChCl. Later, the resultant RF gels underwent carbonization for the generation of monolithic carbons. The RF gels generated in the DESs exhibited identical surface characteristics, which is considered to be an appealing characteristic of the DESs.

The carbon content as well as the conversion rate in the final material was identical to those achieved utilizing the conventional method. Additional appealing characteristic was the capability of DES for dispersing the carbon nanotubes, causing the generation of homogeneous carbon-CNT RF gels. Similar high composite homogeneity is not possible to be achieved in water because of the lower dispersibility of nanotubes in the medium.

Subsequently, Gutierrez *et al.* [56] reported the application of ternary DESs consisting of resorcinol, ChCl and 3-hydroxypyridine as both structure directing agents and precursors for the preparation of nitrogen-doped hierarchic carbon monoliths. The structure of the generated carbons invariably consisted of a bi-continuous porous system of connections composed of immensely cross-linked groups combined into inter-related framework. The DES was illustrated to perform the three distinct roles as (1) a solvent that provides homogeneity of reactants, (2) a source of nitrogen as well as carbon and (3) a structure-directing agent that controls the hierarchical pore framework generation.

In another study, Gutierrez *et al.* [101] illustrated the efficient preparation of hierarchical porous multi-walled CNT (MWCNT) composites in the DESs. The preparation was carried out by means of a furfuryl alcohol (FA) condensation employing the DES ChCl/*p*-toluene sulfonic acid. Further, for attaining a homogeneous dispersion of the CNTs, the DES acted as an excellent medium. The examination of the porosity of the generated carbon-MWCNT composites confirmed that it possessed higher surface areas (400-550 m^2 g^{-1}). In addition, the authors also stated the significant green nature of the recommended procedure as the entire DESs could be used again, because of its complete recovery succeeding the condensation reaction. Thus, it underlined the functional nature of the DES for the generation of advanced materials.

6.4.4 Metal Extraction and Metal Oxides Processing

The metals as well as metal oxides dissolution is essential for a series of extensive procedures like catalyst preparation, remediation of corrosion and metal winning [102]. Processing as well as re-processing act as the source of large quantities of aqueous waste, with treatment of basic as well as acidic byproducts both chemical in nature and energy intensive. In majority of molecular solvents, the metal oxides are insoluble and are usually soluble only in aqueous alkali or acid. The extraction method is dependent on the value of end product, the original matrix and the demanded purity. In accordance with the current standards for environmental conservation, the entire solutes should be separated before the solvent is released to the environment. The four major techniques employed for recovering the metallic solutes from the solution are ion exchange, precipitation, cementation and electro-deposition. For recovering chromium and aluminum, these should be electro-chemically or chemically reduced in higher temperature operations requiring large energy input.

Currently, research is focused towards employing ILs as well as DESs for the solvato-metallurgical processing. Considerable studies have been carried out on the metal oxides' solubility in a range of DESs. It was confirmed that the type III DESs have the ability to solvate a range of metal oxides [103]; ligands like oxalate, urea and thiourea are famous complexants for a series of metals and could be used with DESs. Abbott *et al.* [103] examined the solubility of a series of metal oxides in a DES of ChCl/urea, and noted that using electrodeposition, the dissolved metals could be reclaimed from the mixed metal oxide matrix.

Separation of metals from a complex mixture can be performed employing electro-chemistry [104]. Solubility of 17 metal oxides present in the elemental mass series Ti through to Zn were reported in three DESs based on ChCl [22], with considerate selection of HBD inducing the selectivity for the extraction of some metals from complicated matrices. In aqueous solutions, the redox potentiality could be judiciously shifted by selective choice of a suitable ligand, although, in principle, it can be attained utilizing the same ligand in DESs. Relative strength of metal–anion complex was mostly obscure, however, still remained a critical parameter to quantify. The first commensurable electrochemical series in two DESs was developed recently by Abbott *et al.* [102]. It was proved that the metals could be electro-won into their composite materials by employing the dissimilarity in the metals' redox potentials in a model experiment employing Cu/Zn mixtures. The authors also illustrated the application of iodine as a reversible, stable electro-catalyst used for the dissolution of a range of metals. The redox potential of I_2/I^- in a DES ChCl:ethylene glycol was observed to be more positive, when compared to majority of common metals and was proved to be capable of oxidizing as well as recovering gold.

A hybrid DES of urea, ChCl and ethylene glycol was also reported for use in the large scale extraction of zinc (Zn) and lead (Pb) from the electric arc furnace (EAF) dust [105]. DES exhibited good selectivity in the EAF dust towards lead and zinc oxides only, whereas the aluminum and iron oxides were insoluble. The experiment results indicated that Zn could be latterly electrodeposited with an effectiveness of 75%. Nevertheless, this was not cost-effective for a comparatively cheap metal, hence, the most economically feasible solution was the precipitation of ZnCl using ammonia (NH_3), then filtering off the precipitate, and boiling away the NH_3, thus, leaving behind the DES for reusing. A pilot plant was also developed for testing the productiveness of utilizing the DESs for large scale extraction of metals. For the separation of copper oxides as well as fluorides from post etch residues, the DESs ChCl:urea and ChCl:malonic acid were proved to be effective [51]. Further, for the separation of residues from CF_4/O_2 etching of a Cu coated DUV photoresist, the malonic acid based DESs were observed to be efficient. The DESs have also been reported to be promising candidates for the removal of organic sulfides from fuels [106].

6.5 Summary and Outlook

This review outlines the state of the art on the DES types, properties as well as application in the oil and gas industries ranging from gas absorption and enhanced oil recovery (EOR) to the preparation of materials as well as metal extraction and metal oxides processing. The literature studies revealed majority of the DESs to be based on ChCl, with a few research studies also reporting the generation of DESs from other classes of compounds. The DESs have been revealed to display close physico-chemical characteristics (conductivity, density, viscosity) to those of conventional ILs.

In comparison with the conventional methods of EOR, the DESs are proved to be effective and promising candidates for the EOR process. It was observed that the DESs ChCl:glycerol and ChCl:urea have high potential to be utilized as additives in the thermal EOR processes. The ChCl:glycerol DES exhibited superior comprehensive performance in terms of the *in-situ* heavy oil upgrading. Due to the potential characteristics of DESs as CO_2 and SO_2 absorbents, and primarily as gas separating agents, the information on the physico-chemical properties of DESs with regards to their effect on gas separation mechanisms and gas solubility are also presented in the review. The capturing capabilities and properties of DESs have also been compared with ILs. Overall, DESs have the ability to absorb more CO_2 than the toxic ILs because of their tunable chemical as well as physical properties. Regarding the parameters affecting the solubility of CO_2 and SO_2 in DESs, the literature revealed that at constant pressure, the solubility reduced with enhancing the temperature and increased with increasing the pressure at consistent temperature. The solvent composition as well as the mixture ratio also assists the absorption capacity regardless of the preparation method. The literature studies on gas solubility are restricted to CO_2 and SO_2, but studies on the carbon capture from complex gas mixtures like flue gases is generally non-existent. Further, studies reporting the separation of other gases such as H_2, CO, and methane using DESs are very restricted, and there is a requirement for experimental analyses to generate a database on the nature of DESs in coexistence with these gases.

In the area of materials chemistry, for the iono-thermal preparation of a broad series of materials like MOF, carbon, metal phosphates and phosphites with different frameworks and textures, it is possible that the ILs can be favorably substituted by safe as well as cheap DESs. Despite the fact that an appropriate choice of the DES still continues to be a great challenge, the illustrated studies in the review have clearly revealed that substantial materials, from micro-porous zeo-types to the carbon materials, could be prepared in the DESs. Current application of DESs for the material preparation illustrates the remarkable capability of these solvents for the preparation of novel structures as well as engineered materials. During the preparation process, the DESs might perform multiple roles like solvent, reactant for structure crystallization, structure-directing agent, water inhibitor, etc. Inorganic materials, which are reactive to water, could be effectively synthesized by the iono-thermal method in DESs. Apart from the superior dissolution properties for CO_2, organic molecules and inorganic salts, the DESs could also judiciously dissolve various metal oxides, which signifies capacity for the pure metals recovery, specifically in electro-chemistry.

In conclusion, it can be confirmed that the DESs are an appropriate alternative platform to ILs for technical applications like CO_2 removal, SO_2 removal, EOR and preparation of various materials like MOF, carbon materials as well as metal phosphates and phosphites with open-framework. This is due to the feasibility of tailoring the properties of DES by the proper selection of HBDs, salts and molar ratios. In all the DES-based processes analyzed in this review, it was observed that the usage of DESs not only permitted the design of secure and green processes but also contributed a genuine access to advanced chemicals as well as materials. Even though DESs are not able to currently replace the ILs in entire fields of chemistry, it is envisaged that the attractive price as well as less ecological footprint of DESs will certainly support the industrial development of this advanced medium in a near future.

References

1. Anastas, P. T., and Kirchhoff, M. M. (2002) Origins, current status, and future challenges of green chemistry. *Accounts of Chemical Research*, **35**(9), 686-694.

2. Gani, R., Jimenez-Gonzalez, C., and Constable, D. J. (2005) Method for selection of solvents for promotion of organic reactions. *Computers & Chemical Engineering*, **29**(7), 1661-1676.

3. Clark, J. H., and Tavener, S. J. (2007) Alternative solvents: shades of green. *Organic Process Research & Development*, **11**(1), 149-155.

4. Walden, P. (1914) Molecular weights and electrical conductivity of several fused salts. *Bull Acad Imp Sci*, **1800**, 405-422.

5. Petkovic, M., Seddon, K. R., Rebelo, L. P. N., and Pereira, C. S. (2011) Ionic liquids: a pathway to environmental acceptability. *Chemical Society Reviews*, **40**(3), 1383-1403.

6. Zhang, Q., Vigier, K. D. O., Royer, S., and Jerome, F. (2012) Deep eutectic solvents: syntheses, properties and applications. *Chemical Society Reviews*, **41**(21), 7108-7146.

7. Deetlefs, M., and Seddon, K. R. (2010) Assessing the greenness of some typical laboratory ionic liquid preparations. *Green Chemistry*, **12**(1), 17-30.

8. Romero, A., Santos, A., Tojo, J., and Rodriguez, A. (2008) Toxicity and biodegradability of imidazolium ionic liquids. *Journal of Hazardous Materials*, **151**(1), 268-273.

9. Abo-Hamad, A., Hayyan, M., AlSaadi, M. A., and Hashim, M. A. (2015) Potential applications of deep eutectic solvents in nanotechnology. *Chemical Engineering Journal*, **273**, 551-567.

10. Smith, E. L., Abbott, A. P., and Ryder, K. S. (2014) Deep eutectic solvents (DESs) and their applications. *Chemical Reviews*, **114**(21), 11060-11082.

11. Hadj-Kali, M. K., Al-khidir, K. E., Wazeer, I., El-blidi, L., Mulyono, S., and AlNashef, I. M. (2015) Application of deep eutectic solvents and their individual constituents as surfactants for enhanced oil recovery. *Colloids and Surfaces A: Physicochemical and Engineering Aspects*, **487**, 221-231.

12. Dai, Y., van Spronsen, J., Witkamp, G. J., Verpoorte, R., and Choi, Y. H. (2013) Natural deep eutectic solvents as new potential media for green technology. *Analytica Chimica Acta*, **766**, 61-68.

13. Zhao, B.-Y., Xu, P., Yang, F.-X., Wu, H., Zong, M.-H., and Lou, W.-Y. (2015) Biocompatible deep eutectic solvents based on choline chloride: Characterization and application to the extraction of Rutin from *Sophora japonica*. *ACS Sustainable Chemistry & Engineering*, **3**(11), 2746-2755.

14. Abbott, A. P., Capper, G., Davies, D. L., Rasheed, R. K., and Tambyrajah, V. (2003) Novel solvent properties of choline chloride/urea mixtures. *Chemical Communications*, 70-71.

15. Abbott, A. P., Harris, R. C., Ryder, K. S., D'Agostino, C., Gladden, L. F., and Mantle, M. D. (2011) Glycerol eutectics as sustainable solvent systems. *Green Chemistry*, **13**(1), 82-90.

16. Ru B, C., and Konig, B. (2012) Low melting mixtures in organic synthesis-an alternative to ionic liquids? *Green Chemistry*, **14**(11), 2969-2982.

17. Paiva, A., Craveiro, R., Aroso, I., Martins, M., Reis, R. L., and Duarte, A. R. C. (2014) Natural deep eutectic solvents – Solvents for the 21st century. *ACS Sustainable Chemistry & Engineering*, **2**(5), 1063-1071.

18. Tang, B., and Row, K. H. (2013) Recent developments in deep eutectic solvents in chemical sciences. *Monatshefte für Chemie-Chemical Monthly*, **144**(10), 1427-1454.
19. Maugeri, Z., and de Maria, P. D. (2012) Novel choline-chloride-based deep-eutectic-solvents with renewable hydrogen bond donors: levulinic acid and sugar-based polyols. *RSC Advances*, **2**(2), 421-425.
20. Abbott, A. P., Barron, J. C., Ryder, K. S., and Wilson, D. (2007) Eutectic-based ionic liquids with metal-containing anions and cations. *Chemistry - A European Journal*, **13**(22), 6495-6501.
21. Tang, S., Baker, G. A., and Zhao, H. (2012) Ether-and alcohol-functionalized task-specific ionic liquids: attractive properties and applications. *Chemical Society Reviews*, **41**(10), 4030-4066.
22. Abbott, A. P., Capper, G., Davies, D. L., McKenzie, K. J., and Obi, S. U. (2006) Solubility of metal oxides in deep eutectic solvents based on choline chloride. *Journal of Chemical & Engineering Data*, **51**(4), 1280-1282.
23. Chakrabarti, M. H., Mjalli, F. S., AlNashef, I. M., Hashim, M. A., Hussain, M. A., Bahadori, L., and Low, C. T. J. (2014) Prospects of applying ionic liquids and deep eutectic solvents for renewable energy storage by means of redox flow batteries. *Renewable and Sustainable Energy Reviews*, **30**, 254-270.
24. Abbott, A. P., Boothby, D., Capper, G., Davies, D. L., and Rasheed, R. K. (2004) Deep eutectic solvents formed between choline chloride and carboxylic acids: versatile alternatives to ionic liquids. *Journal of the American Chemical Society*, **126**(29), 9142-9147.
25. Kareem, M. A., Mjalli, F. S., Hashim, M. A., and AlNashef, I. M. (2010) Phosphonium-based ionic liquids analogues and their physical properties. *Journal of Chemical & Engineering Data*, **55**(11), 4632-4637.
26. Abbott, A. P., Capper, G., and Gray, S. (2006) Design of improved deep eutectic solvents using hole theory. *ChemPhysChem*, **7**(4), 803-806.
27. D'Agostino, C., Harris, R. C., Abbott, A. P., Gladden, L. F., and Mantle, M. D. (2011) Molecular motion and ion diffusion in choline chloride based deep eutectic solvents studied by 1 H pulsed field gradient NMR spectroscopy. *Physical Chemistry Chemical Physics*, **13**(48), 21383-21391.
28. Abbott, A. P., Harris, R. C., and Ryder, K. S. (2007) Application of hole theory to define ionic liquids by their transport properties. *The Journal of Physical Chemistry B*, **111**(18), 4910-4913.
29. Yadav, A., and Pandey, S. (2014) Densities and viscosities of (choline chloride+ urea) deep eutectic solvent and its aqueous mixtures in the temperature range 293.15 K to 363.15 K. *Journal of Chemical & Engineering Data*, **59**(7), 2221-2229.
30. Franca, J. M., Nieto de Castro, C. A., Lopes, M. M., and Nunes, V. M. (2009) Influence of thermophysical properties of ionic liquids in chemical process design. *Journal of Chemical & Engineering Data*, **54**(9), 2569-2575.
31. Valderrama, J. O. (2003) The state of the cubic equations of state. *Industrial & Engineering Chemistry Research*, **42**(8), 1603-1618.
32. Shahbaz, K., Mjalli, F. S., Hashim, M., and AlNashef, I. M. (2011) Prediction of deep eutectic solvents densities at different temperatures. *Thermochimica Acta*, **515**(1), 67-72.

33. Yadav, A., Trivedi, S., Rai, R., and Pandey, S. (2014) Densities and dynamic viscosities of (choline chloride+ glycerol) deep eutectic solvent and its aqueous mixtures in the temperature range (283.15–363.15) K. *Fluid Phase Equilibria*, **367**, 135-142.
34. Leron, R. B., and Li, M. H. (2012) High-pressure density measurements for choline chloride: Urea deep eutectic solvent and its aqueous mixtures at T=(298.15 to 323.15) K and up to 50MPa. *The Journal of Chemical Thermodynamics*, **54**, 293-301.
35. Abbott, A. P., Ahmed, E. I., Harris, R. C., and Ryder, K. S. (2014) Evaluating water miscible deep eutectic solvents (DESs) and ionic liquids as potential lubricants. *Green Chemistry*, **16**(9), 4156-4161.
36. Mirza, N. R., Nicholas, N. J., Wu, Y., Kentish, S., and Stevens, G. W. (2015) Estimation of normal boiling temperatures, critical properties, and acentric factors of deep eutectic solvents. *Journal of Chemical & Engineering Data*, **60**(6), 1844-1854.
37. Florindo, C., Oliveira, F. S., Rebelo, L. P. N., Fernandes, A. M., and Marrucho, I. M. (2014) Insights into the synthesis and properties of deep eutectic solvents based on cholinium chloride and carboxylic acids. *ACS Sustainable Chemistry & Engineering*, **2**(10), 2416-2425.
38. Hou, Y., Gu, Y., Zhang, S., Yang, F., Ding, H., and Shan, Y. (2008) Novel binary eutectic mixtures based on imidazole. *Journal of Molecular Liquids*, **143**(2), 154-159.
39. Hurley, F. H., and Wler, T. P., Jr. (1951) The electrodeposition of aluminum from nonaqueous solutions at room temperature. *Journal of the Electrochemical Society*, **98**(5), 207-212.
40. Xu, W. G., Lu, X. M., Zhang, Q. G., Gui, J. S., and Yang, J. Z. (2006) Studies on the thermodynamic properties of the ionic liquid BMIGaCl4. *Chinese Journal of Chemistry*, **24**(3), 331-335.
41. Yang, J. Z., Tian, P., He, L. L., and Xu, W. G. (2003) Studies on room temperature ionic liquid InCl 3–EMIC. *Fluid Phase Equilibria*, **204**(2), 295-302.
42. Abbott, A. P., Capper, G., Davies, D. L., and Rasheed, R. (2004) Ionic liquids based upon metal halide/substituted quaternary ammonium salt mixtures. *Inorganic Chemistry*, **43**(11), 3447-3452.
43. Abbott, A. P., Capper, G., Davies, D. L., Munro, H. L., Rasheed, R. K., and Tambyrajah, V. (2001) Preparation of novel, moisture-stable, Lewis-acidic ionic liquids containing quaternary ammonium salts with functional side chains. *Chemical Communications*, 2010-2011.
44. Mbous, Y. P., Hayyan, M., Hayyan, A., Wong, W. F., Hashim, M. A., and Looi, C. Y. (2017) Applications of deep eutectic solvents in biotechnology and bioengineering - Promises and challenges. *Biotechnology Advances*, **35**(2), 105-134.
45. Abbott, A. P., Capper, G., Davies, D. L., and Rasheed, R. K. (2004) Ionic liquid analogues formed from hydrated metal salts. *Chemistry - A European Journal*, **10**(15), 3769-3774.
46. Abbott, A. P., Capper, G., Davies, D. L., Rasheed, R. K., Archer, J., and John, C. (2004) Electrodeposition of chromium black from ionic liquids. *Transactions of the Institute of Metal Finishing*, **82**, 14-17.
47. Abbott, A. P., Cullis, P. M., Gibson, M. J., Harris, R. C., and Raven, E. (2007) Extraction of glycerol from biodiesel into a eutectic based ionic liquid. *Green Chemistry*, **9**(8), 868-872.

48. Abbott, A. P., Bell, T. J., Handa, S., and Stoddart, B. (2006) Cationic functionalisation of cellulose using a choline based ionic liquid analogue. *Green Chemistry*, **8**(9), 784-786.

49. Gambino, M., Gaune, P., Nabavian, M., Gaune-Escard, M., and Bros, J. P. (1987) Enthalpie de fusion de l'uree et de quelques melanges eutectiques a base d'uree. *Thermochimica Acta*, **111**, 37-47.

50. Gambino, M., and Bros, J. P. (1988) Capacite calorifique de l'uree et de quelques melanges eutectiques a base d'uree entre 30 et 140 C. *Thermochimica Acta*, **127**, 223-236.

51. Taubert, J., and Raghavan, S. (2014) Effect of composition of post etch residues (PER) on their removal in choline chloride–malonic acid deep eutectic solvent (DES) system. *Microelectronic Engineering*, **114**, 141-147.

52. Smiglak, M., Reichert, W. M., Holbrey, J. D., Wilkes, J. S., Sun, L., Thrasher, J. S., Kirichenko, K., Singh, S., Katritzky, A. R., and Rogers, R. D. (2006) Combustible ionic liquids by design: is laboratory safety another ionic liquid myth? *Chemical Communications*, 2554-2556.

53. Biczak, R., Pawłowska, B., Bałczewski, P., and Rychter, P. (2014) The role of the anion in the toxicity of imidazolium ionic liquids. *Journal of Hazardous Materials*, **274**, 181-190.

54. Pham, T. P. T., Cho, C. W., and Yun, Y. S. (2010). Environmental fate and toxicity of ionic liquids: a review. *Water Research*, **44**(2), 352-372.

55. Carriazo, D., Gutiérrez, M. C., Ferrer, M. L., and del Monte, F. (2010) Resorcinol-based deep eutectic solvents as both carbonaceous precursors and templating agents in the synthesis of hierarchical porous carbon monoliths. *Chemistry of Materials*, **22**(22), 6146-6152.

56. Gutierrez, M. C., Carriazo, D., Ania, C. O., Parra, J. B., Ferrer, M. L., and del Monte, F. (2011) Deep eutectic solvents as both precursors and structure directing agents in the synthesis of nitrogen doped hierarchical carbons highly suitable for CO_2 capture. *Energy & Environmental Science*, **4**(9), 3535-3544.

57. Zhu, A., Jiang, T., Han, B., Zhang, J., Xie, Y., and Ma, X. (2007) Supported choline chloride/urea as a heterogeneous catalyst for chemical fixation of carbon dioxide to cyclic carbonates. *Green Chemistry*, **9**(2), 169-172.

58. Hao, G. P., Jin, Z. Y., Sun, Q., Zhang, X. Q., Zhang, J. T., and Lu, A. H. (2013) Porous carbon nanosheets with precisely tunable thickness and selective CO_2 adsorption properties. *Energy & Environmental Science*, **6**(12), 3740-3747.

59. Leron, R. B., and Li, M. H. (2013) Solubility of carbon dioxide in a eutectic mixture of choline chloride and glycerol at moderate pressures. *The Journal of Chemical Thermodynamics*, **57**, 131-136.

60. Leron, R. B., and Li, M. H. (2013) Solubility of carbon dioxide in a choline chloride–ethylene glycol based deep eutectic solvent. *Thermochimica Acta*, **551**, 14-19.

61. Lin, C. M., Leron, R. B., Caparanga, A. R., and Li, M. H. (2014) Henry's constant of carbon dioxide-aqueous deep eutectic solvent (choline chloride/ethylene glycol, choline chloride/glycerol, choline chloride/malonic acid) systems. *The Journal of Chemical Thermodynamics*, **68**, 216-220.

62. Leron, R. B., Caparanga, A., and Li, M. H. (2013). Carbon dioxide solubility in a deep eutectic solvent based on choline chloride and urea at T= 303.15–343.15 K and

moderate pressures. *Journal of the Taiwan Institute of Chemical Engineers*, **44**(6), 879-885.

63. Ali, E., Hadj-Kali, M. K., Mulyono, S., Alnashef, I., Fakeeha, A., Mjalli, F., and Hayyan, A. (2014) Solubility of CO 2 in deep eutectic solvents: experiments and modelling using the Peng–Robinson equation of state. *Chemical Engineering Research and Design*, **92**(10), 1898-1906.

64. Cooper, E. R., Andrews, C. D., Wheatley, P. S., Webb, P. B., Wormald, P., and Morris, R. E. (2004) Ionic liquids and eutectic mixtures as solvent and template in synthesis of zeolite analogues. *Nature*, **430**(7003), 1012-1016.

65. Liao, J. H., Wu, P. C., and Bai, Y. H. (2005) Eutectic mixture of choline chloride/urea as a green solvent in synthesis of a coordination polymer: [Zn (O$_3$PCH$_2$CO$_2$)]·NH$_4$. *Inorganic Chemistry Communications*, **8**(4), 390-392.

66. Li, X., Hou, M., Han, B., Wang, X., and Zou, L. (2008) Solubility of CO$_2$ in a choline chloride+ urea eutectic mixture. *Journal of Chemical & Engineering Data*, **53**(2), 548-550.

67. Sze, L. L., Pandey, S., Ravula, S., Pandey, S., Zhao, H., Baker, G. A., and Baker, S. N. (2014) Ternary deep eutectic solvents tasked for carbon dioxide capture. *ACS Sustainable Chemistry & Engineering*, **2**(9), 2117-2123.

68. Francisco, M., van den Bruinhorst, A., Zubeir, L. F., Peters, C. J., and Kroon, M. C. (2013) A new low transition temperature mixture (LTTM) formed by choline chloride+ lactic acid: characterization as solvent for CO$_2$ capture. *Fluid Phase Equilibria*, **340**, 77-84.

69. Gutierrez Ortiz, F. J., Vidal, F., Ollero, P., Salvador, L., Cortes, V., and Gimenez, A. (2006) Pilot-plant technical assessment of wet flue gas desulfurization using limestone. *Industrial & Engineering Chemistry Research*, **45**(4), 1466-1477.

70. Yang, D., Hou, M., Ning, H., Zhang, J., Ma, J., Yang, G., and Han, B. (2013) Efficient SO$_2$ absorption by renewable choline chloride–glycerol deep eutectic solvents. *Green Chemistry*, **15**(8), 2261-2265.

71. Sun, S., Niu, Y., Xu, Q., Sun, Z., and Wei, X. (2015) Efficient SO$_2$ absorptions by four kinds of deep eutectic solvents based on choline chloride. *Industrial & Engineering Chemistry Research*, **54**(33), 8019-8024.

72. Liu, B., Zhao, J., and Wei, F. (2013) Characterization of caprolactam based eutectic ionic liquids and their application in SO$_2$ absorption. *Journal of Molecular Liquids*, **180**, 19-25.

73. Zhang, K., Ren, S., Hou, Y., and Wu, W. (2017) Efficient absorption of SO$_2$ with low-partial pressures by environmentally benign functional deep eutectic solvents. *Journal of Hazardous Materials*, **324**, 457-463.

74. Guo, B., Duan, E., Ren, A., Wang, Y., and Liu, H. (2009) Solubility of SO$_2$ in caprolactam tetrabutyl ammonium bromide ionic liquids. *Journal of Chemical & Engineering Data*, **55**(3), 1398-1401.

75. Duan, E., Guo, B., Zhang, M., Guan, Y., Sun, H., and Han, J. (2011) Efficient capture of SO$_2$ by a binary mixture of caprolactam tetrabutyl ammonium bromide ionic liquid and water. *Journal of Hazardous Materials*, **194**, 48-52.

76. Benzagouta, M. S., AlNashef, I. M., Karnanda, W., and Al-Khidir, K. (2013) Ionic liquids as novel surfactants for potential use in enhanced oil recovery. *Korean Journal of Chemical Engineering*, **30**(11), 2108-2117.

77. Hezave, A. Z., Dorostkar, S., Ayatollahi, S., Nabipour, M., and Hemmateenejad, B. (2013) Dynamic interfacial tension behavior between heavy crude oil and ionic liquid solution (1-dodecyl-3-methylimidazolium chloride ([C 12 mim][Cl]+ distilled or saline water/heavy crude oil)) as a new surfactant. *Journal of Molecular Liquids*, **187**, 83-89.

78. Xie, X., Weiss, W. W., Tong, Z. J., and Morrow, N. R. (2005) Improved oil recovery from carbonate reservoirs by chemical stimulation. *SPE Journal*, **10**(3), 276-285.

79. Mohsenzadeh, A., Nabipour, M., Asadizadeh, S., Nekouie, M., Ameri, A., and Ayatollahi, S. (2011) Experimental investigation of different steam injection scenarios during SAGD process. *Special Topics & Reviews in Porous Media: An International Journal*, **2**(4), 283-291.

80. Canbolat, S., Akin, S., and Kovscek, A. R. (2004) Non-condensable gas steam-assisted gravity drainage. *Journal of Petroleum Science and Engineering*, **45**(1), 83-96.

81. Nares, R., Schacht-Hernandez, P., Ramirez-Garnica, M. A., and Cabrera-Reyes, M. D. C. (2007) Upgrading Heavy and Extra Heavy Crude Oil with Ionic Liquid. *International Oil Conference and Exhibition*, Mexico. Online: https://www.onepetro.org/conference-paper/SPE-108676-MS [assessed 29th April 2017].

82. Mohsenzadeh, A., Al-Wahaibi, Y., Jibril, A., Al-Hajri, R., and Shuwa, S. (2015) The novel use of deep eutectic solvents for enhancing heavy oil recovery. *Journal of Petroleum Science and Engineering*, **130**, 6-15.

83. Mohsenzadeh, A., Al-Wahaibi, Y., Al-Hajri, R., and Jibril, B. (2015) Effects of concentration, salinity and injection scenario of ionic liquids analogue in heavy oil recovery enhancement. *Journal of Petroleum Science and Engineering*, **133**, 114-122.

84. Mohsenzadeh, A., Al-Wahaibi, Y., Al-Hajri, R., Jibril, B., Joshi, S., and Pracejus, B. (2015). Investigation of formation damage by Deep Eutectic Solvents as new EOR agents. *Journal of Petroleum Science and Engineering*, **129**, 130-136.

85. Al-Weheibi, I., Al-Hajri, R., Al-Wahaibi, Y., Jibril, B., and Mohsenzadeh, A. (2015). Oil Recovery Enhancement in Middle East Heavy Oil Field using Malonic Acid based Deep Eutectic Solvent. *SPE Middle East Oil & Gas Show and Conference*, Bahrain. Online: https://www.onepetro.org/conference-paper/SPE-172592-MS [assessed 20th April 2017].

86. Shuwa, S., Jibril, B., Al-Wahaibi, Y., and Al-Hajri, R. (2014) Heavy-oil-recovery enhancement with choline chloride/ethylene glycol-based deep eutectic solvent. *SPE Journal*, **20**(01), 79-87.

87. Mohsenzadeh, A., Al-Wahaibi, Y., Al-Hajri, R., Jibril, B., and Mosavat, N. (2017) Sequential deep eutectic solvent and steam injection for enhanced heavy oil recovery and in-situ upgrading. *Fuel*, **187**, 417-428.

88. Ramsahye, N. A., Maurin, G., Bourrelly, S., Llewellyn, P. L., Serre, C., Loiseau, T., Serre, C., and Ferey, G. (2008) Probing the adsorption sites for CO_2 in metal organic frameworks materials MIL-53 (Al, Cr) and MIL-47 (V) by density functional theory. *The Journal of Physical Chemistry C*, **112**(2), 514-520.

89. Trung, T. K., Trens, P., Tanchoux, N., Bourrelly, S., Llewellyn, P. L., Loera-Serna, S., Serre, C., Loiseau, T., Fajula, F., & Ferey, G. (2008) Hydrocarbon adsorption in the flexible metal organic frameworks MIL-53 (Al, Cr). *Journal of the American Chemical Society*, **130**(50), 16926-16932.

90. Zhang, J., Wu, T., Chen, S., Feng, P., and Bu, X. (2009) Versatile structure-directing roles of deep-eutectic solvents and their implication in the generation of porosity and open metal sites for gas storage. *Angewandte Chemie International Edition*, **48**(19), 3486-3490.

91. Himeur, F., Stein, I., Wragg, D. S., Slawin, A. M., Lightfoot, P., and Morris, R. E. (2010) The ionothermal synthesis of metal organic frameworks, Ln $(C_9O_6H_3)((CH_3NH)_2CO)_2$, using deep eutectic solvents. *Solid State Sciences*, **12**(4), 418-421.

92. Zhan, C. H., Wang, F., Kang, Y., and Zhang, J. (2011) Lanthanide-thiophene-2,5-dicarboxylate frameworks: ionothermal synthesis, helical structures, photoluminescent properties, and single-crystal-to-single-crystal guest exchange. *Inorganic Chemistry*, **51**(1), 523-530.

93. Chen, S., Zhang, J., Wu, T., Feng, P., and Bu, X. (2010) Zinc (II)-boron (III)-imidazolate framework (ZBIF) with unusual pentagonal channels prepared from deep eutectic solvent. *Dalton Transactions*, **39**(3), 697-699.

94. Parnham, E. R., Drylie, E. A., Wheatley, P. S., Slawin, A. M., and Morris, R. E. (2006) Ionothermal materials synthesis using unstable deep-eutectic solvents as template-delivery agents. *Angewandte Chemie*, **118**(30), 5084-5088.

95. Cammarata, L., Kazarian, S. G., Salter, P. A., and Welton, T. (2001) Molecular states of water in room temperature ionic liquids. *Physical Chemistry Chemical Physics*, **3**(23), 5192-5200.

96. Liu, L., Li, Y., Wei, H., Dong, M., Wang, J., Slawin, A. M., Li, J., Dong, J., and Morris, R. E. (2009) Ionothermal synthesis of zirconium phosphates and their catalytic behavior in the selective oxidation of cyclohexane. *Angewandte Chemie, International Edition*, **48**(12), 2206-2209.

97. Liu, L., Yang, J., Li, J., Dong, J., Sisak, D., Luzzatto, M., and McCusker, L. (2011) Ionothermal synthesis and structure analysis of an open-framework zirconium phosphate with a high CO_2/CH_4 adsorption ratio. *Angewandte Chemie International Edition*, **50**(35), 8139-8142.

98. Lin, Z., Wragg, D. S., Lightfoot, P., and Morris, R. E. (2009) A novel non-centrosymmetric metallophosphate-borate compound via ionothermal synthesis. *Dalton Transactions*, 5287-5289.

99. Paraknowitsch, J. P., Zhang, J., Su, D., Thomas, A., and Antonietti, M. (2010) Ionic liquids as precursors for nitrogen-doped graphitic carbon. *Advanced Materials*, **22**(1), 87-92.

100. Gutierrez, M. C., Rubio, F., and del Monte, F. (2010) Resorcinol-formaldehyde polycondensation in deep eutectic solvents for the preparation of carbons and carbon–carbon nanotube composites. *Chemistry of Materials*, **22**(9), 2711-2719.

101. Gutierrez, M. C., Carriazo, D., Tamayo, A., Jimenez, R., Pice, F., Rojo, J. M., Ferrer, M. L., and del Monte, F. (2011) Deep-eutectic-solvent-assisted synthesis of hierarchical carbon electrodes exhibiting capacitance retention at high current densities. *Chemistry - A European Journal*, **17**(38), 10533-10537.

102. Abbott, A. P., Frisch, G., Hartley, J., and Ryder, K. S. (2011) Processing of metals and metal oxides using ionic liquids. *Green Chemistry*, **13**(3), 471-481.

103. Abbott, A. P., Capper, G., Davies, D. L., Rasheed, R. K., and Shikotra, P. (2005)

Selective extraction of metals from mixed oxide matrixes using choline-based ionic liquids. *Inorganic Chemistry*, **44**(19), 6497-6499.

104. Abbott, A. P., Capper, G., Davies, D. L., and Shikotra, P. (2006) Processing metal oxides using ionic liquids. *Mineral Processing and Extractive Metallurgy*, **115**(1), 15-18.

105. Abbott, A. P., Collins, J., Dalrymple, I., Harris, R. C., Mistry, R., Qiu, F., Scheirer, J., and Wise, W. R. (2009) Processing of electric arc furnace dust using deep eutectic solvents. *Australian Journal of Chemistry*, **62**(4), 341-347.

106. Li, C., Li, D., Zou, S., Li, Z., Yin, J., Wang, A. L., Cui, Y. N., Yao, Z. L., and Zhao, Q. (2013) Extraction desulfurization process of fuels with ammonium-based deep eutectic solvents. *Green Chemistry*, **15**(10), 2793-2799.

7

Graphene for Corrosion Protection

7.1 Introduction

Corrosion is an electrochemical process having deterioration effect on the metal or alloy. For iron, corrosion produces porous and pervious film which is composed of different forms of iron oxide. It can be seen from Figure 7.1 that redox reactions occur on the surface during this process. The sodium and chloride ions act as electrolyte, where chloride ions accelerate the corrosion process by destroying any type of passivity, thus, increasing the active corrosion rate. In this case, the accelerating corrosion process involves the dissolution of iron oxide film and, in addition, sodium and chloride ions also enhance the transportation of electrons [1,2].

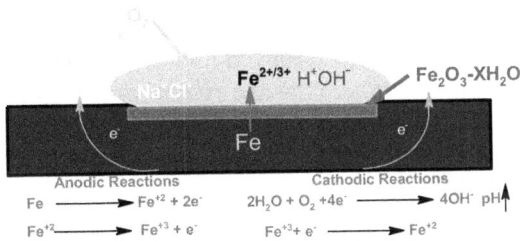

Figure 7.1 Schematic diagram of iron corrosion process; production, and consumption of electrons resulting in corrosion products.

The environmental constraints on using chromium (VI) based coatings promoted the development of non-hazardous organic and inorganic anti-corrosion pigments incorporated in polymer coatings [3]. Graphene sheets are one-atom-thick two-dimensional layers of sp^2-bonded carbon having a variety of remarkable properties and can enhance properties of polymers such as electrical and thermal conductivity, gas impermeability and mechanical properties, etc. [4,5]. Recently, polystyrene/graphene nanocomposites showed superior

Ali U Chaudhry[a,b,], Brajendra Mishra[b,c] and Vikas Mittal[a,**]*
[a]The Petroleum Institute (part of Khalifa University of Science and Technology), Abu Dhabi, UAE; [b]Colorado School of Mines, USA; [c]Worcester Polytechnic Institute, USA
[*]Current address: Texas A&M University, Qatar; [**]Current address: Bletchington, Wellington County, Australia

anti-corrosion properties with the incorporation of 2 wt% of filler owing to excellent barrier properties [6]. Similar results were shown for the composites of silane modified reduced graphene oxide (r-GO)/polyvinylbutyral (PVB) [7] and graphene/pernigraniline/PVB [2,8]. In the same manner, many studies have shown the single time barrier properties of stand-alone graphene films on the surface of aluminum, where excellent protection was shown after 0.5 hours of immersion in chloride environment [9]. Likewise, Raman *et al.* [5,10] measured the anti-corrosion properties of graphene film on copper after one hour of immersion. Further, the barrier properties of composites depend on many factors such as I) nanoscale level dispersion and distribution of fillers, II) interfacial compatibility of polymer and filler phases, and III) polarity match between the filler surface and the polymer chains. The full advantage of nano-fillers can only be achieved by considering above factors which could lead to uniform transfer of chemical, physical and mechanical properties of filler to the host polymer matrix [11-21]. Further, the role of conducting polymers, especially polyaniline (PANI), for corrosion protection of ferrous and non-ferrous metals has been vigorously studied [15]. For this purpose, PANI has been reported to provide corrosion protection of metals either as a neat film or as resin blended coatings [14]. Modification of inorganic pigments with PANI to achieve synergistic anti-corrosion effect has also been reported in literature [14]. Kalendova *et al.* [22] proposed a method to combine the use of inorganic pigments and PANI so as to address the problems associated with resin blended coatings i.e. inefficient PANI distribution, lack of excellent polymer-polymer contact, poor substrate adhesion and change in volume of PANI due to redox reaction. Four pigments specularite (Fe_2O_3), goethite ($FeO(OH)$), talc ($Mg_3(OH)_2$-(Si_4O_{10})) and graphite were surface modified with PANI and subsequently blended with an epoxy binder. Better corrosion resistance was observed in all PANI coated pigments and PANI modified graphite exhibited excellent corrosion inhibition due to improvement in conductivity which promoted redox reactions between iron and PANI or oxygen and PANI. Sathiyanarayanan *et al.* [11,23] modified TiO_2 and Fe_2O_3 with PANI and observed enhanced corrosion protection of steel as compared to pigments without polymer modification. The results were attributed towards the formation of the passive film along with the iron-phosphate salt film on the iron surface. Brodinova *et al.* [24] also reported the presence of PANI filled the pinholes present in the coating and also formed a better interconnection between inorganic pigments and resin. Similarly, Wu *et al.* [25] reported hybrid coating of PANI-layered zinc nickel ferrites and organically modified silicate. The film was deposited using the spin coating method on aluminum alloy. The anti-corrosion performance of the hybrid film was improved due to the denser configuration of organically modified silicate due to the incorporation of nickel zinc ferrites/PANI.

The work presented here deals with the investigation of graphene as anti-corrosion filler for polymer coatings that can be used as a replacement of chromates and other hazardous pigments. Four types of studies have been selected depending on their protection mechanism generated through use of graphene in corrosive solution as well as unmodified graphene and functionalized graphene in polymer coatings. For solution properties, graphene oxide was observed to have no direct effect on the corrosion properties of steel. The effect of unmodified graphene concentration in polymer coatings was also analyzed. Graphene increased the anti-corrosion abilities of coatings when used at higher concentration, but for the shorter periods of time. Further, the effect of the modified graphene nanoplatelets was also studied, which exhibited better anti-corrosion properties of polymer coatings. To impart the electrochemical properties to the graphene, the modification was performed using polyaniline. The long time immersion exhibited that polyaniline-modified graphene had enhanced anti-corrosion properties in coatings owing to the synergistic effect [26].

7.2 Effect of Graphene on Electrochemical Properties of Carbon Steel in a Saline Media

In this work, the electrochemical properties of steel in the presence of suspended nano-graphene oxide in saline media have been studied [2]. The effect of the GO concentration (0–15 ppm) on the electrochemical properties of steel has been evaluated using electrochemical impedance spectroscopy (EIS), in addition to the morphological characterization using microscopy.

7.2.1 Materials and Methods

Nano-graphene oxide aqueous solution (concentration 1g/L, pH 2.9, purity >99%) was purchased from Graphene Supermarket, USA and used as received. The industrial steel used in this study was cut from a pipeline. API-5L X80 steel coupons were machined to $10 \times 10 \times 4$ mm dimensions and a tap and drill hole of 3-48 tpi (threads per inch) was drilled to one long side of the coupon. Machined carbon steel was used as the working electrode and the exposed surface area was 3.4 cm^2. The specimens were surface finished using different grades of SiC grit papers (up to 240 grit) to ensure the same surface roughness [27,28], followed by cleaning and degreasing with industrial grade acetone and ethanol followed by drying in air. To evaluate the protection behavior of nano-GO, solution was prepared in 3.5 wt% NaCl with varying concentration of GO, i.e., 0-15 ppm.

A three-electrode cell assembly consisting of steel coupon as the working electrode (WE), graphite as the counter electrode (CE) and a saturated calomel electrode (SCE) as reference electrode (RE) were used for the electrochemical measurement. Electrochemical testing was performed in a closed system under naturally aerated conditions using a Gamry 600 potentiostat at room temperature. Impedance measurements were performed as a function of open circuit potential after five hours from the time of immersion. The frequency sweep was performed from 10^5 to 10^{-2} Hz at 10 mV AC amplitude. To simulate the electrochemical interface, EIS data was analyzed with Echem analyst using circuit model (Figure 7.2) having electrical equivalent parameters, where R_{ct} is the charge transfer resistance, L is the inductor and CPE is the constant phase element. Accordingly, the impedance can be represented by the following equation:

$$Z(CPE) = Y_0^{-1}[j\omega]^{-n}$$

where Y_0 is the CPE constant, ω is the angular frequency (rad/s), $j=\sqrt{-1}$ and n is another CPE constant that varies from 1 to 0 for pure capacitance and pure resistor, respectively. The double layer capacitance C_{dl} was calculated using the following equation:

$$C_{dl} = Y_0 \left[jw'' \right]^{n-1}$$

where ω'' is the frequency found at the maximum of the imaginary part of the impedance, Z".

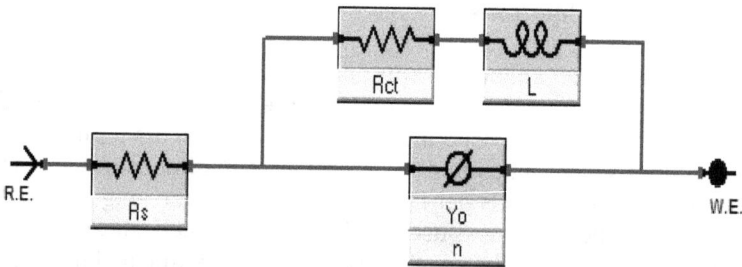

Figure 7.2 A representative circuit model used to model the electrochemical impedance spectroscopy [2].

Steel coupons were carefully disengaged from the cell assembly, dried and observed under the microscope. Field emission scanning electron microscopy (FE-SEM) using JEOL JSM-7000F was performed to evaluate surface morphology. Energy dispersive spectroscopy (EDS) was examined at 5kV under high vacuum at a working distance of 10 mm for elemental composition of the corrosion products.

7.2.2 Results and Discussion

The behavior of electrochemical processes at electrical double layer such as charge transfer resistance and ions adsorption across the electrode/electrolyte interface can be found in Table 7.1 and Figure 7.3. The figure and table depict the typical set of Nyquist plots and modeled EIS plots (Bode and Nyquist) with model circuit where charge transfer resistance was calculated from the diameter of the real part of the semicircle. It can be seen that charge transfer resistance increased with the GO concentration in the solution. It can also be observed from Figure 7.3 that profile of the Nyquist plots remained similar as the concentration of GO increased which indicated that no effect on the corrosion

Table 7.1 Electrochemical impedance spectroscopy parameters in 3.5 wt% after five hours immersion [2]

Conc. ppm	R_{ct} (Ωcm^2)$\times 10^3$	Y_o ($\Omega^{-1}s^n$) $\times 10^{-03}$	n	C_{dl} ($F cm^2$) $\times 10^{-04}$	L ($H cm^2$) $\times 10^{03}$
Blank	1.27	0.72	0.800	6.57	2.9
3	1.50	0.68	0.776	5.86	5.2
9	1.86	1.28	0.773	9.77	8.2
12	2.41	1.50	0.711	9.97	9.8
15	3.27	1.20	0.727	7.63	18.5

mechanism of carbon steel occurred with the addition of GO. Further, the double layer capacitance also increased with the GO concentration and in some cases, it was more than the blank solution, i.e., 9 and 12 ppm. Similarly, the value of n was also observed to decrease as the GO concentration increased, which was also a measure of the surface inhomogeneity; the lower is its value, the higher is the surface roughening of the metal/alloy [29]. Moreover, Y_o value also increased with the increased concentration of GO; the higher Y_o value shows that more surface area is available for the electrochemical reaction [30] due the presence of Cl⁻ ions which increases the film free area [31]. In case of blank solution, a porous corrosion product iron oxide was formed which increased the C_{dl} owing to dielectric effect as given by following equation:

$$C = \frac{\varepsilon\varepsilon_0 r}{d}$$

where d is charge separation distance, ε is relative dielectric constant, ε_0 is permittivity of free space, A is surface area and C is capacitance. The dissolution of the pervious layer is performed by the chloride ions, however, on the addition of GO to the solution, these act as an anionic surfactant which decreases the

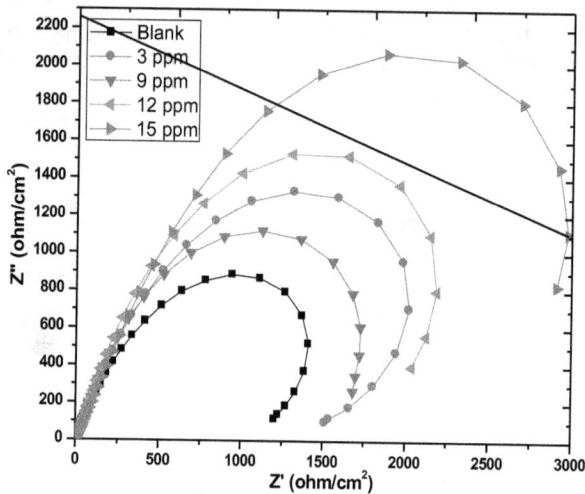

Figure 7.3 Nyquist plots recorded after 5 hour immersion in 3.5 wt% NaCl with blank and different concentrations of GO [2].

solubility of NaCl in solution [32] and hence leads to the precipitation of salt on the working electrode. The precipitation creates porous and inhomogeneous layers which allow the availability of the corrosive solution to the working electrode. However, the precipitation appears to render more or less corrosion protection to the metal below by impeding the transportation of reactants and products among the solution and the metal [33], which results in the increment of charge transfer resistance efficiency up to 70%. Further, there was no second arc seen in the Nyquist plots which indicated that the layer forming on the surface was porous [34,35]. Nyquist plots also showed the inductance loop in the intermediate and low-frequency domain which was mainly ascribed to the occurrence of an adsorbed intermediate on the surface due to chloride ion adsorption on the electrode surface [31]. The total impedance at intermediate and low frequencies was calculated from charge transfer resistance and inductive element in series. The inductive behavior due to adsorption can be defined as $L=R\tau$, where τ represents the relaxation time for adsorbed species at the working electrode [36]. This can be manifested from Table 7.1 that inductance increased with increasing concentration of GO, thus, exhibiting increased absorption of salt at the working surface.

The reactions occurring at anodic and cathodic curves in 3.5 wt% NaCl are given as [37]:

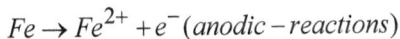

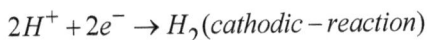

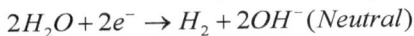

$$Fe \rightarrow Fe^{2+} + e^- (anodic-reactions)$$
$$Fe^{2+} \rightarrow Fe^{3+} + e^-$$
$$2H^+ + 2e^- \rightarrow H_2 (cathodic-reaction)$$
$$2H_2O + 2e^- \rightarrow H_2 + 2OH^- (Neutral)$$

where the formation of a compact and thick layer of precipitates slows down the anodic and cathodic reactions. As a result, decrease in the corrosion current can be seen at higher applied potentials. Figure 7.4 shows the morphology of the outermost corrosion products layer where the surface morphology became more compact on increasing GO concentration. This phenomenon indicated that the presence of GO fostered the formation of NaCl layer on the surface which slowed down the corrosion process. The element analysis for the layers for 9 ppm, 12 ppm and 15 ppm of GO also exhibited that as the amount of added GO increased, NaCl layer became richer and iron oxide reduced.

7.3 Effect of Unmodified Graphene Platelets (UGP) on Electrochemical Properties of Polymer Coatings

In this study, the time-dependent anti-corrosion properties of UGP based PVB composite coating on carbon steel in 0.1M NaCl aqueous solution were measured [5]. Carbon steel samples were coated with thin film of self-crosslinked composites of PVB and UGP. The corrosion properties were measured using electrochemical techniques after an immersion time of 1 and 26 h to differentiate between the corrosion barrier and corrosion promoting phenomena associated with graphene-based composites coatings.

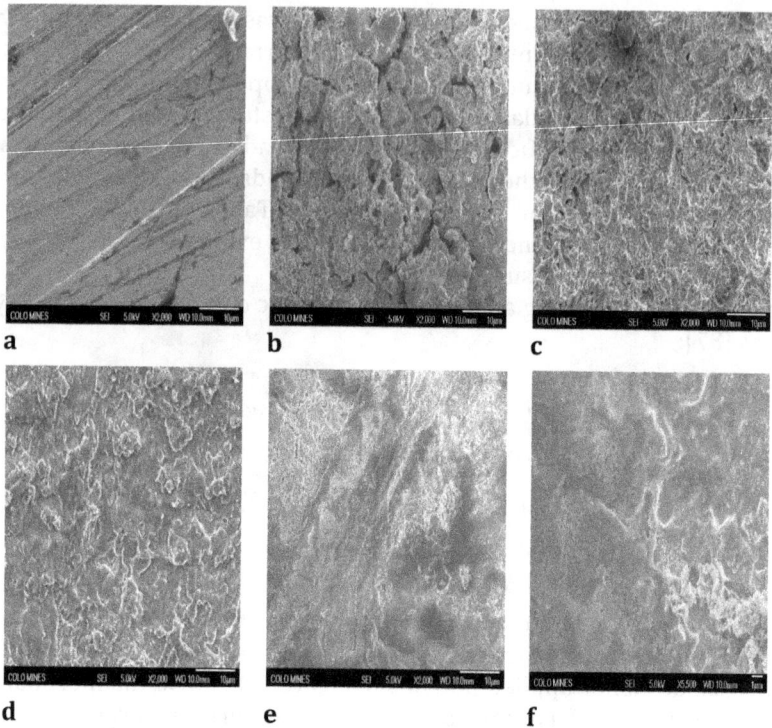

Figure 7.4 SEM micrograph of carbon steel surface after electrochemical testing in 3.5 wt% NaCl, (a) clean surface, (b) blank, different concentrations of GO (c) 3 ppm, (d) 9 ppm, (e) 12 ppm and (f) 15 ppm [2].

7.3.1 Materials and Methods

Model coatings were prepared by dissolving 2000 ppm PVB (Figure 7.5; 0.2 wt% of methanol weight) and 300 ppm SDS (0.03 wt% of methanol weight) in 50 mL (39.6 g) methanol with continuous stirring for 24 h, followed by sonication in a sonicator bath. SDS was used as a dispersant and used in all coatings. Similarly, two different concentrations of UGP powder, i.e., 1000 ppm (0.1 wt% of methanol weight, G-1) and 2000 ppm (0.2 wt% of methanol weight, G-2) were subjected to sonication in 50 mL (39.6 g) methanol with 300 ppm (0.03 wt% of methanol weight) SDS for 1 h. PVB was added to the UGP dispersion and shaken for 72 h to generate a uniform dispersion of UGP in the PVB solution

[38]. Using dip coater, carbon steel substrates were coated with PVB-UGP dispersion by immersion and withdraw speed of 50 and 200 mm/min respectively. Samples were immersed in the solution for 1 min. Three coats were applied for each sample in a similar manner with an interval of 20 min. Further, the samples were dried at room temperature for three days followed by baking in an air circulating oven at 175 °C for 2 h to generate final coating with a thickness in the range of 70±3 μm [12].

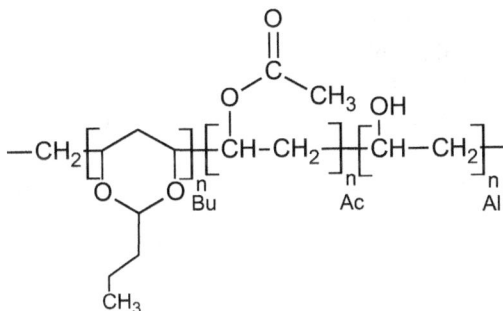

Figure 7.5 Structure of Butvar B-98 (Bu: butyral, Ac: acetate, Al: alcohol)

A flat cell assembly with working volume of 250 mL (Figure 7.6), consisting of carbon steel coupon as the working electrode (WE) with exposing area 2.6 cm², graphite plate as counter electrode (CE) having dimensions of 25×25×5 mm with exposing area 2.6 cm², and a silver/silver chloride electrode as reference electrode (RE), were used for the electrochemical measurements [12]. Carbon steel panels were surface finished using different grades of SiC grit pap-

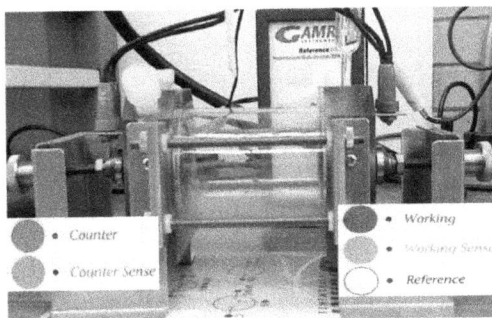

Figure 7.6 Electrochemical flat cell setup.

ers from 240 up to 600 grit, polished to a mirror finish followed by cleaning and degreasing with industrial grade acetone and drying in air. Before coating, specimen were treated with 2% nital for 1 min and used immediately without any further treatment [39].

Electrochemical testing was performed in a closed system under naturally aerated conditions using a Gamry 600 potentiostat/galvanostat/ZRA at room temperature. Corrosion studies were carried out in 0.1 M NaCl conditions [39]. Impedance measurements were performed vs. E_{OCP} at 1 and 26 h from the time of immersion. The frequency sweep was performed from 10^5 to 10^{-2} Hz at 10 mV AC amplitude. The Bode plots were modeled with monophasic circuit model used to fit EIS data as resistor and capacitors as shown in Figure 7.7 (a-b). For the description of a frequency independent phase shift between the applied AC potential and its corresponding current response, a constant phase element (CPE) was used, where impedance of the CPE is given the equation mentioned earlier [3,39]. In the equation, n is a measure of surface inhomogeneity; the lower is its value, the higher is the surface roughening of the metal/alloy [29].

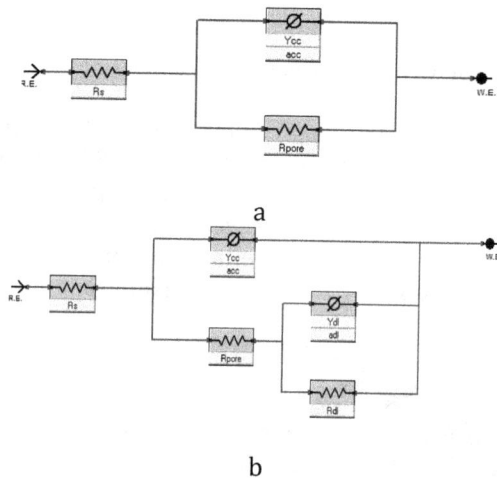

a

b

Figure 7.7 Circuit model after (1) 1 h and (b) 26 h of immersion.

At 1 and 26 h, the alloy exhibits a one-time constant impedance response for all samples. This behavior can be easily noticed in the phase angle Bode curves as a single hump for dominant one-time constant phase. To simulate the electrochemical interface, EIS data was analyzed with Echem Analyst using circuit model having electrical equivalent parameters accordingly. The capacitance C was calculated using the same equation mentioned earlier. The resistance efficiency, ER was calculated as indicated by equation [40]:

$$ER = [R' - R] / R'$$

where R' and R are the resistance value with and without UGP respectively.

Transmission electron microscopy (TEM) imaging was performed to characterize the GP. FEI Philips C200 TEM with 200 kV was used. The samples were prepared by dispersing approximately 1 mg of GP in 10 mL of acetone and sonicating for 30 min in a water bath at room temperature. One drop of the suspension was then deposited on a 400-mesh copper grid covered with a thin amorphous film to view under the microscope [2].

The bulk conductivity of the GP sheet was found to be 5.78 × 10^{+03} S/m which was measured using four probe methods and had good agreement with published literature [41]. The GP sheet was obtained by pressing powder GP in a sheet form on Teflon sheet using 10 lbf for 5 min. The sheet bulk resistance was measured using four probe methods at different areas on GP sheet using 200 mV and 4.53 × 10^{-03} Ampere. The conductivity of sheet was calculated according to following equations:

$$\rho(\Omega.m) = R_s(sheet\ resistance, \frac{\Omega}{\blacksquare}) \times T_s(sheet\ thickness, m)$$

$$\sigma\ (conductivity, \frac{S}{m}) = \frac{1}{\rho(\Omega.m)}$$

7.3.2 Results and Discussion

The conductivity of the sheet indicated that GP had very high conductivity which could also change the conductivity of the coatings. Thus, higher amount of graphene will produce polymer coatings with higher conductivity which may affect the corrosion process happening on the surface. Figure 7.8 shows characteristic low-resolution TEM image of graphene nanoplatelets (sheets) indicating a flaky and transparent structure with wrinkles and folding on the surface [42].

The scheme in Figure 7.9 describes general physicochemical and electrochemical mechanisms of corrosion protection and promotion respectively due to the incorporation of GP in organic coatings [43]. The short-term protection is explained by a physicochemical mechanism which is generally associated with the obstruction of corrosive agents. This effect can be expected to enhance significantly by incorporating plate like reinforcements, thus, increasing the diffusion length for the corrosion agents to reach at the defects through microscopic pores. For reinforcement-free organic coatings, the basic mechanism is a separation of the metal surface from the environment [5].

Figure 7.8 TEM of an aggregate consisting of a folded graphene nanoplatelets.

The permeability of organic coatings also depends on the nature of the binder matrix [5]. In this study, crosslinked film of PVB coating was used in order to have enhanced barrier nature of binder. The long term corrosion promotion effect was explained by the 'active' nature of UGP in the coatings. As water molecules start to accumulate at the interface, which facilitates the corrosion under the coating, graphene nanoplatelets stimulate the electrochemical reaction due to the conductivity of electrons at the interface [44]. This effect was also explained by the addition of carbon black to the zinc filled coating where carbon black was observed to promote the corrosion and acted as a perfect cathode for zinc. The addition of carbon black also improved electrical connections between the zinc particles promoting the galvanic effect [45]. In addition, it was also reported that carbon increased the porosity of the organic coating and increased the absorbance of water, thus, promoting the corrosion [46]. It was also reported that 1 g of reduced graphene (rG) sorbed 14 g of water [47].

EIS was used to examine the characteristics of defects arising in PVB and PVB/UGP composites coatings presented by complex plane plots. The coating exhibited the porous structure and non-ideal capacitive behavior (Figure 7.9

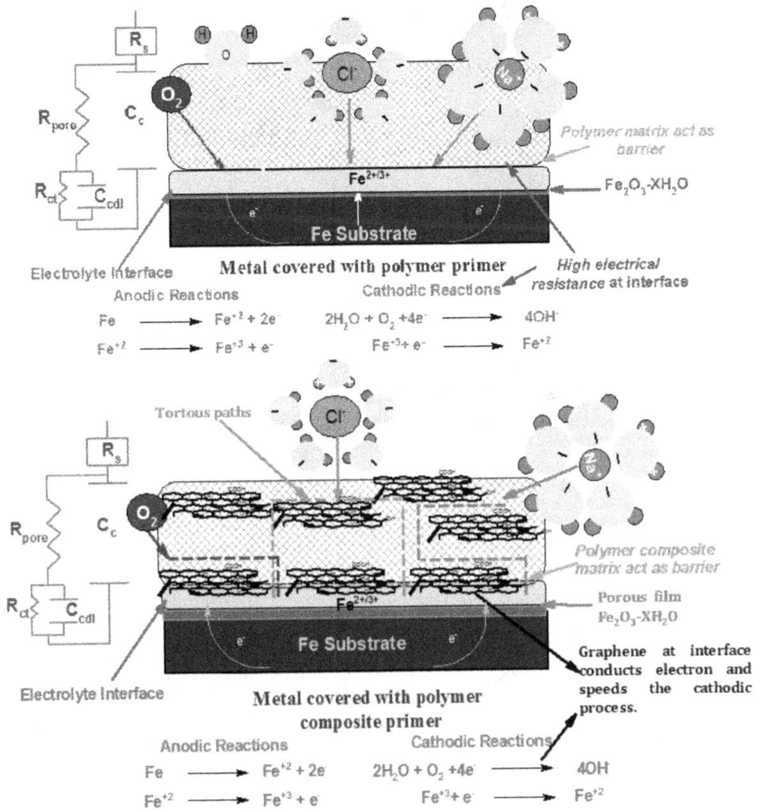

Figure 7.9 Scheme of corrosion protection phenomena in the (a) absence and (b) presence of UGP in PVB with electrochemical corrosion models for 26 h immersion in 0.1 NaCl [5].

and Table 7.2). The pore resistance of coating reduced with the addition of UGP which indicated the increased ionic resistance to current flow between the bulk and interface [48]. At the initial stage of immersion (1 h), G-2 coating exhibited quasi-ideal resistive behavior. The pore resistance of G-2 was observed to be very high as shown by half semicircle of Nyquist plot indicating the strict barrier nature of the coating, resulting in the hindrance of faradic reactions. As shown by the schematic in Figure, this barrier nature could be explained by the low diffusivity of corrosion reactants due to GP sheet-like structure (Figure 7.8). The corrosive solution has to adopt longer diffusion and tortuous paths to reach at the metal surface which in results slows down the corrosion rate. In

Table 7.2 Electrochemical parameters obtained from EIS for different samples [5]

T (h)	Sr. No.	R_{pore} [b] $\Omega.cm^2$ $\times 10^3$	C_C [c] $F.cm^2$ $\times 10^{-04}$	R_{ct} [d] $\Omega.cm^2$ $\times 10^3$	C_{dl} [e] $F.cm^2$ $\times 10^{-4}$
1	PVB	1.17	4.47		
	G-1	2.22	4.11	-	-
	G-2	3.98	4.34	-	-
26	PVB	0.68	4.84	1.05	1.07
	G-1	0.62	6.24	0.82	2.56
	G-2	0.65	9.36	1.04	3.63

case the defects are created by the incorporation filler due to the poor dispersion and agglomerations of fillers, this phenomena may cause inverse results. The pore resistance of coatings was calculated by using the following equation [10]:

$$|Z| = \sqrt{Z_{real}^2 + Z_{imaginary}^2}$$

at lowest frequency, where $Z_{imaginary} \rightarrow 0$ giving $|Z| = Z_{real}$ in Bode plot. It is also evident from Bode plots of Figure 7.10 a that the pore resistance of G-1 and G-2 was at least 50% and 70% respectively greater than that of PVB coated steel after 1 h of immersion. With increasing exposure time, i.e., 26 h of immersion, it was taken into account that corrosive media had accumulated on the carbon steel surface through the coating. EIS data from Figure 7.10 a and Table 7.2 shows the variations in the impedance model as a response to the intact area. The change in the EIS plots indicated that the model used for 1 h of immersion did not satisfy for longer period immersion. It can be noticed that coating capacitance of G-1 and G-2 was increased by 23% and 50% respectively. Similarly, interface capacitance of G-1 and G-2 had an increment of 58% and 71% respectively, showing enhancement of corrosion. Although, there was not much difference in the coating and charge transfer resistance, but these parameters showed decreased value for G-1 and G-2. The presence of water in the pores can change the dielectric constant of the coating or interface and can be determined by the capacitance [48]. The increased capacitance values with increased exposure time for G-1 and G-2 indicated that presence of corrosive reactant and products in coatings and coating/metal interface changed the local effective dielectric constant and water uptake properties according to following equations [13]:

$$C = \frac{\varepsilon \varepsilon_0 A}{d} \text{ and } \theta \text{ (percent)} = \frac{\log (C_t/C_o)}{\log (\varepsilon_w)}$$

where d is the charge separation distance, ε is the relative dielectric constant, ε_0 is the permittivity of free space, A is the surface area, C is the capacitance, θ is the percentage of water uptake, C_t is the coating capacitance at 26 h, C is the initial coating capacitance at 1 h, and ε_w is the relative permittivity of water taken as 80.1 at 20 °C, where relative permittivity is a dimension-less value. It can see from Figure 7.11 that addition of UGP enhanced the water uptake percentage to 9.5% for PVB to UGP ratio of 2:1 and almost double with the 1:1 ratio of PVB to UGP. These results had conformity

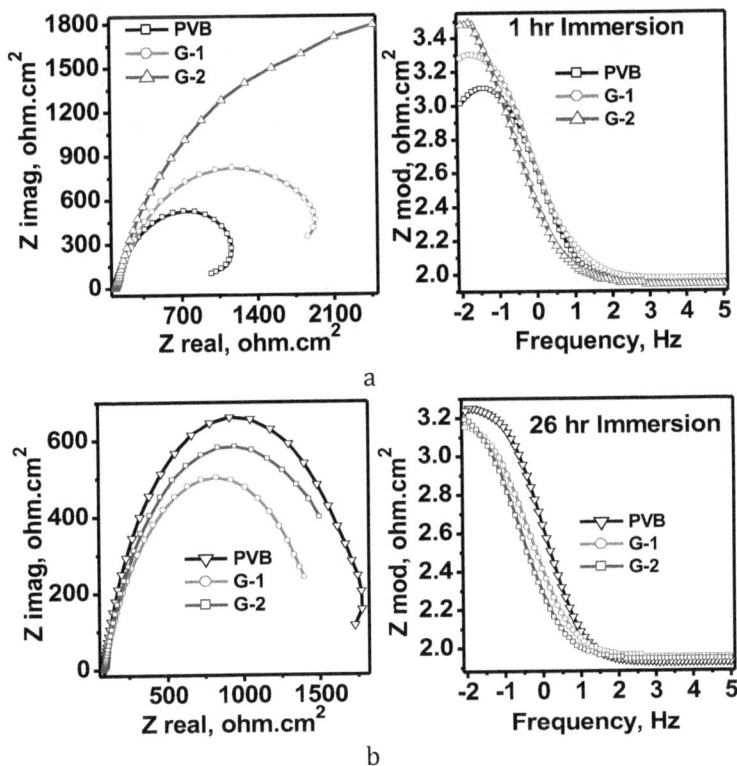

Figure 7.10 EIS magnitude spectra of coated carbon steel after a) 1 hour and b) 26 hour immersion in 0.1 M NaCl [5].

with the previous work reported on coatings containing carbon black [46]. SEM images were taken at the cross-sectional area to study the morphological alterations of coatings at metal/coating interface. The corrosion effects can be confirmed in Figure 7.11 where larger defects could be observed for G-1 and G-2. The approximate defect size of coating for PVB and GP was around 5 µm and 10 µm, respectively. The larger defects in the GP containing coatings indicated that water uptake properties of coatings increased, which in turn produced accelerated coating deterioration.

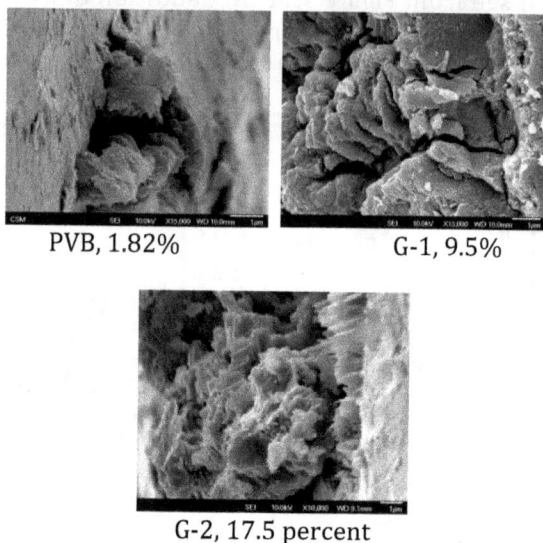

PVB, 1.82%   G-1, 9.5%

G-2, 17.5 percent

Figure 7.11 Cross-sectional area of coating showing coating/metal interface and water uptake θ (percent) after immersion of 26 h [5].

7.4 Effect of Functionalized Graphene Platelets on Electrochemical Properties of Polymer Coatings

This study deals with the comparison of barrier properties of nanocomposites prepared from Haydale's graphene and thermally reduced graphene oxide [49]. In specific, four kinds of graphene, i.e., U-GP and Haydale's processed graphene nanoplatelets produced using three types of process gasses, i.e., fluorine (F-GP), oxygen (O-GP), and argon (A-GP) were used.

7.4.1 Materials and Methods

PVB with trade name Butvar B-98 (molecular weight 40,000-70,000 g/mol) was purchased from Sigma-Aldrich. PVB had 18-20% hydroxyl content and 80% butyral content. M-GP (commercial name HDPlas™), with planer size 0.3-5 μm and platelets thickness < 50nm, was used as purchased from Graphene Supermarket, USA. U-GP (A-12, graphene supermarket, USA), with thickness < 3nm and planer size 2-8 μm, was used for comparison. GP amount was kept constant at 3 wt% of PVB resin (5g resin/50 ml methanol) for all the composites [12]. EIS was performed on coated carbon steel panels at different times of immersion, i.e., 1 and 12 h in 4 wt% percent NaCl. Model coatings were prepared by bath sonication of GP for 4 h in 50 ml methanol followed by 48 h of mixing with PVB in a closely tight glass flask. Carbon steel samples were polished starting from 60 to 600 grit sand paper. Using dip coater, carbon steel substrates were coated with PVB-GP dispersion by immersion and withdraw speed of 50 and 200 mm/min respectively. Samples were immersed in the solution for 1 min. Two coats were applied to each sample in a similar manner with an interval of 15 min. Further, the samples were dried at room temperature for 3 d followed by baking in an air circulating oven at 175 °C for 2 h to generate final coating with a thickness in the range of 270±3 μm. Four carbon steel samples were prepared for each PVB and composite coatings. A flat cell assembly (Figure 7.6) was used for corrosion measurements. The open circuit potential of steel samples was recorded against Ag/AgCl electrode as a reference electrode for 1 and 12 h. After the completion of each step, EIS was measured. Impedance measurements were performed vs. (E_{OCP}) after 1 and 12 h from the time of immersion. The frequency sweep was performed from 10^5 to 10^{-2} Hz at 10 mV AC amplitude.

7.4.2 Results and Discussion

TEM (Figure 7.12) of M-GP indicated different morphological features from that of UGP. Figure 7.13 (a-b) and Table 7.3 depict the phase angle and extracted model circuit parameters respectively for coated carbon steel in 4 wt % NaCl aqueous solution for different immersion time.

Phase angle exhibited distinct behavior at a various range of frequency for PVB and composites coatings. The range of frequency corresponds to the various regions in the coating i.e. starting from coating surface to substrate surface. After 1 h of immersion, the PVB coating at higher frequency ~100 kHz exhibited a mixed capacitive and resistive behavior (~-43°), which indicated the barrier

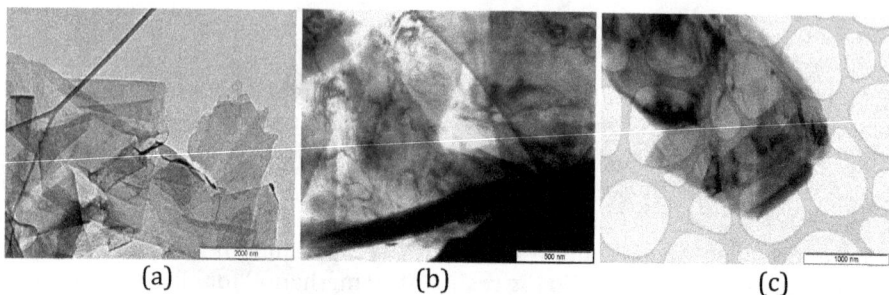

Figure 7.12 TEM images of graphene nanoplatelets; (a) U-GP, (b) A-GP and (c) F-GP [49].

and diffusive nature of the coating. The same behavior continued over intermediate frequency range with a variation of theta around ~-45°±-5°, afterwards it changed to resistive behavior in the low-frequencyspectrum. At frequency ~25 kHz, the slight increase in the angle ~-37° indicated dominant resistive behavior. A little bump could be obserevd at 12.4 Hz corresponding to the phase angle ~-46° between two minima of theta ~-49° at 100 Hz and ~-48.8° at 2.5 Hz. The continuous resistive and capacitive behavior ranging from higher to intermediate frequency could be an indication of constrained diffusion of electrolyte through the coating corresponding to porous bounded Warburg (PB-W) as mentioned in Gamry Echem Analyst™ manual. The major transition from mixed resistive-capacitive behavior to resistive behavior occurs at ~0.9 Hz corresponding to the metal/interface and metal characteristics. Some literature reports (especially Zhang *et al.*) used PB-W element to model the diffusion limited behavior in fuel cell containing meshes of steel [50-52]. The addition of U-GP changed the phase angle behavior considerably. At frequency ~100 kHz, comparatively lower phase angle ~-68° indicated high capacitive nature of coating followed by a dominant resistive behavior owing to higher phase angle ~-30° at 800 Hz. The behavior changed to dominant capacitive behavior ~-57° at ~12.5 Hz. This trend indicated a large amount of trapped electrolyte in the coating which was further unable to respond to the AC frequency at the intermediate frequency. This behavior could be modeled using Gerischer element [53] used for modeling a porous electrode as mentioned in Gamry Echem Analyst™ manual. Table 7.3 shows that addition of U-GP did not significantly improve the pore resistance and admittance of coating. The admittance value was very high as compared to PVB coating. However, there was a significant improvement in the R_{ct} indicating that corrosive electrolyte was although already in the coating, but unable to reach the interface. Upon adding graphene

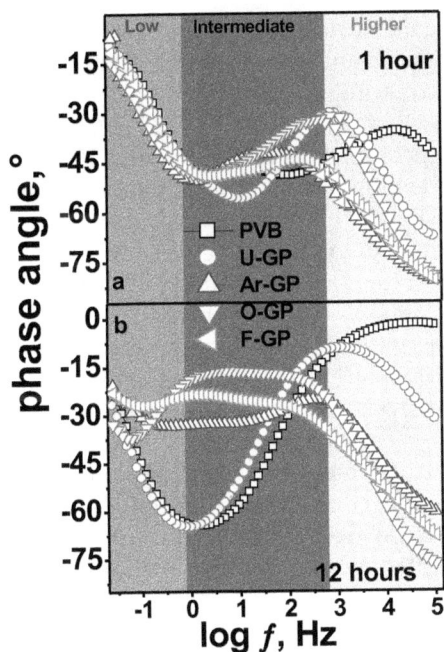

Figure 7.13 Phase angle behavior for PVB and GP-PVB nanocomposites coating on carbon steel recorded after 1 h and 12 h of immersion in 4 wt% NaCl [49].

Table 7.3 EIS parameters extracted using different circuit models

Model Circuits		EIS parameters	PVB	U-GP -PVB	O-GP -PVB	Ar-GP -PVB	F-GP -PVB
	PVB 1h	$R_{pore} \times 10^{03}$	2.93	2.75	26.9	87.4	49.0
	M-GP 1h	$R_{cl} \times 10^{04}$	2.67	21.6	618	456	326
		$Y_{admit.} \times 10^{-05}$	0.32[W]	8.52[G]	0.18[G]	0.09[G]	0.13[G]
	PVB 12h	$R_{pore} \times 10^{04}$	1.01	1.28	1.36	1.52	4.40
	M-GP 12h	$R_{cl} \times 10^{03}$	4.51	12.6	10.4	24.6	34.1
		$Y_{admit.} \times 10^{-05}$	-	79.6	2.87	2.22	9.22

Resistance Ohm.cm^2, Y(admit.) S*s$^{(1/2)}$/cm^2

produced from Haydale's process to PVB, the composite coating showed remarkable enhancement in anti-corrosion properties owing to improved diffusion behavior indicating improved dispersion and interaction with the PVB resin.

In the case of nanocomposites produced from F-GP/PVB, at ~100 kHz frequency, the coating exhibited a highly capacitive behavior indicated by low phase angle, i.e., ~-81° which indicated improved barrier properties. The behavior gradually changed to capacitive-resistive behavior at ~400 Hz frequency. Afterwards, frequency independent behavior could be seen over a range of frequency from 400 Hz to 1.6 Hz with constant phase angle ~-45°±3° indicating a diffusion limited process, a trend shown by all of other cases. This kind of impedance behavior enhanced the RC by 28, 21 and 15 times for O-GP, A-GP, and F-GP respectively, as compared to U-GP after 1 h of immersion. After 12 h of immersion, PVB coating exhibited an entirely different behavior. At the higher frequency, it depicted resistive behavior indicating loss of barrier properties. On the other hand, at intermediate and low-frequency range, dominant capacitive hump could be seen indicating the appearance of larger blisters in the coatings [54]. Similar was the case for U-GP at intermediate and lower frequency, but it exhibited lesser capacitive behavior at a higher frequency as compared to 1 h indicating reduced barrier and adhesion. M-GP nanocomposites still exhibited very high capacitive behavior at higher frequency indicating that the coating still retained barrier properties and was in contact with the substrate. The capacitive behavior gradually changed to more resistive (~-33° for A-GP, ~-23°for F-GP and ~-17°for O-GP) as compared to 1 h immersion and extended over a broad range of low frequencies. For O-GP, a little capacitive hump could be observed at very low frequency indicating appearance of tiny blisters. Further, the continuous resistive behavior also indicated that delaminated area was small where the corrosion was taking place. Table 7.3 also indicated that compared to U-GP, M-GP nanocomposites had very low admittance value determined from Gerischer element after 12 h.

7.5 Effect of Functionalized Graphene Platelets with Polyaniline on Electrochemical Properties of Polymer Coatings

In the current study, PANI and the nano-hybrids based on PANI-thermally reduced graphene (r-GO) were used as the anti-corrosion pigments incorporated in poly(vinyl butyral) (PVB) resin [13]. The PANI/r-GO hybrids were generated with varying r-GO/aniline ratios during the hybrid generation, so as to relate the effect of graphene content to the coating performance. The coatings were applied on carbon steel panels and the corrosion performance of these pigments was investigated in 4% w/v NaCl solution.

7.5.1 Materials and Methods

PVB (M_W 70,000-100,000) resin with broad molecular weight distribution, containing ~20% vinyl alcohol and ~2% vinyl acetate, was obtained as granules from Polysciences, Inc. (USA). Carbon steel coupons (elemental composition: C 0.07 wt%, Mn 1.36 wt%, Ti 0.008 wt%, S 0.003 wt%, P 0.004 wt% and rest Fe) were purchased locally and were machined to 30 x 16 x 5 mm dimensions. Polyaniline modified r-GO was prepared using *in-situ* polymerization. To generate PANI/r-GO composites with different weight ratios, 0, 10, 20, 30, 40 and 50 mg of r-GO nanoparticles were added to a mixture of 1 ml of aniline and 90 ml of 1N HCl in a reaction vessel (Figure 7.14) [11]. The mixture was stirred in ice water bath (5 °C) for 1 h to obtain a uniform dispersion of r-GO. To the mixture, 100 ml pre-cooled 1N HCl solution containing 2.5 g ammonium persulfate (APS) was added dropwise. The reaction was allowed to proceed in the ice bath for 4 h. The resulting product was washed with distilled water several times and at last washed with methanol in order to eliminate oligomers and other impurities [11]. Subsequently, the sample was dried in a vacuum oven for 12 h. Based on the amount of r-GO, these hybrids were named as PANI/10r-GO, PANI/20r-GO, PANI/30r-GO, PANI/40r-GO and PANI/50r-GO.

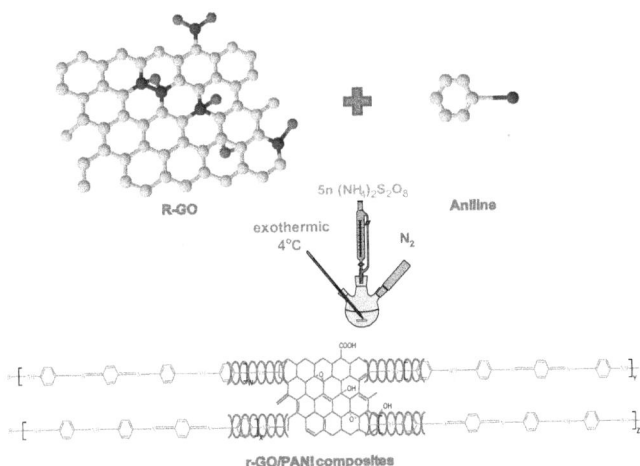

Figure 7.14 Schematic diagram of generation of PANI/r-GO composite particles.

PVB was dissolved in methanol (20 ml) under stirring for 6 h, followed by sonication in a bath sonicator for 2 h. PANI/r-GO nanoparticles of different compositions were finely powdered with a small quantity of methanol to obtain a uniform slurry. The slurry was added and mixed in the PVB solution and soni-cated for 24 h to achieve uniform dispersion of nanoparticles in the PVB solu-tion [38]. The amount of PANI/r-GO particles was fixed to 10 wt % of the amount of dry PVB. For coating of the generated PANI/r-GO/PVB formulations, carbon steel coupons with rounded corners and edges were polished by emery paper, washed with acetone, dried and weighed with an accuracy of ±1 mg. The substrates were brush coated with the PANI/r-GO/PVB formulations and dried at room temperature for 30 min. Subsequently, the coating was baked in an air circulating oven at 60 °C for 6 h and coatings with a thickness in the range of 15-20 μm were obtained.

Transmission electron microscopy (TEM) of hybrid samples was performed using EM 912 Omega (Zeiss, Oberkochen BRD) electron microscope at 120 kV and 200 kV accelerating voltage. Thin flakes of PANI/r-GO platelets were sup-ported on 100 mesh grids sputter coated with a 3 nm thick carbon layer.

The anti-corrosion properties of the coatings were studied by immersion test (NaCl 4% w/v) according to ASTM G31. The coated coupons were placed in a specially designed set up for 800 h at 25 °C. The function of this set-up was to maintain temperature, air flow, and water level so as to attain similar condi-tions throughout the experiment [14]. After completing the immersion period, the samples were taken out of salt solution, cleaned and weighed again. Clean-ing of the samples was achieved by removing the coating using acetone as a solvent and subsequently removing the corrosion products by a solution of 3.5 g of hexamethylenetetramine in 500 ml of hydrochloric acid diluted to 1000 ml with distilled water. The samples were dipped for 10 min in the amine solution at room temperature. Weight loss method was used to evaluate corrosion rate (CR), using the following equation:

$$Corrosion\,rate, CR = \frac{87.6\,W}{DAT}$$

where CR is corrosion rate in mm/year, W is weight loss in mg, D is the sub-strate density, A is the exposed surface area and T is exposure time in the salt solution in hours. The area of the exposed surface of the coated substrate was 9.60 cm², whereas the density of substrate was taken as 7.85 g/cm³. The corro-sion protection efficiency (CE) of anti-corrosion pigment was calculated using following equation:

$$CE = \left[\frac{CR - CR'}{CR} \right]$$

where CR and CR' are the corrosion rates without and with the anti-corrosion pigment respectively.

7.5.2 Results and Discussion

In the current study, PANI/r-GO hybrids with varying amount of graphene were incorporated in PVB matrix so as to achieve anti-corrosion coatings where both PANI and graphene synergistically contributed to the coating's performance. The morphology of the graphene platelets altered completely with PANI intercalation in the interlayers. PANI was observed to uniformly cover the surface of the platelets, which was also the reason for the absence of any graphene diffraction signal in the X-ray diffractograms. Even increasing the amount of graphene in the hybrid exhibited similar morphology. PANI chains are expected to physically adsorb on the surface of graphene nanoparticles due to interactions of N atoms with functional groups on graphene surface as well as due to van der Waals forces between PANI chains and graphene surface. This is, thus, also expected to affect the polymer crystallinity as the PANI chains would adsorb on the surface of platelets simultaneously during their synthesis, which will hinder their effective crystallization [11].

From the corrosion rate analysis, it was observed that under the test conditions used, pure PANI pigment in PVB did not provide any superior corrosion protection than pure PVB coating itself, though both improved the protection of bare metal substrate. The corrosion protection efficiency (CE) was observed to increase with increasing amount of r-GO in the coatings. These results can be attributed to the additional protection provided by r-GO, along with the intrinsic protection provided by PANI. For instance, coating with PANI/40r-GO was nearly 52% more effective than the PVB coating without pigment. Further increase in graphene content in PANI/50r-GO exhibited a marginal increase in protection efficiency to 58%, which indicated that the CE may have reached a plateau.

To further confirm these findings, optical microscopy was employed to gain insights into the change in the surface topography of the substrates after the corrosion test. Figure 7.15 shows the surface of the coatings after the corrosion test as well as removal of the coating. Substrates coated with pure PVB (Figure 7.15 (a)) and pure PANI in PVB (Figure 7.15 (b)) formulations exhibited extensive surface corrosion. It indicated that the corrosion media could penetrate faster through the coatings, thus, initiating the corrosion process. The substrate coated with PANI/10r-GO (Figure 7.15 (c)) in PVB exhibited improved anti-corrosion performance as the extent of surface corrosion was markedly reduced in comparison with Figure 7.15 (a) and (b). A further increase in the graphene

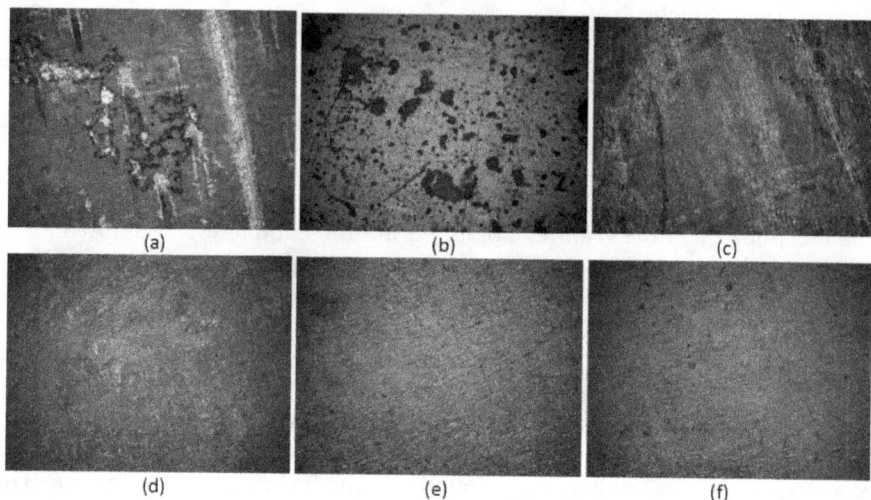

Figure 7.15 Surface of the substrates after corrosion testing and coating removal; (a) blank (pure PVB), (b) PANI, (c) PANI/10r-GO, (d) PANI/20r-GO, (e) PANI/30r-GO and (f) PANI/40r-GO. The width of the images equals 200 μm.

concentration enhanced coatings performance further. PANI/20r-GO, PANI/30r-GO, and PANI/40r-GO coated substrates optically exhibited a high level of stability to the immersion conditions. Interestingly, the conductivity of the coatings decreased as the amount of reduced graphene was increased. It has been reported earlier that the lower conductivity of PANI/graphene composites results probably due to decrease in the degree of doping in PANI and change in the morphology of the composites [55, 56]. Similarly, Sun *et al.* [8] have also reported graphene/pernigraniline composite (GPCs) coatings in PVB with reduced conductivity for the corrosion protection of copper. The GPCs were generated by in-situ polymerization-reduction/doping process and the authors concluded that pernigraniline modified graphene in PVB was able to provide effective corrosion protection.

The corrosion protection by the coating is due to the generation of disconnection between bare metal and corrosive environment, thus, impeding the transportation of corrosive material. The protection behavior of PANI at the interface of metal/electrolyte has been described earlier in the literature [57]. In this phenomenon, the formation of metal oxide layer occurs with the aid of more noble emeraldine salt (ES) of PANI. This form of PANI further leads to lower energy state and reduces to different forms such as non-conducting leucoemeraldine base (LEB) and emeraldine base (EB). This cycle is continuous and LE form turns to ES via EB form by oxygen. This cycle can be continuous only if the barrier properties of applied coating are able to avoid the removal of H$^+$ (maintained acidic pH) by the surroundings. This process slows the formation of rust at the iron surface by providing passive layers of different forms of PANI. However, as observed in the current study, pure PANI present in the PVB matrix was unable to significantly slow down the corrosion rate probably due to the lower extent of PANI-metal interface formation. In addition, the process of corrosion protection could be specifically enhanced by the addition of r-GO, as it resulted in the tortuous path for the corrosive media, thus, delaying the corrosion process. This was also confirmed from the improving corrosion protection efficiency as the amount of graphene was enhanced in the coatings. Such barrier effect of graphene-based reinforcement is called a 'passive' role. Further, the 'active' behavior of such material was realized in the coatings by the formation of 'Schottky barrier' at the interface, thus, leading to the depletion of electrons, which slowed down the corrosion half-cell reactions. Composites of PANI acted like hetro-junction, where PANI behaves as p-type while r-GO being n-type, thus, hindering the anodic and cathodic reactions respectively. The decreased conductivity of composites also impedes the transport of electron from graphene to PANI [38].

Furthermore, due to the physical adsorption of PANI on the graphene surface, the effect of graphene incorporation was synergistically enhanced by the presence of PANI due to its better interface with the metal. It has also to be noted that with these coatings systems, the corrosion reaction was not completely eliminated, however, it was significantly slowed down, thus underlining the need for the functional coatings to achieve superior material performance [6,58].

7.6 Conclusions

Graphene represents a very useful nanomaterial for the development of anti-corrosion coatings. The studies presented in this chapter confirm the high potential of graphene and its derivatives in inhibiting and delaying corrosion on metal surfaces in various corrosion environments. In this category, the specific applications or performance enhancements can also be generated by suitably modifying the graphene surface as well as optimizing its interaction with the polymers, along with nano-scale dispersion. In addition, graphene significantly impacts both short-term and ling-term anti-corrosion behavior of the coatings.

References

1. Jones, D. A. (2013) *Principles and Prevention of Corrosion,* Pearson Education Limited, USA.
2. Chaudhry, A. U., Mittal, V., and Mishra, B. (2015) Effect of graphene oxide nanoplatelets on electrochemical properties of steel substrate in saline media. *Materials Chemistry and Physics,* **163**, 130-137.
3. Chaudhry, A. U., Bhola, R., Mittal, V., and Mishra, B. (2014) $Ni_{0.5}Zn_{0.5}Fe_2O_4$ as a potential corrosion inhibitor for API 5L X80 steel in acidic environment. *International Journal of Electrochemical Science,* **9**, 4478-4492.
4. Kim, J., Cote, L. J., Kim, F., Yuan, W., Shull, K. R., and Huang, J. (2010) Graphene oxide sheets at interfaces. *Journal of the American Chemical Society,* **132**(23), 8180-8186.
5. Chaudhry, A.U., Mittal, V., and Mishra, B. (2015) Inhibition and promotion of electrochemical reactions by graphene in organic coatings. *RSC Advances,* **5**(98), 80365-80368.
6. Yu, Y.-H., Lin, Y.-Y., Lin, C.-H., Chan, C.-C., and Huang, Y.-C. (2014) High-performance polystyrene/graphene-based nanocomposites with excellent anti-corrosion properties. *Polymer Chemistry,* **5**(2), 535-550.
7. Sun, W., Wang, L., Wu, T., Wang, M., Yang, Z., Pan, Y., and Liu, G. (2015) Inhibiting the corrosion-promotion activity of graphene. *Chemistry of Materials,* **27**(7), 2367-2373.
8. Sun, W., Wang, L., Wu, T., Pan, Y., and Liu, G. (2014) Synthesis of low-electrical-conductivity graphene/pernigraniline composites and their application in corrosion protection. *Carbon,* **79**, 605-614.
9. Liu, J., Hua, L., Li, S., and Yu, M. (2015) Graphene dip coatings: An effective anticorrosion barrier on aluminum. *Applied Surface Science,* **327**, 241-245.
10. Singh Raman, R. K., Chakraborty Banerjee, P., Lobo, D. E., Gullapalli, H., Sumandasa, M., Kumar, A., Choudhary, L., Tkacz, R., Ajayan, P. M., and Majumder, M. (2012) Protecting copper from electrochemical degradation by graphene coating. *Carbon,* **50**(11), 4040-4045.
11. Mittal, V., Chaudhry, A. U., and Matsko, N. B. (2016) Organic functionalization of thermally reduced graphene oxide nanoplatelets by adsorption: structural and morphological characterization. *Philosophical Magazine,* **96**(20), 2143-2160.
12. Mishra, B., Chaudhry, A., and Mittal, V. (2017) Development of polymer-based composite coatings for the gas exploration industry: Polyoxometalate doped conducting polymer based self-healing pigment for polymer coatings, *Materials Science Forum,* in print.
13. Chaudhry, A. U., and Mittal, V. (2017) Polyaniline-graphene composite nanoparticle pigments for anti-corrosion coatings, in preparation.

14. Khan, M. I., Chaudhry, A. U., Hashim, S. and Iqbal, M. Z. (2010) Investigation of corrosion-protective performance of polyaniline covered inorganic pigments. *Nucleus*, **47**(4), 287-293.
15. Khan, M. I., Chaudhry, A. U., Hashim, S., Zahoor, M. K., and Iqbal, M. Z. (2010) Recent developments in intrinsically conductive polymer coatings for corrosion protection. *Chemical Engineering Research Bulletin*, **14**(2), 73-86.
16. Luckachan, G. E., and Mittal, V. (2015) Anti-corrosion behavior of layer by layer coatings of cross-linked chitosan and poly(vinyl butyral) on carbon steel. *Cellulose*, **22**, 3275-3290.
17. Mittal, V., and Chaudhry, A. U. (2015) Polymer-graphene nanocomposites: effect of polymer matrix and filler amount on properties. *Macromolecular Materials and Engineering*, **300**(5), 510-521.
18. Mittal, V., and Chaudhry, A. U. (2015) Effect of amphiphilic compatibilizers on the filler dispersion and properties of polyethylene-thermally reduced graphene nanocomposites. *Journal of Applied Polymer Science*, **132**(35), DOI: 10.1002/app.42484.
19. Mittal, V., Chaudhry, A. U., and Khan, M. I. (2011) Comparison of anti-corrosion performance of polyaniline modified ferrites. *Journal of Dispersion Science and Technology*, **33**(10), 1452-1457.
20. Mittal, V., Chaudhry, A. U., and Luckachan, G. E. (2014) Biopolymer-thermally reduced graphene nanocomposites: Structural characterization and properties. *Materials Chemistry and Physics*, **147**(1-2), 319-332.
21. Mittal, V., Chaudhry, A. U., and Matsko, N. B. (2014) "True" biocomposites with biopolyesters and date seed powder: Mechanical, thermal, and degradation properties. *Journal of Applied Polymer Science*, **131**(19), DOI: 10.1002/app.40816.
22. Kalendova, A., Sapurina, I., Stejskal, J., and Vesely, D. (2008) Anticorrosion properties of polyaniline-coated pigments in organic coatings. *Corrosion Science*, **50**(12), 3549-3560.
23. Sathiyanarayanan, S., Azim, S. S., and Venkatachari, G. (2007) A new corrosion protection coating with polyaniline-TiO_2 composite for steel. *Electrochimica Acta*, **52**(5), 2068-2074.
24. Brodinova, J., Stejskal, J., and Kalendova, A. (2007) Investigation of ferrites properties with polyaniline layer in anticorrosive coatings. *Journal of Physics and Chemistry of Solids*, **68**(5-6), 1091-1095.
25. Wu, K. H., Chao, C. M., Liu, C. H., and Chang, T. C. (2007) Characterization and corrosion resistance of organically modified silicate–NiZn ferrite/polyaniline hybrid coatings on aluminum alloys. *Corrosion Science*, **49**(7), 3001-3014.
26. Usman, C. A. (2016) *Anti-corrosion Behaviour of Barrier, Electrochemical and Self-healing Fillers in Polymer Coatings for Carbon Steel in a Saline Environment*, Ph.D. Thesis, Colorado School of Mines, USA.
27. *New Trends in Electrochemical Impedance Spectroscopy (EIS) and Electrochemical Noise Analysis (ENA)*, Mansfeld, F., Huet, F., and Mattos, O. (eds.), The Electrochemical Society, USA.
28. Asma, R. N., Yuli, P., and Mokhtar, C. (2011) Study on the effect of surface finish on corrosion of carbon steel in CO2 environment. *Journal of Applied Sciences*, **11**(11), 2053-2057.

29. Chongdar, S., Gunasekaran, G., and Kumar, P. (2005) Corrosion inhibition of mild steel by aerobic biofilm. *Electrochimica Acta*, **50**(24), 4655-4665.
30. Mora-Mendoza, J. L., and Turgoose, S. (2002) Fe_3C influence on the corrosion rate of mild steel in aqueous CO_2 systems under turbulent flow conditions. *Corrosion Science*, **44**(6), 1223-1246.
31. Fekry, A. (2011) Electrochemical corrosion behavior of magnesium alloys in biological solutions. In: *Magnesium Alloys - Corrosion and Surface Treatments*, Czerwinski, F. (ed.), Intech, Croatia, doi: 10.5772/13027.
32. Zhou, X., and Hao, J. (2011) Solubility of NaBr, NaCl, and KBr in surfactant aqueous solutions. *Journal of Chemical & Engineering Data*, **56**(4), 951-955.
33. Lopez, D. A., Simison, S. N., and de Sanchez, S. R. (2003) The influence of steel microstructure on CO_2 corrosion. EIS studies on the inhibition efficiency of benzimidazole. *Electrochimica Acta*, **48**(7), 845-854.
34. Jiang, X., and Nesic. S. (2008) Electrochemical Investigation of the Role of Cl-on Localized CO_2 Corrosion of Mild Steel. *17th International Corrosion Congress, USA.* Online: http://www.corrosioncenter.ohiou.edu/documents/publications/8209.pdf [assessed 23rd April 2017].
35. Hernandez, J., Munoz, A., and Genesca, J. (2012) Formation of iron-carbonate scale-layer and corrosion mechanism of API X70 pipeline steel in carbon dioxide-saturated 3% sodium chloride. *Afinidad*, **69**(560), 251-258.
36. Fekry, A., Ghoneim, A., and Ameer, M. (2014) Electrochemical impedance spectroscopy of chitosan coated magnesium alloys in a synthetic sweat medium. *Surface and Coatings Technology*, **238**, 126-132.
37. Sarkar, P. P., Kumar, P., Manna, M. K., and Chakraborti, P. C. (2005) Microstructural influence on the electrochemical corrosion behaviour of dual-phase steels in 3.5% NaCl solution. *Materials Letters*, **59**(19-20), 2488-2491.
38. Radhakrishnan, S., Siju, C. R., Mahanta, D., Patil, S., and Madras, G. (2009) Conducting polyaniline-nano-TiO_2 composites for smart corrosion resistant coatings. *Electrochimica Acta*, **54**(4), 1249-1254.
39. Chaudhry, A. U., Mittal, V., and Mishra, B. (2015) Nano nickel ferrite ($NiFe_2O_4$) as anti-corrosion pigment for API 5L X-80 steel: An electrochemical study in acidic and saline media. *Dyes and Pigments*, **118**, 18-26.
40. Sathiyanarayanan, S., Jeyaprabha, C., Muradidharan, S., and Venkatachari, G. (2006) Inhibition of iron corrosion in 0.5 M sulphuric acid by metal cations. *Applied Surface Science*, **252**(23), 8107-8112.
41. Kundhikanjana, W., Lai, K., Wang, H., Dai, H., Kelly, M. A., and Shen, Z.-x. (2009) Hierarchy of electronic properties of chemically derived and pristine graphene probed by microwave imaging. *Nano Letters*, **9**(11), 3762-3765.
42. Shao, Y., Zhang, S., Wang, C., Nie, Z., Liu, J., Wang, Y., and Lin, Y. (2010) Highly durable graphene nanoplatelets supported Pt nanocatalysts for oxygen reduction. *Journal of Power Sources*, **195**(15), 4600-4605.
43. Funke, W. (1986) How organic coating systems protect against corrosion. In: *Polymeric Materials for Corrosion Control*, American Chemical Society, USA, pp. 222-228.

44. Schriver, M., Regan, W., Gannett, W. J., Zaniewski, A. M., Crommie, M. F., and Zettl, A. (2013) Graphene as a long-term metal oxidation barrier: Worse than nothing. *ACS Nano*, **7**(7), 5763-5768.

45. Marchebois, H., Touzain, S., Joiret, S., Bernard, J., and Savall, C. (2002) Zinc-rich powder coatings corrosion in sea water: influence of conductive pigments. *Progress in Organic Coatings*, **45**(4), 415-421.

46. Nazeri, M. F. M., Suan, M. S. M., Masri, M. N., Alias, N., and Mohamed, A. A. (2012) Corrosion studies of conductive paint coating using battery cathode waste material in sodium chloride solution. *International Journal of Electrochemical Science*, 7, 6976-6987.

47. Iqbal, M., and Abdala, A. (2013) Oil spill cleanup using graphene. *Environmental Science and Pollution Research*, **20**(5), 3271-3279.

48. Macdonald, D. D., and McKubre M. C. H. (1981) Electrochemical impedance techniques in corrosion science, ASTM, USA. Online: https://www.astm.org/DIGITAL_LIBRARY/STP/PAGES/STP28030S.htm [assessed 23rd April 2017].

49. Chaudhry, A. U., Mittal, V., and Mioshra, B. (2017) Impedance response of nanocomposite coatings comprising of polyvinyl butyral and Haydale's plasma processed graphene. *Progress in Organic Coatings*, **110**, 97-103.

50. Zhang, F., Merrill, M. D., Tokash, J. C., Saito, T., Cheng, S., Hickner, M. A., and Logan, B. E. (2011) Mesh optimization for microbial fuel cell cathodes constructed around stainless steel mesh current collectors. *Journal of Power Sources*, **196**(3), 1097-1102.

51. Lamaka, S. V., Zheludkevich, M. L., Yasakau, K. A., Montemor, M. F., and Ferreira, M. G. S. (2007) High effective organic corrosion inhibitors for 2024 aluminium alloy. *Electrochimica Acta*, **52**(25), 7231-7247.

52. Kuzum, D., Takano, H., Shim, E., Reed, J. C., Juul, H., Richardson, A. G., de Vries, J., Bink, H., Dichter, M. A., Lucas, T. H., Coulter, D. A., Cubukcu, E., and Litt, B. (2014) Transparent, flexible, low noise graphene electrodes for simultaneous electrophysiology and neuroimaging. Nature Communications, **5**, 5259.

53. Gonzalez-Cuenca, M., Zipprich, W., Boukamp, B. A., Pudmich, G., and Tietz, F. (2001) Impedance studies on chromite-titanate porous electrodes under reducing conditions. *Fuel Cells*, **1**(3-4), 256-264.

54. Oliveira, C. G., and Ferreira, M. G. S. (2003) Ranking high-quality paint systems using EIS. Part I: intact coatings. *Corrosion Science*, **45**(1), 123-138.

55. Zhang, K., Zhang, L. L., Zhao, X. S., and Wu, J. (2010) Graphene/polyaniline nanofiber composites as supercapacitor electrodes. *Chemistry of Materials*, **22**(4), 1392-1401.

56. Das, T. K., and Prusty, S. (2013) Graphene-based polymer composites and their applications. *Polymer-Plastics Technology and Engineering*, **52**(4), 319-331.

57. Wessling, B. (1999) Scientific engineering of anti-corrosion coating systems based on organic metals (polyaniline). *The Journal of Corrosion Science & Engineering*, **1**, paper 15.

58. Lu, W.-K., Elsenbaumer, R. L., and Wessling, B. (1995) Corrosion protection of mild steel by coatings containing polyaniline. *Synthetic Metals*, **71**(1-3), 2163-2166.

8

Theoretical Insights into Gas Molecular Adsorption on Graphene: Implications for Gas Sensing Applications

8.1 Introduction

Graphene, a two-dimensional hexagonal sheet of carbon atoms, has grabbed immense research and industrial interest since its first isolation in 2004 [1] owing to its potentially tuneable and exotic structural, electronic, mechanical and optical properties [2-11]. Graphene has emerged as an attractive candidate for applications such as energy storage, electronics, field emission, photovoltaics, catalysis, biomedicine etc. [12-20]. It has a very large surface area of 2630 m^2/g (300 times higher than graphite and two times higher than single-walled carbon nanotubes), which provides large active area for sensing applications [21-27].

The unique two-dimensional nature which maximizes the interactions of gas molecules with graphene [28], metallic conductivity, low Johnson noise [2,3,8,28,29] and the limited crystal defects [2,3,8,28] contribute to the excellent gas sensing behavior of graphene. A little change of carrier concentration of graphene can cause a notable variation of electrical conductivity [28]. Apart from the above features, graphene monocrystals allow the fabrication of four-probe devices that can avoid the influence of the contact resistance and can extend the limit of sensitivity [30].

The linear energy-momentum dispersion near the Dirac point and the zero-energy band gap in the electronic spectrum of pure graphene [8] make graphene more accessible to doping with either electrons or holes by the gas molecules adsorbed on the surface of graphene [28,31]. Thus, the electronic properties of graphene are strongly influenced by the adsorption of the gas molecules in the surrounding environment, which makes this material attractive for gas sensing applications. Graphene can detect minute concentrations of gases present in the environment by measuring the change of graphene's electrical conductivity, caused by the change in the carrier concentration of graphene induced by the adsorbed gas molecules. The possibility of using graphene as a highly sensitive gas sensor has been experimentally demon-

Seba S. Varghese[a,b], Sundaram Swaminathan[c], Krishna K. Singh[b] and Vikas Mittal[a,*]
[a]The Petroleum Institute (part of Khalifa University of Science and Technology), Abu Dhabi, UAE; [b]Birla Institute of Technology and Science, Dubai, UAE; [c]DIT University, India
*Current address: Bletchington, Wellington County, Australia

strated by Schedin *et al.* [28] with detection limit of the order of 1 parts per billion (ppb) for NO_2, H_2O, NH_3 and CO. The ultimate sensitivity of an individual molecule of NO_2 was also reported.

The excellent gas sensing properties of graphene have inspired theoreticians to investigate the mechanisms involved in the adsorption of gas molecules on graphene's surface. Numerous theoretical researches based on first-principles have been performed to analyze the interactions between gas molecules and graphene, so as to exploit the potential of graphene as gas sensors. In this chapter, the state of the art research on first-principles theoretical studies on gas molecular adsorption on graphene and chemically functionalized graphene has been reviewed. In addition, the effect of heteroatom dopants and defects on the adsorption potential of graphene in the light of reported simulation based studies has been discussed. Before discussing the latest computer simulation based works on adsorption properties of doped and defected graphene for various gas molecules, research studies on gas molecular adsorption on intrinsic graphene (IG), graphene nanoribbon (GNR) and graphene oxide (GO) are also described. Finally, the existing challenges and future perspectives of gas molecular adsorption on chemically functionalized graphene are also provided.

8.2 Theoretical Studies on Graphene-gas Molecular Interaction

Understanding the interactions between gas molecules and graphene are of great significance in developing graphene based gas sensors. Quantum mechanical simulations on the gas molecular adsorption of graphene initially decide on the most favorable adsorption configuration of the gas molecule on the surface of graphene and calculate the adsorption energy to understand the nature of interactions. The adsorption mechanisms are also discussed from charge transfers and density of states (DOS). The changes in the electronic structures of graphene caused by the adsorption of gas molecules are determined from charge transfer calculations. The effect of gas molecules caused by the physi- or chemi-sorption of gas molecules on the electronic conductivity of graphene are verified from the DOS of gas molecule-graphene adsorption systems. The properties such as adsorption energies of gas molecules on graphene surfaces and the charge transfers between gas molecules and graphene are studied from the simulated gas molecule-graphene adsorption systems. The results from the theoretical calculations are useful for application of graphene as gas sensors. The results can direct in designing effective gas sensors with optimized sensing performances.

8.2.1 Gas Molecules on Intrinsic Graphene

There have been many reports on the adsorption of gas molecules on IG. Wehling *et al.* [31], for the first time, presented a combined experimental and theoretical work on doping of graphene by gas adsorbates and its relevance to gas sensor applications. The relationship between the adsorbate-induced doping strength and the shell configuration of the adsorbate was demonstrated by considering the open- and closed-shell configurations of NO_2. The authors observed that single, open shell NO_2 molecule acts as a strong acceptor that results in strong doping of graphene, whereas its closed shell dimer N_2O_4 acts as a weak dopant. The effect of NO_2 and N_2O_4 on the electronic structure of graphene was well understood from the DOS shown in Figure 8.1, which reveals a

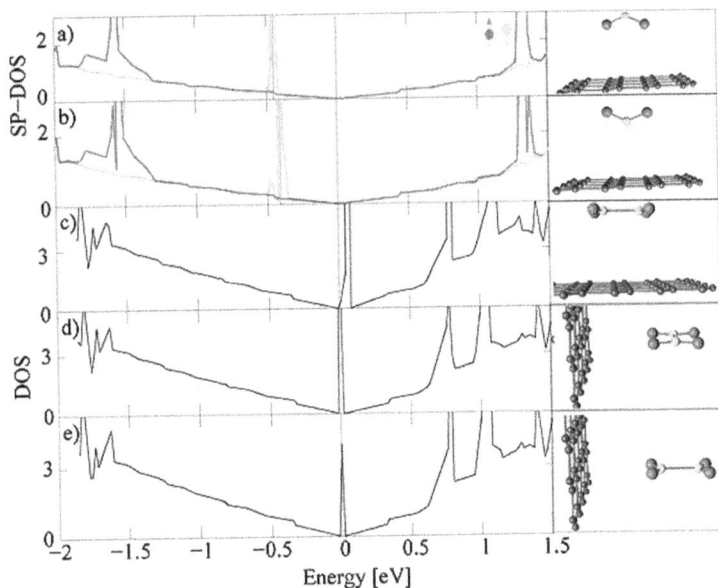

Figure 8.1 Left: Spin-polarized DOS of the graphene supercells with adsorbed NO_2 (a, b) and DOS of graphene with N_2O_4 (c-e), in various adsorption geometries. The energy of the Dirac points is defined as $E_D = 0$. In the case of NO_2, the Fermi level E_f of the supercell is below the Dirac point, directly at the energy of the spin down partially occupied molecular orbital, whereas for N_2O_4 E_f is directly at the Dirac points. Right: Adsorption geometries obtained with generalized gradient approximation (GGA). The carbon atoms are shown in blue, nitrogen is shown in green, and oxygen is shown in red. Reproduced from Reference 31 with permission from American Chemical Society.

strong acceptor level at 0.4 eV below the Dirac point in the NO_2 adsorption geometries (Figure 8.1 a,b). The authors pointed out that as the graphene's DOS spectrum has zero gap, a small chemical potential mismatch between the adsorbate and graphene is sufficient to induce a donor or acceptor level in the spectrum, as compared to conventional semiconductors. Hence, gas sensors made from graphene are expected to be more sensitive than those made from other semiconductors [31].

Later, Dan *et al.* [32] measured the intrinsic chemical responses of graphene devices by removing the resist residue from the lithography step and observed sharp decrease in sensing responses with the clean graphene samples (Figure 8.2). The electrical response of graphene devices was found to be small even upon exposure to strong analytes such as ammonia vapor. I-V characteristics and the device current showed minimal change after exposure to ammonia concentrations as high as 1000 ppm. The authors observed that the contamination layer from the conventional nanolithography process chemically dopes graphene and acts as an adsorbent layer which enhances the

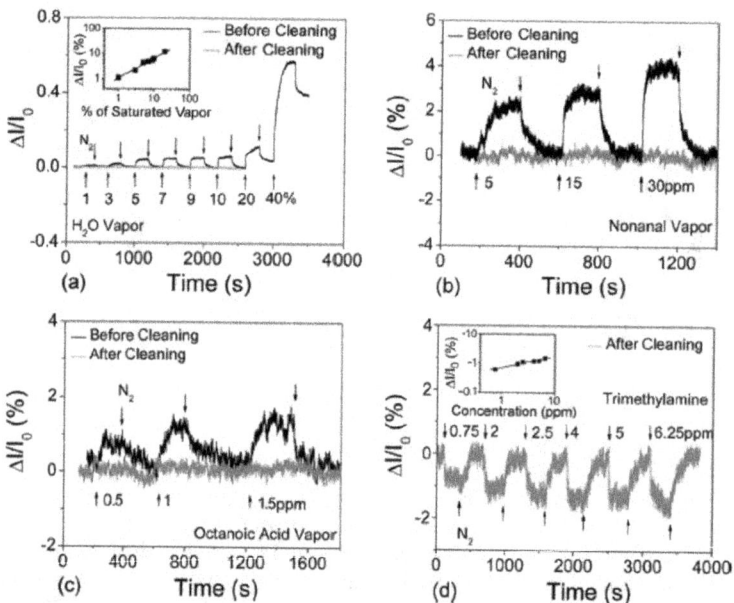

Figure 8.2 Measured sensor responses, before (black) and after (red) sample cleaning, to vapors of (a) water, (b) nonanal, (c) octanoic acid, and (d) trimethylamine. Reproduced from Reference 32 with permission from American Chemical Society.

interactions between the gas molecules and graphene surface, thus, leading to the observed ultrahigh sensitivity of IG. This work confirmed the potential of graphene for vapor sensing applications due to the demonstrated desirable properties of graphene vapor sensors such as fast response and recovery (tens of seconds) and reversibility without heating or other refreshing methods. The authors also suggested the possibility of tuning the chemical and electronic properties of graphene through doping with impurities which could result in enhanced reactivity of graphene towards gas molecules [32].

The adsorption of H_2O, NH_3, CO, NO_2 and NO on IG studied from first-principles based on density functional theory (DFT) calculations have shown physisorption of these gas molecules on IG surface [33]. The signs of the calculated charge transfer between the gas adsorbates and IG were found to be in good agreement with the experimental results reported by Schedin *et al.* [28]. The authors compared the doping strength induced by the paramagnetic gas adsorbates, NO_2 and NO from which they observed that NO_2 induces strong doping, whereas NO induces only weak doping. The result for NO was found to be in contrary to the claim by Wehling *et al.* [33] that paramagnetic molecules induces strong doping.

The adsorption of NH_3 molecule onto IG has been investigated using molecular mechanics and DFT based calculations [34]. The most stable site for the adsorption of NH_3 on graphene surface was observed to be the top of the center of the carbon hexagon (H) from both DFT and molecular mechanics based studies. The calculated adsorption energy of ammonia molecule on the hydrogen terminated graphene flake was 0.01 eV, which indicates weak physisorption. The DOS plots of graphene flake with and without NH_3 molecule revealed that the interactions between NH_3 and graphene flake does not change the electronic structure of graphene flake significantly [34].

Lee and Kim [35] studied the chemisorption and physisorption of CO_2 on graphene based on both DFT and second-order Møller-Plesset (MP2) methods, since MP2 is more accurate than DFT for geometric relaxation of physisorbed molecules. The determined adsorption energies for both types of adsorption showed reasonably good agreement with experimental findings and previously reported theoretical works.

The adsorption mechanism of hydrogen onto various graphene flakes was investigated using first-principles calculations where non-local van der Waals (vdW) density functional (B3LYP-D3) method was used for structural relaxation and total energy estimations [36]. Both conventional and vdW corrected DFT based calculations for a hydrogen molecule above a coronene (a graphene model) surface were carried out to analyze the effect of using these approaches on the hydrogen molecule-graphene adsorbed system. The obtained adsorption energy (5.013 kJ mol^{-1}) for physisorbed hydrogen on coronene surface using B3LYP-D3 method was found to be in agreement with the experimental findings. It was found that non-local dispersion interactions are important in getting accurate results for hydrogen adsorption on graphene surface [36].

Theoretical works on gas molecular adsorption on IG showed very low adsorption energies and negligible charge transfers, as it has highly stable carbon atoms in the honeycomb lattice and has no dangling bonds on its surface. Also the adsorption of most of the gaseous species on IG does not change its electronic properties, as the interactions of gas molecules on IG are very weak [37,38]. Due to the weak binding of gas molecules with IG, significant change in the electrical conductivity of the system does not occur. Thus IG even being a 2D structure is not an ideal material as gas sensor.

8.2.2 Gas Molecules on Graphene Nanoribbons

Theoreticians have also explored the feasibility of employing GNR having free reactive edges, compared to graphene for the detection of NH_3 [38,39], NO_2 [38,39], CO [38-40], O_2 [38,39], N_2 [38], CO_2 [38-41] and NO [38]. DFT investigation of the adsorption of various gas molecules around the sites of dangling bond defects on semiconducting GNRs with armchair-shaped edges (AGNRs) indicated energetically favourable adsorption of CO, NO, NO_2, O_2, CO_2, and NH_3 on AGNR, except that the N_2 adsorption is endothermic in nature [38]. Figure 8.3 also demonstrates the optimized structure of AGNRs with adsorbed CO, NO, NO_2, O_2, CO_2, and NH_3 [38]. The adsorption energies of CO, NO, NO_2, and O_2 on AGNR were larger than 1 eV, which indicates strong chemisorption. For CO_2 and NH_3 adsorption, the adsorption energies were -0.31 eV and -0.18 eV respectively which indicate that the adsorptions lie between strong physisorption and weak chemisorption. The authors also exhibited that all gas molecules influence the electronic structure of AGNRs. Quantum transport calculations indicated that semiconducting AGNRs can be employed for selective detection of NH_3 out of the other gas molecules due to the modification of the conductance of AGNR by NH_3 adsorption, while all others have little effect on conductance. The charge transfer analysis on the adsorbed configurations showed that CO, NO, NO_2, O_2, and CO_2 act as electron acceptors whereas NH_3 act as electron donor. The adsorption energies and the charge transfers of gas molecules on GNR edges are found to be larger than that observed on the surface of graphene or GNR [33,37]. These results led to the conclusion that the interactions of the gas molecules with GNR edges are much stronger than that with graphene or GNR surface [38].

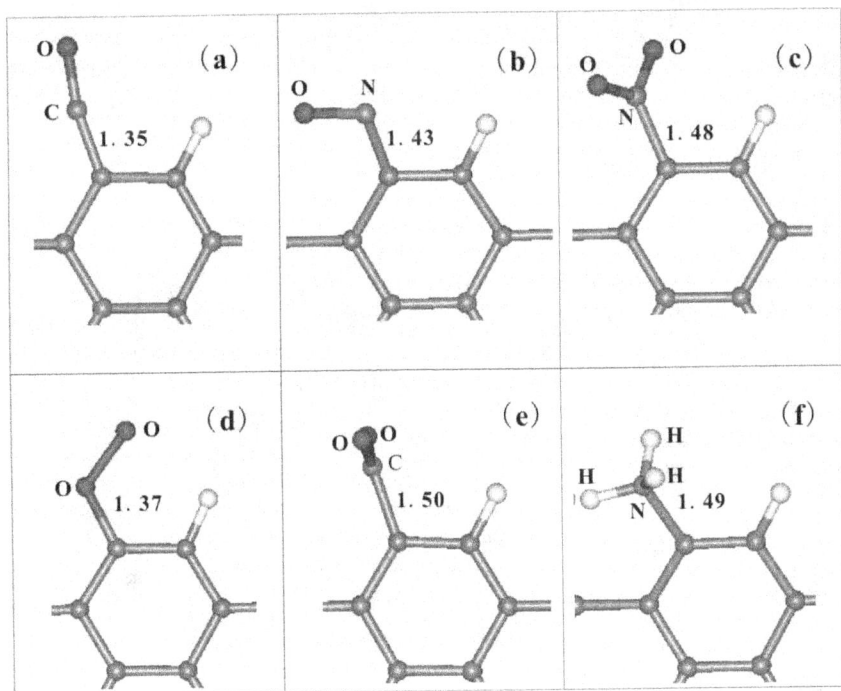

Figure 8.3 Optimized structure of AGNRs with gas molecule adsorption: (a) CO, (b) NO, (c) NO_2, (d) O_2, (e) CO_2, and (f) NH_3. Reproduced from Reference 38 with permission from American Chemical Society.

Paulla and Farajian [40] used MP2 calculations and DFT to examine the detection capability of AGNR having ~1 nm width for CO and CO2 gas molecules. These gas molecules undergo physisorption on AGNR with binding energies of -0.35 eV for CO2 and -0.252 eV for CO with low charge transfer values ranging from -0.005 e- to +0.005 e- respectively. Quantum conductance calculations on AGNR based nanosensor showed shift in the conductance characteristics on the order of few meV, compared to pristine AGNR for the adsorption of one and two molecules per two unit cells of AGNR. The effect of concentration of O_2 molecule adsorption on the electronic properties of semiconducting AGNRs with hydrogenated edges (HEAGNR) has been investigated via DFT. The adsorption energy calculations for different positions of gas molecule on HEAGNRs indicated that the carbon atoms at the edges are the most favorable centers for adsorption. The results showed decrease in the energy gap with increase in the concentration of O_2 molecule and the energy gap disappears at gas concentration of 0.02. This dependence of the electronic properties of HEAGNR on gas concentration could be used for sensing O_2 gas [42].

8.2.3 Gas Molecules on Graphene Oxide

Compared to graphene, GO serves as a suitable candidate for high performance gas molecular sensing under practical situations [43]. The enhanced interactions between the gas adsorbates and the reactive oxygen rich functional groups on GO improves the sensitivity of GO towards gas molecules. Several studies have focused on using GO as gas molecular sensors [43,44]. Peng *et al.* [43] investigated the adsorption of NH_3 on IG and GO using first-principles calculations. The results have shown that doping of graphene with -O and -OH species could increase the adsorption energy and the charge transfer of the NH_3 adsorbate. The calculated adsorption energy and the charge transfer for NH_3 on IG's surface were small, whereas for GO, the surface epoxy and hydroxyl groups promote the interactions between NH_3 and GO. The enhancement of the adsorption energy for the hydroxyl groups was found to be greater than that for the epoxy groups on the GO surface. Adsorption of NH_3 on epoxy or hydroxyl groups that exist on the basal plane of graphene does not show an appreciable increase in the adsorption energy [43].

Another DFT study on the interactions of nitrogen oxides NO_x (x = 1, 2, 3) and N_2O_4 with IG and GO also showed stronger adsorption of nitrogen oxides on GO than that on IG due to the presence of active defect sites such as hydroxyl, carbonyl functional groups and the carbon atom near these groups [44]. The sufficient active sites on GO increase the binding energies and enhance the charge transfer from nitrogen oxides to GO, which confirms the chemisorption of these gas molecules on the surface of GO. The DOS graphs of gas molecule-adsorbed GO showed strong hybridization of frontier orbitals of NO_2 and NO_3 with the electronic states around the Fermi level of GO, which resulted in strong acceptor doping of GO by these gas molecules [44].

Mattson *et al.* [45] reported *in-situ* infrared (IR) microspectroscopy investigation of NH_3 adsorption on reduced graphene oxide (RGO) to study the dynamics of adsorption under various realistic working conditions (i.e. ambient pressure) along with DFT results. The authors observed that the residual oxygen functional groups and structural defects in the RGO lattice act as highly active centers for NH_3 adsorption. The results suggested that the interactions of the epoxide groups and carbon vacancies of RGO with NH_3 will lead to the molecularly physisorbed NH_3 and also a variety of chemisorbed species arising from the dissociation of NH_3. The calculations have shown that the chemisorbed fragments such as NH_2, OH, and CH can produce small electron donor effect (Figure 8.4) [45].

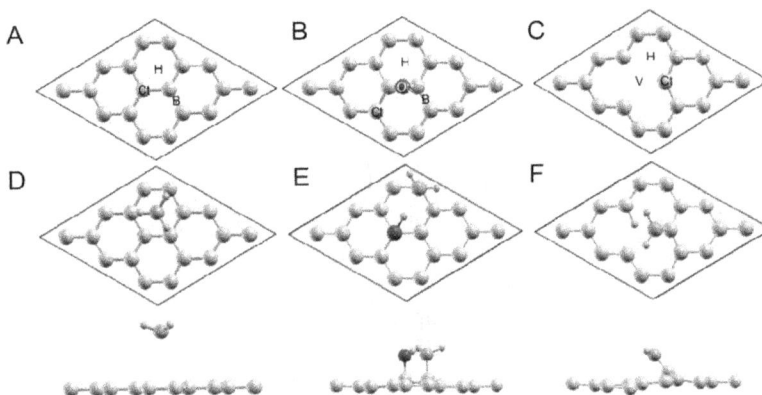

Figure 8.4 Structural models of RGO and NH_3 adsorption on graphene and RGO. (A-C) 3 × 3 supercells used to model bare RGO containing (A) pristine graphene, (B) graphene with a single epoxide group, and (C) graphene with a carbon vacancy. (D) NH_3 molecules adsorb molecularly on pristine graphene, (E) NH_3 molecules that adsorb near epoxide groups dissociate, resulting in NH_2 and OH groups, and (F) NH_3 molecules adsorbed at defects are dissociated into NH_2 and H bonded to next-nearest-neighbor carbon atoms. Reproduced from Reference 45 with permission from American Chemical Society.

8.2.4 Gas Molecules on Doped and Defected Graphene

Heteroatom doping has proved to be an efficient method for tailoring the physical, chemical and electronic properties of graphene [46]. Heteroatom doping of graphene can significantly alter the electronic structure of graphene [47] and make it appropriate to use as a highly sensitive gas sensor. As heteroatom doping allows tuning of the chemical reactivity of graphene, this approach has been verified for improving the sensing efficiency using quantum mechanical simulations.

For instance, Zhang *et al.* [37] investigated the interactions of intrinsic, boron-, nitrogen-doped and defective graphenes (hereafter abbreviated as IG,

BG, NG and DG respectively) with small gas molecules such as CO, NO, NO_2 and NH_3 using density functional computations. The authors observed enhancement in the interactions between gas molecules and doped or defective graphenes due to the much stronger adsorption of gas molecules on graphene modified with dopants and defects than that on IG. NG had weak interactions with CO, NO and NH_3, but had strong binding with NO_2. B-doping of graphene improved its adsorption properties for NO, NO_2 and NH_3, out of which NH_3 had tightest binding with BG. The adsorption energies, charge transfers and the electronic total charge density plots for systems of CO, NO and NO_2 adsorbed on DG showed that vacancy on DG acts as strong chemisorption binding site for CO, NO and NO_2 molecules. Therefore, the authors concluded that DG can be used as a suitable sensing platform for the detection of CO, NO and NO_2 as compared to other graphenes such as IG, BG and NG. The calculations also showed dramatic changes to graphene's electronic properties due to the strong interactions between adsorbed molecules and modified graphenes. It was concluded from the DOS plots of gas molecule-graphene adsorption systems that DG, BG and NG are appropriate for sensing CO, NO and NO_2, respectively. The current-voltage curve of gas sensor based on BG showed considerable increase of current on NO_2 adsorption and sensitivity of nearly two orders of magnitude higher than that of IG was observed at an optimum bias voltage of 1.0 V for NO_2. BG was found to be more sensitive to NO_2 than NH_3 as the observed sensitivity to NH_3 was only one order of magnitude higher than that of IG, when the bias voltage is greater than 1.0 V. The work revealed that the sensitivity of graphene based chemical gas sensors could be drastically improved by introducing appropriate dopant or defect [37].

Recently, Rad *et al.* [48] used DFT calculations to investigate the adsorption of SO_x ($x = 2, 3$) molecules on the surface of IG and NG. The study showed that the binding energy and the net charge transfer of the molecules on graphene got increased by doping with N. The DOS plots displayed major orbital hybridization between SO_x and NG, while there was no hybridization between SO_x and IG molecules. Physical and chemical adsorption of SO_2 and SO_3 molecules were observed on the surface of NG with binding energies of -27.5 and 65.2 kJmol^{-1}, respectively. From the results, the authors concluded that NG can be used as sensitive sensor for the detection of SO_2 and SO_3 [48]. Another work on the adsorption properties of O_3, SO_2 and SO_3 on BG reported strong chemisorption for O_3 with dissociation on the surface of BG [49]. The very low adsorption energies of SO_2 and SO_3 on BG indicated weak interactions of these molecules with BG. Rad *et al.* [49] pointed out that the n-type semiconducting property of BG is responsible for the improved sensitivity of BG to O_3 molecule. The observed high sensitivity of BG to O_3 can be exploited for highly selective O_3 sensors.

The enhanced sensitivity of graphene to CO through aluminium (Al) doping has been demonstrated theoretically [50]. DFT based calculations showed strong chemisorption of CO molecule on Al-doped graphene (hereafter abbreviated as AlG) through the formation of a strong Al-C bond with significant electron transfer from AlG to CO, whereas weak physisorption of CO was observed on IG. The charge difference caused by CO adsorption in AlG ($Q = 0.027$ e^-) was almost one order larger than that in IG ($Q = 0.003$ e^-), due to the modified electronic properties of graphene by Al doping. The noticeable increase in the DOS value of AlG near the Fermi level and the closure of energy band gap after CO adsorption indicated the introduction of extra number of shallow acceptor states in AlG through CO interaction. The large increase in the electrical conductivity of AlG upon chemisorption of CO molecule can be exploited for developing highly sensitive CO sensors [50].

Wang *et al.* [51] recently studied the adsorption of CO on IG, NG and AlG and observed very low adsorption energies of CO molecule on IG and NG. Similar to the findings in reported in Ref. [50], large adsorption energy and large charge transfer were observed for CO molecule adsorption on AlG. Although N-doping enhances the interactions of CO molecule with graphene, the adsorption still remained to be weak. The calculated band structures, DOS, and the charge density difference for the doped graphene sheets with and without adsorbed CO molecule showed significantly different adsorption on AlG compared to that on IG and NG. The study concluded that AlG is much more sensitive than IG and NG to the adsorption of CO [51].

AlG was also suggested to be a promising sensing material for detecting formaldehyde (H_2CO) from DFT based study [52]. The chemisorption of H_2CO molecule on the surface of AlG was confirmed by the calculated high adsorption energy value, short binding distance and the observed orbital hybridization between H_2CO and AlG, compared to that calculated for IG. The Mulliken charge analysis of H_2CO-adsorbed IG and AlG systems showed a charge transfer of 0.085 e^- from AlG to H_2CO molecule, which was about 4 times more than the charge transfer of 0.019 e^- from IG to H_2CO molecule [52].

High sensitivity of AlG to NO_2 and N_2O has also been reported [53]. The larger adsorption energy, significant net charge transfer and small binding distance of NO_2 and N_2O on the surface of AlG compared to that calculated for IG indicated strong chemical interactions of these gas molecules with AlG. The calculated DOS plots showed orbital hybridization between NO_2 as well as N_2O and AlG during adsorption, while no hybridization between the gas molecules and IG was observed [53]. Another work on the comparison of the adsorption behaviors of CO, CO_2 and H_2O on the surface of IG and AlG reported very high adsorption on AlG in contrast to relatively low adsorption for IG [54]. The weak physisorption of the gas molecules on IG and the strong chemisorption on AlG was confirmed from natural bond orbital analysis and DOS plots apart from the analysis of adsorption energies and charge transfers. The enhancement in the adsorption of the gases on graphene through Al-doping could be attributed to the modification in the electronic properties of graphene with increased electron density on carbon atoms surrounding the Al-dopant and decreased electron density on Al-dopant. This charge relocation makes the dopant atom as the reactive site for gas molecules [54].

In another study, Milowska and Majewski [55] reported the transport properties of sensors based on covalently modified graphene monolayer (GML) functionalized by $-CH_3$, $-CH_2$, $-NH_2$, $-NH$ and $-OH-CH_3$, $-CH_2$, $-NH_2$, $-NH$ and $-OH$ Green's function (NEGF). For instance, Figure 8.5 demonstrates the current-voltage characteristics of pure and functionalized GML. The authors observed that carbene introduces significant changes in the transmission

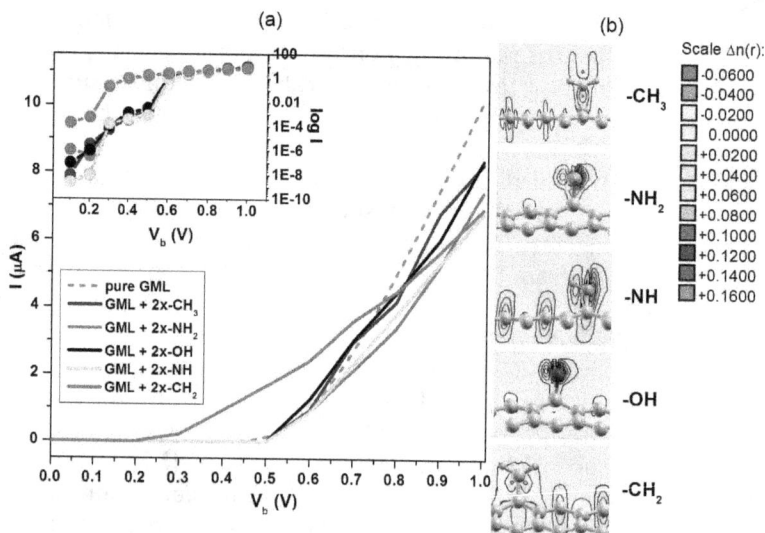

Figure 8.5 (a) The current-voltage characteristics of pure and functionalized GML; (b) the difference of the valence pseudocharge density and the superposition of the atomic valence pseudocharge densities. Reproduced from Reference 55 with permission from American Chemical Society.

spectrum and current-voltage characteristics, whereas the methyl groups were observed to cause minimal change. These findings provide fundamental insights into the design of sensors using doped and functionalized graphene for various gases.

A systematic theoretical investigation of the adsorption of common gas molecules such as H_2, H_2O, O_2, CO_2, CO, NO_2, NO, SO_2, NH_3, and N_2 on graphene doped with substitutional impurities such as boron, nitrogen, aluminium and sulfur has also been reported [56]. It was found that BG possesses the ability to chemically bind NO and NO_2, whereas S-doped graphene (hereafter abbreviated as SG) can only bind NO_2 molecule. The Lowdin charge analysis showed a charge transfer of about 0.35 e^- from BG to NO_2, and of about 0.76 e^- from SG to NO_2, which could produce large changes in the conductivity of these doped graphenes. Out of the considered doped graphenes, AlG was found to be highly reactive and binds all the considered gases except H_2 through strong Al–X (X = O, N, C) bonds, of length around 2 Å, which renders it as a highly sensitive gas sensing material. The large local curvature induced by the Al atom in the graphene lattice tends to make AlG more chemically reactive. The authors suggested that BG and SG could be used as good sensors for detecting polluting gases such as NO and NO_2. The results also predict the suitability of using SG as a selective sensor for NO_2 detection in the presence of other gas molecules such as NO, NH_3, CO, CO_2, H_2O, SO_2, O_2, H_2 and N_2 [56].

Dai and Yuan [57] evaluated the effect of O_2 on doped graphene for enabling practical applications of gas sensors, electronic and spintronic devices by simulating the adsorption of molecular oxygen on BG, NG, AlG, Si-, P-, Cr- and Mn-doped graphene (hereafter abbreviated as SiG, PG, CrG and MnG respectively) using DFT. Physisorption was observed for O_2 adsorption on BG and NG due to the weak interactions of O_2 molecule with the dopant atom (B and N). O_2 molecule was found to be chemisorbed on AlG, SiG, PG, CrG and MnG with induced obvious changes in the electronic structures due to the enhanced reactivity of these doped graphenes towards O_2. The results from the simulations indicate that the presence of O_2 molecule in air greatly affects the sensitivity of the gas sensors based on AlG, SiG, PG, CrG and MnG, which limit their usage as efficient gas sensors [57]. The authors, in another study, also investigated the adsorption property of various gases such as H_2, H_2O, CO_2, CO, NO, NO_2, SO_2, N_2, O_2 and NH_3 on PG. The calculations proved that PG can only physisorb gas molecules such as H_2, H_2O, CO_2, CO, N_2 and NH_3, whereas firm adsorption of NO, NO_2, SO_2 and O_2 was observed with large adsorption energies and short graphene-gas molecule distances through the formed P–X (X = O, N, S) bonds [58].

Zou *et al.* [59] investigated the adsorptions of small gas molecules such as CO, O_2, NO_2 and H_2O on IG and SiG using *ab initio* calculations. IG was found to be sensitive to the presence of O_2 and NO_2 molecules, as the electronic properties of IG got influenced by the adsorption of these gas molecules. However, there was no change in the electronic properties of IG upon the adsorption of CO and H_2O molecules. Doping of graphene with silicon makes it more reactive to the adsorption of CO, O_2, NO_2 and H_2O molecules and hence could be used as a good sensor for CO, O_2, NO_2 and H_2O. The strong interactions between SiG and the adsorbed molecules induced dramatic changes to the electronic properties of SiG [59]. Another DFT investigation of the adsorption of NO, N_2O and NO_2 on IG and SiG predicted the possibility of using SiG as an excellent candidate for the detection of NO and NO_2, as well as a metal-free catalyst for N_2O reduction [60]. The results indicated significant enhancement in the interactions of these gas molecules with graphene sheet upon Si doping. It was found that the adsorption of NO and NO_2 result in large changes in the electronic properties of SiG [60]. Zhang *et al.* [61] reported that H_2S adsorption was found to be much stronger on SiG through the formation of Si–S bond, compared to the weak adsorption of H_2S molecule observed on IG, BG and NG.

Lv *et al.* [62] carried out first-principles study of N_2O interaction with IG, AlG and Gallium (Ga) doped graphene (GaG). N_2O was found to be chemisorbed on the surface of AlG and GaG by the stronger covalent bond formed between the dopant atom and the N_2O molecule via the *p* orbitals coupling. The authors observed that the N_2O molecule adsorbed on the surface of AlG and GaG can be easily decomposed to N_2 and O_2 with the application of a perpendicular electric field. Therefore, both AlG and GaG can be used as excellent candidates for the detection and the dissociation of N_2O, thus protecting the ozone layer and controlling the global warming effects [62].

Sharma *et al.* [63] evaluated the H_2S adsorption potential of IG and group III (B, Al and Ga)-doped graphene using DFT modelling. It was found that H_2S weakly binds with IG and BG, on the other hand, chemisorption was observed for AlG and GaG with large adsorption energies and short binding distances. The authors also reported that the increased reactivity of AlG and GaG towards H_2S is due to the local curvature in these doped graphenes formed by the long X-C (X = Al or Ga) bonds compared to C-C bonds [63]. DFT study of phenol adsorption on IG and group III (B, Al and Ga)-doped graphene by Avila *et al.* [64] showed weak physical interactions of phenol molecule with IG and BG. The structural and electronic properties of IG and graphene doped with B atom remain unaffected by phenol adsorption. Much stronger interaction was seen between the phenol molecule and graphene doped with Al and GaG, corresponding to chemical adsorption [64]. Based on the report of Shao *et al.* [65] on the adsorption of SO_2 on IG and heteroatom (B, N, Al, Si, Cr, Mn, Ag, Au, and Pt)-doped graphene sheets using first-principles approach, AlG, SiG, CrG, MnG, Ag-, Au- and Pt-doped graphene (hereafter abbreviated as AgG, AuG and PtG respectively) can be used as good candidates as SO_2 gas sensors due to the strong chemisorption of SO_2 on these doped graphenes. SO_2 molecule had weak adsorption on IG, BG and NG. The authors observed that CrG and MnG are the best choices for SO_2 adsorption due to large adsorption energies and charge transfers of SO_2-adsorbed CrG and MnG systems, compared to that calculated for other doped graphene-SO_2 adsorbed systems [65]. Another study on the adsorption of H_2CO on BG, NG, SiG, AlG, CrG, MnG, and AuG suggested that BG and NG are not sensitive to H_2CO, whereas SiG, AlG, CrG, MnG, and AuG are sensitive to H_2CO as they can strongly adsorb H_2CO molecule [66]. Al and Mn were found to be the most appropriate dopants on graphene for the detection of H_2CO molecule because of their relatively large adsorption energies, large charge transfers and significant changes of DOS before and after adsorption.

In another study, Pramanik and Kang [67] reported the adsorption of paramagnetic O_2 and NO on pristine as well as N-doped and P-doped graphene (GrS) using DFT calculations. The authors observed that the van der Waals interaction significantly affects the physisorption energy along with the adsorption geometry of the gases in pristine and N-doped graphene. The authors also observed adsorption of the gases on P-doped graphene to be stronger than that on pristine or N-doped graphene. It was concluded that the O_2-adsorption is a normal chemisorption, whereas NO-adsorption was weak chemisorption. For instance, Figure 8.6 also demonstrates the optimized geometry of P-doped GrS-O_2.

In another study, Choudhuri *et al.* [68] performed DFT calculations to study the gas (CO, CO_2, NO, and NO_2) sensing mechanisms prevailing in pure and doped (B@, N@, and B–N@) graphene surfaces. Figure 8.7 represents the optimized structures of (a) CO, (b) CO_2, (c) NO, and (d) NO_2 molecules adsorbed on 6.25% B–N@graphene surface. The authors observed that the doping improves adsorption energy and selectivity. The electronic properties of the B–N@graphene surfaces were reported to change significantly in comparison with pure and B@ and N@graphene surfaces, when the gas molecules were adsorbed. It was also reported that the electronic properties changed maximum for NO and NO_2, but not for CO and CO_2. The authors reported that the B–N@graphene is a promising semiconductor based gas sensor, which is a significant finding.

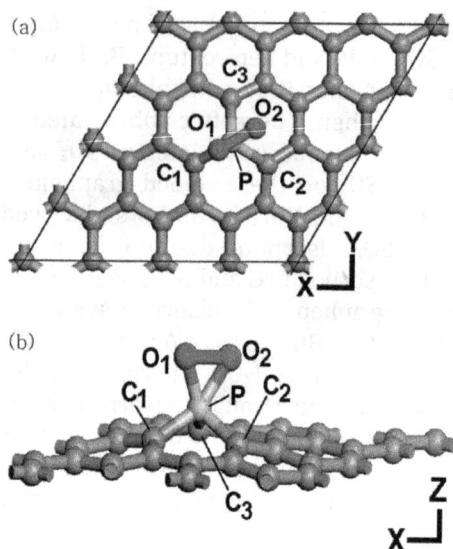

Figure 8.6 Optimized geometry of P-doped GrS-O_2 in XY (a) and XZ (b) planes. Reproduced from Reference 67 with permission from American Chemical Society.

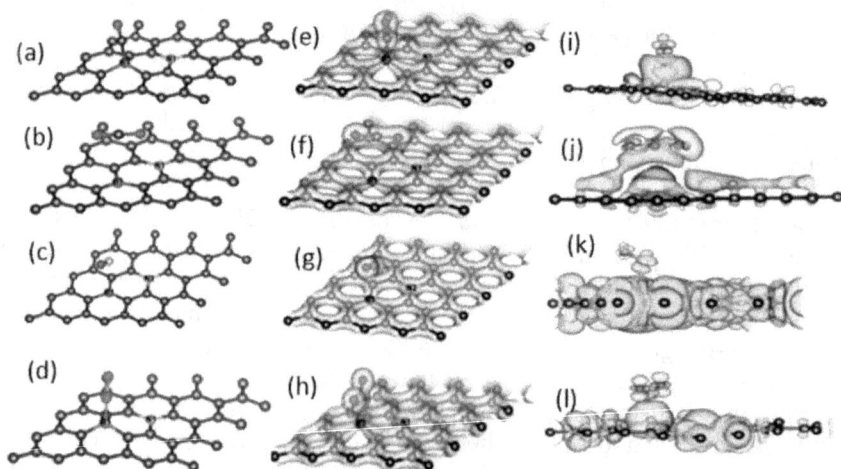

Figure 8.7 Optimized structures of (a) CO, (b) CO_2, (c) NO, and (d) NO_2 molecules adsorbed on 6.25% B–N@graphene surface. Reproduced from Reference 68 with permission from American Chemical Society.

Jiang *et al.* [69] also reported first principles density functional theory calculations to analyze the effect of designed sub nanometer pores on the permeabilityand selectivity of graphene sheets. High selectivity for H_2 was observed for H_2/CH_4 gas pair, along with high permeability for H_2 a nitrogen-functionalized pore. On the other hand, very high selectivity for H_2 was observed for an all-hydrogen passivated pore. Figure 8.8 also demonstrates the interaction energy between H_2 and the all-hydrogen passivated porous graphene as a function of adsorption height. Though the developed porous graphene was useful as a gas separation membrane, however, the use of designer nanopores can also find potential applications towards gas sensing.

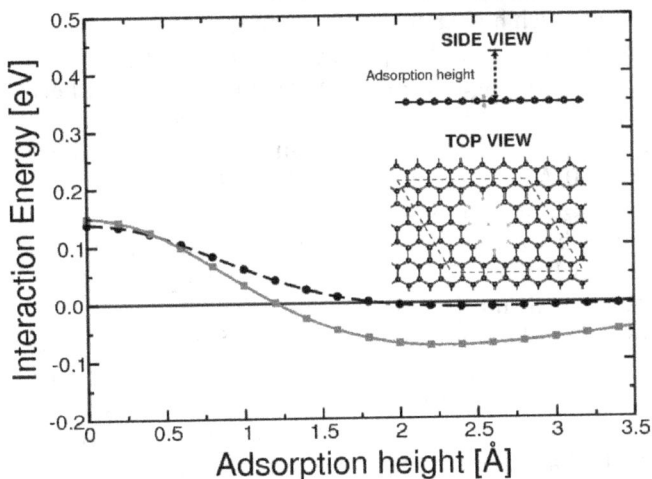

Figure 8.8 Interaction energy between H_2 and the all-hydrogen passivated porous graphene as a function of adsorption height. Squares and solid line: Rutgers-Chalmers van der Waals density functional (vdW-DF); circles and dashed line: Perdew, Burke, and Erzenhoff (PBE). Reproduced from Reference 69 with American Chemical Society.

Lee *et al.* [70] investigated the effect of adsorption of N_2, O_2, NO_2 and SO_2 on the magnetic properties of PtG. N_2, O_2, NO_2 and SO_2 were observed to be chemisorbed on PtG. The authors observed four different types of magnetic properties of PtG based on the chemisorbed molecules on the surface of PtG. Even though there was no spin polarization for N_2 adsorption, other gas molecules changed the magnetic properties of PtG. O_2 adsorption on PtG led to local polarization on the gas molecule, whereas the adsorption of NO_2 and SO_2 on PtG introduced two types of complete polarization on PtG, which differs in the spin direction of PtG and gas adsorbate. These results suggest that the differences in magnetic properties of the PtG based on the type of adsorbed gas molecules could be exploited for developing graphene-based gas sensors [70].

The H_2S sensing behaviour of graphene sheet has been found to be significantly enhanced by the substitution of graphene with transition metal (TM) atoms such as Ca, Co, Fe and modification with single vacancy defect from DFT calculations and non-equilibrium Green's function formalism [61]. The observed dramatic changes of the electronic and magnetic properties of TM-doped graphene upon H_2S adsorption suggested strong interactions of H_2S molecule with TM-doped graphene. The electron transport properties of gas sensors using Fe-doped graphene (abbreviated as FeG) sheets as sensing materials exhibited much higher sensitivity than that of the devices made with IG. Similar study on the effect of doped Fe atom on the H_2S sensing behavior of graphene also showed significantly improved interactions between H_2S and FeG sheet from the results of adsorption energy, electron density difference, and DOS plots [71]. Different H_2S adsorbed FeG systems with distinct binding distances (2.5-7.0 Å) were simulated to investigate the influence of the distance between H_2S and the FeG sheet on their interactions. The adsorption energies approach to zero for binding distances larger than 5.0 Å, which is similar to that of H_2S on IG. The DFT calculations proved that the adsorbed H_2S dissociates into S and H_2 [71].

The analyses of the adsorption energies of molecular halogens, CH_3OH, CH_3SH, H_2O, and H_2S on graphene, SG and 2SG showed that the SG is the best adsorbent, fluorine is the best adsorbate, and all molecular halogens adsorb on graphene better than other molecules. The enhancement in the electrical conductivity of SG after the interactions with molecular halogens was predicted from the molecular orbital results and DOS graphs [72].

Wanno *et al.* [73] used DFT calculations to investigate the use of VIIIB TM (Fe, Ru, Os, Co, Rh, Ir, Ni, Pd, and Pt)-doped graphene sheets for CO sensing and observed that CO molecule forms a strong chemisorption bond to VIIIB TM-doped graphene. Among the considered VIIIB TM-doped graphene sheets, Os- and Fe-doped graphene displayed strong chemical interactions with C and O atoms of CO molecule respectively [73]. Later Nasehnia and Seifi [74] considered the effect of O_2 molecule on VIIIB TM-doped graphene structures to evaluate the suitability of these TM-doped graphenes as toxic gas sensing materials in air. O_2 molecule was found to be chemisorbed on these doped graphene sheets with large adsorption energies (−0.653 eV to −1.851 eV) and appreciable amounts of charge transfers (0.375 e^- to 0.650 e^-) from doped graphene sheets to O_2 molecule. The enhanced reactivity of VIIIB TM-doped graphene structures to O_2 could be attributed to the metallic doping and the pyramidalization of the doped graphene sheets. The adsorption energies for both O_2 and CO molecules on VIIIB TM-doped graphene sheets have been found to be similar. Hence, these materials may not be considered as a good choice for toxic gas sensing application [74].

Ma *et al.* [75] analyzed the sensitivity of IG and Pd-doped graphene (abbreviated as PdG) toward small gas molecules such as CO, NH_3, O_2 and NO_2 using first-principles. The authors observed weak adsorption of the gases on IG and significant increase in the interactions of graphene with gas molecules by the introduction of Pd dopants. The authors proposed that PdG is more appropriate for the detection of CO, NH_3, O_2 and NO_2 compared to IG owing to the dramatically increased adsorption energy and elevated charge transfers of gas molecule-Pd doped graphene systems [75].

Cr-doped zigzag graphene nanoribbons (Cr-ZGNRs) have shown high sensitivity to SO_2 molecule than pure ZGNRs from the investigation of the effect of SO_2 molecule on the structural, electronic and transport properties of pure and Cr-ZGNR using DFT and non-equilibrium Green's functional formalism [76]. The adsorption of SO_2 did not induce any change in the I-V curves of pure ZGNRs, which shows the insensitivity of pure ZGNRs towards SO_2. The observed significant increase of molecular current through Cr-ZGNR after SO_2 molecule adsorption revealed the potential of using Cr-ZGNR as a good SO_2 gas sensor [76].

DFT studies have been carried out by Aguila *et al.* [77] to elucidate the interactions between three different graphene structures (intrinsic, divacancy and the inclusion of N_2 in the graphene) and four H_2S concentrations (1 H_2S, 2 H_2S, 3 H_2S and 4 H_2S). H_2S binds to IG layer, graphene layer modified with defects and impurities with adsorption energies in the range of 43-433 meV, that correspond to the physisorption of H_2S on the considered graphene structures. Largest binding energies were observed for all the three graphene structures upon the adsorption of 3 H_2S molecules, which indicates the concentration of 3 H_2S adsorbed molecules saturate both intrinsic and modified graphene sheets [77].

DFT based study of the adsorption of H_2S molecules onto defective graphene as a function of vacancy concentration showed stronger interactions of the molecule with the carbon atoms surrounding the vacancy than with the carbon atoms in a perfect arrangement, which corresponds to chemisorption process followed by the release of H_2 molecule. The DOS of IG, DG and DG with chemisorbed sulfur demonstrated changes in the electrical conductivity. The most favorable adsorption configuration was found to be the one in which the sulfur atom of the H_2S molecule is at the center of the vacancy defect, which facilitates covalent binding with three carbon atoms through unsaturated bonds [78].

Using first-principles calculations, Ma *et al.* [79] showed chemisorption of CO, NO and O_2 gas molecules on vacancy-defected graphene (VG) and hence VG cannot selectively detect CO in air, as the chemisorbed NO and O_2 produce charge transfer which strongly disturb the sensing signals of VG for CO. From the investigation of the adsorption of NO, CO, N_2 and O_2 molecules on pyri-dinic-like N-doped graphene (NG), it was found that NG has high selectivity for CO in the presence of N_2, O_2 and NO due to chemisorption of CO along with large charge transfer compared to the physical adsorption of N_2, O_2 and NO on the surface of NG [79].

The adsorption studies of H_2CO on graphene modified with a combination of vacancy and dopants such as B, N and S suggested that VG is more sensitive to the presence of H_2CO molecule that IG [80]. However, H_2CO molecule had physisorption on VG with small adsorption energy. The calculations have shown chemisorption of H_2CO on B-doped, N-doped and S-doped vacancy-defected graphene (abbreviated as BVG, NVG and SVG respectively) for which the adsorption energies and net charge transfer are larger than that of VG without dopants. The enhanced adsorption of H_2CO molecule on the dopant-defect combination due to the strong orbital interaction between the *p* orbit-als of dopant atoms and carbon atoms around the vacancy was also evident from the results of partial DOS [80].

TM-doped Stone-Wales (SW) defected graphene has also been verified as a promising sensing material for H_2CO detection due to the enhanced adsorp-tion on its surface [81]. H_2CO molecule undergoes chemisorption on TM atoms (Cr, Mn and Co) doped perfect graphene (abbreviated as Cr-PG, Mn-PG and Co-PG respectively) and SW defective graphene structures. The larger binding energy and shorter binding distance of H_2CO molecule on TM-doped defected graphene (Cr-SWG, Mn-SWG and Co-SWG) than that on TM-doped perfect graphene structures showed that the presence of SW-defect improved the ad-sorption of H_2CO. It was found that the adsorption of H_2CO affect the electron-ic conductance of the Cr-doped and Mn-doped defective graphene, and hence by measuring the change in electronic conductance, the presence of H_2CO molecule can be detected sensitively. These works theoretically revealed the combined effect of defect and doping on the sensitivity of graphene towards H_2CO gas, for developing future H_2CO sensing devices [81].

Hussain *et al.* [82] investigated the adsorption capability of hydrogenated graphene (graphane) sheet for H_2S and NH_3 gases using DFT. Both pristine graphane sheet and defected graphane sheet (obtained by removing a few surface H atoms from the graphane sheet) have been found to have low affinity for both H_2S and NH_3 gas molecules. The sensing affinity of graphane sheet to these gas molecules was found to be stronger on doping the graphane sheet with Li adatoms. Li-doped graphane sheet showed higher affinity for NH_3 gas compared to H_2S gas due to the stronger hybridization of s orbitals of Li adatom with p orbitals of nitrogen atom of NH_3 compared to that of with the p orbitals of sulfur atom of H_2S. The adsorption energy calculations indicated that Li-doped graphane sheet is suitable for the detection of H_2S and NH_3 gases [82].

In another work by Tanveer *et al.* [83], pristine graphane sheet showed less affinity towards CO, H_2O and NO_2 gas molecules. Defected graphane sheet had strong affinity towards these gas molecules. However, CO and H_2O were found to be weakly physisorbed to the defected graphane sheet. Defected graphane sheet chemically binds NO_2 molecule due to the strong hybridization of the N (p) and O (p) states with the most active C (p) states which lie at the defected site of the graphane sheet. The observed increase in the trend of the work function of the defected graphane sheet on the adsorption of CO, H_2O and NO_2 gas molecules opens up possibilities for next generation gas sensors [83].

Tawfik *et al.* [84] examined the sensitivity of ZGNR with titanium or tin (Sn) atom adsorbed in a double carbon vacancy (DV ZGNR) to H_2S and SO_2 gas molecules. Both Ti- and Sn-doped DV ZGNRs (abbreviated as Ti/DV ZGNR and Sn/DV ZGNR respectively) have shown high sensitivity to H_2S gas. The observed weak sensitivity of Sn/DV ZGNR and the mild sensitivity of Ti/DV ZGNR to SO_2 indicate that both these sensor structures are poor sensors of SO_2. The strong interaction of O_2 molecule with Ti/DV ZGNR through chemisorption process with associated dissociation resulted in reduction of the adsorption energy of both H_2S and SO_2 on Ti/DV ZGNR. In the case of Sn/DV ZGNR, there was no significant reduction in the adsorption energy of subsequent H_2S, while the adsorption energy of SO_2 was found to be significantly reduced after the adsorption of O_2. These results propose the reusability of oxidized Ti/DV ZGNR for both gases, whereas Sn/DV ZGNR exhibited reusability only with respect to SO_2. The authors observed that oxidation does not affect the sensitivity of Ti/DV ZGNR to H_2S, but prevents the dissociation of the H–S bond during the adsorption of H_2S on Ti/DV ZGNR, which makes Ti/DV ZGNRs reusable for this gas [84].

Recently, Xie *et al.* [85] explored the gas sensing behavior of molecular devices constructed with an azulene-like dipole molecule sandwiched between two GNR electrodes using DFT and nonequilibrium Green's function method. The authors showed that toxic gas molecules such as CO, NO and NO_2 would adsorb on the doped B atoms of the AGNR. Different gas molecule adsorption on molecular devices with symmetric B-doped AGNR electrodes resulted in different change in the current–voltage profile and hence these devices can be employed for the detection of these gases. The adsorption of these gas molecules induced changes in the subbands of B-doped AGNR near the Fermi energy, which depends on the type of the adsorbed molecule. These changes led to obvious variation in the current profile and were employed for specific detection of gas molecules [85].

As shown in Figure 8.9 [86], a transient voltage across the rGO was applied

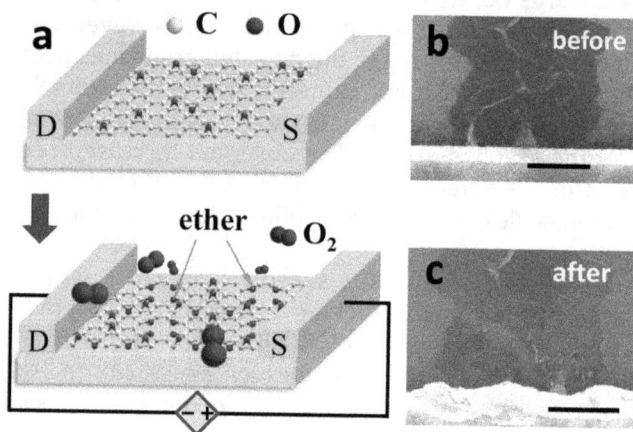

Figure 8.9 Schematic illustration of the voltage activation process for rGO in the air. Reproduced from Reference 86 with permission from American Chemical Society.

to generate defects in the rGO. The authors observed that the voltage activation process led to an increase in the number of epoxide and ether functional groups in the rGO. The activated rGO was observed to be sensitive to NO_2. The sensitivity of the activated rGO was 500% higher than that of the original rGO, with the detection limit to very low concentration of 50 ppb. The authors observed through density functional theory (DFT) calculations that the high sensitivity to NO_2 and sensing was due to charge transfer from ether groups to NO_2.

Kumar *et al.* [87] also studied the effect of external defects in sensing the performance of graphene field-effect transistors (chemFETs). The authors observed that the sensitivity of graphene chemFETs was not solely due to graphene, and can be generated by external defects in the insulating substrate (Figure 8.10). These external defects in the insulating substrate can help to tune the electronic properties of graphene.

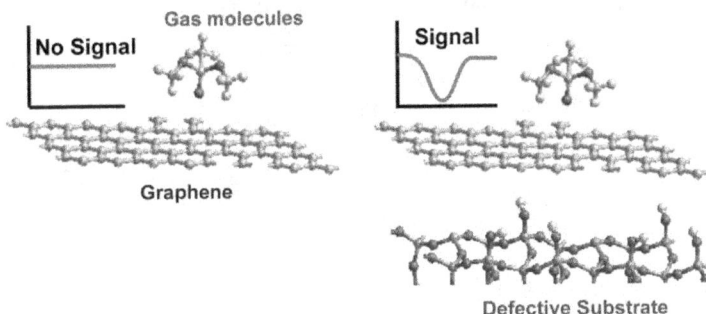

Figure 8.10 Representation of the sensing performance in the presence of extremal defects. Reproduced from Reference 87 with permission from American Chemical Society.

To verify the effect of different dopant atoms such as N, Al, Zn, Ti, Zr on the interactions between hydrogen molecules and graphene sheets, Zhang *et al.* [88] performed DFT based calculations on the hydrogen molecule-intrinsic and doped graphene adsorbed systems. The interaction energy with the hydrogen molecule was found to be the largest for Ti-doped graphene (abbreviated as TiG), which is followed by the Zn-doped graphene (abbreviated as ZnG) and then AlG. N-doping did not enhance the adsorption characteristics of graphene sheet with H_2 molecule [88]. The authors, in another report, studied the effect of doped N and Ti on the interactions of CO, NO, SO_2, and HCHO with graphene and observed that Ti-doping can greatly improve the interactions of these gas molecules with graphene with the order of interactions as SO_2 > NO > HCHO > CO from the analyses of adsorption energy, electron density difference and DOS [89]. TiG also enabled selective gas adsorption, whereas NG did not exhibit selective adsorption behavior. Thus, TiG sheet can be used as an effective sensor material than IG and NG sheets for the detection and removal of these harmful gases [89]. TiG was also found to be very promising for HF detection [88]. The calculated charge transfers and the band structure plots of TiG with and without HF molecule indicated that the adsorption of HF on TiG will lead to large increase in the electrical conductivity of TiG sheet [90].

Table 8.1 also summarizes the results of the first-principles simulation studies on different graphene-gas systems.

Table 8.1 Adsorption energy E_a (eV), equilibrium graphene-molecule distance (d) and charge transfer (Q) for gas molecules adsorbed on intrinsic, doped and defected graphene

Type of graphene	Gas mole-cule	E_a	d (Å)	Q (e)	Ref.
IG	CO	-0.12	3.02	-0.01	
	NO	-0.30	2.43	0.04	[37]
	NO_2	-0.48	2.73	-0.19	
	NH_3	-0.11	2.85	0.02	
	CO	-0.0102	3.8561	-0.0011	[51]
	H_2CO	-0.083	3.127	0.085	[52]
	CO	-0.05	2.80	-0.01	
	O_2	-0.18	2.34	-0.15	[59]
	NO_2	-0.38	2.80	-0.24	
	H_2O	-0.14	2.31	-0.14	
	H_2S	-0.16	2.64	0.01	[61]
	SO_2	0.012	3.279	-0.077	[65]
BG	CO	-0.14	2.97	-0.02	
	NO	-1.07	1.99	0.15	
	NO_2	-1.37	1.67	-0.34	[37]
	NH_3	-0.50	1.66	0.40	
	O_2	-0.0232	3.5099	-0.0863	[57]
	H_2S	-0.11	3.26	0.01	[61]
	SO_2	0.205	3.162	-0.110	[65]
	H_2CO	-0.018	3.845	0.002	[66]
NG	CO	-0.14	3.15	0	
	NO	-0.40	2.32	0.01	[37]
	NO_2	-0.98	2.87	-0.55	
	NH_3	-0.12	2.86	0.04	
	CO	-0.0261	3.5540	-0.0027	[51]
	O_2	-0.1228	3.3196	0.0235	[57]
NG	H_2S	-0.14	3.22	0.01	[61]
	SO_2	0.172	3.478	-0.263	[65]
	H_2CO	-0.046	3.193	-0.073	[66]
DG	CO	-2.33	1.33	0.26	
	NO	-3.04	1.34	-0.29	[37]
	NO_2	-3.04	1.42	-0.38	
	NH_3	-0.24	2.61	0.02	
AlG	H_2S	-0.91	1.77	0.74	[61]
	CO	-2.6928	2.0623	0.2346	[51]
	H_2CO	-3.721	1.876	0.019	[52]
	O_2	-1.5589	1.8770	0.4367	[57]
	SO_2	1.262	1.825	-0.744	[65]
	H_2CO	-1.262	1.858	0.502	[66]

Table 8.1 (*contd.*)

Type of graphene	Gas molecule	E_a	d (Å)	Q (e)	Ref.
	O_2	-1.3132	1.7109	0.9184	[57]
	CO	-0.58	2.08	0.24	
SiG	O_2	-3.99	1.71	-1.10	[59]
	NO_2	-3.89	1.76	0.12	
	H_2O	-0.59	2.01	0.24	
	H_2S	-0.94	2.53	0.26	[61]
	SO_2	0.902	1.737	-0.959	[65]
	H_2CO	-0.769	1.722	0.772	[66]
	O_2	-1.0359	1.6275	0.8181	[57]
	O_2	-1.0895	1.6279	0.8181	
	H_2	-0.0143	3.3625	-	
	H_2O	-0.0469	3.4650	-	
PG	N_2	-0.0091	4.2252	-	[58]
	CO_2	-0.0107	4.0019	-	
	CO	-0.0685	4.3139	-	
	NH_3	-0.0146	4.5685	-	
	NO_2	-1.8875	1.5364	0.6198	
	NO	-0.5093	1.8074	0.4995	
	SO_2	-0.3216	1.6580	0.5007	
	O_2	-2.6098	1.7935	0.2166	[57]
CrG	SO_2	1.675	1.927	-0.672	[65]
	H_2CO	-1.215	1.960	0.275	[66]
	O_2	-2.0918	1.8641	0.1865	[57]
MnG	SO_2	1.729	1.905	-0.599	[65]
	H_2CO	-1.025	1.859	0.357	[66]
AgG	SO_2	0.968	2.173	-0.454	[65]
AuG	SO_2	1.284	2.167	-0.479	[65]
	H_2CO	-0.748	2.111	0.211	[66]
	SO_2	1.018	2.229	-0.550	[65]
	N_2	0.414	2.084	-	
PtG	O_2	1.487	2.038	-	
	NO_2	2.210	2.234	-	[70]
	SO_2	1.060	2.250	-	
	CO_2	0.086	2.197	-	
	H_2O	0.095	2.914	-	
CaG	H_2S	-0.66	4.81	0.04	
CoG	H_2S	-1.80	3.77	1.16	[61]
FeG	H_2S	-1.92	3.70	1.23	

8.2.5 Gas Molecules on Other Modified Graphene

Omidvar and Mohajeri [91] investigated the interactions between small gas molecules (O_2, N_2, CO and NO) and different armchair graphene nanoflakes (GNFs) such as intrinsic, B- or N-doped as well as functionalized GNFs with COOH, CN, and NO_2. In this study, pristine, B-, N-doped and functionalized GNF were found to be not sensitive to N_2, O_2 and CO. Pristine GNFs exhibited high sensitivity and reactivity towards NO molecule. The edge functionalization of GNFs with electron withdrawing groups such as COOH-groups significantly enhanced the adsorption of NO molecule which enables highly sensitive and selective detection of NO. The authors observed that the stronger interactions between NO and edge COOH-functionalized GNFs result in large variation of the energy band gap of about 555% after NO adsorption [91].

The adsorption energies and the charge transfer calculations of H_2S adsorption on Cu_2O modified graphene (Cu_2O-GS) showed strong adsorption of H_2S on the surface of Cu_2O-GS. DOS of Cu_2O-GS exhibited considerable variations after H_2S adsorption and also the electrical conductance of the nanosensor at room temperature got significantly enhanced by H_2S adsorption. From the obtained results, it was predicted that Cu_2O-GS nanosensor can be exploited as a nanosensor for H_2S detection. The obtained results were found to be in excellent agreement with previous experimental results [92].

The adsorption of various gas molecules on graphene based materials have been implemented in quantum mechanical simulation software packages such as Atomistix Toolkit [93], Materials Studio DMol^3 [94], CASTEP [95], Gaussian [96], VASP [97], Quantum Espresso [98], SIESTA [99] and ABINIT [100].

8.3 Conclusion and Future Perspectives

Graphene has been considered as one of the most promising candidates for gas sensing applications owing to its excellent structural and electronic properties. Graphene based gas sensors have already demonstrated to have the highest sensitivities and lowest detection limits compared to sensors based on one-dimensional carbon nanotubes and semiconductor nanowires due to the lower electrical noise of graphene arising from its two-dimensional crystal structure.

Theoretical studies based on quantum mechanical calculations enable us to gain a deep understanding of the interactions of gas molecules with graphene, before going for expensive device fabrication. Modeling of graphene-gas molecule adsorbed systems has verified weak physisorption of gas molecules on intrinsic graphene, which poses a major limitation for future applications of graphene based gas sensors. Graphene oxide and reduced graphene oxide have shown highly sensitive sensing response compared to intrinsic graphene, by virtue of the surface rich oxygen-containing functional groups. Theoretical studies have revealed that the interactions of graphene with gas molecules could be improved by heteroatom doping, introducing defects or by the combination of dopants and defects, thereby enhancing the sensitivity of graphene gas sensors. The predictions from quantum mechanical simulations can stimulate interest in experimental works on suitable modification of graphene and future development of novel micro devices based on modified graphene for enabling high sensitivity detection of gas molecules.

In spite of the great advancements in the field of graphene based gas sensors during the past 10 years, there are some major challenges that need to overcome to meet the increasing demands of the industry for high performance gas sensors. Stronger binding of gas molecules required for gas sensing applications, achieved by oxidized, doped or defected graphene would result in noticeable changes in electrical conductivity of oxidized, doped and defected graphene upon exposure to gases. The main drawback with these modified graphene based gas sensors is the long recovery period, as the thermal energy at room temperature is not sufficient to overcome the activation energy needed for releasing the strongly adsorbed gas molecules from the surface. The adsorption and desorption processes on intrinsic graphene are energetically viable as compared to the chemisorption on modified graphene. Strong adsorption on the surface of modified graphene makes gas desorption extremely difficult without the assistance of ultra-violet light illumination or annealing at high temperatures or by application of an electric field. These treatments make the adsorption process reversible in practical time periods. Thus, a proper balance should be maintained between sensitivity and recovery of the modified graphene based gas sensors. Modified graphene based gas sensors are best choices for ultrahigh sensitivity detection of gas molecules, but as just "on off" sensors. Hence, future work needs to focus on functionalization schemes that could control the binding energy of the gas molecules on graphene which could enable fast regeneration of the sensor at room temperature. The realization of highly sensitive and reversible gas sensors based on graphene can ultimately lead to single molecule detection of toxic or polluting gases in practical situations.

As graphene gets strongly influenced by a range of different gas species and mixtures, specific identification of gases by selective adsorption is challenging. The adsorption of electron withdrawing gases such as O_2 and NO_2 on intrinsic graphene will lead to an increase in electrical conductivity of graphene for both gases, whereas under controlled environment, which makes graphene difficult to distinguish between these oxidizing gases. Similarly, the adsorption of electron donating gases will result in decrease in the electrical conductivity of graphene and thus specific identification of electron donating gases is also difficult. Apart from this, a mixture containing equal amounts of oxidizing gases (for example, O_2 and NO_2) and reducing gases (for example, NH_3 and CO), will not create any change in the conductivity of graphene as the p-type and n-type doping induced by oxidizing and reducing gases cancel out each other. Thus, future work needs to concentrate on functionalization of graphene with appropriate capture agents that specifically binds the target gas to graphene's surface.

Modified graphene has been widely studied as highly sensitive gas sensors. The inherent sensitivity of modified graphene to most of the gases makes it uninteresting for developing selective gas sensing devices. For example, Al-doped graphene has shown high chemical reactivity to almost all gases and hence Al-doped graphene based sensor is not a suitable choice for the detection of a particular gas of interest in atmospheric air, as the other constituent gases of air strongly affect the sensing signal of Al-doped graphene for the desired gas. Similar is the case with other modified graphene materials, barring the few exceptions. Several modified graphene materials have demonstrated high selectivity. Among these, S-doped graphene is highly selective to NO_2, B-doped graphene is selective to NO and NO_2 in the presence of other common gas molecules, pyridinic-like N-doped graphene has shown high selectivity for CO in the presence of N_2, O_2 and NO. Until now, most of the previous studies have centered on different modifications of graphene for improving the interactions between gas molecules and graphene. In addition to the focus on increased reactivity, theoretical works need to focus on selectivity aspects of modified graphene based gas sensors and also on the suitability of using them under practical environments. Overall, the potential of the use of graphene for functional and efficient gas sensors is immense and is expected to be realized in greater details in the coming years.

References

1. Novoselov, K. S., Geim, A. K., Morozov, S. V., Jiang, D., Zhang, Y., Dubonos, S. V., Grigorieva, I. V., and Firsov, A. A. (2004) Electric field effect in atomically thin carbon films. *Science*, **306**, 666-669.

2. Novoselov, K. S., Geim, A. K., Morozov, S. V., Jiang, D., Katsnelson, M. I., Grigorieva, I. V., Dubonos, S. V., and Firsov, A. A. (2005) Two-dimensional gas of massless Dirac fermions in graphene. *Nature,* **438,** 197-200.

3. Zhang, Y., Tan, Y.-W., Stormer, H. L., and Kim, P. (2005) Experimental observation of the quantum Hall effect and Berry's phase in graphene. *Nature,* **438,** 201-204.

4. Lee, C., Wei, X., Kysar, J. W., and Hone, J. (2008) Measurement of the elastic properties and intrinsic strength of monolayer graphene. *Science,* **321,** 385-388.

5. Nair, R. R., Blake, P., Grigorenko, A. N., Novoselov, K. S., Booth, T. J., Stauber, T., Peres, N. M., and Geim, A. K. (2008) Fine structure constant defines visual transparency of graphene. *Science,* **320,** 1308.

6. Katsnelson, M. I. (2007) Graphene: carbon in two dimensions. *Materials Today,* **10,** 20-27.

7. Geim, A. K., and Novoselov, K. S. (2007) The rise of graphene. *Nature Materials,* **6,** 183-191.

8. Castro Neto, A. H., Guinea, F., Peres, N. M. R., Novoselov, K. S., and Geim, A. K. (2009) The electronic properties of graphene. *Reviews of Modern Physics,* **81,** 109-162.

9. Geim, A. K. (2009) Graphene: Status and prospects. *Science,* **324,** 1530-1534.

10. Craciun, M. F., Russo, S., Yamamoto, M., and Tarucha, S. (2011) Tuneable electronic properties in graphene. *Nano Today,* **6,** 42-60.

11. Novoselov, K. S., Falko, V. I., Colombo, L., Gellert, P. R., Schwab, M. G., and Kim, K. (2012) A roadmap for graphene. *Nature,* **490,** 192-200.

12. Singh, V., Joung, D., Zhai, L., Das, S., Khondaker, S. I., and Seal, S. (2011) Graphene based materials: Past, present and future. *Progress in Materials Science,* **56,** 1178-1271.

13. Xiang, Q., Yu, J., and Jaroniec, M. (2012) Graphene-based semiconductor photocatalysts. *Chemical Society Reviews,* **41,** 782-796.

14. Zhao, G., Wen, T., Chen, C., and Wang, X. (2012) Synthesis of graphene-based nanomaterials and their application in energy-related and environmental-related areas. *RSC Advances,* **2,** 9286-9303.

15. Zhu, J., Yang, D., Yin, Z., Yan, Q., and Zhang, H. (2014) Graphene and graphene-based materials for energy storage applications. *Small,* **10,** 3480-3498.

16. Weiss, N. O., Zhou, H., Liao, L., Liu, Y., Jiang, S., Huang, Y., and Duan, X. (2012) Graphene: An emerging electronic material. *Advanced Materials,* **24,** 5782-5825.

17. Tao, L., Wang, D., Jiang, S., Liu, Y., Xie, Q., Tian, H., Deng, N., Wang, X., Yang, Y., and Ren, T.-L. (2016) Fabrication techniques and applications of flexible graphene-based electronic devices. *Journal of Semiconductors,* **37,** 041001.

18. Lahiri, I., Verma, V. P., and Choi, W. (2011) An all-graphene based transparent and flexible field emission device. *Carbon,* **49,** 1614-1619.

19. Bonaccorso, F., Sun, Z., Hasan, T., and Ferrari, A. C. (2010) Graphene photonics and optoelectronics. *Nature Photonics,* **4,** 611-622.

20. Yang, Y., Asiri, A. M., Tang, Z., Du, D., and Lin, Y. (2013) Graphene based materials for biomedical applications. *Materials Today,* **16,** 365-373.

21. Shao, Y., Wang, J., Wu, H., Liu, J., Aksay, I. A., and Lin, Y. (2010) Graphene based electrochemical sensors and biosensors: A review. *Electroanalysis,* **22,** 1027-1036.

22. Basu, S., and Bhattacharyya, P. (2012) Recent developments on graphene and

graphene oxide based solid state gas sensors. *Sensors and Actuators B: Chemical,* **173**, 1-21.

23. Liu, Y., Dong, X., and Chen, P. (2012) Biological and chemical sensors based on graphene materials. *Chemical Society Reviews,* **41**, 2283-2307.

24. Yuan, W., and Shi, G. (2013) Graphene-based gas sensors. *Journal of Materials Chemistry A,* **1**, 10078-10091.

25. Varghese, S. S., Lonkar, S., Singh, K. K., Swaminathan, S., and Abdala, A. (2015) Recent advances in graphene based gas sensors. *Sensors and Actuators B: Chemical,* **218**, 160-183.

26. Wang, T., Huang, D., Yang, Z., Xu, S., He, G., Li, X., Hu, N., Yin, G., He, D., and Zhang, L. (2016) A review on graphene-based gas/vapor sensors with unique properties and potential applications. *Nano-Micro Letters,* **8**, 95-119.

27. Varghese, S., Varghese, S., Swaminathan, S., Singh, K., and Mittal, V. (2015) Two-dimensional materials for sensing: Graphene and beyond. *Electronics,* **4**, 651-687.

28. Schedin, F., Geim, A. K., Morozov, S. V., Hill, E. W., Blake, P., Katsnelson, M. I., and Novoselov, K. S. (2007) Detection of individual gas molecules adsorbed on graphene. *Nature Materials,* **6**, 652-655.

29. Danneau, R., Wu, F., Craciun, M. F., Russo, S., Tomi, M. Y., Salmilehto, J., Morpurgo, A. F., and Hakonen, P. J. (2008) Shot noise in ballistic graphene. *Physical Review Letters,* **100**, 196802.

30. Hill, E. W., Vijayaragahvan, A., and Novoselov, K. (2011) Graphene sensors. *IEEE Sensors Journal,* **11**, 3161-3170.

31. Wehling, T. O., Novoselov, K. S., Morozov, S. V., Vdovin, E. E., Katsnelson, M. I., Geim, A. K., and Lichtenstein, A. I. (2008) Molecular doping of graphene. *Nano Letters,* **8**, 173-177.

32. Dan, Y., Lu, Y., Kybert, N. J., Luo, Z., and Johnson, A. T. C. (2009) Intrinsic response of graphene vapor sensors. *Nano Letters,* **9**, 1472-1475.

33. Leenaerts, O., Partoens, B., and Peeters, F. M. (2008) Adsorption of H_2O, NH_3, CO, NO_2, and NO on graphene: A first-principles study. *Physical Review B,* **77**, 125416.

34. Seyed-Talebi, S. M., and Beheshtian, J. (2013) Computational study of ammonia adsorption on the perfect and rippled graphene sheet. *Physica B: Condensed Matter,* **429**, 52-56.

35. Lee, K.-J., and Kim, S.-J. (2013) Theoretical investigation of CO_2 adsorption on graphene. *Bulletin of the Korean Chemical Society,* **34**, 3022-3026.

36. Ganji, M. D., Hosseini-Khah, S., and Amini-Tabar, Z. (2015) Theoretical insight into hydrogen adsorption onto graphene: a first-principles B3LYP-D3 study. *Physical Chemistry Chemical Physics,* **17**, 2504-2511.

37. Zhang, Y.-H., Chen, Y.-B., Zhou, K.-G., Liu, C.-H., Zeng, J., Zhang, H.-L., and Peng, Y. (2009) Improving gas sensing properties of graphene by introducing dopants and defects: a first-principles study. *Nanotechnology,* **20**, 185504.

38. Huang, B., Li, Z., Liu, Z., Zhou, G., Hao, S., Wu, J., Gu, B.-L., and Duan, W. (2008) Adsorption of gas molecules on graphene nanoribbons and its implication for nanoscale molecule sensor. *The Journal of Physical Chemistry C,* **112**, 13442-13446.

39. Rostam, M., Yawar, M., and Nader, G. (2008) Investigation of gas sensing properties of armchair graphene nanoribbons. *Journal of Physics: Condensed Matter,* **20**, 425211.

40. Paulla, K. K., and Farajian, A. A. (2013) Concentration effects of carbon oxides on sensing by graphene nanoribbons: Ab initio modeling. *The Journal of Physical Chemistry C,* **117**, 12815-12825.

41. Akbari, E., Yousof, R., Ahmadi, M. T., Kiani, M. J., Rahmani, M., Feiz Abadi, H. K., and Saeidmanesh, M. (2014) The effect of concentration on gas sensor model based on graphene nanoribbon. *Neural Computing and Applications,* **24**, 143-146.

42. Fathalian, A., Jalilian, J., and Shahidi, S. (2013) Effects of gas adsorption on the electronic properties of graphene nanoribbons. *Physica B: Condensed Matter,* **417**, 75-78.

43. Peng, Y., and Li, J. (2013) Ammonia adsorption on graphene and graphene oxide: a first-principles study. *Frontiers of Environmental Science & Engineering,* **7**, 403-411.

44. Tang, S., and Cao, Z. (2011) Adsorption of nitrogen oxides on graphene and graphene oxides: Insights from density functional calculations. *The Journal of Chemical Physics,* **134**, 044710.

45. Mattson, E. C., Pande, K., Unger, M., Cui, S., Lu, G., Gajdardziska-Josifovska, M., Weinert, M., Chen, J., and Hirschmugl, C. J. (2013) Exploring adsorption and reactivity of NH_3 on reduced graphene oxide. *The Journal of Physical Chemistry C,* **117**, 10698-10707.

46. Wang, X., Sun, G., Routh, P., Kim, D.-H., Huang, W., and Chen, P. (2014) Heteroatom-doped graphene materials: syntheses, properties and applications. *Chemical Society Reviews,* **43**, 7067-7098.

47. Denis, P. A. (2010) Band gap opening of monolayer and bilayer graphene doped with aluminium, silicon, phosphorus, and sulfur. *Chemical Physics Letters,* **492**, 251-257.

48. Shokuhi Rad, A., Esfahanian, M., Maleki, S., and Gharati, G. (2016) Application of carbon nanostructures toward SO_2 and SO_3 adsorption: a comparison between pristine graphene and N-doped graphene by DFT calculations. *Journal of Sulfur Chemistry,* **37**, 176-188.

49. Rad, A. S., Shabestari, S. S., Mohseni, S., and Aghouzi, S. A. (2016) Study on the adsorption properties of O_3, SO_2, and SO_3 on B-doped graphene using DFT calculations. *Journal of Solid State Chemistry,* **237**, 204-210.

50. Ao, Z. M., Yang, J., Li, S., and Jiang, Q. (2008) Enhancement of CO detection in Al doped graphene. *Chemical Physics Letters,* **461**, 276-279.

51. Wang, W., Zhang, Y., Shen, C., and Chai, Y. (2016) Adsorption of CO molecules on doped graphene: A first-principles study. *AIP Advances,* **6**, 025317.

52. Chi, M., and Zhao, Y.-P. (2009) Adsorption of formaldehyde molecule on the intrinsic and Al-doped graphene: A first principle study. *Computational Materials Science,* **46**, 1085-1090.

53. Rad, A. S. (2015) First principles study of Al-doped graphene as nanostructure adsorbent for NO_2 and N_2O: DFT calculations. *Applied Surface Science,* **357**, Part A, 1217-1224.

54. Rad, A. S., and Foukolaei, V. P. (2015) Density functional study of Al-doped graphene nanostructure towards adsorption of CO, CO_2 and H_2O. *Synthetic Metals,* **210**, Part B, 171-178.

55. Milowska, K. Z., and Majewski, J. A. (2014) Graphene based sensors: Theoretical study. *Journal of Physical Chemistry C,* **118**(31), 17395-17401.

56. Dai, J., Yuan, J., and Giannozzi, P. (2009) Gas adsorption on graphene doped with B, N, Al, and S: A theoretical study. *Applied Physics Letters,* **95,** 232105.

57. Dai, J., and Yuan, J. (2010) Adsorption of molecular oxygen on doped graphene: Atomic, electronic, and magnetic properties. *Physical Review B,* **81,** 165414.

58. Dai, J., and Yuan, J. (2010) Modulating the electronic and magnetic structures of P-doped graphene by molecule doping. *Journal of Physics: Condensed Matter,* **22,** 225501.

59. Zou, Y., Li, F., Zhu, Z. H., Zhao, M. W., Xu, X. G., and Su, X. Y. (2011) An ab initio study on gas sensing properties of graphene and Si-doped graphene. *The European Physical Journal B,* **81,** 475-479.

60. Chen, Y., Gao, B., Zhao, J.-X., Cai, Q.-H., and Fu, H.-G. (2012) Si-doped graphene: an ideal sensor for NO- or NO_2-detection and metal-free catalyst for N_2O-reduction. *Journal of Molecular Modeling,* **18,** 2043-2054.

61. Zhang, Y.-H., Han, L.-F., Xiao, Y.-H., Jia, D.-Z., Guo, Z.-H., and Li, F. (2013) Understanding dopant and defect effect on H_2S sensing performances of graphene: A first-principles study. *Computational Materials Science,* **69,** 222-228.

62. Lv, Y.-a., Zhuang, G.-l., Wang, J.-g., Jia, Y.-b., and Xie, Q. (2011) Enhanced role of Al or Ga-doped graphene on the adsorption and dissociation of N_2O under electric field. *Physical Chemistry Chemical Physics,* **13,** 12472-12477.

63. Sharma, S., and Verma, A. S. (2013) A theoretical study of H_2S adsorption on graphene doped with B, Al and Ga. *Physica B: Condensed Matter,* **427,** 12-16.

64. Avila, Y., Cocoletzi, G. H., and Romero, M. T. (2014) First principles calculations of phenol adsorption on pristine and group III (B, Al, Ga) doped graphene layers. *Journal of Molecular Modeling,* **20,** 1-9.

65. Shao, L., Chen, G., Ye, H., Wu, Y., Qiao, Z., Zhu, Y., and Niu, H. (2013) Sulfur dioxide adsorbed on graphene and heteroatom-doped graphene: a first-principles study. *The European Physical Journal B,* **86,** 1-5.

66. Liu, X.-Y., and Zhang, J.-M. (2014) Formaldehyde molecule adsorbed on doped graphene: A first-principles study. *Applied Surface Science,* **293,** 216-219.

67. Pramanik, A., and Kang, H. S. (2011) Density functional theory study of O_2 and NO adsorption on heteroatom-doped graphenes including the van der Waals interaction. *Journal of Physical Chemistry C,* **115,** 10971-10978.

68. Choudhuri, I., Patra, N., Mahanta, A., Ahuja, R., Pathak, B. (2015) B-N@Graphene: Highly sensitive and selective gas sensor. *Journal of Physical Chemistry C,* **119**(44), 24827-24836.

69. Jiang, D.-e., Cooper, V. R., and Dai, S. (2009) Porous graphene as the ultimate membrane for gas separation. *Nano Letters,* **9**(12), 4019-4024.

70. Lee, Y., Lee, S., Hwang, Y., and Chung, Y.-C. (2014) Modulating magnetic characteristics of Pt embedded graphene by gas adsorption (N_2, O_2, NO_2, SO_2). *Applied Surface Science,* **289,** 445-449.

71. Zhang, H.-p., Luo, X.-g., Song, H.-t., Lin, X.-y., Lu, X., and Tang, Y. (2014) DFT study of adsorption and dissociation behavior of H_2S on Fe-doped graphene. *Applied Surface Science,* **317,** 511-516.

72. Tavakol, H., and Mollaei-Renani, A. (2014) DFT, AIM, and NBO study of the interaction of simple and sulfur-doped graphenes with molecular halogens, CH_3OH, CH_3SH, H_2O, and H_2S. *Structural Chemistry,* **25,** 1659-1667.

73. Wanno, B., and Tabtimsai, C. (2014) A DFT investigation of CO adsorption on VIIIB transition metal-doped graphene sheets. *Superlattices and Microstructures,* **67**, 110-117.

74. Nasehnia, F. and Seifi, M. (2014) Adsorption of molecular oxygen on VIIIB transition metal-doped graphene: A DFT study. *Modern Physics Letters B,* **28**, 1450237.

75. Ma, L., Zhang, J.-M., Xu, K.-W., and Ji, V. (2015) A first-principles study on gas sensing properties of graphene and Pd-doped graphene. *Applied Surface Science,* **343**, 121-127.

76. Shao, L., Chen, G., Ye, H., Niu, H., Wu, Y., Zhu, Y., and Ding, B. (2014) Sulfur dioxide molecule sensors based on zigzag graphene nanoribbons with and without Cr dopant. *Physics Letters A,* **378**, 667-671.

77. Castellanos Aguila J. E., Cocoletzi, H. H., and Cocoletzi, G. H. (2013) A theoretical analysis of the role of defects in the adsorption of hydrogen sulfide on graphene. *AIP Advances,* **3**, 032118.

78. Borisova, D., Antonov, V., and Proykova, A. (2013) Hydrogen sulfide adsorption on a defective graphene. *International Journal of Quantum Chemistry,* **113**, 786-791.

79. Ma, C., Shao, X., and Cao, D. (2014) Nitrogen-doped graphene as an excellent candidate for selective gas sensing. *Science China Chemistry,* **57**, 911-917.

80. Zhou, Q., Yuan, L., Yang, X., Fu, Z., Tang, Y., Wang, C., and Zhang, H. (2014) DFT study of formaldehyde adsorption on vacancy defected graphene doped with B, N, and S. *Chemical Physics,* **440**, 80-86.

81. Zhou, Q., Wang, C., Fu, Z., Tang, Y., and Zhang, H. (2014) Adsorption of formaldehyde molecule on Stone–Wales defected graphene doped with Cr, Mn, and Co: A theoretical study. *Computational Materials Science,* **83**, 398-402.

82. Hussain, T., Panigrahi, P., and Ahuja, R. (2014) Enriching physisorption of H_2S and NH_3 gases on a graphane sheet by doping with Li adatoms. *Physical Chemistry Chemical Physics,* 16, 8100-8105.

83. Hussain, T., Panigrahi, P., and Ahuja, R. (2014) Sensing propensity of a defected graphane sheet towards CO, H_2O and NO_2. *Nanotechnology,* **25**, 325501.

84. Tawfik, S. A., Cui, X., Carter, D., Ringer, S., and Stampfl, C. (2015) Sensing sulfur-containing gases using titanium and tin decorated zigzag graphene nanoribbons from first-principles. *Physical Chemistry Chemical Physics,* **17**, 6925-6932.

85. Xie, Z., Zuo, X., Zhang, G.-P., Li, Z.-L., and Wang, C.-K. (2016) Detecting CO, NO and NO_2 gases by Boron-doped graphene nanoribbon molecular devices. *Chemical Physics Letters,* **657**, 18-25.

86. Cui, S., Pu, H., Mattson, E. C., Wen, Z., Chang, J., Hou, Y., Hirschmugl, C. J., and Chen, J. (2014) Ultrasensitive chemical sensing through facile tuning defects and functional groups in reduced graphene oxide. *Analytical Chemistry,* **86**, 7516-7522.

87. Kumar, B., Min, K., Bashirzadeh, M., Farimani, A. B., Bae, M.-H., Estrada, D., Kim, Y. D., Yasaei, P., Park, Y. D., Pop, E., Aluru, N. R., and Salehi-Khojin, A. (2013) The role of external defects in chemical sensing of graphene field-effect transistors. *Nano Letters,* **13**, 1962-1968.

88. Zhang, H.-p., Luo, X.-g., Lin, X.-y., Lu, X., and Leng, Y. (2013) Density functional theory calculations of hydrogen adsorption on Ti-, Zn-, Zr-, Al-, and N-doped and intrinsic graphene sheets. *International Journal of Hydrogen Energy,* **38**, 14269-14275.

89. Zhang, H.-p., Luo, X.-g., Lin, X.-y., Lu, X., Leng, Y., and Song, H.-t. (2013) Density

functional theory calculations on the adsorption of formaldehyde and other harmful gases on pure, Ti-doped, or N-doped graphene sheets. *Applied Surface Science,* **283**, 559-565.

90. Song, E. H., Zhu, Y. F., and Jiang, Q. (2013) Density functional theory calculations of adsorption of hydrogen fluoride on titanium embedded graphene. *Thin Solid Films,* **546**, 124-127.

91. Omidvar, A., and Mohajeri, A. (2014) Edge-functionalized graphene nanoflakes as selective gas sensors. *Sensors and Actuators B: Chemical,* **202**, 622-630.

92. Mohammadi-Manesh, E., Vaezzadeh, M., and Saeidi, M. (2015) Theoretical study on electronic structure, and electrical conductance at room temperature of Cu_2O–GS nanosensors and detection of H_2S gas. *Computational Materials Science,* **97**, 181-185.

93. Brandbyge, M., Mozos, J.-L., Ordejon, P., Taylor, J., and Stokbro, K. (2002) Density-functional method for nonequilibrium electron transport. *Physical Review B,* **65**, 165401.

94. Delley, B. (2000) From molecules to solids with the DMol3 approach. *The Journal of Chemical Physics,* **113**, 7756-7764.

95. Milman, V., Winkler, B., White, J. A., Pickard, C. J., Payne, M. C., Akhmatskaya, E. V., and Nobes, R. H. (2000) Electronic structure, properties, and phase stability of inorganic crystals: A pseudopotential plane-wave study. *International Journal of Quantum Chemistry,* **77**, 895-910.

96. Frisch, M. J., Trucks, G. W., Schlegel, H. B., Scuseria, G. E., Robb, M. A., Cheeseman, J. R., Scalmani, G., Barone, V., Mennucci, B., Petersson, G. A., Nakatsuji, H., Caricato, M., Li, X., Hratchian, H. P., Izmaylov, A. F., Bloino, J., Zheng, G., Sonnenberg, J. L., Hada, M., Ehara, M., Toyota, K., Fukuda, R., Hasegawa, J., Ishida, M., Nakajima, T., Honda, Y., Kitao, O., Nakai, H., Vreven, T., Montgomery, Jr., J. A., Peralta, J. E., Ogliaro, F., Bearpark, M., Heyd, J. J., Brothers, E., Kudin, K. N., Staroverov, V. N., Kobayashi, R., Normand, J., Raghavachari, K., Rendell, A., Burant, J. C., Iyengar, S. S., Tomasi, J., Cossi, M., Rega, N., Millam, J. M., Klene, M., Knox, J. E., Cross, J. B., Bakken, V., Adamo, C., Jaramillo, J., Gomperts, R., Stratmann, R. E., Yazyev, O., Austin, A. J., Cammi, R., Pomelli, C., Ochterski, J. W., Martin, R. L., Morokuma, K., Zakrzewski, V. G., Voth, G. A., Salvador, P., Dannenberg, J. J., Dapprich, S., Daniels, A. D., Farkas, O., Foresman, J. B., Ortiz, J. V., Cioslowski, J., Fox, D. J. (2009) *Gaussian 09,* Gaussian, Inc., USA.

97. Kresse, G., and Furthmuller, J. (1996) Efficient iterative schemes for *ab initio* total-energy calculations using a plane-wave basis set. *Physical Review B,* **54**, 11169-11186.

98. Giannozzi, P., Baroni, S., Bonini, N., Calandra, M., Car, R., Cavazzoni, C., Ceresoli, D., Chiarotti, G. L., Cococcioni, M., Dabo, I., Corso, A. D., de Gironcoli, S., Fabris, S., Fratesi, G., Gebauer, R., Gerstmann, U., Gougoussis, C., Kokalj, A., Lazzeri, M., Martin-Samos, L., Marzari, N., Mauri, F., Mazzarello, R., Paolini, S., Pasquarello, A., Paulatto, L., Sbraccia, C., Scandolo, S., Sclauzero, G., Seitsonen. A. P., Smogunov, A., Umari, P., and Wentzcovitch, R. M. (2009) QUANTUM ESPRESSO: a modular and open-source software project for quantum simulations of materials. *Journal of Physics: Condensed Matter,* **21**, 395502.

99. Soler, J. M., Artacho, E., Gale, J. D., Garcia, A., Junquera, J., Ordejon, P., and Sanchez-Portal, D. (2002) The SIESTA method for ab initio order- N materials simulation. *Journal of Physics: Condensed Matter,* **14**, 2745.

100.Gonze, X., Amadon, B., Anglade, P. M., Beuken, J. M., Bottin, F., Boulanger, P., Bruneval, F., Caliste, D., Caracas, R., Cote, M., Deutsch, T., Genovese, L., Ghosez, P., Giantomassi, M., Goedecker, S., Hamann, D. R., Hermet, P., Jollet, F., Jomard, G., Leroux, S., Mancini, M., Mazevet, S., Oliveira, M. J. T., Onida, G., Pouillon, Y., Rangel, T., Rignanese, G. M., Sangalli, D., Shaltaf, R., Torrent, M., Verstraete, M. J., Zerah, G., and Zwanziger, J. W. (2009) ABINIT: First-principles approach to material and nanosystem properties. *Computer Physics Communications,* **180**, 2582-2615.

DFT based Investigation of N$_2$O Adsorption on Graphene and Heteroatom-doped Graphene

9.1 Introduction

Graphene, the novel two-dimensional carbon based nanomaterial, has aroused great interest in numerous application areas including electronics [1, 2], energy storage [3-6], biomedicine [7, 8], catalysis [9], etc. The physicochemical properties of graphene such as large surface area (2630 m^2/g for single layer graphene), excellent electrical conductivity, low Johnson noise, and limited crystal defects along with the high mechanical characteristics [1,10-13] make graphene one of the most promising candidates for gas sensing applications [14-16]. Experimental research studies have shown that graphene can be used as a good sensing material to detect various gas molecules by measuring the change in the conductivity induced by the gas adsorbates, which act as either electron donors or acceptors [14,17]. Previous studies that focused on pure or intrinsic graphene (hereafter abbreviated as IG) showed weak physical adsorption of gas molecules on its surface [18,19]. It has been found that the gas sensing response of graphene towards gas molecules could be improved by using appropriate dopants [20,21].

The detection of N$_2$O, commonly known as laughing gas, is very important since increasing atmospheric concentrations of N$_2$O leads to global warming and stratospheric ozone-layer destruction [22-24]. N$_2$O is considered as the single greatest ozone-depleting substance and is expected to remain the dominant ozone-depleting substance throughout the 21st century [25]. The global warming potential of N$_2$O is 298 times higher than that of CO$_2$, with an atmospheric lifetime of about 120 years [26]. N$_2$O is estimated to contribute about 6% of the global warming effect due to greenhouse gases [27]. Significant efforts have been made to find an effective method for detecting N$_2$O gas. The sensing affinity of graphene or doped graphene to N$_2$O may provide new insights into the development of future N$_2$O gas sensors for environmental monitoring and control.

In this study, the adsorption of N$_2$O on B-, N-, Si-, P-, Ga-, Cr- and Mn-doped graphene (hereafter abbreviated as BG, NG, SiG, PG, GG, CG and MG) sheets have been investigated by performing first-principles calculations

Seba S. Varghese[a,b], Sundaram Swaminathan[c], Krishna K. Singh[b] and Vikas Mittal[a,*]
[a]The Petroleum Institute (part of Khalifa University of Science and Technology), Abu Dhabi, UAE; [b]Birla Institute of Technology and Science, UAE; [c]DIT University, India
*Current address: Bletchington, Wellington County, Australia

based on density functional theory (DFT). The adsorption of N_2O on IG is also studied for comparison. The energetically favorable adsorption configurations, adsorption energies and charge transfers of the graphene-N_2O adsorbed systems were calculated and analyzed to exploit the potential application of graphene or doped graphene as N_2O gas sensors.

9.2 Computational Methods

All the calculations have been carried out using DFT framework implemented in ABINIT simulation package [28]. The exchange correlation functional of generalized gradient approximation (GGA) in the Perdew-Burke-Ernzerhof (PBE) form is employed for the study [29]. The interactions of the valence electrons with atomic core are represented using norm-conserving Troullier-Martins pseudopotentials [30]. The single-layer graphene sheet is modelled using a 4×4 graphene supercell with a single molecule of N_2O adsorbed onto it. Different doped graphenes (abbreviated as XG, where X corresponds to the dopant atom), are modelled by substituting a carbon (C) atom with the dopant atom (B, N, Si, P, Ga, Cr or Mn). To minimize the interactions between adjacent graphene layers, a space width of 16 Å is taken in the direction normal to the graphene plane. A planewave basis set with the converged kinetic energy cutoff value of 816 eV is used. The Brillouin zone is sampled using a $5 \times 5 \times 1$ Monkhorst-Pack k-point [31]. Structural optimization has been performed for all considered systems using the Broyden-Fletcher-Goldfard-Shanno minimization (BFGS) [32] until the residual forces on each atom were smaller than 5×10^{-4} Hartree/Bohr.

The adsorption energy, E_{ad} of N_2O on graphene is calculated as:

$$E_{ad} = E_{total} - (E_{sheet} + E_{N_2O})$$ (1)

where E_{total} , E_{sheet} and E_{N_2O} represent the energies of the relaxed graphene system with the adsorbed N_2O molecule, isolated graphene sheet and isolated N_2O molecule respectively. The charge transfer analysis have been carried out using the Hirshfeld method [33].

9.3 Results and Discussion

The atomic geometries of isolated N_2O molecule and IG were first optimized. The structural parameters for N_2O molecule are determined with N-N bond length of 1.13 Å, N-O bond length of 1.19 Å and the N–N–O angle measures 180°, which are in good agreement with experimental values (N-N bond length of 1.1282 Å, N-O bond length of 1.1842 Å and N–N–O angle of 180° respectively). For graphene, the calculated C-C bond length of 1.42

Å (Figure 9.1) is observed to be in good agreement with the experimental value [34,35].

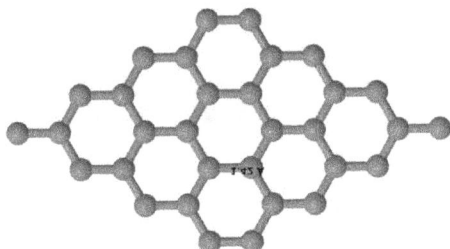

Figure 9.1 Optimized structure of IG sheet.

9.3.1 Adsorption of N_2O on IG

The interaction of N_2O with the graphene sheet was investigated in two modes based on the binding atom of N_2O, labelled as N-end and O-end. In N-end binding mode, the N atom of N_2O is close to the graphene sheet, whereas in O-end binding mode, the O atom of the gas molecule is close to the graphene sheet. The adsorption mechanism of N_2O on IG is discussed here. The nature of interactions of N_2O molecule with graphene is studied from the calculations of adsorption energies of N_2O molecule adsorbed on IG *via* N-end and O-end binding mode. After N_2O adsorption, the atomic geometry of IG remains unchanged (Figure 9.2). Figure 9.2 (a) and (b) show the optimized structures of N_2O adsorbed on IG through the N-end and O-end binding mode respectively. N_2O molecule either N-end (E_{ad} = -0.011 eV) or O-end (E_{ad} = -0.015 eV) was adsorbed on top of the carbon atom on IG (Figure 9.2), and the IG-N_2O binding distance is 3.5 Å in both cases, which indicates weak interaction between N_2O molecule and IG, in agreement with the results reported by Lv *et al.* [36].

9.3.2 Adsorption of N_2O on BG, NG, SiG, PG and GG

The adsorption of N_2O on graphene doped with B, N, Si, P and Ga is considered next. Initially, the doped graphene sheets are allowed to relax. After relaxation, both BG and NG retain the planar form of undoped graphene as seen in Figure 9.3 (a) and (b). The carbon-dopant atom distance is found to be 1.49 Å (B-C bond) for BG and 1.41 Å (N-C bond) for NG. The substitution of a single carbon atom by other dopant atoms such as Si, P and Ga in SiG, PG and GG respectively results in a distorted geometry compared to IG, due

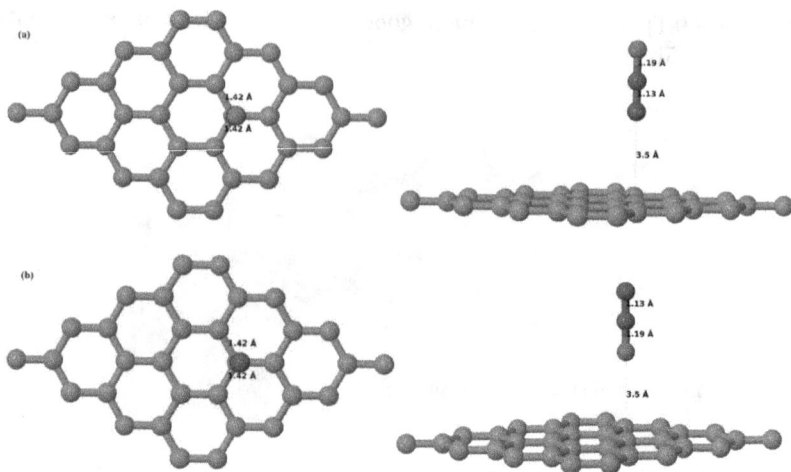

Figure 9.2 Top (left) and side (right) views of the adsorption structures of N_2O on IG *via* (a) N-end and (b) O-end.

to the stress introduced by the bigger sized dopant atoms compared to the host carbon atoms. Figure 9.3 (c-e) present the relaxed structures of SiG, PG and GG sheet. The dopant atom protrudes instead out of the plane, at a distance d from the plane of $d = 0.989$ Å for Si, $d = 1.106$ Å for P, $d = 1.220$ Å for Ga. The bond around the dopant atom expands to 1.74 Å (Si-C bond) in SiG and 1.76 Å (P-C bond) in PG and 1.83 Å (Ga-C bond) in GG from the ideal C–C bond length of 1.42 Å (Figure 9.3 (c-e)). These results are in good agreement with previous reported theoretical works [37-40].

Having determined the energetically most stable configuration of BG, NG, SiG, PG and GG sheets, the next step is to investigate the N_2O adsorption process. N_2O molecule is initially placed on the top site of the dopant atom on the above mentioned doped graphene sheets *via* two binding modes (N-end and O-end).

Figure 9.4 (a) and (b) present the relaxed structures of N_2O adsorbed on BG and NG system respectively. There is no significant change in the structure of the BG and NG after N_2O adsorption. For both N_2O adsorbed on BG and NG, the energetically favourable geometry is the one in which the O-end binding mode is tilted with the oxygen atom of N_2O facing towards the dopant atom (Figure 9.4 (a), (b)). The carbon-dopant atom distance remains unchanged in both BG and NG even after N_2O interaction. The low absorption energy (-0.027 eV for BG and -0.022 eV for NG) and long binding distance (3.72 Å for BG and 3.58 Å for NG) indicate weak forces and suggest that the interaction involved physisorption.

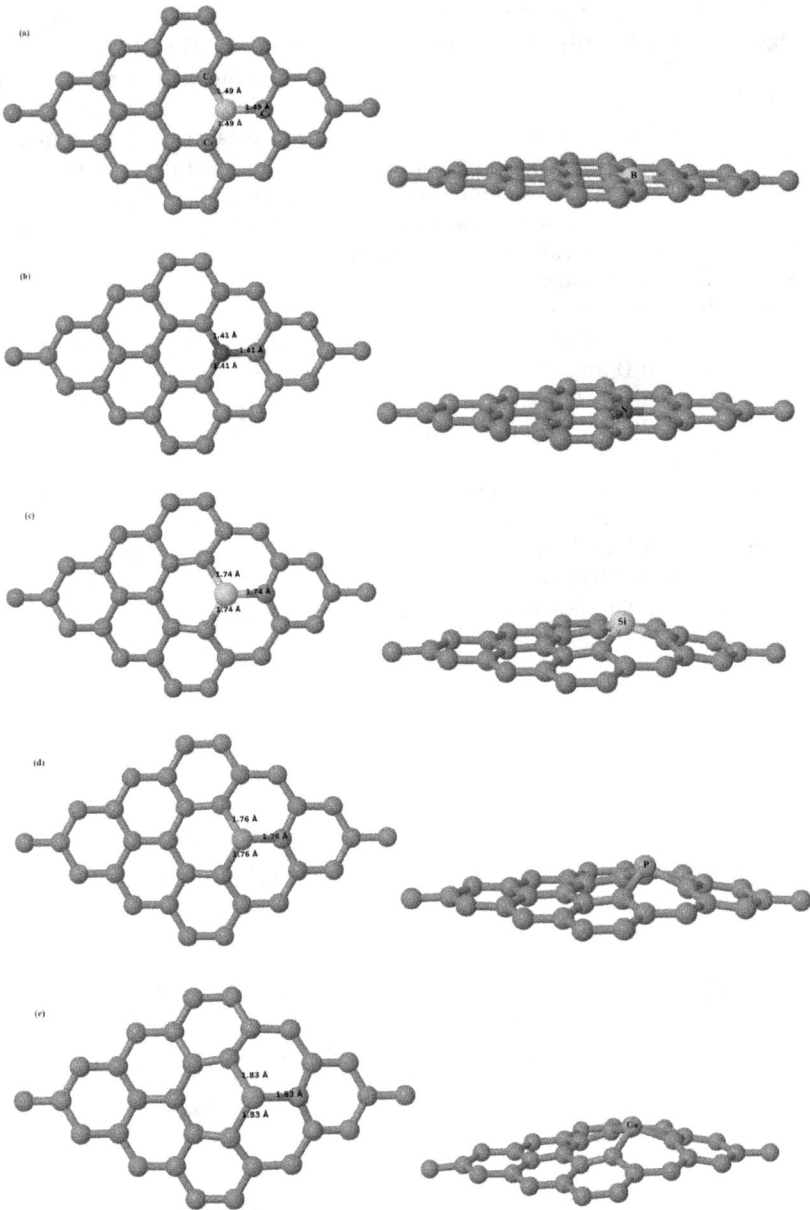

Figure 9.3 Top (left) and side (right) views of (a) BG, (b) NG, (c) SiG (d) PG, and (e) GG.

The optimized structures of N₂O adsorbed on SiG and PG are shown in Figure 9.4 (c) and (d) respectively. In N₂O adsorbed-SiG and -PG systems, N₂O molecule is far away from the doped graphene sheets with distances of 3.41 Å in SiG and 4.28 Å in PG (Figure 9.4 (c-d)). Upon N₂O adsorption, the elevation of Si atom decreases to 0.863 Å with decreased Si–C bond length of 1.73 Å (Figure 9.4 (c)). In the case of PG, the elevation of P atom decreases to 1.01 Å with the same P-C bond length of 1.76 Å (Figure 9.4 (d)). These structural parameters indicate the repulsive force between the N₂O molecule and the Si (or P) atom. The calculated adsorption energies for N₂O on SiG and PG are found to be almost the same (-0.030 eV), which clearly shows physisorption of N₂O on SiG and PG.

N₂O was adsorbed onto the top site of the Ga atom of GG via N-end (E_{ad} = -0.239 eV) and O-end (E_{ad} = -0.164 eV) and the corresponding dopant atom-gas molecule distance are 2.28 Å (Figure 9.4 (e)) and 2.43 Å (Figure 9.4 (f)) respectively, which shows that the interaction of Ga atom with N₂O is stronger for N-end adsorbed on GG. Figure 9.4 (e)) depicts the optimized structure of N₂O adsorbed on GG *via* the N-end with a strong chemical bond between Ga atom and the adsorbed N atom. The elevation of Ga atom and the Ga–C bond length increase to 1.245 Å and 1.84 Å on N₂O adsorption (Figure 9.4 (e)). These results are consistent with previous reported values [36]. The obtained adsorption energies are presented in Table 9.1.

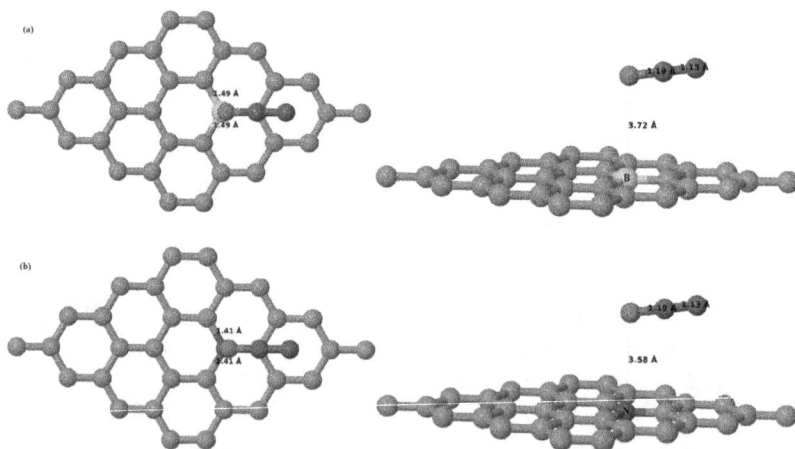

Figure 9.4 Top (left) and side (right) views of optimized adsorption structures of N₂O molecule on (a) BG, (b) NG. (The figure continues to the other page)

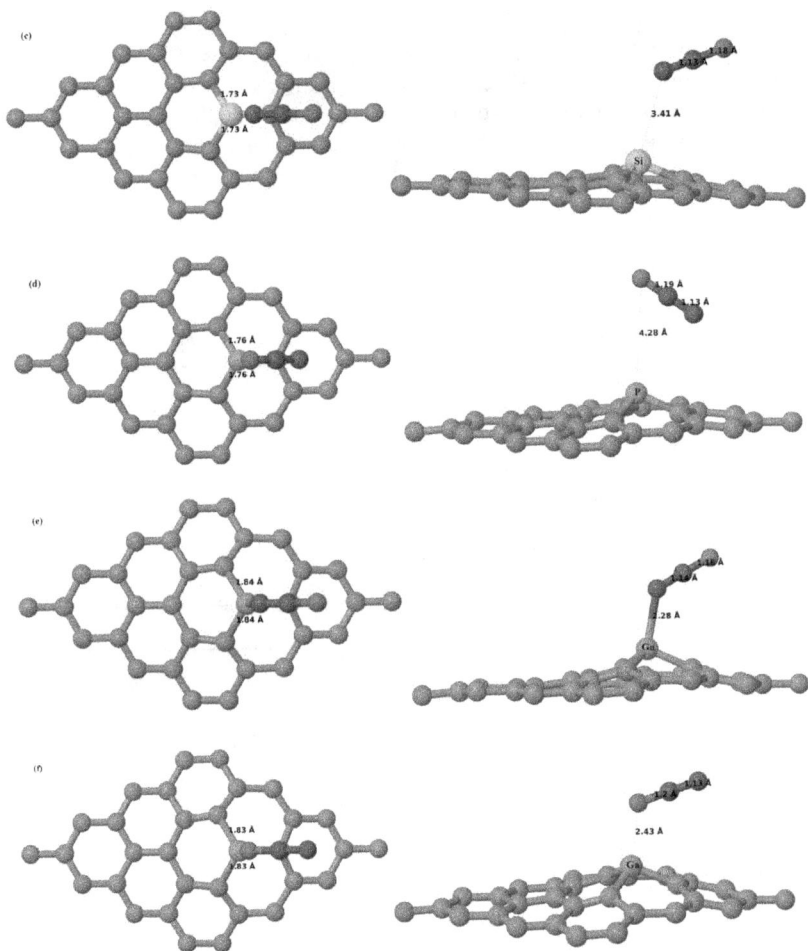

Figure 9.4 (contd.) Top (left) and side (right) views of optimized adsorption structures of N$_2$O molecule on (c) SiG (d) PG, (e) GG *via* N-end and (f) GG *via* O-end.

Table 9.1 Values of adsorption energy and distance of N$_2$O above graphene surface for BG, NG, SiG, PG and GG sheets

System	BG	NG	SiG	PG	GG
E_{ad} in eV	-0.027	-0.022	-0.030	-0.030	-0.239
x in Å	3.72	3.58	3.41	4.28	2.28

Functional Nanomaterials & Nanotechnologies

Table 9.2 shows the results from the Hirshfeld charge population analysis of BG, NG, SiG, PG, GG and N_2O adsorbed-XG systems. The charge states of the dopant atoms, three nearest carbon atoms surrounding the dopant atom (C_1, C_2 and C_3) and the adsorbed N_2O molecule are presented in Table 9.2. In IG, all the C atoms are charge neutral and have a charge state of zero. In BG, the three neighbouring C atoms have a charge state of -0.061 and the

Table 9.2 Hirshfeld charge population analysis of doping atom, three nearest carbon atoms surrounding doped atom and the N_2O molecule on the surface of BG, NG, SiG, PG and GG sheet[a]

System	C_1	C_2	C_3	B	N	Si	P	Ga	N	O
IG	0	0	0	-	-	-	-	-	-	-
BG	-	-	-	0.023	-	-	-	-	-	-
	0.061	0.061	0.061							
NG	0.045	0.045	0.045	-	0.030	-	-	-	-	-
SiG	-	-	-	-	-	0.205	-	-	-	-
	0.084	0.084	0.084							
PG	-	-	-	-	-	-	0.172	-	-	-
	0.046	0.046	0.046							
GG	-	-	-	-	-	-	-	0.020	-	-
	0.028	0.028	0.028							
O-end N_2O-BG	-	-	-	0.018	-	-	-	-	-	-
	0.061	0.061	0.065							0.0944
O-end N_2O-NG	0.044	0.044	0.041	-	0.028	-	-	-	-	0.0999
N-end N_2O-SiG	-	-	-	-	-	0.222	-	-	0.086	-
	0.085	0.085	0.088							
O-end N_2O-PG	-	-	-	-	-	-	0.170	-	-	-0.105
	0.046	0.046	0.046							
N-end N_2O-GG	-	-	-	-	-	-	-	0.006	-	-
	0.033	0.033	0.040						0.030	

[a] A negative sign indicates electrons gained, whereas positive sign implies electron lost by the atom (the unit of charge is electron)

B-dopant has a charge state of +0.023, since the neighbouring C atoms attract electrons from the B atom. Similarly in SiG, the C neighbours attain a charge state of -0.084 and the Si-dopant lose electronic charge of +0.205. Similar behaviour of the charge state of C neighbours and the dopant atom in BG, SiG, PG and GG is due to the fact that the electronegativity of C atom is higher than B, Si, P, and Ga atoms. As N atom is more electronegative than C atom, the neighbouring C atoms possess a charge state of +0.045 and the N-dopant has a charge state of +0.030 in NG.

In BG, NG, SiG and PG, the charge state of the dopant atom and the C neighbours does not show any significant change after N_2O adsorption. These results indicate that there is negligible charge transfer between these doped graphene systems and N_2O. But when N_2O molecule is adsorbed on the top site of GG, the three carbon atoms neighboring the Ga atom attain electrons and the Ga atom loses valence electrons featuring 0.006 for N-end, which suggests that a chemical bond is formed between the Ga and N_2O.

9.3.3 Adsorption of N_2O on CG and MG

The adsorption of N_2O on CG and MG sheets are discussed in this section. The optimized atomic structures of CG and MG are shown in Figure 9.5. In CG and MG, the dopant atom is displaced outward forming a bump (Figure 9.5) with an elevation of 1.318 Å and 1.304 Å respectively. The carbon-dopant atom distances of 1.85 Å (Cr-C bond) and 1.82 Å (Mn-C bond), are in good agreement with the results reported in [41].

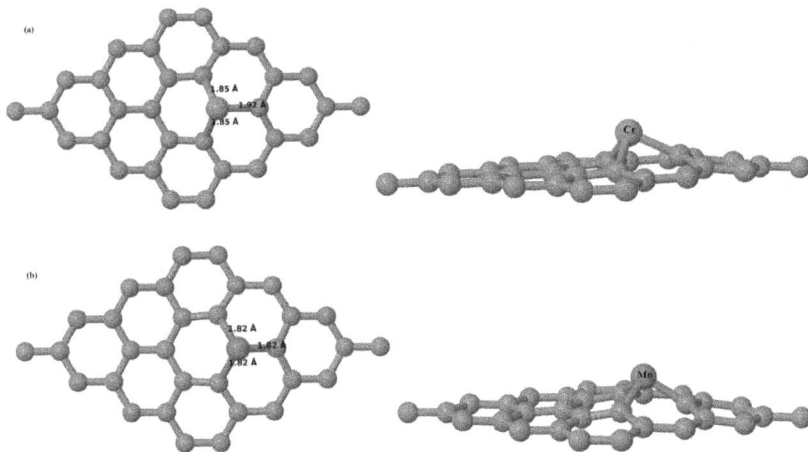

Figure 9.5 Top (left) and side (right) views of optimized structures of (a) CG and (b) MG sheets.

After N_2O adsorption, the structures are changed dramatically. The stable adsorption configurations of N_2O molecule adsorbed on CG and MG *via* N-end and O-end are presented in Figure 9.6. From the figure, it can be seen that N_2O molecule can strongly bond to CG and MG sheets by the O atom or by the two N atoms with dopant atom-gas molecule distance of around 2 Å. The calculated adsorption energies of these structures for CG and MG are listed in Table 9.3, suggesting that, for CG, the structure bonded by O and N atoms is the most stable one, whereas for MG, the structure bonded by N atoms is the most stable one, as these have the largest E_{ad}. Figure 9.6 (b) presents the most energetically stable adsorption configurat-

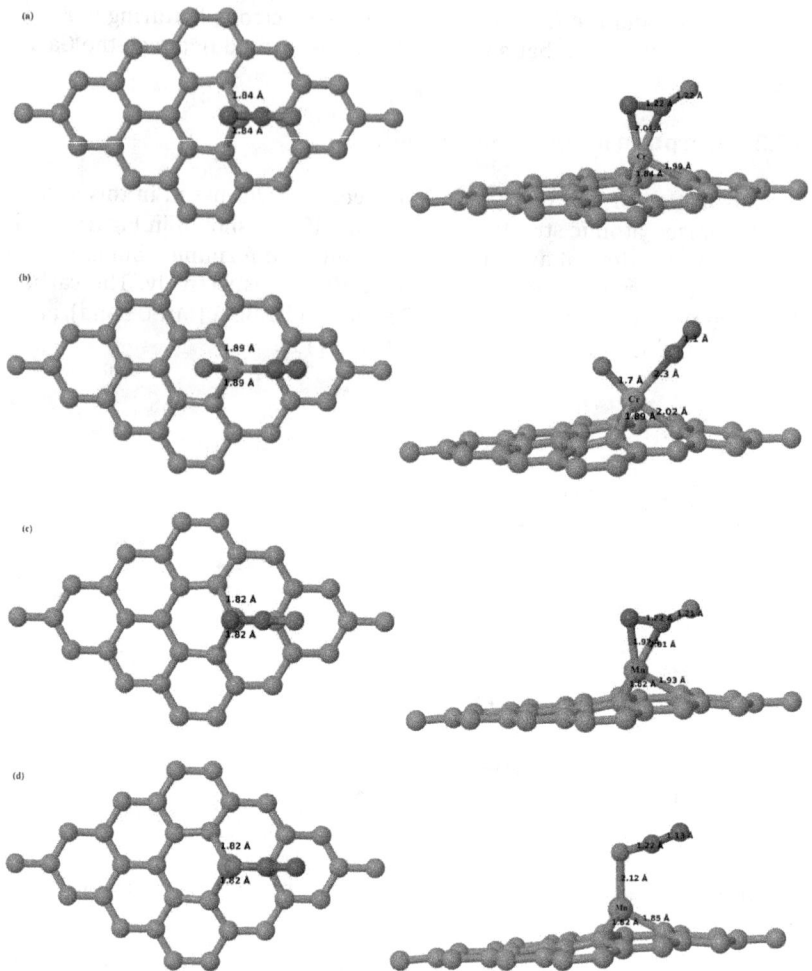

Figure 9.6 Top (left) and side (right) views of the adsorption structures of (a) N_2O on CG *via* N-end (b) N_2O on CG *via* O-end (c) N_2O on MG *via* N-end, and (b) N_2O on MG *via* O-end.

ion of N_2O on CG, where a partial dissociation of N_2O into N_2 and O-species on the Cr-dopant can be seen. The resulting NNCrO species remains bonded to the doped graphene sheet through the Cr dopant. In both CG and MG, the carbon-dopant atom distance got extended after N_2O adsorption (Figure 9.6). The elevation of Cr and Mn above the graphene sheet has extended

significantly to 1.550 Å and 1.396 Å, upon N_2O interaction. The above results show that N_2O is chemisorbed on both CG and MG with large adsorption energies (Table 9.3) and small binding distances.

Table 9.3 Adsorption energies released during N_2O chemisorption on CG and MG sheets

System	CG via N-end	CG via O-end	MG via N-end	MG via O-end
E_{ad} in eV	-1.168	-2.406	-1.088	-0.478

Table 9.4 shows the results of Hirshfeld charge distribution analysis performed on CG, MG and N_2O adsorbed-CG and MG systems. As the C atoms
around the dopant atom attract electrons from the dopant atom, the C
neighbors attain a charge state of -0.067 and -0.037 in CG and MG respectively. The dopant atoms lose electronic charge and attain a charge state of
+0.327 and +0.151in CG and MG respectively.

Table 9.4 Hirshfeld charge distribution analysis of the dopant atom, three C atoms
around dopant atom and the N_2O molecule on the surface of CG and MG sheet

System	C_1	C_2	C_3	Cr	Mn	N	O
CG	-0.067	-0.067	-0.076	0.327	-	-	-
MG	-0.037	-0.037	-0.037	-	0.151	-	-
N-end N_2O-CG	-0.062	-0.062	-0.055	0.277	-	-0.160	-
O-end N_2O-CG	-0.052	-0.052	-0.100	0.303	-	-	-0.293
N-end N_2O-MG	-0.025	-0.025	-0.022	-	0.064	-0.116	-
O-end N_2O-MG	-0.035	-0.035	-0.040	-	0.084	-	-0.049

Upon N_2O adsorption on both CG and MG, the charge state of Cr and Mn
reduces for both binding ends of the N_2O molecule. For N-end adsorbing on
CG and MG, the carbon neighbors loses electrons. For the scheme of N_2O
adsorbed on CG and MG through the O-end, both C_1 and C_2 neighbors
around the Cr and Mn-dopant loses electronic charge, whereas the third C
neighbor attain more electrons. The adsorption of N_2O on CG and MG result in significant change in the charge state of the dopant atoms, which
could be attributed to the charge transfer between the Cr- or Mn-dopant
and the N_2O molecule. The charge of O atom for O-end adsorbing on the CG
is larger than the charge of N for N-end binding mode, whereas in the case

of N_2O adsorbed on MG, the charge of N atom for N-end is larger. The binding atoms of N_2O get charge state of -0.160 for N-end and -0.293 for O-end on CG, and -0.116 for N-end and -0.049 for O-end on MG respectively, which implies that a chemical bond exists in the region between N_2O and the Cr- and Mn-dopants.

9.4 Summary

We have carried out first-principles studies to investigate the interaction of an isolated N_2O molecule on intrinsic and heteroatom-doped (B, N, Si, P, Ga, Cr and Mn) graphene. The stable adsorption geometries and adsorption energies are obtained on the basis of density functional theory calculations. Our calculations indicate that intrinsic graphene, B-, N-, Si- and P-doped graphene are not sensitive for N_2O molecule, due to weak adsorption. In contrast, Ga-doped graphene was found to be sensitive towards N_2O molecule, due to the strong interaction of N_2O with Ga-dopant. Cr- and Mn-doped graphene show chemisorption of N_2O on their surface with comparatively high adsorption energies and short binding distances. The charge population analysis show that the electronic properties of Ga-, Cr- and Mn- doped graphene are sensitive to the adsorption of a single N_2O molecule. The results also indicate that Cr- and Mn-doped graphene can be more suitable for the detection of N_2O gas.

References

1. Geim, A. K. and Novoselov, K. S. (2007) The rise of graphene. *Nature Materials,* **6**, 183-191.
2. Jang, H., Park, Y. J., Chen, X., Das, T., Kim, M.-S., and Ahn, J.-H. (2016) Graphene-based flexible and stretchable electronics. *Advanced Materials,* **28**, 4184-4202.
3. Tozzini, V., and Pellegrini, V. (2013) Prospects for hydrogen storage in graphene. *Physical Chemistry Chemical Physics,* **15**, 80-89.
4. Gadipelli, S., and Guo, Z. X. (2015) Graphene-based materials: Synthesis and gas sorption, storage and separation. *Progress in Materials Science,* **69**, 1-60.
5. Pumera, M. (2011) Graphene-based nanomaterials for energy storage. *Energy & Environmental Science,* **4**, 668-674.
6. Brownson, D. A. C., Kampouris, D. K., and Banks, C. E. (2011) An overview of graphene in energy production and storage applications. *Journal of Power Sources,* **196**, 4873-4885.
7. Goenka, S., Sant, V., and Sant, S. (2014) Graphene-based nanomaterials for drug delivery and tissue engineering. *Journal of Controlled Release,* **173**, 75-88.
8. Feng, L., and Liu, Z. (2011) Graphene in biomedicine: opportunities and challenges. *Nanomedicine,* **6**, 317-324.
9. Machado, B. F., and Serp, P. (2012) Graphene-based materials for catalysis. *Catalysis Science & Technology,* **2**, 54-75.

10. Zhang, Y., Tan, Y.-W., Stormer, H. L., and Kim, P. (2005) Experimental observation of the quantum Hall effect and Berry's phase in graphene. *Nature,* **438**, 201-204.
11. Novoselov, K. S., Geim, A. K., Morozov, S. V., Jiang, D., Katsnelson, M. I., Grigorieva, I. V., Dubonos, S. V., and Firsov, A. A. (2005) Two-dimensional gas of massless Dirac fermions in graphene. *Nature,* **438**, 197-200.
12. Danneau, R., Wu, F., Craciun, M. F., Russo, S., Tomi, M. Y., Salmilehto, J., Morpurgo, A. F., and Hakonen, P. J. (2008) Shot noise in ballistic graphene. *Physical Review Letters,* **100**, 196802.
13. Lee, C., Wei, X., Kysar, J. W., and Hone, J. (2008) Measurement of the elastic properties and intrinsic strength of monolayer graphene. *Science,* **321**, 385-388.
14. Schedin, F., Geim, A. K., Morozov, S. V., Hill, E. W., Blake, P., Katsnelson, M. I., and Novoselov, K. S. (2007) Detection of individual gas molecules adsorbed on graphene. *Nature Materials,* **6**, 652-655.
15. Yavari, F., and Koratkar, N. (2012) Graphene-based chemical sensors. *The Journal of Physical Chemistry Letters,* **3**, 1746-1753.
16. Wang, T., Huang, D., Yang, Z., Xu, S., He, G., Li, X., Hu, N., Yin, G., He, D., and Zhang, L. (2016) A review on graphene-based gas/vapor sensors with unique properties and potential applications. *Nano-Micro Letters,* **8**, 95-119.
17. Ko, G., Kim, H. Y., Ahn, J., Park, Y. M., Lee, K. Y., and Kim, J. (2010) Graphene-based nitrogen dioxide gas sensors. *Current Applied Physics,* **10**, 1002-1004.
18. Leenaerts, O., Partoens, B., and Peeters, F. M. (2008) Adsorption of H_2O, NH_3, CO, NO_2, and NO on graphene: A first-principles study. *Physical Review B,* **77**, 125416.
19. Giannozzi, P., Car, R., and Scoles, G. (2003) Oxygen adsorption on graphite and nanotubes. *The Journal of Chemical Physics,* **118**, 1003-1006.
20. Yong-Hui, Z., Ya-Bin, C., Kai-Ge, Z., Cai-Hong, L., Jing, Z., Hao-Li, Z., and Yong, P. (2009) Improving gas sensing properties of graphene by introducing dopants and defects: a first-principles study. *Nanotechnology,* **20**, 185504.
21. Shao, L., Chen, G., Ye, H., Wu, Y., Qiao, Z., Zhu, Y., and Niu, H. (2013) Sulfur dioxide adsorbed on graphene and heteroatom-doped graphene: a first-principles study. *The European Physical Journal B,* **86**, 1-5.
22. Wuebbles, D. J. (2009) Nitrous oxide: No laughing matter. *Science,* **326**, 56-57.
23. Kramlich, J. C., and Linak, W. P. (1994) Nitrous oxide behavior in the atmosphere, and in combustion and industrial systems. *Progress in Energy and Combustion Science,* **20**, 149-202.
24. Forster, P., Ramaswamy, V., Artaxo, P., Berntsen, T., Betts, R., Fahey, D. W., Haywood, J., Lean, J., Lowe, D. C., Myhre, G., Nganga, J., Prinn, R., Raga, G., Schultz, M., and Van Dorland, R. (2007) Changes in atmospheric constituents and in radiative forcing, Cambridge University Press, United Kingdom.
25. Ravishankara, A. R., Daniel, J. S., and Portmann, R. W. (2009) Nitrous oxide (N_2O): The dominant ozone-depleting substance emitted in the 21st century. *Science,* **326**, 123-125.
26. Perez-Ramírez, J. (2007) Prospects of N_2O emission regulations in the European fertilizer industry. *Applied Catalysis B: Environmental,* **70**, 31-35.
27. Rapson, T. D., and Dracres, H. (2014) Analytical techniques for measuring nitrous oxide. *TrAC Trends in Analytical Chemistry,* **54**, 65-74.
28. Gonze, X., Amadon, B., Anglade, P. M., Beuken, J. M., Bottin, F., Boulanger, P., Bruneval, F., Caliste, D., Caracas, R., Cote, M., Deutsch, T., Genovese, L., Ghosez, P., Giantomassi, M., Goedecker, S., Hamann, D. R., Hermet, P., Jollet, F., Jomard, G., Leroux, S., Mancini, M., Mazevet, S., Oliveira, M. J. T., Onida, G., Pouillon, Y., Rangel,

T., Rignanese, G. M., Sangalli, D., Shaltaf, R., Torrent, M., Verstraete, M. J., Zerah, G., and Zwanziger, J. W. (2009) ABINIT: First-principles approach to material and nanosystem properties. *Computer Physics Communications,* **180,** 2582-2615.

29. Perdew, J. P., Burke, K., and Ernzerhof, M. (1996) Generalized gradient approximation made simple. *Physical Review Letters,* **77,** 3865-3868.

30. Troullier, N., and Martins, J. L. (1991) Efficient pseudopotentials for plane-wave calculations. *Physical Review B,* **43,** 1993-2006.

31. Monkhorst, H. J., and Pack, J. D. (1976) Special points for Brillouin-zone integrations. *Physical Review B,* **13,** 5188-5192.

32. Schlegel, H. B. (1982) Optimization of equilibrium geometries and transition structures. *Journal of Computational Chemistry,* **3,** 214-218.

33. Hirshfeld, F. L. (1977) Bonded-atom fragments for describing molecular charge densities. *Theoretica Chimica Acta,* **44,** 129-138.

34. Dresselhaus, M. S., Dresselhaus, G., Saito, R., and Jorio, A. (2005) Raman spectroscopy of carbon nanotubes. *Physics Reports,* **409,** 47-99.

35. Castro Neto, A. H., Guinea, F., Peres, N. M. R., Novoselov, K. S., and Geim, A. K. (2009) The electronic properties of graphene. *Reviews of Modern Physics,* **81,** 109-162.

36. Lv, Y.-a., Zhuang, G.-l., Wang, J.-g., Jia, Y.-b., and Xie, Q. (2011) Enhanced role of Al or Ga-doped graphene on the adsorption and dissociation of N_2O under electric field. *Physical Chemistry Chemical Physics,* **13,** 12472-12477.

37. Dai, J., Yuan, J., and Giannozzi, P. (2009) Gas adsorption on graphene doped with B, N, Al, and S: A theoretical study. *Applied Physics Letters,* **95,** 232105.

38. Zou, Y., Li, F., Zhu, Z. H., Zhao, M. W., Xu, X. G., and Su, X. Y. (2011) An ab initio study on gas sensing properties of graphene and Si-doped graphene. *The European Physical Journal B,* **81,** 475-479.

39. Jiayu, D., and Jianmin, Y. (2010) Modulating the electronic and magnetic structures of P-doped graphene by molecule doping. *Journal of Physics: Condensed Matter,* **22,** 225501.

40. Sharma, S., and Verma, A. S. (2013) A theoretical study of H_2S adsorption on graphene doped with B, Al and Ga. *Physica B: Condensed Matter,* **427,** 12-16.

41. Dai, J., and Yuan, J. (2010) Adsorption of molecular oxygen on doped graphene: Atomic, electronic, and magnetic properties. *Physical Review B,* **81,** 165414.

10

Nano-catalysts: Advances in Synthesis and Applications in the Field of Energy

10.1 Introduction

Globally, catalyst synthesis solely accounts around US$10 billion in sales in different chemical processes [1]. Catalysis is a process during which the catalyst lowers the activation energy and increases the rate of a reaction without affecting the state of equilibrium of the reaction and regenerates at the last step of reaction. The field of catalysis has been developed in many directions like electro (e.g. ethanol powered fuel cells), photo (e.g. self-cleaning glass), enzymatic (e.g. Baeyer–Villiger oxidation), homogeneous (e.g. hydroformylation) and heterogeneous (e.g. automotive catalytic converter) [2]. Catalysts play vital role in many industrial processes such as olefin polymerization, hydrogenation, cracking, reforming, gas synthesis, fermentation, etc. Since the discovery of catalysts, the improvement in the catalytic activity and performance is continuously under focus in order to reduce the major cost associated with the large-scale chemical processes [3]. This includes many aspects of the catalytic process such as elimination of catalyst poisoning, re-generation, enhancing the useful lifetime, etc.

Nano-scale materials refer to those which have at least one external dimension or internal structure between 1-100 nanometers. These materials may have different physico-chemical properties than coarser materials of same type and can be used to produce state-of-the-art technologies and end-products [4]. Example of such technologies and end-products containing novel nanomaterials are clinical neuroscience [5,6], bio-labelling [7], rechargeable batteries [8], antimicrobial surfaces [9], tissue engineering and regenerative medicine [10], etc. With the advent of nanotechnology in technological and industrial areas, nano-science has also made considerable and revolutionary advances in the field of catalysis (late 1990s) for different applications. Nano-catalysts refer to nano-materials that deal with the unconventional material properties for innovative applications. According to recent reports, nano-catalyst is a *"hierarchical system of basic structural units (active sites) with defined assembling strategies*

Ali U. Chaudhry* and Vikas Mittal**, The Petroleum Institute (part of Khalifa University of Science and Technology), Abu Dhabi, UAE
*Current address: Texas A&M University, Qatar; **Current address: Bletchington, Wellington County, Australia

for multiple length scales" [11,12]. Nano-catalysts are used in several environmental, chemical and process industries such as energy conversion and conservation [13], pharmaceutical, production of chemicals and petrochemicals, control emission of hazardous gases like CO, NO [14], etc. Considering the benefits of catalysts, there is always a need to develop catalysts with improved efficiency, recycling, selectivity and stability. Nano-science has made it possible to produce nano-catalysts with high activity and selectivity which is highly unlikely to be achieved with micro dimensional materials [2]. At nano scale, different chemical and physical properties of materials may appear due to different extent of delocalized electrons. Also, the presence of large intrinsic surface reactivity (catalytic activity) owing to smaller size makes nano-catalysts promising candidates for catalysis. For instance, Figure 10.1 demonstrates the TEM image of multi-armed nano-star colloidal platinum nano-catalyst particles [15]. For the synthesis of nano-star catalyst particles, the authors employed tetrahedral platinum nanocrystals as seeds, which avoided the need of organic solvents, templates, substrates, etc.

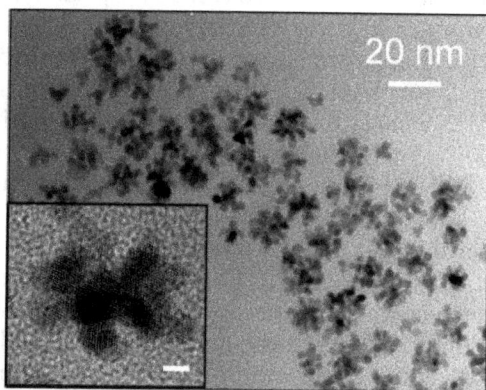

Figure 10.1 TEM image of the multi-armed nano-star colloidal platinum nano-catalyst particles. Inset represents the HRTEM image of the nano-star displaying multiple arms (scale bar 2 nm). Reproduced from Reference 15 with permission from American Chemical Society.

10.2 Applications of Nano-catalysts in the Field of Energy

In the twenty-first century, the nano-catalysts have gained significant research attention for exploring large scale applications, especially in the field of energy due to ecological and environmental friendly operation and low cost synthesis. Also, the depleting fossil fuel resources indicate the need to reassess our energy needs and optimize the efficiency of existing processes [16]. According to a recent report, around 85% of the energy used globally comes from fossil fuels which in turn discharges 30 billion tons of carbon dioxide [17]. The grand challenge of clean environment and efficiently control processes may be met earnestly with the help of precisely controlled composition and nano-catalyst structure [18]. In addition, the implications of nanostructure materials should not only encircle the efficient conversion (clean) of fossil fuel to useful products, but should also lead to a vital role in developing new catalytic routes for renewable energy. Development of new catalytic routes for non-conventional energy is specifically important to open new avenues so as to reduce the reliance on the conventional fossil based fuels, however, the increment in the efficiency of the existing processes would equally benefit the process industries.

This review reports recent advances in nanostructured materials as catalysts for energy applications. The review covers not only the research and development of nano-catalysts in already developed processes for fossil fuel conversion, but also focuses on the alternative means of energy using nano-catalysts (Table 10.1). More specifically, the review focuses on some of the energy related applications mentioned in Table 10.1.

10.2.1 Oil Refining

Oil refinery uses various kind of processes including catalyzed reactions for the separation of individual compounds present in the crude oil. During these operations, many environmentally hazardous gases like CO_2, hydrogen sulfide, ammonia, etc., are liberated. Refineries have to treat/process all kind of harmful products to make them harmless for the environment. Moreover, in order to meet the current and future demands of market in terms of fuel quality and quantity, efficient conversion of heaviest compounds of crude oil is required. In these scenarios, the contributions by the nano-catalysts with increased activity and selectivity are expected to be beneficial, especially in following refining processes [19]:

Hydrotreating

Hydrotreating (HT) is a major reductive process to remove oxygen, sulfur, nitrogen, halogen atoms and metals from the oil streams. Mainly, the catalysts containing hydrogen reduced palladium, platinum (low H_2S environments),

Table 10.1 Application of nano-catalysis in the field of Energy [1,17,19-33]

	Catalytic Processes
Energy from Fossil & Renewable Resources	Hydrotreating
	Hydrodemetalation
	Catalytic Reforming
	Fluid Catalytic Cracking
	Hydrocracking
	Isomerization
	Alkylation of Isobutene with Light Alkenes
	Oligomerization
	Etherification
	Water Gas shift and COS removal
	Steam Reforming
	Methanol Synthesis
	Methanol to Hydrocarbons
	The Fischer-Tropsch (FT) Synthesis Process
	Gas to Liquids
	Oxidative Coupling of Methane
	Direct Coal Liquefaction
	Coal and Carbon gasification
	Fuel Cell
	Biomass gasification and syngas conversion
	Fast pyrolysis and bio-oil upgrading
	Liquid-phase/aqueous-phase catalytic processing
	Vegetable Oil Conversion to Bio-Fuel
	Photocatalytic water splitting
	Hydrogenation
	Reforming
	Dehydration
	Energy Storage

sulfides of nickel, cobalt, tungsten or molybdenum metals supported on alumina, amorphous aluminosilicates or zeolites (X.Y or morenite) are used in the treatment process [25,34]. Hydrodesulfurization (HDS) is one of the vital processes of hydrotreating which is used to reduce the sulfur compounds from the crude oil. The performance of the catalysts during the treatment process mainly depends on the support on which catalyst resides, sulfidation conditions and promoters. The main function of the support is to provide good dispersion of the active part of the catalysts whereas promoters improve the performance. In a recent report, trimetallic nano-catalysts composed of Ni–Mo–W supported on multiwalled carbon nanotubes (MWCNT) were synthesized using ultrasound assisted co-precipitation method. The high energy environment (ultrasound) was used to avoid the metallic agglomeration. The incorporation of MWCNT was also useful due to weak interaction of metallic part and nanotubes [35].

Nano-catalysts prepared with conventional methods result in samples with polydisperse particles and inhomogeneous compositions. Further, metallic sulfide results in low active site densities which are unable to generate ultralow level of sulfur in petroleum products. Previous reports showed that silica supported metallic phosphide showed excellent activity, selectivity and resistance to sulfur poisoning. Danforth *et al.* [36] prepared robust system of mono-disperse nanoparticles composed of $Ni_{2-x}Co_xP$ and $Ni_{2-x}Fe_xP$ using solution-phase arrested precipitation. The nanoparticles were encapsulated in mesoporous silica ($mSiO_2$). Nano-catalyst composed of $Ni_{1.92}Co_{0.08}P@mSiO_2$ exhibited highest activity HDS on mass basis, whereas highest turnover frequency (TOF) was observed for $Ni_{2-x}Fe_xP@mSiO_2$ owing to increased electron donation from Fe to Ni than from Co to Ni [36].

Hydrodemetalation

Hydrodemetalation is among the basic steps of petroleum refining in which heteroatoms such as O, N, S and specifically metals are removed. With the reducing amount of light crude, refineries are now focusing on the heavier crude containing concentrated metals. The metals have damaging consequences on the catalysts and refinery equipment [28,37]. The fundamental process to remove the heteroatoms and metals is HM in the presence of catalyst and high hydrogen pressure. The heteroatoms usually convert to oxides (H_2O, SO_2) and hydrogen pnictides (NH_3), whereas the metals such as nickel and vanadium (as porphyrins and complexes) become sulfides. These sulfides usually deactivate the expensive catalyst by plugging the active sites on the catalyst surface and cause underutilization. The main catalysts of hydrodemetalation process are hydrogen (gas) and oxides of Co, Mo, etc., supported on γ-Al_2O_3 [19]. In order to overcome the problem of deactivation of catalyst due to diffusion limited (transport) reactions, Rao *et al.* [38] adopted a mathematical approach for the design of nano-catalyst with optimized broad pore network. This hierarchically structured catalyst (CoMo/AlO), using modeling, exhibited activity for a longer period of time as compared to optimized mesosphere industrial catalysts [38].

The use of nano-sized MoS_2 or MoS_2/NiS for the conversion of extra heavy crude oil and petroleum residue was also reported [39]. During this process V, Ni, and Mo metals were quantitatively passed into high boiling fractions.

Thus, the development of nanostructured nano-catalysts leads to process enhancement as well as alleviation of the existing challenges.

Catalytic Reforming

Catalytic reforming (CR) deals with the crude oil distillate in order to make high-octane reformates. The feed of CR is usually composed of aromatics (6-12 carbon atoms), naphthenes and paraffins. Generally, impregnated conventional catalysts based on platinum (noble metals) with promoters such as Re, Sn, and Ir and chloride alumina support are mainly used in this process [19,24,40]. The advances in the nanoparticles chemistry have made it possible to design nano-catalysts for CR process with improved resistance to deactivation, high catalytic activity and selectivity at high temperature [41]. In a recent research study, bi-metallic nanoparticles based on Pt (catalyst) and Rh (promoter) supported on mesoporous silica were used for isomerization reaction. The study showed that turnover frequency (TOF) and selectivity were mainly affected by the change in composition and size of the nano-catalyst [42]. The high temperature usage of nano-catalyst is limited owing to unwanted sintering and Ostwald ripening due to non-equilibrium state of nano-catalyst and support. This problem can be solved by introducing isolate nanoparticles concept using core-shell structure approach [11]. Recently, An *et al.* [43] studied the effect of thermal stability of designed nano-catalyst and the role of the support on CR. The authors synthesized four different types of nano-catalysts from platinum nanoparticles (Figure 10.2) supported on mesoporous SiO_2 or TiO_2. The selectivity of the catalysts was highly improved due to improved charge transfer at the interface in case of Pt-TiO_2 for n-Hexane reforming reactions [43]. These findings underlined the role of nano-catalytic processes in enhancing the efficiency and yield of conventional processes.

Fluid Catalytic Cracking

Fluid catalytic cracking (FCC) is a major refinery process and used to convert heavy vacuum distillates (like vacuum gas oil), atmospheric and refinery residues into lighter ones like motor fuels and gasoline. The main catalysts of FCC are mainly composed of functional matrix (zeolite Y), binder (NaY) and kaolin clay. Owing to the new environmental regulations on the release of CO, NO_x and SO_x, there is need to develop new effective FCC catalysts [19,23]. The main challenge associated with the catalysts is limited transport phenomena (poor mass transfer) within the catalysts pore structures, as the zeolite having pores struc-

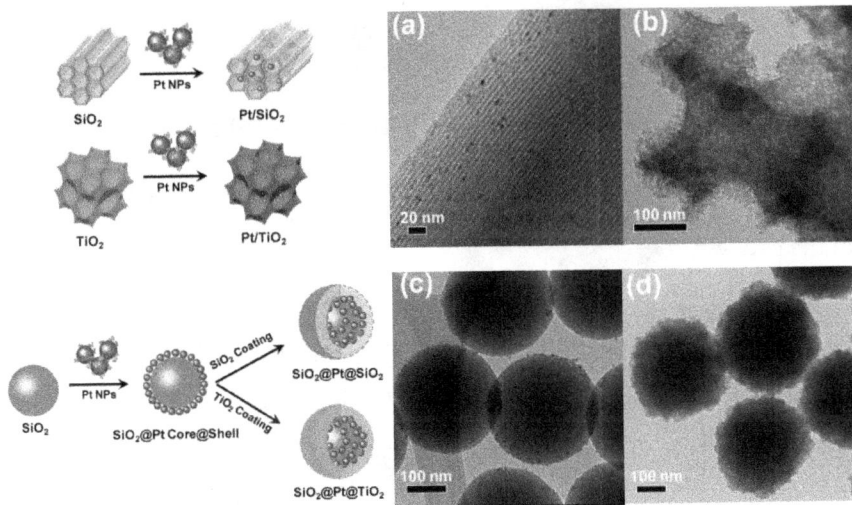

Figure 10.2 Preparation of supported Pt nanoparticle catalysts and sandwich-type Pt core@shell catalysts and TEM images of supported Pt nanoparticle catalysts (a,b) and sandwich catalysts (c,d): (a) Pt/SiO$_2$, (b) Pt/TiO$_2$, (c) SiO$_2$@Pt@SiO$_2$, and (d) SiO$_2$@Pt@TiO$_2$. Reproduced from Reference 43 with permission from American Chemical Society.

ture less than 1 nm lead to limited diffusion reactions. Extensive research efforts have been dedicated to improve the pore accessibility through nano-sized zeolites [44-46]. Moreover, transport of reactants within catalyst pores can occur through single or combination of following three mechanisms or models, i.e., Knudsen diffusion [47], intercrystalline diffusion [48], and molecular diffusion [49]. Knudsen and molecular diffusion mechanism describe the trasnporatation of molecules within the micropores on the basis of molecular sizes and are often called rate-limiting step of a reaction [50,51]. On the other hand, intercrystalline diffusion is more active at smaller pore structure. Garcia-Martinez *et al.* [51] introduced a novel, well-controlled and direct surfactant-assisted synthesis method for controlled mesoporosity in Y zeolites (low Si/Al ratios) for FCC applications. The synthesized FCC catalysts (mesostructured zeolites) exhibited improved hydrothermal stability and selectivity in terms of useful fuels. In another research study, well-organized zeolitic nanocrystal aggregates were synthesized [52]. The aggregates system was composed of interconnected hierarchically micro–meso–macro porous. Catalytic testing validated their supremacy as the cracking catalysts compared to the conventional zeolite catalysts. Similar kind of structures were also reported in the other literature studies, such as delaminated zeolite [53] and stable single-unit-cell nanosheets of zeolite as nano-catalysts.

In another study, Dejhosseini *et al.* [54] also studied catalytic cracking reaction of heavy oil in the presence of cerium oxide nano-catalyst (Figure 10.3) in supercritical water. The authors concluded that the rate of conversion could be enhanced by increasing the exposed surface area and reducing the particle size of the catalyst.

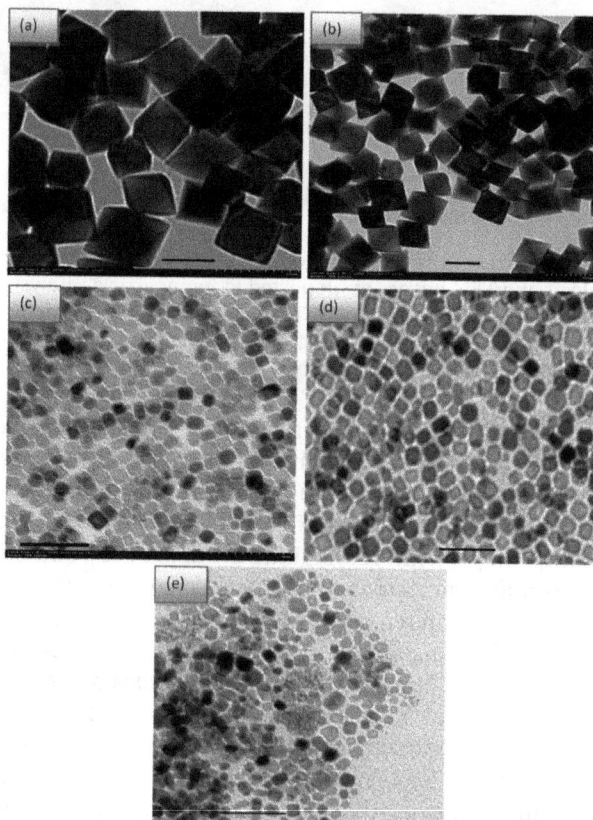

Figure 10.3 Transmission electron micrographs of the CeO_2 nanoparticles (a,b) without hexanoic acid, (c) with hexanoic acid, (d) after calcination at 300 °C, and (e) after calcination at 650 °C. Reproduced from Reference 54 with permission from American Chemical Society.

Isomerization

Isomerization reaction in refinery industry deals with the skeleton isomerization of alkanes, alkenes, isomerization of double bond of alkenes, alkyl aromatics, saturated hydro-carbon (bi/poly-cyclic) and inter-conversion of alkyl cyclo-pentanes and cyclo-hexanes. In the various isomerization processes, two types of catalysts are mainly used, i.e., mono-functional acidic and bi-functional catalysts [19]. Mostly, mono-functional are consisted of Friedel–Crafts type of acid $AlCl_3$ with HCl whereas $HSbF_6$ is used as liquid acid catalyst. Bi-functional catalysts are advantageous, promising and mainly composed of combination of acid with a metal such as zeolites (acid forms) with platinum or palladium [26]. For the isomerization of hexane, nano-catalysts system comprising of Pd and zeolites HY, HZSM-5 and mixtures of zeolites with γ-Al_2O_3 were synthesized [55]. The increased activity was observed in case of Pd/HY due to improved dispersity and reduction in Pd cluster size. Recently Aghdam *et al.* [56] used membrane reactor packed with Pt/ZSM-5 nano-catalyst for the isomerization of n-C5. The membrane reactor containing nano-catalyts exhibited improved i-C5 yield due to selectivity of reactants in reaction zone. In another study, An *et al.* [57] synthesized colloidal Pt nanoparticles of different controlled sizes and supported onto ordered macroporous oxides such as SiO_2, Al_2O_3, TiO_2, Nb_2O_5, Ta_2O_5, and ZrO_2 for n-hexane isomerization. The results revealed that Nb_2O_5 and Ta_2O_5 exhibited the highest selectivity when the nanoparticles size increased from 1.7 to 5.5 nm due to combined effect, i.e., charge transfer at Pt-oxide interfaces and smaller size of catalysts [57].

Steam Reforming

Steam reforming (SR), another key industrial process, is used to synthesize hydrogen and products like ammonia, methanol, CO-rich gas, etc., from steam, CO_2 and hydrocarbons. Main catalysts used of SR are usually based on Cu, Ni, Co, Fe Ru, Rh, etc., and perform better than non-metal catalysis [29]. There are certain disadvantages associated with the usage of catalysts for SR process as reported in literature such as reduction in the surface area due to sintering of particles, sulfur poisoning, carbon formation, etc. [19]. The requirements for good steam reforming catalysts were also reported like high selectivity to CO_2, minimum production of CO, low active energy, low cost, resistance to sintering and coke formation, etc. [58]. Attempt were made to improve the diffusion, i.e., the rate-limiting step, as well as to induce high activity and long-lasting stability of catalysts. Koo *et al.* [59] developed a monolith catalyst using locally developed coating technique, i.e., deposition-precipitation for hydrogen production. Nickel as nano-layer was deposited on Fe-Cr alloy flat rolled and corrugated metal strips using nickel nitrate as the precursor. The prepared catalysts exhibited improved activity and long-term stability due to high dispersion and large active surface area. Similarly for improved activity, stability, and reduced CO formation for steam reforming of methanol, novel nano-structured copper supported ZrO_2 was synthesized [60]. A polymer templating technique was used to control the morphology and porosity of the supported material. During catalysis, copper particles stayed apart due to ZrO_2 support, thus, resulting in improved sintering process and higher available surface area for carrying out the reaction.

Methanol Synthesis

According to a recent report, the global annual consumption of methanol was about of 65 million tons/year during 2013 [61]. Methanol is synthesized from synthesis gas, hydrogen mixture, CO and CO_2 using different technologies such as reforming of natural gas, adiabatic pre-, fired, tubular, auto-thermal or oxygen-fired processes. Main catalysts for the process are composed of ZnO/Cr_2O_3, $Cu/ZnO/Al_2O_3$, Cu/ZrO_2, as well as Pd-based and sulfide-based. The problems associated with the catalyst have been reported in the literature such as sintering, poisoning of sulfur, chlorine, carbonyls of iron and by-product formation [19,30]. In order to improve the poor sintering behavior of Cu nano-catalyst and metal-support interaction, Berg *et al.* [62] synthesized Cu catalysts on SiO_2 via precipitation and incipient wetness impregnation method. The catalysts system created through precipitation exhibited entrapment of Cu particles in the plate-like silica structure which, in turn, limited the growth of Cu particles during catalysis [62]. Another attempt was made to avoid Ostwald ripening of nano-particles during copper methanol catalysis by producing amine-functionalized support material (Figure 10.4) [63]. The improved sintering behavior was owing to the functionalization which, in turn, increased the distance between the Cu nanoparticles due to limiting transport of copper atoms over the support surface.

The function of a promoter element in main catalyst particle is to enhance or suppress the reactivity or selectivity for certain reaction. Recently, Kuld *et al.* [61] discussed the complex promotional interactions of ZnO with Cu during catalysis of methanol from synthesis gas. The authors showed with the help of experiments and density functional theory calculations that promotion of catalysis is due to Zn atoms migration in the Cu surface during catalysis [61].

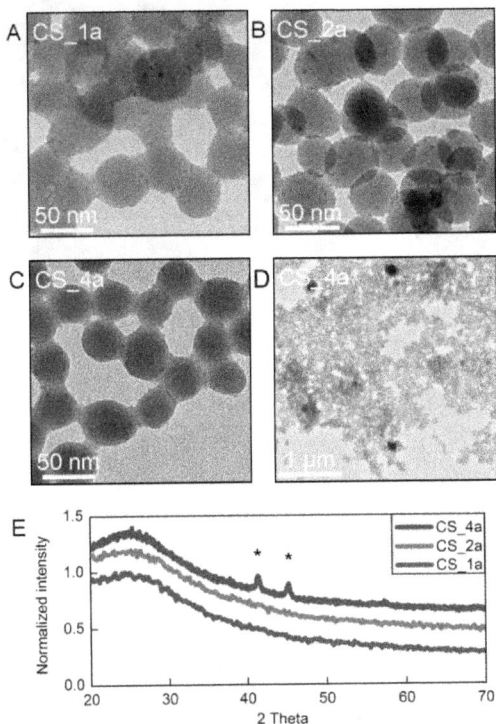

Figure 10.4 TEM images of various nanoparticles and X-ray diffractograms of the samples (peaks (*) correspond to CuO). Reproduced from Reference 63 with permission from American Chemical Society.

Similarly, Berg *et al.* [64] also reported about the structure sensitivity of Cu and CuZn for methanol synthesis. It was revealed that activity of the catalyst was highly depending on structure. It was considerably decreased for smaller particles (8 nm) due to sole atomic configuration like step-edge sites, which is usually missing for smaller particles [64]. In another study, Rungtaweevoranit *et al.* [65] reported a catalyst system (100% selectivity and improved activity) consisting of Cu nanoparticles encapsulated in a metal-organic framework (MOF) for methanol conversion from CO_2 [65]. Figure 10.5 shows the morphological features of the the Cu encapsulated in MOF, in comparison with the Cu on MOF. Ternary nano-sized $Cu/ZnO/Al_2O_3$ was also reported in another study, which was prepared from alternatives novel sol and colloidal primary solutions and exhibited improved catalytic activity [66].

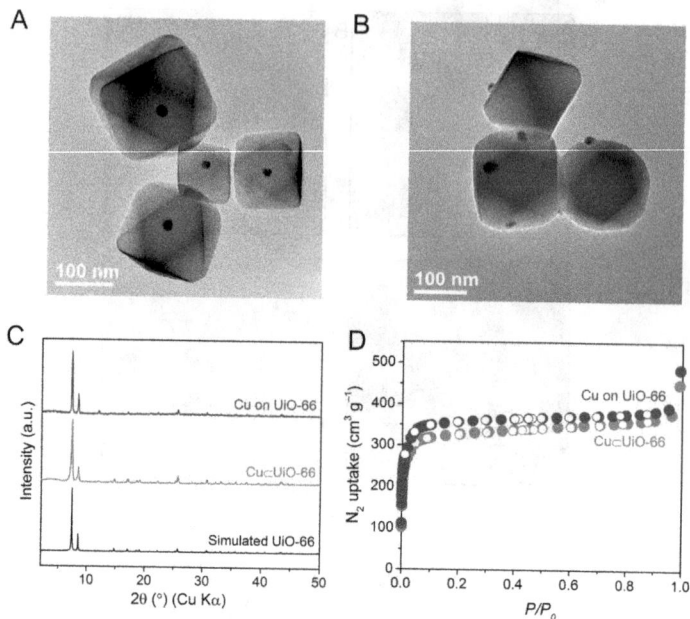

Figure 10.5 TEM images of (A) Cu⊂UiO-66 (single Cu NC inside UiO-66), (B) Cu on UiO-66, (C) XRD patterns in comparison with simulated pattern from single crystal X-ray diffraction data, and (D) N_2 adsorption–desorption isotherms at 77 K (adsorption and desorption points represented by closed circles and open circles, respectively). Reproduced from Reference 65 with permission from American Chemical Society.

Fischer–Tropsch Synthesis Processes

Fischer-Tropsch synthesis process (FT) is a highly exothermic reaction and used to produce wide-range of hydrocarbons such as high quality diesel fuel, gasoline and linear chemicals from catalytic conversion of mixtures of CO and H_2 (syngas, catalytic CO hydrogenation) through surface polymerization. The syngas is usually produced from coal or methane (non -crude oil feedstock) by treating with oxygen and steam at high temperatures. Main catalysts used in FT process are iron, nickel, cobalt, ruthenium or osmium [11,67]. Many attempts have been made to enhance the activity of the FT catalysts. Using promoter along with the main catalyst system is one of the method to increase the activity, selectivity, degree of reduction or the dispersion of the supported catalyst. Recently, Pt was used as promoter with Co/Nb_2O_5 catalyst system supported on γ-Al$_2$O$_3$- and α-Al$_2$O$_3$ [68]. Experimental data showed that Pt promoted the activity and active sites of catalyst system by a factor of 2.4 and 1.7 respectively. Similarly, different kinds of noble metals promoters such as Ag, Pt, Ru, and Re were added through (co-)impregnation to Co/TiO$_2$ catalysts system. Temperature programmed reduction (TPR) studies showed that Pt and Ru decreased the reduction temperatures of cobalt oxide more than Re and Ag [69].

An attempt was also made to improve the low porosity limits of promoter like niobia by synthesizing niobia modified silica by impregnation and subsequent use as support for Co and Pt-Co catalysts [70]. The selectivity of the niobia modified catalysts system was found to be pressure dependent due to higher intrinsic CO coverage. It was also found that combination of Nb nanocrystals and Pt increased the cobalt-weight normalized activity by a factor of 3-4 due to higher number of active sites and increased TOF. In another study, Wang *et al.* [71] reported platinum-modulated cobalt nano-catalysts for low-temperature Fischer–Tropsch process. The authors observed high activity of the catalyst for CO dissociation, which was attributed to the formation of Co overlayer structures on Pt NPs or Pt–Co alloy NPs. As observed in Figure 10.6, the higher activity of the supported Co layer resulted from the formation of a more favorable transition state, which thus facilitates the C–O dissociation on supported Co layers.

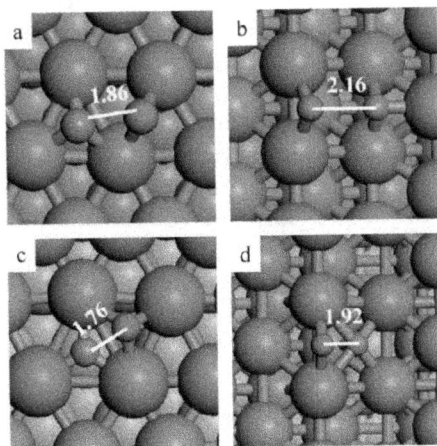

Figure 10.6 Optimized transition states for CO dissociation on: (a) Co(111), (b) Co(311), (c) 1 ML Co/Pt(111), and (d) 2 ML Co/Pt(311). Blue, orange, gray, and red spheres represent Co, Pt, C, and O atoms, respectively. Reproduced from Reference 71 with permission from American Chemical Society.

Usually parameters associated with support in catalyst system like pore size, symmetry, geometry, connectivity and morphology, textures, specific surface area, stability, metal dispersion and reducibility play a vital role during catalysis. Use of ordered mesoporous substrates for FT catalysts have plenty of advantageous due to symmetry, long range order and narrow pore size distribution. Oschatz *et al.* [72] compared mesoporous ordered particles of silica, carbon and silicon carbide as supports for Fe-based Ft catalysts. The catalyst system was synthesized by encapsulating iron nanoparticles and sodium and sulphur (promoters) within the selected supports which, in turn, increased the stability of the system. The results showed that carbon had highly selective behavior for lower olefins and low methane production [72].

Etherification

Etherification is a useful exothermic and acid catalytic process in refining industry which is used to convert branched alkene into important products such as tertiary ethers (gasoline as methyl tert-butyl ether (MTBE), ethyl tert-butyl ether (ETBE) and tert-amyl methyl ether (TAME)), linear ethers (diesel fuel as dimethyl ether (DME), di-*n*-pentyl ether (DNPE)), etc. It is a simply the addition reaction of alcohol which is mainly catalyzed by acidic ion-exchange resins (sulfonated organic polymers), zeolites (molecular sieves), clays, solid hetro-polyacids, sulfided zirconia, etc. The main feedstock for etherification process is reactive streams of iso-butene or pentenes, or alcohol (methanol or ethanol). During catalysis, deactivation of catalysts usually happens due to poisoning of acidic sites by Na^{+2} and NH_4^+ ions, nitrogen containing compounds, dienes, metallic cations, etc. [31]. Other limitations of the catalysts have also been reported in the literature such as high solubility of the mineral acids and high temperature stability of hetero-polyacid. Yee *et al.* [73] discussed the catalysis of ETBE synthesis from tert-butyl alcohol (TBA) using functionalized multiwalled carbon nanotube (F-MCNT) having Lewis acid sites. The proposed mechanism of etherification was as follows: 1) creation of tert-butyl carbonium ion (TBC), 2) interaction of TBC ion with ethanol forming intermediate, 3) rearrangement of TBC to towards stable molecule of ETBE and 4) desorbance of ETBE molecule from the surface. The associated advantage of this process was to avoid the formation of side product like diethyl ether. The catalyst system improved degradation, reusability and regeneration behavior, selectivity and conversion (64% & 68%) at 3 wt% of catalytic loading [73]. Catalytic performance of ion-exchange resin during catalysis has critical importance for producing analogous ether. In a recent study sixteen sulfonic ion-exchange resins were evaluated for catalysis of iso-butene with four different alcohols, i.e., 1-butanol, 1- propanol, ethanol or methanol [74]. It was observed that reaction rates were higher as the chain length of the alcohol increased. It was also concluded that the highly acidic resins with inflexible morphology showed higher activity for the reaction.

Molecular sieves catalysts are known as "electron transfer stations" and can be used as alternatives to acidic catalysts due to many benefits, e.g., thermal stability, modification with different ions, amendable pore configuration and acidity and easy regeneration, even though these display reduced activity at low range of temperature. These catalysts have also other properties like empty outermost (5d) orbital, different valency state and ability to form thermally stabilized structure. Modifications and treatments of zeolites with metallic ions, acidic leaching or steam treatment are also reported in literature for improved catalytic performance of these molecular sieves. Recently, Yan *et al.* [75] modified the Hβ zeolite with a combination of low steam treatment and lanthanum ion exchange. The comparative study with single modification process showed that combination modification processes exhibited best results in terms of etherfication activity and conversion. The combined modification process did not destroy the crystal phase and framework structure, while generated new Brønsted acid sites and enlarged pore structure, which was useful for the etherification [75]. Excellent activity of the sulfated zirconia is also reported for etherification by Jaworski *et al.* [76]. The authors reported the modification of zirconia with sulfuric acid which resulted in tetragonal phase of zirconia. The authors used the modified ZrO_2 for etherification of glycerol with benzyl alcohol at different temperatures (120-140° C), where the conversion grew with increasing sulfuric acid content and temperature.

10.2.2 Alternative/Renewable Energy

According to The International Energy Agency (IEA) World Energy Outlook 2009, the CO_2 target of 450 ppm up to year 2030 cannot be achieved without considerable involvement of renewable energy, and carbon capture and sequestration (CCS) [77]. Brining renewable energy into the main stream, thus, would be the significant challenge for the current century. Alternative or renewable energy refers to the energy which is evolved from sources other than earth's natural resources or does not damage the environment. In this category, mainstream technologies could be based on extraction of power from wind, hydro, etc., as well as the generation of energy from solar, geothermal, biomass, etc. In this respect, nano-catalysts due to their large surface to volume ratio and efficient catalytic capabilities have a pivotal role by improving the efficiency of clean energy production, energy conversion, energy storage and transport. Specifically, the role of nanotechnology in electro-catalysis of alternative resources can be beneficial in terms of improved efficiency and selectivity of current processes and permitting new technologies which are not viable till now [78].

Fuel Cells

Fuel cells are considered as ultra-low/zero noise and pollutant emission devices for the production of electrical energy directly converted from chemical energy. The main problem associated with the commercialization of fuel cells is the costly material used for manufacturing and its insufficient short life span of operation. Different kinds of low and high operating temperature fuel cells with different electrolytes have been commercialized/researched, like RMFC (reformed methanol), PEMFC (proton/polymer exchange/electrolyte membrane), UMFC (up-flow microbial), DMFC (direct methanol), MHFC (metal hydride), DCFC (direct carbon), MCFC (molten carbonate), PCFC (protonic ceramic), EGFC (electro-galvanic), SOFC (solid oxide), EBFC (enzymatic biofuel), PAFC (phosphoric acid), MFC (microbial), DBFC (direct boron-hydride), DFAFC (direct formic acid), AFC (alkaline), RFC (regenerative), TSOFC (tubular solid oxide), etc. The main working mechanism of the fuel cell depends on the reaction of positively charges H^+ ions with O_2 or some other oxidizing agent [33]. AFC is low operating temperature (80° C) FC and uses aqueous alkaline (30% KOH) solution in a porous matrix. The electrical efficiency and power density of cell is about 60% and 0.2-0.4 W/cm^2 respectively using H_2 and O_2 as fuel. The main catalyst reported in the literature is Pt as cathodic catalyst (oxygen reduction reaction (ORR)) and its foremost disadvantage is the high cost and scarcity. Bhandary *et al.* [79] proposed a green and cheaper method of synthesis of an electro-catalyst based on nano-flakes of AgCd alloy for ORR in AFC. The electro-activity of the catalyst was determined using cyclic voltammetry. The catalyst exhibited outstanding activity for ORR, producing <15% of peroxide in the process which facilitated it to tolerate its electro-catalytic activity for more than 3 hours in alkaline conditions. Similarly, Nagai *et al.* [80] proposed a complex system of non-precious metal electro-catalyst system based on perovskite type oxide-carbon mixture ($La_{0.6}Sr_{0.4}MnO_3/C$, $La_{0.6}Sr_{0.4}CoO_3/C$, and $La_{0.6}Sr_{0.4}Mn_{0.7}Co_{0.3}O_3/C$) and an iron phthalocyanine complex (Fe-Pc) for ORR in AFC. The activity of the catalysts were determined using rotating ring-disk electrode (RRDE) which showed improved ORR activities in all of the cases. Specifically, Fe-Pc/$La_{0.6}Sr_{0.4}Mn_{0.7}Co_{0.3}O_3/C$ catalyst was observed to exhibit the maximum catalytic activity, which in comparison was similar to that of the Pt/C catalyst.

The other type of low operating temperature (60-90° C) FC is PEMFC with high electrical efficiency between 50-60% and power density 0.5-1 W/cm^2 and appropriate for many applications. The main reactants are H_2, H_2 rich reformates with CO less than 30 ppm and oxygen/air. The main electrolyte is proton conducting polymer membrane whereas mostly Pt, Pt alloy or Pt/C are used as catalysts. The reactants are separated by the polymer membrane. Hydrogen and O_2 react at anode and cathodic side respectively and flow independently in the bipolar plate. The catalyst layer is loaded with supported metallic particles where the hydrogen and oxygen react to form water and byproducts. The electrons from H_2 dissociation flow through external circuit [81]. Along with many advantages like highest power density and fast start and stop sequence, catalyst poisoning (inactivation of catalytic sites) by CO is a major problem in commercialization of these promising PEMFCs. In order to overcome the drawback associated with the CO poisoning so as to increase the tolerance of the anodic catalyst, many efforts have been made. One of the solutions suggested the use of binary or ternary alloyed catalysts with Pt to improve the adsorption and oxidation of CO. Further, the cost of the Pt based catalysts in PEMFC can also be decreased by replacement with other cheaper metals or lower loadings. Kheradmandinia *et al.* [82] used low cost non-noble metals nano-catalysts composed of C/SnO$_2$, C/Ni and C/CoO. These nano-catalysts exhibited improved catalytic activities by lowering the potentials of CO electro-oxidation [33,82]. Another attempt was also made to reduce the cost of the Pt catalysts by developing nano-alloy Pt catalysts on high surface area carbon and Vulcun with different loadings of Pt, Ni and Co [83].

Another type of low operating temperature (20-90° C) and proton conducting polymer electrolyte membrane containing FC is DMFC where liquid methanol is utilized as direct fuel instead of H_2. The cell efficiency and power density is around 20-30% and 0.05-0.1 W/cm^2 respectively. The working of DMFC depends on the methanol oxidation at anode where generated protons transfer towards the cathode through a membrane followed by oxygen reduction at cathode. The electricity is produced through the transfer of electrons through an external circuit. Similar to the PEMFC, DMFCs also use carbon supported Pt metal based catalysts system [33]. In this regard, Pt has the same disadvantage of cost-effectiveness and much attention has been given to decrease the cost. The electro-catalysts should have good methanol tolerance, activity and selectivity for ORR. Along with this, other problems related to DEMFC have also been reported in the literature like lethargic kinetics of reactions at both electrodes, reduced durability, electrode flooding and methanol crossover. The overall performance of the DMFCs can be improved by modifying the structure of membrane electrode assembly (MEA), i.e., backing, diffusion and catalysts layers. Also, in DMFC, transport of methanol through membrane causes efficiency loss and drop in cell voltage owing to cathodic oxidation of methanol. To overcome this, studies have focused on replacement of Pt with other metals like Pd (low cost and activity for ORR). Efforts have been made to expand the electro-activity of Pd by using transition metals such as Co, Fe, and Ni as alloying agent. These studies have exhibited feasibility of Pd as electro-catalysts and have resolved the methanol crossover problem for DMFCs. For instance, Golmohammadi et al. [84] used Pd_3Co as nano-catalyst on binary carbon supports (MWCNTs and Vulcan) for fabrication of MEA for a passive DMFC and showed that the elector-catalytic properties for the ORR were best when MWCNTs and Vulcan at a mass ratio of 25:75 were used. Introduction of MWCNTs into the support provided high surface area, good electronic conductivity, and fast ORR kinetics as shown by improved electrode kinetic parameters [84].

In another study, Wang et al. [85] reported an electro-catalyst with low Pt mole fraction, consisting of intermetallic nanoparticles of Cu_3Pt for oxygen reduction reaction activity. A simple impregnation-reduction method was used for the catalyst synthesis. The authors used electrochemical and chemical dealloying methods for the synthesis of the catalyst particles. It was observed that the electrochemical dealloying method resulted in the formation of a core-shell morphology with 1 nm thin Pt skin and ordered Cu_3Pt core. On the other hand, the chemical dealloying method resulted in a spongy structure (Figure 10.7). The authors observed that both dealloying methods led to enhanced specific and mass activities toward the ORR. The mass activity was also observed to still enhance even after 5000 potential cycles, which indicated the functional nature of the developed electro-catalysts with improved ORR.

Vegetable Oil Conversion to Bio-fuel

The importance of energy from alternative resources is progressively vital owing to shrinking crude oil reserves and environmental hazards of liberated gases. Recently, vegetable oil (VO) conversion to bio-diesel (BD) has fascinated the researchers and attracted significant attention. BD is the colloquial name for "fatty acid methyl ester" and usually produced by trans-esterification of VO and methanol in the presence of catalysts. It can also be produced by other methods like fermentation, ecofining, thermal and catalytic cracking [86]. Since it is synthesized from the renewable resources, a lesser amount of pollutants

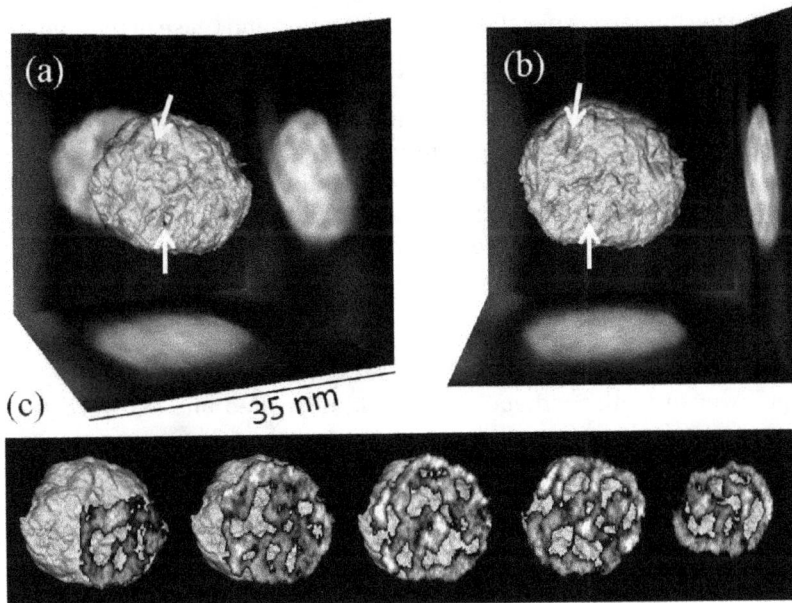

Figure 10.7 (a,b) Tomographic reconstruction of the spongy nanoparticle. The arrows indicate channels connecting to the exterior surfaces. (c) Consecutive slices through the surfaces to   demonstrate the porous networks. Reproduced from Reference 85 with permission from American Chemical Society.

and hazardous gases (CO_2, sulfur and nitrogen) is released due to clean burning, biodegradability, nontoxicity, absence of sulfur and aromatic compounds, etc. VO represents a large family of resources with oil from many sources includingcanola, avocado, palm, mustard, soybean, olive sunflower, peanut rapeseed, recycled oil, etc. In the trans-esterification reaction with alcohol for the synthesis of BD, diglycerides are formed from triglycerides during the first stage, followed by the formation of monoglycerides. Finally, glycerol is synthesized producing one methyl ester molecule at every stage. The reactions are catalyzed by conventional homogeneous or heterogeneous based alkali (faster) or acidic (slower) based materials, enzymatic catalysts or supercritical alcoholysis conditions. The use of conventional homogenous catalysts like soluble Na and KOH make this process longer in duration and due to improved solubility, the catalysts are difficult to be removed. Heterogeneous catalysts such as ZrO_2, titania, sodium hydroxide, sodium on alumina, potassium chloride, potassium carbonate, potassium nitrate, potassium iodide, potassium bromide, potassium fluoride have numerous benefits like regeneration and reuse of the catalyst. Different nanomaterials have also been employed for VO conversion to BD like calcium oxides, nano-γ-Al$_2$O$_3$, MgO, TiO$_2$, zeolites [87,88]. Verziu *et al.* [89] compared the performance of three dissimilar MgO nano-catalysts, different in terms of sizes and morphologies, for the conversion of sunflower and rapeseed oils. In order to prepare nanoparticles with different surfaces like (111) (110) and (100), different preparation methods were employed such hydrolysis, precipitation, and calcination. Under microwave conditions, the studies indicated that the activation temperature depended on the exposed face. It was lowest for (110) and (100) faces due to density of the basic sites [89]. Bet-Moushoul *et al.* [90] used five different types of gold nanoparticles supported calcium oxide-based nano-catalysts systems for biodiesel synthesis using trans-esterification of sunflower oil with methanol. Nano-gold deposited on CaO were prepared from AuNps solution using the impregnation method. The conversion process produced higher quality biodiesel and showed that the efficiency of the reaction was around 90–97% with 3.9-4.3 mg/kg amount of glycerol for all the samples. The reusability of the catalyst was also found to be improved as it could be re-employed up to ten times without impairment of activity [90].

Bio-gasoline (BG) is another kind of biofuel where exploration is being conducted in both academic and private sectors. Catalytic cracking process can be used to produce bio-based fuels like gasoline, kerosene and diesel from cooking or non-cooking oils at higher yields and comparatively lower temperature range. Common zeolites based catalysts are used for catalytic cracking due to their several advantages such as ordered structures and acidic sites. Ahmad *et al.* [86] prepared zeolite (ZSM-5) based nano-catalyst system loaded with different concentration of hetero-metallic (Fe–Zn–Cu) nano-oxides and used for the catalytic cracking of palm oil for the production of bio-gasoline. The synthesis of nano-catalysts initially started with the preparation of metallic organic frameworks with hetro-metallic complexes. Higher efficiency (56%) was obtained due to higher loading of loaded metal oxides catalysts [86].

Photo-catalytic Water Splitting

Photo-catalytic water splitting (PWS) is a promising technology to produce energy in the form of H$_2$ as fuel by artificial photosynthesis of water using artificial or natural light. PWS converts solar energy into chemical energy by photo-catalysis/photo-electro-catalysis in a photochemical/photo-electrochemical cell. The reactants (sunlight and H$_2$O) of PWS process are found as abundant resources on the globe. With increasing environmental concerns caused by fossil fuel burning, hydrogen as a fuel can be an alternative due to environmental friendliness and high energy yield. Among other water splitting methods like electrolysis, steam reforming of hydrocarbons, thermochemical and photobiological mwthods, PWS has numerous advantageous such as low cost, comparatively high efficiency, ability to separate hydrogen and oxygen streams, and appropriate technology for small scale production. The hydrogen production mechanism from H_2O splitting in 2:1 ratio is as follows: during initial stage, electron-hole pairs are generated after absorption of photons followed by migration of these charge carriers on the catalyst surface and finally electrons produce H_2 through water reduction, whereas holes produce oxygen by an oxidizing reaction [91]. Main catalysts reported in the literature for PWS are TiO_2, surface and structural modified TiO_2, metal sulfides and oxides, nitrides, oxysulfides and oxy-nitrides. The main principals associated with the use of photocatalyst for PWS are their energy band gap which should be at least 1.23 eV and anti-photo-corrosion. The major challenge associated with the use of semiconductor as photo-catalyst is ultraviolet (UV) light sensitivity. Attempts have been made to improve the sensitivity and efficiency of the photo-catalysts for visible light response by developing visible-light-driven reactors. The use of conductive surface containing photo-catalyst particles can be useful for PWS process. Addition of reduced graphene oxide with photo-catalysts can create new paths for electron migration towards the electrode. Iwase *et al.* [92] fabricated composites of RGO and $CuGaS_2$ in order to achieve enhanced cathodic photo-current under sunlight irradiation. Composite electrode depicted improvements in terms of cathodic photo-current under visible light irradiation than the pristine photo-electrode. On coupling with a CoO_x loaded $BiVO_4$ based photo-anode, the composite electrode exhibited that hydrogen evolution can be achieved without applying external bias under visible irradiation [92].

Corrosion of the photo-catalysts is another major problem for photo-cells. Hence, it is vital to develop photo-catalyst with improved corrosion resistance behavior. This can be achieved by either surface modification or doping with other materials more resistant to corrosion. For instance, nano-crystalline zinc oxide is a useful catalyst for PWS process owing to its relatively higher electron and holes mobility, and easy fabrication of nano-structures with large surface-to-volume ratios. The photo-degradation of ZnO in presence of aggressive solutions is a serious hurdle in using ZnO for PWS. Nano-coatings on ZnO electrodes of suitable material like TiO_2 could be beneficial for water splitting photo-anode. Liu *et al.* [93] stabilized the ZnO nanowire photo-anodes with ultrathin shell of TiO_2 through atomic layer deposition (ALD) (Figure 10.8). The chemically grown thin shells of TiO_2 were observed to exhibit stable operation for water splitting in aggressive solution with 25% higher water splitting activity [93].

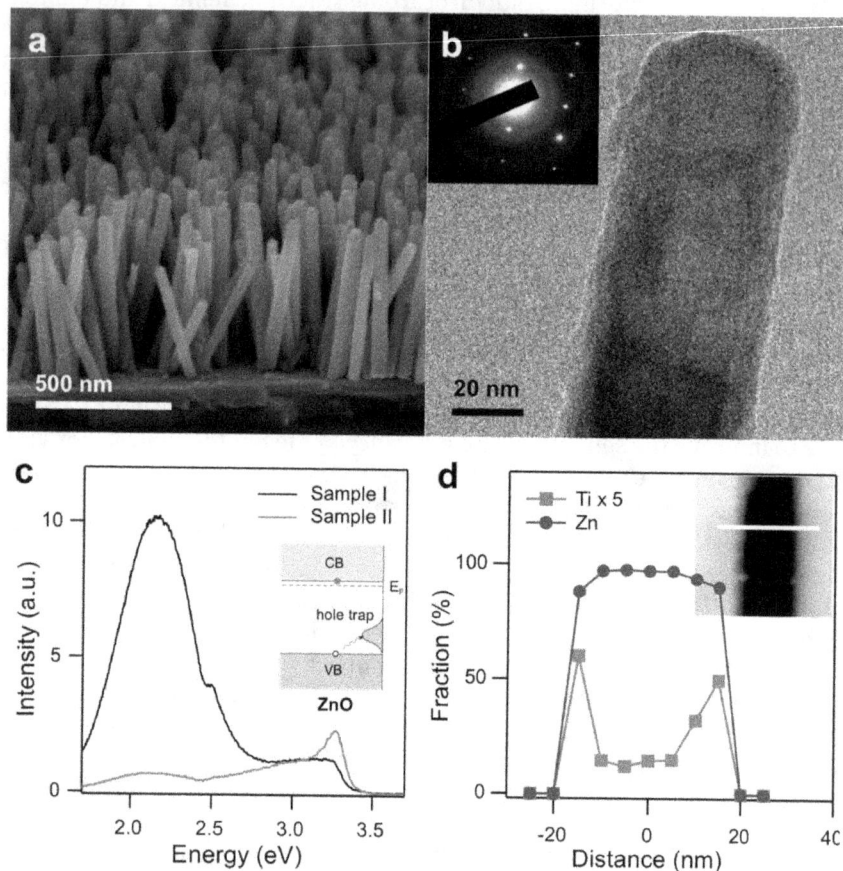

Figure 10.8 (a) SEM image of a ZnO nanowire array; (b) TEM image of a ZnO nanowire; (c) PL spectra of the ZnO nanowire array (sample I, higher intensity curve) and the ZnO/TiO$_2$ core/shell nanowire array (sample II, lower intensity curve) and (d) EDX elemental scans of a ZnO/TiO$_2$ core/shell nanowire. Reproduced from Reference 93 with permission from American Chemical Society.

Enhancing visible light absorption characteristics can improve the efficiency of the photo-catalytic reactor. Engineering of nano-structured materials to produce different nano-architectures is a promising strategy to improve the antireflection and absorption properties. Wei *et al.* [94] synthesized low cost, ordered hetero-nano-structure bismuth vanadate (BiVO₄) nano-pyramid arrays through aqueous deposition (60 °C for 6 hours) and annealed (400 °C for 30 min). The tips of the nano-pyramid were capped with CoPi nanoparticles (electro-catalyst) through photo-deposition at room temperature. The resulted structure exhibited low-cost solar water splitting without the use of a sacrificial agent [94].

Biomass Gasification

Energy from biomass gasification (BG), a flow based system, depends on continuous supply of sustainable feedstock. It is the most technically established fields of biofuel synthesis. During the thermochemical process, solid biomass such as lignocellulosic convert into combustible gases. The major application of BG process could be cogeneration of heat, power, bio-fuel, bio-tar, electricity and fertilizers. On the other hand, there are deficiencies and bottlenecks in the current technology, which are main hurdles in BG commercialization such as economically unviability and the presence of condensates, tar, dust and carbon burnout in the product streams. Catalysis research in these areas is requisite for appropriate inexpensive and innovative technology which can improve the commercial feasibility of BG [22,95].

Gasification of biomass usually occurs through many reactions such as pyrolysis, thermal cracking, dehydration, water-gas shift, and steam reforming. Supercritical water gasification (SWG) is a beneficial process to produce compressed and unpolluted gas with high H_2. One of the important advantages of using catalyst along SWG process is the lowering down of the temperature of the BG process, thus, resulting in reduced cost. Major catalysts systems used for SWG processes are usually alkaline salts (homogenous) and bimetallic, transitional metals, activated carbon (heterogeneous), etc. Kang *et al.* [96] catalyzed cellulose and lignin from feedstock of wheat straw, canola meal, and timothy grass. The gasification was performed by using by K_2CO_3 and 20Ni-0.36Ce/Al₂O₃ (homo/heterogeneous) catalyst system. The study concluded that higher catalyst loading encouraged the hydrogen production [96]. It was also reported that impregnation of biomass with catalyst systems is another suitable and innovative method to improve the selectivity and activity of the reaction. Similarly, in another attempt, lignocellulosic feedstocks from wheat straw and pinewood were impregnated with Ni nanoparticles for SWG [97]. Hydrothermal gasification of impregnated biomass exhibited significant gas yields (hydrogen, carbon dioxide, and methane, etc.) as compared to non-catalytic gasification. Enhanced catalysis of biomass was due to the active sites present on nickel nanoparticles improving H_2 production [97].

Marine biomass like algal, known as third generation of biofuels, is another kind of biomass which is rich in C and H_2 and can be used as alternative to agricultural waste for the production of green fuel. Around 28 million tons of marine biomass presents enormous potential to produce renewable and environmentally pleasant biofuels and chemicals, which will also lower the burden on agricultural feed. The structure of algal biomass presents an ability to hydrolyze easily due to the low content of hemicelluloses and lignin. Norouzi *et al.* [98] investigated the potential of biofuel production using SWG of algal biomass from southern coast of Caspian Sea in a batch reactor. The nano-catalyst system used in this study was composed of un-promoted and Ru promoted Ni-Fe/γ-Al_2O_3. Nano-catalysts were synthesized via reverse micro-emulsion technique using 12 wt% of Ni, 6 wt% of Fe and 0.5-2 wt% of Ru. The best performance was depicted by promoted Ni–Fe/c-Al_2O_3/2 wt% Ru, during SWG of Enteromorpha intestinalis (algal) [98]. Another main bottleneck which obstructs the commercialization of BG process is the formation of tar with the gas products. Many studies have been carried out to alleviate the problem of the tar production in biomass conversion. In this respect too, nano-catalysis has played a vital role in improving the cracking behavior of tar. In a recent study, nanocomposites comprising of Ni/SiO_2 and Ni-Ce/SiO_2 were generated by deposition-precipitation method and applied as a nano-catalyst in the cracking of tar [99]. It was reported that the utilization of nanocomposites for tar cracking increased the catalytic cracking by 93% and 98.5% for Ni/SiO_2 and Ni-Ce/SiO_2, respectively. Catalytic cracking of tar also affected the quality of the product gas by increasing the carbon monoxide and hydrogen contents, whereas carbon dioxide and methane contents were decreased.

10.2.3 Other Systems

Among other functional uses of nano-catalysts for energy, Hussain *et al.* [100] reported heterogeneous ruthenium–molybdenum disulfide nano-catalyst for the selective aerobic oxidation of amines (Figure 10.9). The generated catalyst yielded with high selectivity nitriles from different aromatic, aliphatic, allylic, and heteroatomic amines. The advanced nano-catalyst reported in the study can be also potentially applied in conventional and new processes for energy applications. The authors also reported that the oxidation reactions could be achieved with a very small amount of the catalyst, which indicated the environmentally friendly and economically efficient process. In another study, Liu *et al.*

[101] reported ultrafine metal oxide nano-catalysts for high-performance lith-ium-oxygen batteries. An electrochemical prelithiation process was used for the catalyst synthesis, which reduces the size of $NiCo_2O_4$ (NCO) particles from 20-30 nm to a uniformly distributed domain of ~2 nm, thereby significantly improving the catalytic activity. A $Li\text{-}O_2$ battery employing the developed

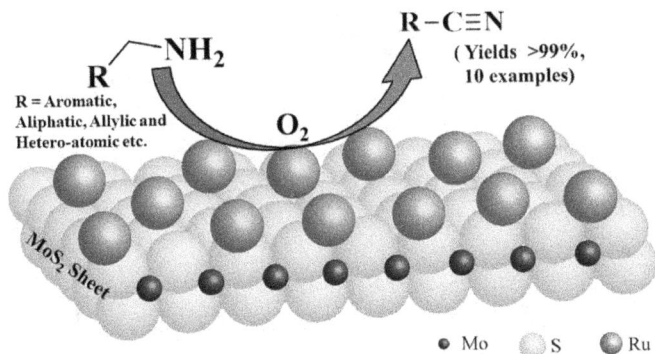

Figure 10.9 Schematic representation of the amine oxidation process. Reproduced from Reference 100 with permission from American Chemical Society.

catalyst was observed to have an initial capacity of 29,280 mAh g^{-1}. In addition, the battery retained the capacity of >1000 mAh g^{-1} after 100 cycles. Figure 10.10 also demonstrates the morphology of prelithiated (PL)-NCO nanowires (NW)/carbon fabric (CF) electrodes after prelithiation process.

10.3 Conclusion

Catalyst systems based on nano-structured materials have created interesting research areas for efficient catalysis of fossil and renewable resources. Nano-structured nano-catalyst materials present promising progress avenues in the current and future energy areas due to their higher activity and improved se-lectivity. Due to many current experimental, technological and environmental constraints, there is a need to improve the quality and quantity of the products generated from fossil or renewable resources from the perspective of both en-vironment and energy requirements. In this respect, efficient and selective nano-catalysts are attractive alternatives to the conventional catalyst systems for the sustainable energy, clean environment, efficient use of resources and cost effectiveness.

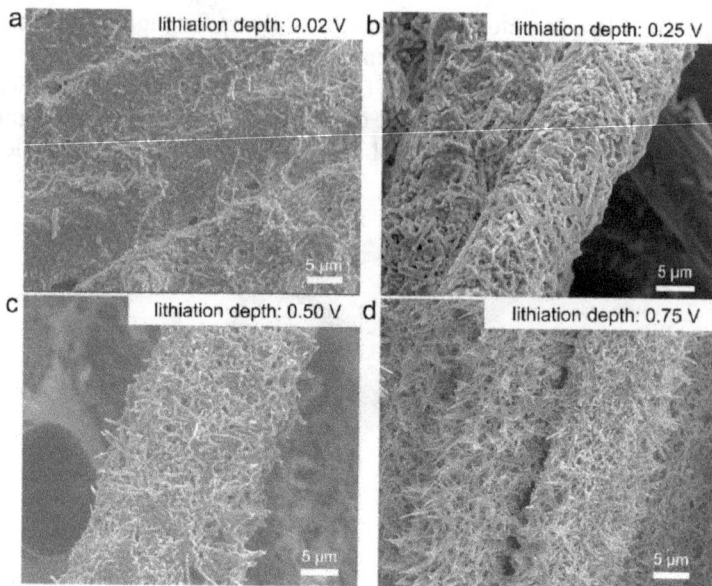

Figure 10.10 SEM images depicting the morphology of PL-NCO NWs/CF electrodes after prelithiation process at depth of (a) 0.02, (b) 0.25, (c) 0.50, and (d) 0.75 V. Reproduced from Reference 101 with permission from American Chemical Society.

References

1. Shiju, N. R., and Guliants, V. V. (2009) Recent developments in catalysis using nanostructured materials. *Applied Catalysis A: General*, **356**(1), 1-17.
2. Philippot, K., and Serp, P. (2013) Concepts in nanocatalysis. In: *Nanomaterials in Catalysis*, Serp, P., and Philippot, K. (eds.), Wiley-VCH, Germany, pp. 1-54.
3. Machado, R. M., Heier, K. R., and Broekhuis, R. R. (2001) Developments in hydrogenation technology for fine-chemical and pharmaceutical applications. *Current Opinion in Drug Discovery And Development*, **4**(6), 745-755.
4. Lovestam, G., Rauscher, H., Roebben, G., Kluttgen, B. S., Gibson, N., Putaud, J.-P., and Stamm, H. (2010) Considerations on a Definition of Nanomaterial for Regulatory Purposes. *Joint Research Centre (JRC) Reference Reports*, European Union. Online: https://ec.europa.eu/jrc/sites/jrcsh/files/jrc_reference_report_201007_nano-materials.pdf [assessed 29th April 2017].
5. Kaur, R., and Badea, I. (2013) Nanodiamonds as novel nanomaterials for biomedical applications: drug delivery and imaging systems. *International Journal of Nanomedicine*, **8**, 203-220.

6. Gilmore, J. L., Yi, X., Quan, L., and Kabanov, A. V. (2008) Novel nanomaterials for clinical neuroscience. *Journal of Neuroimmune Pharmacology*, **3**(2), 83-94.
7. Cordier, S., Dorson, F., Grasset, F., Molard, Y., Fabre, B., Haneda, H., Sasaki, T., Mortier, M., Ababou-Girard, S., Perrin, C. (2009) Novel nanomaterials based on inorganic molybdenum octahedral clusters. *Journal of Cluster Science*, **20**(1), 9-21.
8. Garbarczyk, J. E., Wasiucionek, M., Jozwiak, P., Nowinski, J. L., and Julien, C. M. (2009) Novel nanomaterials based on electronic and mixed conductive glasses. *Solid State Ionics*, **180**(6-8), 531-536.
9. Li, Q., Mahendra, S., Lyon, D. Y., Brunet, L., Liga, M. V., Li, D., and Alvarez, P. J. J. (2008) Antimicrobial nanomaterials for water disinfection and microbial control: Potential applications and implications. *Water Research*, **42**(18), 4591-4602.
10. Zhang, L., and Webster, T. J. (2009) Nanotechnology and nanomaterials: Promises for improved tissue regeneration. *Nano Today*, **4**(1), 66-80.
11. Bonnemann, H., Khelashvili, G., Hormes, J., Kuhn, T. J., and Richter, W.-J. (2013) Nanostructured metal particles for catalysts and energy-related materials. In: *Nanomaterials in Catalysis*, Serp, P., and Philippot, K. (eds.), Wiley-VCH, Germany, pp. 123-201.
12. Schlogl, R., and Abd Hamid, S. B. (2004) Nanocatalysis: Mature science revisited or something really new? *Angewandte Chemie, International Edition*, **43**(13), 1628-1637.
13. Li, Y., and Somorjai, G. A. (2010) Nanoscale advances in catalysis and energy applications. *Nano Letters*, **10**(7), 2289-2295.
14. Chaturvedi, S., Dave, P. N., and Shah, N. K. (2012) Applications of nano-catalyst in new era. *Journal of Saudi Chemical Society*, **16**(3), 307-325.
15. Mahmoud, M. A., Tabor, C. E., El-Sayed, M. A., Ding, Y., and Wang, Z. L. (2008) A new catalytically active colloidal platinum nanocatalyst: The multiarmed nanostar single crystal. *Journal of the American Chemical Society*, **130**, 4590-4591.
16. Basic Research Needs: Catalysis for Energy (2008) *Report from the U.S. Department of Energy, Office of Basic Energy Sciences Workshop*, USA. Online: http://www.iact.anl.gov/publications/CAT_rpt.pdf [assessed 23rd April 2017].
17. Thompson, L. (2011) Applications: Energy from fossil resources. In *International Assessment of Research and Development in Catalysis by Nanostructured Materials*, Davis, R. (ed.), Imperial College Press, UK, pp. 151-183.
18. Hu, E. L., Davis, S. M., Davis, R., and Scher, E. (2011) Applications: Catalysis by nanostructured materials. In: *Nanotechnology Research Directions for Societal Needs in 2020: Retrospective and Outlook*, Roco, M. C., Mirkin, C. A., and Hersam, M. C. (eds.), Springer, Netherlands, pp. 445-466.
19. *Handbook of Heterogeneous Catalysis: 8 Volumes*, Ertl, G., Knozinger, H., Schuth, F., and Weitkamp, J. (eds.), 2nd edition, Wiley-VCH, Germany (2008).
20. Guliants, V. V. (2011) Applications: Chemicals from fossil resources. In: *International Assessment of Research and Development in Catalysis by Nanostructured Materials*, Davis, R. (ed.), Imperial College Press, UK, pp. 185-237.
21. Huber, G. (2011) Applications: Renewable fuels and chemicals. In: *International Assessment of Research and Development in Catalysis by Nanostructured Materials*, Davis, R. (ed.), Imperial College Press, UK, pp. 239-262.

22. Davis, R. J. (2011) Overview of catalysis by nanostructured materials. In: *International Assessment of Research and Development in Catalysis by Nanostructured Materials*, Davis, R. (ed.), Imperial College Press, UK, pp. 1-24.

23. Cheng, W.-C., Habib, Jr., E. T., Rajagopalan, K., Roberie, T. G., Wormsbecher, R. F., and Ziebarth, M. S. (2008) Fluid catalytic cracking. In: *Handbook of Heterogeneous Catalysis*, Ertl, G., Knozinger, H., Schuth, F., and Weitkamp, J. (eds.), 2nd Edition, Wiley-VCH, Germany, pp. 2741-2775.

24. Moser, M. D., and Bogdan, P. L. (2008) Catalytic reforming. In: *Handbook of Heterogeneous Catalysis*, Ertl, G., Knozinger, H., Schuth, F., and Weitkamp, J. (eds.), 2nd Edition, Wiley-VCH, Germany, pp. 2728-2740.

25. Prins, R. (2008) Hydrotreating. In: *Handbook of Heterogeneous Catalysis*, Ertl, G., Knozinger, H., Schuth, F., and Weitkamp, J. (eds.), 2nd Edition, Wiley-VCH, Germany, pp. 2695-2716.

26. Sie, S. T. (2008) Isomerization. In: *Handbook of Heterogeneous Catalysis*, Ertl, G., Knozinger, H., Schuth, F., and Weitkamp, J. (eds.), 2nd Edition, Wiley-VCH, Germany, pp. 2809-2828.

27. van Veen, J. A. R., Minderhoud, J. K., Huve, L. G., and Stork, W. H. J. (2008) Hydrocracking and catalytic dewaxing. In: *Handbook of Heterogeneous Catalysis*, Ertl, G., Knozinger, H., Schuth, F., and Weitkamp, J. (eds.), 2nd Edition, Wiley-VCH, Germany, pp. 2778-2805.

28. Wei, J. (2008) Hydrodemetalation. In: *Handbook of Heterogeneous Catalysis*, Ertl, G., Knozinger, H., Schuth, F., and Weitkamp, J. (eds.), 2nd Edition, Wiley-VCH, Germany, pp. 2718-2727.

29. Rostrup-Nielsen, J. R. (2008) Steam reforming. In: *Handbook of Heterogeneous Catalysis*, Ertl, G., Knozinger, H., Schuth, F., and Weitkamp, J. (eds.), 2nd Edition, Wiley-VCH, Germany, pp. 2882-2903.

30. Hansen, J. B., and Hojlund Nielsen, P. E. (2008) Methanol synthesis. In: *Handbook of Heterogeneous Catalysis*, Ertl, G., Knozinger, H., Schuth, F., and Weitkamp, J. (eds.), 2nd Edition, Wiley-VCH, Germany, pp. 2920-2949.

31. Krause, A. O. I., and Keskinen, K. I. (2008) Etherification. In: *Handbook of Heterogeneous Catalysis*, Ertl, G., Knozinger, H., Schuth, F., and Weitkamp, J. (eds.), 2nd Edition, Wiley-VCH, Germany, pp. 2864-2879.

32. Peters, R. (2008) Fuel processors. In: *Handbook of Heterogeneous Catalysis*, Ertl, G., Knozinger, H., Schuth, F., and Weitkamp, J. (eds.), 2nd Edition, Wiley-VCH, Germany, pp. 3045-3076.

33. Gasteiger, H. A., and Garche, J. (2008) Fuel cells. In: *Handbook of Heterogeneous Catalysis*, Ertl, G., Knozinger, H., Schuth, F., and Weitkamp, J. (eds.), 2nd Edition, Wiley-VCH, Germany, pp. 3081-3117.

34. Moqadam, S. I., and Mahmoudi, M. (2013) Advent of nanocatalysts in hydrotreating process: Benefits and developements. *American Journal of Oil and Chemical Technologies*, **1**,13-21.

35. Nejad, D. M., Rahemi, N., and Allahyari, S. (2017) Effect of tungsten loading on the physiochemical properties of nanocatalysts of Ni–Mo–W/carbon nanotubes for the hydrodesulfurization of thiophene. *Reaction Kinetics, Mechanisms and Catalysis*, **120**(1), 279-294.

36. Danforth, S. J., Liyanage, D. R., Hitihami-Mudiyanselage, A., Ilic, B., Brock, S. L., and Bussell, M. E. (2016) Probing hydrodesulfurization over bimetallic phosphides using monodisperse Ni$_{2-x}$M$_x$P nanoparticles encapsulated in mesoporous silica. *Surface Science*, **648**, 126-135.

37. Jenifer, A. C., Sharon, P., Prakash, A., and Sande, P. C. (2015) A review of the unconventional methods used for the demetallization of petroleum fractions over the past decade. *Energy & Fuels*, **29**(12), 7743-7752.

38. Rao, S. M., and Coppens, M.-O. (2012) Increasing robustness of deactivating nanoporous catalysts by optimizing the pore network - application to hydrodemetalation. *Chemical Engineering Science*, **83**, 66-76.

39. Visaliev, M. Y., Shpirt, M. Y., Kadiev, K. H., Dvorkin, V. I., Magomadov, E. E., and Khadzhiev, S. N. (2012) Integrated conversion of extra-heavy crude oil and petroleum residue with the recovery of vanadium, nickel, and molybdenum. *Solid Fuel Chemistry*, **46**(2), 100-107.

40. *Catalysis: An Integrated Approach to Homogeneous, Heterogeneous and Industrial Catalysis*, Moulijn, J. A., van Leeuwen, P. W. N. M., and van Santen, R. A. (eds.), Volume 79, 1st edition, Elsevier, USA (1993).

41. Park, J. Y., and Somorjai, G. A. (2014) Bridging materials and pressure gaps in surface science and heterogeneous catalysis. In: *Current Trends of Surface Science and Catalysis*, Park, J. Y. (ed.), Springer, USA, pp. 3-17.

42. Musselwhite, N., Alayoglu, S., Melaet, G., Pushkarev, V. V., Lindeman, A. E., An, K., and Somorjai, G. A. (2013) Isomerization of n-hexane catalyzed by supported monodisperse PtRh bimetallic nanoparticles. *Catalysis Letters*, **143**(9). 907-911.

43. An, K., Zhang, Q., Alayoglu, S., Musselwhite, N., Shin, J.-Y., and Somorjai, G. A. (2014) High-temperature catalytic reforming of n-hexane over supported and core–shell Pt nanoparticle catalysts: Role of oxide-metal interface and thermal stability. *Nano Letters*, **14**(8), 4907-4912.

44. Madsen, C., Madsen, C., and Jacobsen, C. J. H. (1999) Nanosized zeolite crystals-convenient control of crystal size distribution by confined space synthesis. *Chemical Communications*, **1999**(8), 673-674.

45. Fan, W., Snyder, M. A., Kumar, S., Lee, P.-S., Yoo, W. C., McCornick, A. V., Penn, R. L., Stein, A., and Tsapatsis, M. (2008) Hierarchical nanofabrication of microporous crystals with ordered mesoporosity. *Nature Materials*, **7**(12), 984-991.

46. Tosheva, L., and Valtchev, V. P. (2005) Nanozeolites: Synthesis, crystallization mechanism, and applications. *Chemistry of Materials*, **17**(10), 2494-2513.

47. Koros, W. J., and Fleming, G. K. (1993) Membrane-based gas separation. *Journal of Membrane Science*, **83**(1), 1-80.

48. Koros, W. J., and Chern, R. T. (1987) Separation of gaseous mixtures using polymer membranes. In: *Handbook of Separation Process Technology*, Rousseau, R. W. (ed.), Wiley, USA, pp. 862-953.

49. Koros, W. J., Fleming, G. K., Jordan, S. M., Kim, T. H., and Hoehn, H. H. (1988) Polymeric membrane materials for solution-diffusion based permeation separations. *Progress in Polymer Science*, **13**(4), 339-401.

50. Murase, K., Fujiwara, T., Umemura, Y., Suzuki, K., Iino, R., Yamashita, H., Saito, M., Murakoshi, H., Ritchie, K., and Kusumi, A. (2004) Ultrafine membrane compartments

for molecular diffusion as revealed by single molecule techniques. *Biophysical Journal*, **86**(6), 4075-4093.

51. Garcia-Martinez, J., Johnson, M., Valla, J., Li, K., and Ying, J. Y. (2012) Mesostructured zeolite Y-high hydrothermal stability and superior FCC catalytic performance. *Catalysis Science & Technology*, **2**(5), 987-994.

52. Yang, X. Y., Tian, G., Chen, L. H., Li, Y., Rooke, J. C., Wei, Y. X., Liu, Z. M., Deng, Z., Van Tendeloo, G., Su, B. L. (2011) Well-organized zeolite nanocrystal aggregates with interconnected hierarchically micro–meso–macropore systems showing enhanced catalytic performance. *Chemistry - A European Journal*, **17**(52), 14987-14995.

53. Corma, A., Fornes, V., Pergher, S. B., Maesen, T. L. M., and Buglass, J. G. (1998) Delaminated zeolite precursors as selective acidic catalysts. *Nature*, **396**(6709), 353-356.

54. Dejhosseini, M., Aida, T., Watanabe, M., Takami, S., Hojo, D., Aoki, N., Arita, T., Kishita, A., and Adschiri, T. (2013) Catalytic cracking reaction of heavy oil in the presence of cerium oxide nanoparticles in supercritical water. *Energy & Fuels*, **27**(8), 4624-4631.

55. Luu, C. L., Dao, T. K. T., Nguyen, T., Bui, T. H., Dang, T. N. Y., Hoang, M. N., and Ho, S. T. (2013) Effect of carriers on physico-chemical properties and activity of Pd nano-catalyst in n-hexane isomerization. *Advances in Natural Sciences: Nanoscience and Nanotechnology*, **4**(4), 045001.

56. Aghdam, N. C., Ejtemaei, M., Babaluo, A. A., Tavakoli, A., Bayati, B., and Bayat, Y. (2016) Enhanced i-C5 production by isomerization of C5 isomers in BZSM-5 membrane reactor packed with Pt/ZSM-5 nanocatalyst. *Chemical Engineering Journal*, **305**, 2-11.

57. An, K., Alayoglu, S., Musselwhite, N., Na, K., and Somorjai, G. A. (2014) Designed catalysts from Pt nanoparticles supported on macroporous oxides for selective isomerization of n-hexane. *Journal of the American Chemical Society*, **136**(19), 6830-6833.

58. Behrens, M., and Armbruster, M. (2012) Methanol steam reforming. In: *Catalysis for Alternative Energy Generation*, Guczi, L., and Erdohelyi, A. (eds.), Springer, USA, pp. 175-235.

59. Koo. K. Y., Eom, H. J., Jung, U. H., and Yoon, W. L. (2016) Ni nanosheet-coated monolith catalyst with high performance for hydrogen production via natural gas steam reforming. *Applied Catalysis A: General*, **525**, 103-109.

60. Purnama, H., Girgsdies, F., Ressler, T., Schattka, J. H., Caruso, R. A., Schomacker, R., and Schlogl, R. (2004) Activity and selectivity of a nanostructured CuO/ZrO_2 catalyst in the steam reforming of methanol. *Catalysis Letters*, **94**(1), 61-68.

61. Kuld, S., Thorhauge, M., Falsig, H., Elkjaer, C. F., Helveg, S., Chorkendorff, I., and Sehested, J. (2016) Quantifying the promotion of Cu catalysts by ZnO for methanol synthesis. *Science*, **352**(6288), 969-974.

62. van den Berg, R., Zecevic, J., Sehested, J., Helveg, S., de Jongh. P. E., and de Jong, K. P. (2016) Impact of the synthesis route of supported copper catalysts on the performance in the methanol synthesis reaction. *Catalysis Today*, **272**, 87-93.

63. van den Berg, R., Parmentier, T. E., Elkjaer, C. F., Gommes, C. J., Sehested, J., Helveg, S., de Jongh, P. E., and de Jong, K. P. (2015) Support functionalization to retard Ostwald ripening in copper methanol synthesis catalysts. *ACS Catalysis*, **5**(7), 4439-4448.

64. van den Berg, R., Prieto, G., Korpershoek, G., van der Wal, L. I., van Bunningen, A. J., Laegsgaard-Jorgensen, S., de Jongh, P. E., and de Jong, K. P. (2016) Structure sensitivity of Cu and CuZn catalysts relevant to industrial methanol synthesis. *Nature Communications*, **7**, 13057.
65. Rungtaweevoranit, B., Baek, J., Araujo, J. R., Archanjo, B. S., Choi, K. M., Yaghi, O. M., and Somorjai, G. A. (2016) Copper nanocrystals encapsulated in Zr-based metal-organic frameworks for highly selective CO_2 hydrogenation to methanol. *Nano Letters*, **16**(12), 7645-7649.
66. Bahmani, M., Vasheghani Farahani, B., and Sahebdelfar, S. (2016) Preparation of high performance nano-sized $Cu/ZnO/Al_2O_3$ methanol synthesis catalyst via aluminum hydrous oxide sol. *Applied Catalysis A: General*, **520**, 178-187.
67. Dry, M. E. (2008) The Fischer–Tropsch (FT) synthesis processes. In: *Handbook of Heterogeneous Catalysis*, Ertl, G., Knozinger, H., Schuth, F., and Weitkamp, J. (eds.), 2nd Edition, Wiley-VCH, Germany, pp. 2965-2992.
68. den Otter, J. H., Yoshida, H., Ladesma, C., Chen, D., and de Jong, K. P. (2016) On the superior activity and selectivity of $PtCo/Nb_2O_5$ Fischer Tropsch catalysts. *Journal of Catalysis*, **340**, 270-275.
69. Eschemann, T. O., Oenema, J., and de Jong, K. P. (2016) Effects of noble metal promotion for Co/TiO_2 Fischer-Tropsch catalysts. *Catalysis Today*, **261**, 60-66.
70. den Otter, J. H., Nijveld, S. R., and de Jong, K. P. (2016) Synergistic promotion of Co/SiO_2 Fischer–Tropsch catalysts by niobia and platinum. *ACS Catalysis*, **6**(3), 1616-1623.
71. Wang, H., Zhou, W., Liu, J.-X., Si, R., Sun, G., Zhong, M.-Q., Su, H.-Y., Zhao, H.-B., Rodriguez, J. A., Pennycook, S. J., Idrobo, J.-C., Li, W.-X., Kou, Y., and Ma, D. (2013) Platinum-modulated cobalt nanocatalysts for low-temperature aqueous-phase Fischer.Tropsch synthesis. Journal of the American Chemical Society, **135**, 4149-4158.
72. Oschatz, M., Lamme, W. S., Xie, J., Dugulan, A. I., and de Jong, K. P. (2016) Ordered mesoporous materials as supports for stable iron catalysts in the Fischer-Tropsch synthesis of lower olefins. *ChemCatChem*, **8**(17), 2846-2852.
73. Yee, K. F., Ng, E.-P., Mohamed, A. R., Adam, F., and Tan, S. H. (2016) Functionalized multi-walled carbon nanotubes as heterogeneous Lewis acid catalysts in the etherification reaction of tert-butyl alcohol and ethanol. *Chemical Engineering Communications*, **203**(10), 1385-1394.
74. Badia, J. H., Fite, C., Bringue, R., Ramirez, E., and Iborra, M. (2016) Relevant properties for catalytic activity of sulfonic ion-exchange resins in etherification of isobutene with linear primary alcohols. *Journal of Industrial and Engineering Chemistry*, **42**, 36-45.
75. Yan, X., Ke, M., Song, Z., Jiang, Q., and Yu, P. (2016) Etherification over H[small beta] zeolite modified by lanthanum ion exchange combined with low-temperature steam treatment. *RSC Advances*, **6**(18), 14799-14808.
76. Jaworski, M. A., Vega, S. R., Siri, G. J., Casella, M. L., Salvador, A. R., and Lopez, A. S. (2015) Glycerol etherification with benzyl alcohol over sulfated zirconia catalysts. *Applied Catalysis A: General*, **505**, 36-43.

77. World Energy Outlook (2009), The International Energy Agency, France. Online: http://www.worldenergyoutlook.org/media/weowebsite/2009/WEO2009.pdf [assessed 30th April 2017].

78. Centi, G., Lanzafame, P., and Perathoner, S. (2012) Introduction and general overview. In: *Catalysis for Alternative Energy Generation*, Guczi, L., and Erdohelyi, A. (eds.), Springer Science & Business Media, USA, pp. 1-28.

79. Bhandary, N., Basu, S., and Ingole, P. P. (2016) Rudimentary simple, single step fabrication of nano-flakes like AgCd alloy electro-catalyst for oxygen reduction reaction in alkaline fuel cell. *Electrochimica Acta*, **212**, 122-129.

80. Nagai, T., Yamazaki, S.-i., Fujiwara, N., Asahi, M., Siroma, Z., and Ioroi, T. (2016) Non-precious metal catalyst that combines perovskite-type oxide and iron phthalocyanine for use as a cathode catalyst in an alkaline fuel cell. *Journal of The Electrochemical Society*, **163**(5), F347-F352.

81. Devanathan, R. (2008) Recent developments in proton exchange membranes for fuel cells. *Energy & Environmental Science*, **1**(1), 101-119.

82. Kheradmandinia, S., Khandan, N., and Eikani, M. H. (2016) Synthesis and evaluation of CO electro-oxidation activity of carbon supported SnO_2, CoO and Ni nano catalysts for a PEM fuel cell anode. *International Journal of Hydrogen Energy*, **41**(42), 19070-19080.

83. Van Schalkwyk, F. Pattrick, G., Olivier, J., Conrad, O., and Blair. S. (2016) Development and scale up of enhanced ORR Pt-based catalysts for PEMFCs. *Fuel Cells*, **16**(4), 414-427.

84. Golmohammadi, F., Gharibi, H., and Sadeghi, S. (2016) Synthesis and electrochemical characterization of binary carbon supported Pd_3Co nanocatalyst for oxygen reduction reaction in direct methanol fuel cells. *International Journal of Hydrogen Energy*, **41**(18), 7373-7387.

85. Wang, D., Yu, Y., Xin, H. L., Hovden, R., Ercius, P., Mundy, J. A., Chen, H., Richard, J. H., Muller, D. A., DiSalvo, F. J., and Abruna, H. D. (2012) Tuning oxygen reduction reaction activity via controllable dealloying: A model study of ordered Cu_3Pt/C intermetallic nanocatalysts. *Nano Letters*, **12**, 5230-5238.

86. Ahmad, M., Farhana, R., Rahman, A. A. A., and Bhargava, S. K. (2016) Synthesis and activity evaluation of heterometallic nano oxides integrated ZSM-5 catalysts for palm oil cracking to produce biogasoline. *Energy Conversion and Management*, **119**, 352-360.

87. Wang, L., and Yang, J. (2007) Transesterification of soybean oil with nano-MgO or not in supercritical and subcritical methanol. *Fuel*, **86**(3), 328-333.

88. Boz, N., Degirmenbasi, N., and Kalyon, D. M. (2009) Conversion of biomass to fuel: Transesterification of vegetable oil to biodiesel using KF loaded nano-γ-Al_2O_3 as catalyst. *Applied Catalysis B: Environmental*, **89**(3-4), 590-596.

89. Verziu, M., Cojocaru, B., Hu, J., Richards, R., Ciuculescu, C., Filip, P., and Parvulescu, V. I. (2008) Sunflower and rapeseed oil transesterification to biodiesel over different nanocrystalline MgO catalysts. *Green Chemistry*, **10**(4), 373-381.

90. Bet-Moushoul, E., Farhadi, K., Mansourpanah, Y., Nikbakht, A. M., Molaei, R., and Forough, M. (2016) Application of CaO-based/Au nanoparticles as heterogeneous nanocatalysts in biodiesel production. *Fuel*, **164**, 119-127.

91. Ahmad, H., Kamarudin, S. K., Minggu, L. J., and Kassim, M. (2015) Hydrogen from photo-catalytic water splitting process: A review. *Renewable and Sustainable Energy Reviews*, **43**, 599-610.

92. Iwase, A., Ng, Y. H., Amal, R., and Kudo, A. (2015) Solar hydrogen evolution using a $CuGaS_2$ photocathode improved by incorporating reduced graphene oxide. *Journal of Materials Chemistry A*, **3**(16), 8566-8570.

93. Liu, M., Nam, C.-Y., Black, C. T., Kamcev, J., and Zhang, L. (2013) Enhancing water splitting activity and chemical stability of zinc oxide nanowire photoanodes with ultrathin titania shells. *The Journal of Physical Chemistry C*, **117**(26), 13396-13402.

94. Wei, Y., Su, J., Wan, X., Guo, L., and Vayssieres, L. (2016) Spontaneous photoelectric field-enhancement effect prompts the low cost hierarchical growth of highly ordered heteronanostructures for solar water splitting. *Nano Research*, **9**(6), 1561-1569.

95. Ahrenfeldt, J., Thomsen, T. P., Henriksen, U., and Clausen, L. R. (2013) Biomass gasification cogeneration - A review of state of the art technology and near future perspectives. *Applied Thermal Engineering*, **50**(2), 1407-1417.

96. Kang, K., Azargohar, R., Dalai, A. K., and Wang, H. (2016) Hydrogen production from lignin, cellulose and waste biomass via supercritical water gasification: Catalyst activity and process optimization study. *Energy Conversion and Management*, **117**, 528-537.

97. Nanda, S., Reddy, S. N., Dalai, A. K., and Kozinski, J. A. (2016) Subcritical and supercritical water gasification of lignocellulosic biomass impregnated with nickel nanocatalyst for hydrogen production. *International Journal of Hydrogen Energy*, **41**(9), 4907-4921.

98. Norouzi, O., Safari, F., Jafarian, S., Tavasoli, A., and Karimi, A. (2017) Hydrothermal gasification performance of Enteromorpha intestinalis as an algal biomass for hydrogen-rich gas production using Ru promoted Fe–Ni/γ-Al_2O_3 nanocatalysts. *Energy Conversion and Management*, **141**, 63-71.

99. Shanmuganandam, K., and Ramanan, M. V. (2016) Ni-Ce/SiO_2 nanocomposite: Characterization and catalytic activity in the cracking of tar in biomass gasifiers. *Energy Sources, Part A: Recovery, Utilization, and Environmental Effects*, **38**(16), 2418-2425.

100. Hussain, M. A., Yang, M.-H., Jang, H.-S., Hwang, S.-Y., Um, B.-H., Choi, B. G., and Kim, J. W. (2016) Two-dimensional heterogeneous ruthenium-molybdenum disulfide nanocatalyst for the selective aerobic oxidation of amines. *Industrial & Engineering Chemistry Research*, **55**(25), 7043-7047.

101. Liu, B., Yan, P., Xu, W., Zheng, J., He, Y., Luo, L., Bowden, M. E., Wang, C.-M., and Zhang, J.-G. (2016) Electrochemically formed ultrafine metal oxide nanocatalysts for high-performance lithium-oxygen batteries. *Nano Letters*, **16**(8), 4932-4939.

11

Carbon Based Coatings

11.1 Introduction

During the past few decades, carbon based coatings have gained immense importance for a wide range of industrial, commercial and domestic applications. Carbon exhibits wide variety of allotropes and structures because of its unique ability of hybridization which imparts it the status of a special element in the periodic table. Carbon-based materials have attracted immense interest, especially by invention of fullerenes, carbon nanotubes and graphene. By modifying bonding ratio of carbon from sp^3 to sp^2 and mixing or blending with other elements, the resultant unique physical and mechanical characteristics of carbon make it feasible for use in various demanding applications. For example, graphite coatings have been used since long as solid lubricants because of their low-friction coefficient. This is possible because of the sp^2 bonding present in them. In a similar way, the sp^3 bonding arrangement of "diamond like coatings" (DLC), which are high hardness coatings, has been used since early 1950s. Combining the properties of graphite coatings and DLC has also resulted in very efficient coatings with high order of mechanical resistance [1].

In 1990s, the application of carbon based coatings found its way into more technological applications like fuel injection system in the diesel engines which provided safety against corrosion, wear and tear effects of the lubricant system. In the 21st century, carbon based coatings have become even more profound in their use. Almost every industry like pharmaceuticals, food packaging, and petroleum industry utilizes carbon based coatings for a wide spectrum of functional applications. Carbon coatings also find great utilization in mechanical industries where these are used to protect the surfaces from corrosion resulting from wear and tear of the metal surface; some of the examples include compressors and pumps, gears, shafts, etc. [2]. Besides this, carbon coatings are added on various metallic and non-metallic substrates which enhances the properties of the material; a common example being the application of carbon based coatings on lithium electrodes.

Ahmad Tabish and Vikas Mittal*
The Petroleum Institute (part of Khalifa University of Science and Technology), Abu Dhabi, UAE
*Current address: Bletchington, Wellington County, Australia

Carbon exists in various forms like carbon black, carbon nanomaterials, mesoporous carbon, amorphous carbon, microporous carbon, etc. The following sections review the various types of coatings generated with active carbon component in various structural forms. The most recent findings in the synthesis, characterization and application of these carbon-based coatings are highlighted.

11.2 Mesoporous Carbon Coatings

Mesoporous materials have pores whose diameter ranges usually from 2 to 50 nm. Porous carbon materials are of interest in many applications because of their physio-chemical properties and high surface area. Due to their ordered mesostructures and high surface areas, these materials are widely used as catalysts, advanced electronic materials, separation media, etc. The most recent type of porous carbon materials are ordered mesoporous carbons (OMCs). The BET specific surface area of these materials is found to be around 2200 m^2/g. Moreover, these materials exhibit immense stability at high temperatures and have high resistance in acidic or basic media.

Recently, the mesoporous carbon based coatings have gained importance in various applications. The below sections summarize various research studies on the subject.

11.2.1 Fabrication

In a study by Mittal *et al.* [3], a mesoporous carbon nanocapsule (MCC)/polyvinylidene fluoride (PVDF) polymer composite based free-standing film was fabricated with superhydrophobic nature and multifunctionalities. Mesoporous carbon (MCC) capsules which were in micron range were synthesized by impregnating carbon precursor into the silica spheres which were later carbonized leading to the removal of silica from the latter. The PVDF films were synthesized by first dissolving 10 wt% of PVDF in N-methyl-2-pyrolidone (NMP) solvent. Later, 1H,1H,2H,2H-perfluorodecyltriethoxysilane (PFS) was added in the above solution. MCC was subsequently added to the polymer mixture by manually stirring. It was followed by dipping of metal substrate in the polymer solution and drying at 120 °C for around 2 hours.

Wan *et al.* [4] studied the adsorption application of ordered mesoporous carbon based coating over cordierite with honeycomb structure. Here, the mesoporous carbon was coated on cordierite as a carrier and the surfactant-templating method was used. Carbon sources used for the work were formaldehyde and some phenols. Moreover, copolymer F127 having triblock shape was utilized as a structure dictating agent. In a very similar fashion, another film of mesoporous carbon was synthesized with silicon wafer as substrate, instead of cordierite and the spin-coating method was employed instead of templating for the purpose of comparison of resulting films by the two methods.

Sol-gel spin-coating technique has also been utilized by Pang *et al.* [5] for synthesizing thin films based on mesoporous carbon. In this study, precursors used were tetraethyl orthosilicate (TEOS) and cheaply available sucrose. Thus, films based on the blend of sucrose and silica nanocomposites were synthesized by reacting TEOS in a solution having acidic sucrose, followed by spin coating the formed solution having both sucrose and silica. Later, the carbonization of the films was carried out which resulted in the formation of carbon/silica nanocomposite thin films. Finally, the removal of silica network was carried out using HF which resulted in the thin films based on mesoporous carbon.

In another study, coating-etching approach was used by Feng *et al.* [6]. In this study, similar mesoporous carbon based free-standing thin films were fabricated with arranged pore structures as well as varying diameters (Figure 11.1). For the synthesis of the film, preoxidized silicon wafer was used as a substrate, on which resol precursors/pluronic copolymer were coated. It resulted in the formation of polymeric mesostructures of highly ordered nature. Finally, carbonization was carried out at 600 °C which was followed by etching the layer

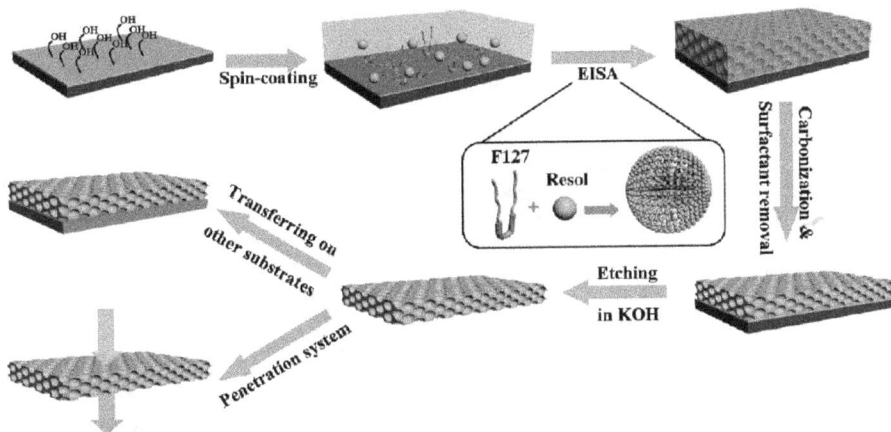

Figure 11.1 Schematic of the preparation of free-standing mesoporous carbon thin films. Reproduced from Reference 6 with permission from American Chemical Society.

of oxide formed between the silicon substrate and carbon film.

Another recent study on the generation of mesoporous carbon coatings has been reported by Mittal *et al.* [7]. In this study, mesoporous carbon nanocapsules (MCC) based super hydrophobic coatings have been fabricated which are multifunctional in nature. Just like previous studies, PVDF was used as a binder. A facile brush-on process was used on multiple substrates like cotton based fabric, metals and glass, etc. Solid mesoporous carbon coatings have also been synthesized in a similar manner in other research studies [8-10]. For instance, Figure 11.2 demonstrates metal-organic framework (IRMOF-1) after controlled pyrolysis for generating carbon-coated ZnO quantum dots [9].

Figure 11.2 Metal organic framework (IRMOF-1; top) and IRMOF-1 after controlled pyrolysis for generating carbon-coated ZnO QDs without agglomeration. Reproduced from Reference 9 with   permission from American Chemical Society.

Another example of the mesoporous coating framework has been demonstrated by Ji *et al.* [11] in which centimeter-scale free standing thin films have been synthesized from 2D nanocrystal superlattices self-assembled at the solid– or liquid–air interface and are collectively called mesoporous graphene frameworks (MGFs). MGF thin films were obtained based on the transformation of 2D nanocrystal superlattices by ligand carbonization, nanocrystal etching, and framework graphitization [12].

11.2.2 Characterization and Testing

Mittal *et al.* [3] carried out the transmission electron microscopy (TEM) analysis of MCC with and without silica core for morphological and structural characterization. Moreover, field emission scanning electron microscopy (SEM) was also used for studying the free standing films (FSFs). SEM revealed homogeneously distributed membrane resembling a porous film with sub-micron sized pores varying from 0.5 to 3 um approx. in size on the top film surface. Higher magnification exhibited that mesoporous carbon nanocapsules were mostly around 200-300 nm in size. The SEM images of the edge view of the MCC/polymer based superhydrophobic FSF showed that the FSFs had a thickness of around 120 um.

Wan *et al.* [4] characterized the mesoporous carbon coating on cordierite by TEM, high-resolution scanning electron microscopy (HRSEM), and nitrogen sorption techniques which exhibited the presence of carbon coating pore arrays ordered hexagonally over cordierite. Also, ordered mesoporous carbon was used to coat the honeycomb monolith adsorbents [13]

which exhibited adsorption capacities for chlorinated organic pollutants in water with 178 mg/g for *p*-chloroaniline and 200 mg/g for *p*-chlorophenol (with respect to the net carbon coating), along with adsorption of lower concentration contaminants, handling of high processing volumes and recyclability. Similarly, Pang *et al.* [5] used atomic force microscopy (AFM) and SEM images of mesoporous carbon films after the removal of silica template which indicated the appearance of smooth, homogeneous and continuous crack free thin films. Further, SEM indicated the film thickness to be 1 micron on an average. TEM study of mesoporous carbon films suggested that mesoporous carbon contained a non-ordered, but homogeneously sized mesoporous structure.

Mittal *et al.* [7] also characterized the morphology and structure of MCC based coatings by microscopy and optical profilometer. SEM exhibited the 3D structure of structured MCC with a narrow size dispersion (radius around 100-150 nm) revealing presence of pores formed within. Moreover, TEM demonstrated the presence of capsules of thin carbon shell before and after silica was etched out from the core. Fourier transform infrared (FTIR) spectroscopy was utilized to study the chemical nature of superhydrophobic coating. The spectra obtained after annealing at 100 °C for 10 hours exhibited peaks at 895, 1017, and 1138 cm^{-1}, which were attributed C–H bending modes, C–F stretching modes and definite variations of CF_2 and CF groups present in polyvinylidene fluoride.

Nitrogen sorption technique has been utilized by Feng *et al.* [6] to evaluate the mesoporosity of carbon films, which demonstrated a large BET surface area of 700 m^2/g and large homogeneous mesopores with average size of ~4.4 nm. The SEM analysis exhibited a homogenous and smooth free-standing film layer with no visible cracks. The film thickness was observed as 500 nm and was uniform in nature, as confirmed by both the surface profiler and SEM. The mesopore architecture of the synthesized free-standing mesoporous carbon films was investigated by cross-section transmission electron microscopy, 2-D small-angle X-ray scattering (SAXS), and high-resolution scanning electron microscopy (HR-SEM). Moreover, TEM analysis of the cross-section of mesoporous carbon films showed a distorted hexagonal lattice. Similarly, Ji *et al.* [11] used HRSEM and TEM for characterization of mesoporous graphene framework (MGF) coatings which established that MGF films possessed a 3D ordered, interconnected mesoporous structure with a pore size of ~10 nm (Figure 11.3). The slightly reduced pore size as compared to the initial diameter of Fe_3O_4 nanocrystals (NCs) was attributed to the framework contraction caused by heat treatment. TEM analysis identified the lattice projections such as (111) and (110) which suggested that MGF films possessed the same fcc symmetry inherited from NC superlattice films. For the capacitive performance of MGF films,

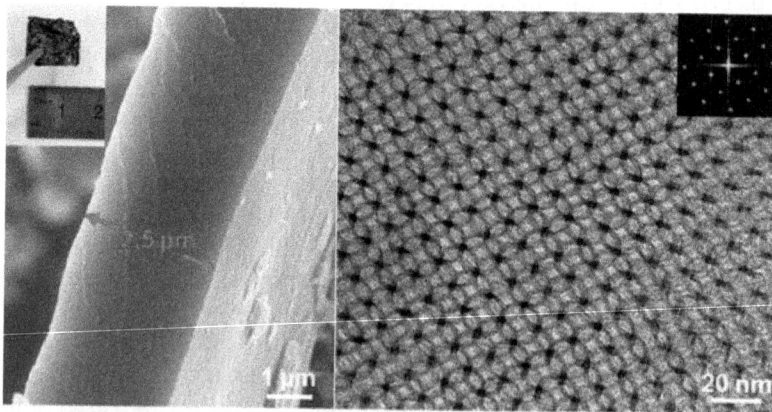

Figure 11.3 Ordered mesoporous few-layer graphene framework films. Reproduced from Reference 11 with permission from American Chemical Society.

H$_2$SO$_4$ and HNO$_3$ were used for acid treatment and later FTIR spectra was carried out which indicated a much stronger peak corresponding to OH stretching (broad peak beyond 3000 cm^{-1}) and the significant peak of carbonyl (C=O) extending (~1716 cm^{-1}) after the acid treatment, indicating the presence of a large portion of hydroxyl (–OH) and carboxyl (–COOH) groups on the graphene framework surface after acid treatment.

11.2.3 Applications

Superhydrophobic free-standing films synthesized by Mittal *et al.* [3] find many applications requiring high temperatures like automobile parts or cooking wares along with capability of long exposure to various corrosive environments. These are even suitable for various applications in microelectromechanical systems, optics along with enhanced resistance to the solvents. Moreover, utilization of MCC for synthesis of superhydrophobic and conductive films assists in release/encapsulation of varying functional materials and tuning the properties like electrical conductivity. Therefore, the carbon nanocontainers embedded in matrices have paved the way for wide range of applications of hydrophobic films based smart coatings.

Fabrication of mesoporous carbon coatings on cordierite, as shown by Wan *et al.* [4], can be used as a reusable adsorbent with high volume processing rates, and recyclability. Similarly, mesoporous carbon films synthesized by spin coating technique by Pang *et al.* [5] find application in sensors, separation processes, membrane based reactors, fuel cells, and many other commercial products. The mesoporous carbon based free-standing thin films, as fabricated by Feng *et al.* [6], have wide range of commercial use in electrochemical and biomedical fields. The films were used to make an electrochemical supercapacitor device which exhibited a capacitance around 136 F/g at 0.5 A/g. Also, nanofilters derived from the above films of carbon demonstrated high rate of filtration of bovine serum albumin and cytochrome c based on size.

Multifunctional superhydrophobic coatings fabricated by Mittal *et al.* [7] based on mesoporous carbon nanocapsules (MCC) with polyvinylidene fluoride (PVDF) find application in bio-medical field with reduced bacterial adhesion. Moreover, the free-standing ordered mesoporous few-layer graphene framework films, synthesized by Ji *et al.* [11], have potential applications in energy storage. The synthesized films were used as electrode materials to build supercapacitors, which exhibited high specific capacitances with excellent cycling stabilities in both aqueous and organic electrolytes, with the capacitive performance comparable to or higher than that of most graphene-based materials.

11.3 Amorphous Carbon Based Coatings

With the improvement in thin film synthesis and development strategies in the last few years such as dip coating, vapor deposition (chemical based), and cathodic arc deposition, fabrication of pure amorphous carbon (a-C) based coatings has been extensively studied. As a result, applications based on amorphous carbon coatings have also enhanced. Figure 11.4 and 11.5 depict the examples of amorphous carbon-coated tin anode material as well as amorphous carbon-coated silicon nanocomposites [14,15]. This section reviews the recent advances in the synthesis, characterization and applications of such functional coatings.

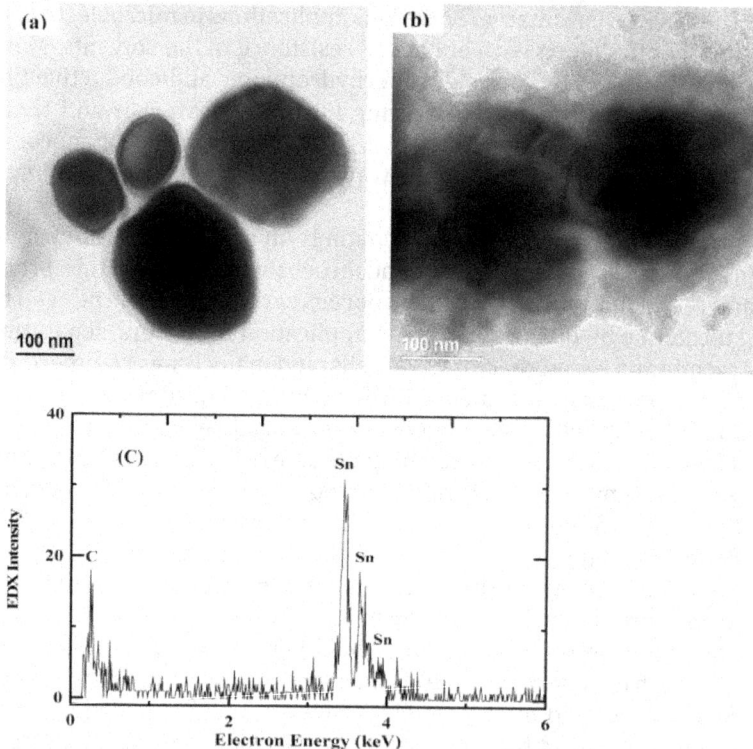

Figure 11.4 TEM images of (a) tin particles and (b) amorphous carbon-coated tin particles, and (c) EDAX spectra of the carbon coated particles. Reproduced from Reference 14 with permission from American Chemical Society.

11.3.1 Fabrication

Kim *et al.* [16] coated the $LiNi_{1/3}Mn_{1/3}Co_{1/3}O_2$ cathode material with the amorphous carbon. $LiNi_{1/3}Mn_{1/3}Co_{1/3}O_2$ and sugar were stirred well for 3 hours using ethanol as solvent. The resulting well mixed solution underwent sintering at 350 °C for 1 h in a furnace which resulted in the coating layer being developed. Alakoski *et al.* [17] used the pulsed arc discharge (FPAD) technique to synthesize the diamond-like carbon (DLC) coatings called tetrahedral amorphous carbon (ta-C) coatings on the identical silicon wafers in a vacuum of 100 μPa using high energy unit with high anode-cathode voltage (6 kV), high peak current (13 kA) and short pulses (15 μs). The average carbon ion energies were 600 eV. Due to the large plasma ion energy, low quality of the coatings was observed with sp^3 fraction around 40%.

In another study, Cao *et al.* [18] coated the surface of $LiCoO_2$ with a nanolayer of amorphous carbon using chemical means without reduction of the original surface. Das *et al.* [19] fabricated amorphous carbon based coating by chemical vapor deposition technique on the synthesized Si nanowires using acetylene as the source of carbon. A current density of 12.5 mA cm^{-2} and a working potential of 1.5 kV was used. Also, 0.4 mbar was the working pressure of acetylene and electrode separation was kept at 2.5 cm. A thin layer of a-C was deposited on Si wafers without application of any extra substrate temperature. In addition, Show [20] synthesized a a-C film on Ti bipolar plates at various growth temperatures using radio frequency plasma enhanced chemical vapor deposition (RF-PECVD), where ethylene served as the gas source. Moreover, flow rate of ethylene, RF power and deposition pressure were 10 sccm, 150 W and 0.5 Pa respectively. The growth time was settled for 3 h and the growth temperature was varied between room temperature to 600 °C in order to study the effect of temperature variation.

Liu *et al.* [21] fabricated carbon coated core shell nanorods of magnetite (Fe_3O_4) using hydrothermal technique with Fe_2O_3 as precursor and citric acid as carbon source. 0.2 g citrus extract was broken down in 5 mL ethanol to produce a homogeneous and transparent mixture. At that point, 0.2 g Fe_2O_3 nanorods was also mixed into the above solution and stirred in a sealed fixed container for 24 h. Drying and grounding were carried out followed by sintering for 2 h at 300 °C. Perez-Huerta and Cusack [22] used amorphous carbon to coat on the surface of two carbonates namely biogenic aragonite and calcite to analyze the effect of the carbon thickness on the crystallographic orientation of the substrate. Carbon coating thickness was controlled utilizing a precision etching-coating system.

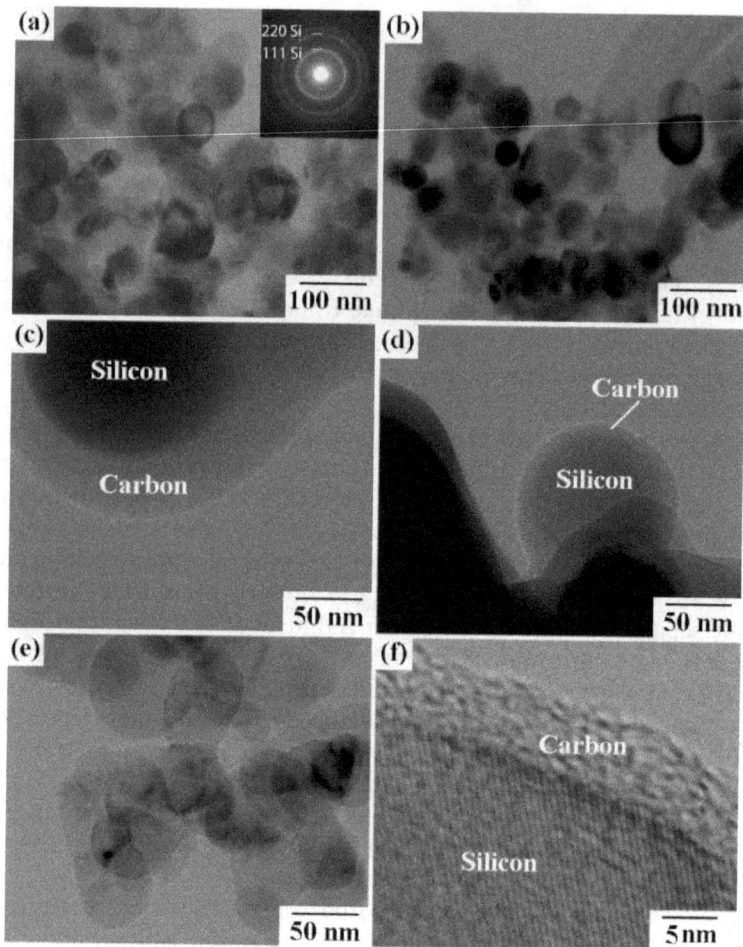

Figure 11.5 TEM images of nanocrystalline Si (a) and carbon-coated Si nanocomposites (b-f). Reproduced from Reference 15 with permission from American Chemical Society.

11.3.2 Characterization and Testing

Das *et al.* [19] employed electron microscopy technique to analyze amorphous carbon (a-C) coated silica (Si) nanowires (NWs) and observed that the whole surface of Si NWs was covered with amorphous carbon. Crystalline as well as homogeneous nanowires were observed in the lattice images and same appearance on the coating confirmed its amorphous nature. The FTIR study exhibited an absorbance peak at 2900 cm^{-1} assigned to C–H$_n$ vibrational bonds, which also indicated the presence of amorphous carbon coating on Si NWs. The absorbance peak nearby 1070 cm^{-1} was designated to the Si–O bond obtained from the residual SiO$_2$ and the one at 3400 cm^{-1} was assigned to O-H bond. Moreover, high vacuum field emission set up was used to study various field emission properties like the enhancement factors and turn-on field which were significantly enhanced due to amorphous carbon coating on the Si NWs.

Show [20] carried out the Raman spectroscopy of the a-C films deposited on the surface of Ti bipolar plates at various growth temperatures and intense Raman signals were observed at around 1410 and 1560 cm^{-1}, attributed as D and G lines. Sharp D and G lines observed for a-C films deposited at 500 and 600 °C had the peak position displaced to 1350 and 1590 cm^{-1}. Impedance measurement for the fuel cells was also carried out and the Cole-Cole plot exhibited the presence of a curve of semicircular form in the negative axis region of impedance (-Z$_{img}$) for all cells. Moreover, fuel cells assembled from bipolar plates and coated with a-C film at 600 °C had the bottommost position on Z$_{real}$ axis (real impedance region).

In other studies, Kim *et al.* [16] used the X-Ray diffraction for the structural analysis of carbon-coated LiNi$_{1/3}$Mn$_{1/3}$Co$_{1/3}$O$_2$. No peaks due to the impurities were seen but the hexagonal crystal structure of α-NaFeO$_2$ with a space group of *R3m* was visible. DSC was used to measure the thermal stability of the fully charged cell (4.3V versus Li/Li+). Heat release from the uncoated sample was observed to be 161.3 J g^{-1} and the maximum temperature of the exothermic reaction were found as 362.4 and 457.8 °C, respectively. The authors inferred that the layer of coated carbon on the surface of LiNi$_{1/3}$Mn$_{1/3}$Co$_{1/3}$O$_2$ active material suppressed the oxygen release and hence the thermal stability was improved. Kim *et al.* [16] also evaluated the rate capability of the bare- and carbon-coated powder for which cells underwent charging and discharging at varied cut off voltages. It was observed that the carbon coating improved the rate ability of the material as compared to uncoated material. Moreover, the capacity holding ability of 1 wt% carbon coated material at the rate of 5 °C was found to be 87.4% as compared to that of 0.2 °C rate. Lastly, discharge holding capacity after 50 charge-discharge cycles of same material as above was observed as nearly the same to that of bare material (98% similarity).

Alakoski *et al.* [17] used the simple visual inspection method to determine the quality of tetrahedral amorphous carbon coating. The authors incorporated the use of X-ray photoelectron spectroscopy and profilometer for determining the sp³ fraction and coating thickness respectively. It was observed that the tetrahedrally bonded amorphous carbon (ta-C) layer was thickest in the center of the silicon wafer. Cao *et al.* [18] used various techniques to characterize the carbon coated $LiCoO_2$. From SEM, a very smooth surface of original $LiCoO_2$ was observed. On the other hand, for coated $LiCoO_2$, the surface was found to be rough, non-homogeneous with some small particles found stuck to the surface which seemed like small scattered carbon particles. On the other hand, the TEM analysis exhibited a homogeneously coated nanolayer of carbon on the surface of $LiCoO_2$. X-ray Diffraction patterns were also analyzed for both uncoated and carbon-coated $LiCoO_2$. The main diffraction peaks indicated that both samples had rhombohedral structure. It could also be inferred that there was a weakening and reduction in the intensities of XRD peaks because of amorphous carbon coating on the surface of $LiCoO_2$. The authors also plotted typical Nyquist plots of the coated and the original $LiCoO_2$ composite electrodes. Both the plots exhibited a straight line and a semicircle in low and high frequency regions, respectively.

Liu *et al.* [21] used the high resolution TEM (HRTEM) to characterize the coating of carbon layer on the surface of Fe_3O_4 nanorods. Moreover, ac impedance spectroscopy, cyclic voltammetry and galvanostatic charge/discharge techniques were employed to evaluate the electrochemical properties of Fe_3O_4/carbon nanorods as anodes in lithium-ion cells. X-ray diffraction was used to characterize the crystal structure of Fe_3O_4/carbon nanorods using Cu Kα radiation which indicated that the structure was similar to standard hematite (a-Fe_2O_3). Moreover, the impurity of α-Fe_2O_3 phase could not be observed which indicated that the α-Fe_2O_3 was totally converted to magnetite Fe_3O_4. On the other hand, TEM analysis of prepared carbon coated Fe_3O_4 nanorods exhibited the agglomeration of individual nanorods into bundles and all the nanorods had a coating of amorphous carbon. The HRTEM inferred the presence of Fe_3O_4 lattice and a layer of amorphous carbon having thickness of 2-5 nm. Lastly, energy dispersive X-ray (EDX) study confirmed the presence of amorphous layer of carbon.

Perez-Huerta and Cusack [22] utilized electron backscatter diffraction (EBSD) to study crystallographic introduction in biogenic carbonates (aragonite and calcite) in the basic blue mussel, Mytilus edulis, utilizing distinctive sorts of gum and thicknesses of carbon covering. It was observed that there was no appearance of diffraction in aragonite and calcite. EBSD examinations were additionally performed in the electron magnifying instrument in low vacuum mode to analyze the impact of carbon covering thickness with various weight conditions.

11.3.3 Applications

Carbon coated Si nanowires synthesized by Das *et al.* [19] find applications in anti-corrosion coatings. Ti bipolar plates coated with amorphous carbon (a-C) film synthesized by Show [20] are useful for polymer electrolyte membrane fuel cells (PEMFC) as it increased the output power and reduced the internal resistance of the fuel cell. $LiNi_{1/3}Mn_{1/3}Co_{1/3}O_2$ cathode material coated with amorphous carbon is commercially used for lithium batteries because of enhanced electrochemical performance and thermal stability. Hence, it has the potential to be useful for future promising applications in hybrid electric vehicles as an electrode material.

Carbon-coated $LiCoO_2$ lithium ion batteries synthesized by Cao *et al.* [18] have promising applications where charging and discharging at large rates with high capacity are needed. Carbon coated magnetite (Fe_3O_4) core-shell nanorods fabricated by Liu *et al.* [21] are indispensable in ferromagnetic, biomedical, and catalysis applications. Perez-Huerta and Cusack [22] generated the coating on biogenic carbonates which are used as impregnating agents in the resins like epoxy resin for grinding and polishing applications.

11.4 Coatings Based on Pyrolytic Carbon and Carbon Black

Some of the characteristics of pyrolytic carbon resemble with that of graphite, but different in the sense that it has covalent bonding present on its surface in the graphene sheets because of various defects and imperfections generated during its manufacturing. Pyrolytic carbon has vast range of applications in automobiles, nuclear reactors, and spectrometry. Its extensive use in biomedical applications has made it very relevant in current perspective. On the other hand, carbon black is obtained from the combustion of various types of petroleum products. Properties like high surface to volume ratio make carbon black very appealing for research and development. The below section summarizes some of the carbon coatings synthesized using the two materials of pyrolytic carbon and carbon black.

11.4.1 Fabrication

Kim *et al.* [23] synthesized carbon coating layers comprising of silicon carbide and pyrolytic carbon on UO_2 pellets by utilizing combustion reaction technique between silicon and carbon. For the coating purpose, propane underwent thermal decomposition reaction in a chemical vapor deposition unit at 1250 °C and microwave pulsed electron cyclotron resonance plasma enhanced chemical vapor deposition (ECR PECVD) at 500 °C by utilizing silane. The carbon deposition unit comprised of a 1500 kW resistance heater, a preheater of the source gasses, a gas blender, a temperature controller, and a gas stream control framework.

Lopez-Honorato *et al.* [24] fabricated silicon carbide and pyrolytic carbon coatings using time domain thermoflectance on fuel particles (simulated). Three variety of samples were synthesized utilizing fluidized bed chemical vapor deposition. Two single layer PyC coatings and a triple layered coating comprising of a buffer, silicon carbide (SiC) and inward pyrolytic carbon (IPyC) were spread on alumina particles 250 μm in radius. The single layer coatings were deposited with 50% v/v acetylene concentration at 1250 °C (1250-PyC) and 1450 °C (1450-PyC). For the triple layered particle, a low density PyC (buffer) was synthesized with 40% v/v acetylene at 1450 °C, while the inward high density PyC was deposited with a blend of 33% v/v propylene/acetylene at 1300 °C.

In another study, Kim *et al.* [25] fabricated coatings using carbon black on the surface of $LiCoO_2$ cathode and utilized surfactants for high-density Li-ion cell. The coating strategy contained two stages: (i) scattering of amassed carbon black utilizing orotan®, a polyacrylate dispersant; (ii) carbon-black covering of the cathode material utilizing a surfactant gelatin which is amphoteric in nature. The procedure diminished the carbon content in the electrode and had no effect on the cycle-life working and efficiency of the cell. Furthermore, it also enhanced the rate ability of the Li-ion particle cell. For coating purpose, around 2 g of gelatin was mixed in 30 ml of DI water for 10 min at 50 °C. Finally, the carbon-dispersed orotan® solution was added in the above mixture and gelatin coating was accomplished on the dispersed carbon black. The pH of gelatin was required to be brought close to its isoelectric point of pH ~ 4-5 by utilizing acetic acid, as it was observe to help in the direct coating of carbon on the cathode surface.

Ponrouch *et al.* [26] fabricated carbon coatings by physical deposition of carbon through dissipation under vacuum (around 10^{-4} mbar) performed at normal room temperature along with dry conditions. A commercially available carbon evaporator was used for the coating purpose on electrode active material powders. A specimen holder was fabricated for accomplishing uniform covering on all particles and comprised of a glass petri dish fixed to the traditional pivoting/tilting plate of the device. For achieving a very uniform and smooth coating, persistent mixing of the sample along with rotation with the sample holder was carried out.

In another study, Rodriguez-Mirasol *et al.* [27] also formed pyrolytic carbon within a microporous zeolite template using chemical vapor infiltration, as shown in Figure 11.6. Propylene, in a nitrogen stream, was used as the carbon precursor. The process resulted in the microporous carbon with high surface area, with wide microporosity, well-developed mesoporosity and high adsorption capacity.

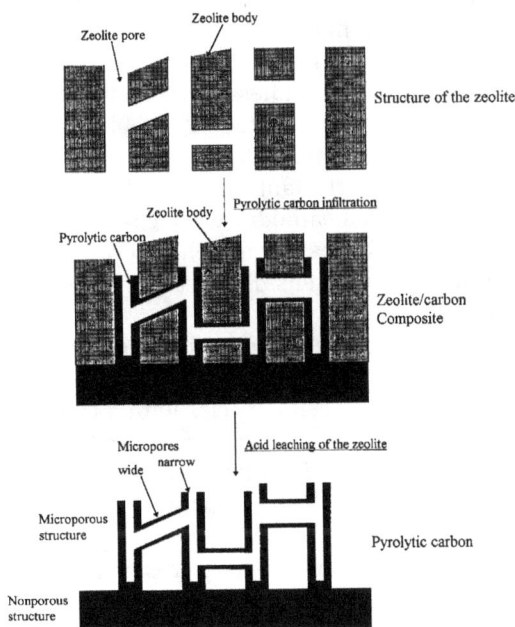

Figure 11.6 Scheme of pyrolytic carbon infiltration of zeolite, followed by removal of the zeolitic substrate. Reproduced from Reference 27 with permission from American Chemical Society.

11.4.2 Characterization and Testing

Kim *et al.* [23] carried out the chemical analyses of the multi-layer coating with X-ray diffraction and Auger electron spectroscopy (AES) along with TEM and SEM. SEM exhibited uniform layers of coating having silicon carbide and pyrolytic carbon due to the combustion reaction. The X-ray spectra indicated β-SiC was the product with negligible presence of unreacted silicon. Moreover, AES and XRD studies confirmed the presence of pyrolytic carbon and silicon carbide at the inner and outer coating layers respectively.

Lopez-Honorato *et al.* [24] used the time-domain thermoreflectance for pyrolytic carbon and silicon carbide coatings on fuel particles and performed thermal conductivity mapping to obtain the line profiles of the coatings. Later, SEM of these coatings was carried out which led to the inference of differences between density and porosity. Values of thermal conductivity obtained from the maps for PyC were 4.2 W/m K for 1250-PyC and 3.4 W/m K for 1450-PyC and corresponding density values were 2.12 and 1.41 g/cm^3 respectively.

Kim *et al.* [25] carried out the studies of volumetric limits in Li-ion particle cells utilizing uncovered or carbon-coated $LiCoO_2$ electrodes. Discharge capacities of $LiCoO_2$ were seen from the anode/cathode dimensional proportion. To further study the impact of surfactant on the stability of the Li cell, the cells having bare or carbon-covered cathodes were charged to 4.2 V after each 50 cycles and the changes in thickness of the cell at room temperature was calculated. It was concluded that the cells displayed comparable swelling with expanding the number of cycles.

Ponrouch *et al.* [26] carried out TEM examination to analyze the carbon deposition on the electrode materials. TEM demonstrated a clear carbon layer at the surface of the dynamic material particles with a thickness ranging from 1 nm up to 4 nm. Additionally, TGA for carbon covered NTO powder with a layer of carbon having thickness around 2.5 nm were analyzed. Weight loss due to the decay of the carbon layer into CO and CO_2 was observed. Raman spectroscopy was utilized to get further knowledge into the graphitization level of the deposited carbon layer. Apart from the sharp crest at around 1000 cm^{-1} and the more extensive and less visible peak near 1100 cm^{-1}, the range comprised of a several broad bands which overlapped on one another, the majority of which were related to C vibrations. The effect of carbon coating on the kinetics of electrolyte reduction and SEI development was examined through the investigation of Co_3O_4 based electrodes. Electrodes synthesized with uncoated and carbon coated nanosized Co_3O_4 were cycled in galvanostatic mode with potential restriction (GCPL) with a progression of 10 cycles.

11.4.3 Applications

The multilayer coatings of pyrolytic carbon and silicon carbide over the surface of UO_2 pellets synthesized by Kim *et al.* [23] represented the coating process which is beneficial to study the feasibility of the above coatings for fuels of light water and heavy water reactors (LWR, HWR). Lopez-Honorato *et al.* [24] also developed pyrolytic carbon and silicon carbide coatings on simulated fuel particles were helpful for nuclear purpose. Plotting of the conductivity of coated fuel particles provided valuable information to optimize fuel efficiency amid the operation of atomic reactors. Direct carbon-black coating of the cathode material using a surfactant fabricated by Kim *et al.* [25] improved the rate capacity of the Li-ion cell and, thus, has excellent application in technology like mobiles, PCs, cameras, laptops, etc.

Carbon coatings synthesized by Ponrouch *et al.* [26] are largely utilized to fabricate conductive carbon coatings with a thickness of ca. 1 nm and upwards on insulating specimens to empower charge free imaging in electron microscopy and microprobe examination. Also, it has the benefit of delivering coatings covering uneven and non-homogeneous surfaces.

11.5 Coatings Based on 'Carbon Alcohols' as Source

Various coatings have been synthesized in the recent past with carbon based alcohols as source like ethylene glycols, ethyl alcohol, ethanol, etc. This section summarizes the fabrication techniques and performance results of such coatings.

11.5.1 Fabrication

Lin *et al.* [28] synthesized carbon coating using ethylene glycol as source using the substrate LiFePO$_4$, with various molar ratios of high electron conductive iron phosphide phase by an aqueous sol–gel technique in a reductive sintering environment. Different parameters were utilized for modifying the microstructure and varying phase concentration of products. The LiFePO$_4$ cathode material was prepared using the sol-gel method utilizing various constituents as FeC$_2$O$_4$·2H$_2$O (ferrous oxalate), LiNO$_3$ (lithium nitrate) and NH$_4$H$_2$PO$_4$ (dihydrogen ammonium phosphate). In another study, Klebanoff *et al.* [29] utilized ethyl alcohol as the carbon source for manufacturing radiation-instigated protective carbon coating on extraordinary bright optics. A 5 Angstrom carbon coating was kept on EUV Mo/Si optics by means of co-exposure to radiation (EUV photons, electrons) and ethanol vapor. A 2 kV electron beam was used as a EUV/water exposure. Moreover, a sputtered carbon sample was used and prepared in a custom-built magnetron sputtering system which was operated at the base pressure of less than $3*10^{-7}$ Torr using argon at a constant pressure of 0.9 mTorr as a sputtering gas. As an experimental setup, two substrate holders were spun during the entire deposition process.

Wang *et al.* [30] fabricated a nano-carbon coating on the surface of LiFePO$_4$ using ethanol as carbon source using spray pyrolysis system. In this study, effect of corrosion was studied by the addition of carbon layer at olivine LiFePO$_4$ stored in moisture-contaminated electrolyte. Effect of iron dissolution on corrosion was also studied.

11.5.2 Characterization and Testing

Lin *et al.* [28] studied the release/discharge abilities to analyze the impact of the carbon covering and iron phosphides on the electrochemical properties of the LiFePO$_4$/C electrodes at rates of 0.1-5C (1C = 170mAhg^{-1}) and examining the CV bends. It was observed that carbon covering in a ratio of 1.5 wt% significantly diminished the molecule size of LiFePO$_4$, and enhanced the rate ability of LiFePO$_4$. Additionally, the impact of the weight of FeP on the capacity of the carbon coated LiFePO$_4$ varied at various release rates. Incrementing the concentration of FeP from 1.2 to 3.7 wt% somewhat diminished the capacity of LiFePO$_4$/C at low release rate (0.1C and 1C). Likewise, an excessive amount of iron phosphides brought down the release limit of the electrode since these were not reactive for the disinsertion/insertion of lithium particle. XRD patterns of the products synthesized by various blend parameters demonstrated that the principle phase was LiFePO$_4$, along with iron phosphides peaks. Additionally, SEM analysis of the LiFePO$_4$ affirmed that the molecule size of the carbon free LiFePO$_4$ synthesized from gel without ethylene glycol was substantially bigger than that of the LiFePO$_4$ particles coated with carbon.

Klebanoff *et al.* [29] carried out the sputter Auger depth profiling, Auger electron spectroscopy, and EUV reflectivity calculations of the synthesized radiation-induced carbon coating on EUV optics. It was observed in a nutshell that the coating was void free and protected the optic from water-induced oxidation at the water partial pressures which were used in the experiment. Moreover, the coating was resistant to atmospheric degradation along with gasification with a 0.5% reduction in the relative reflectivity of the optic.

Wang *et al.* [30] used various testing methods to describe the morphology and structure of olivine LiFePO$_4$ after the application of nanocarbon coating and iron dissolution. The SEM images of carbon coated LiFePO$_4$ showed that the conductivity was greatly improved after coating and a homogeneous layer of nanocarbon was visible. Using FIB along with HRTEM, the thickness was determined to be 10 nm and a clear Au protection film was also observed. Also, the corrosion extent of coated LiFePO$_4$ was greatly reduced after aging process, along with reduced loss of iron, hence, surface corrosion was greatly reduced.

Further, ToF-SIMS testing was performed for coated LiFePO$_4$ which showed much less corrosion degradation by acidic species during aging process, as carbon coating was helpful in preventing attack from HF on LiFePO$_4$ surface and hence provided a protective layer. Lastly, X-ray absorption near-edge structure (XANES) spectra was studied for coated LiFePO$_4$ and was found to be nearly same for uncoated sample, except indication of partial oxidation of iron. On the whole, materials stability and protection was enhanced by addition of carbon coating based on ethanol on the surface of LiFePO$_4$.

11.5.3 Applications

The carbon coating on LiFePO$_4$ prepared by Lin *et al.* [28] find enormous applicability in increasing the intrinsic electronic conductivity, hence, making it a promising future commercial alternative cathode material for lithium ion battery. Radiation-induced protective coating fabricated by Klebanoff *et al.* [29] are commercially viable as these are resistant to atmospheric degradation. Moreover, these coatings improve the properties of EUV optics which are extremely important for the manufacturing of next-generation semiconductor chips.

The nanocarbon coatings synthesized by Wang *et al.* [30] on the surface of olivine LiFePO$_4$ greatly enhanced its performance and, hence, make it a potential candidate as a safe and efficient cathode material for advanced LIBs applied in commercial electric vehicles, representing a next generation safe technology for future.

11.6 Coatings Based on Graphene and Other Related Materials

Graphene has been extensively used in the recent years for various research studies. As a result, graphene based materials have been used in a wide variety of protective coatings. For instance, Figure 11.7 shows the partially covered platinum surface with graphene [31]. The graphene coated surface exhibited effective protection against CO and O$_2$. Another example of graphene based surface protection is also demonstrated in Figure 11.8 for Gr/SiO$_2$, Cu, and Gr/Cu samples, where graphene was coated on the surface for corrosion protection [32].

The below section summarizes the recent studies in which graphene and other related functional nanomaterials have been used to synthesize the coatings with tuned structure and properties, along with their performance in diverse areas.

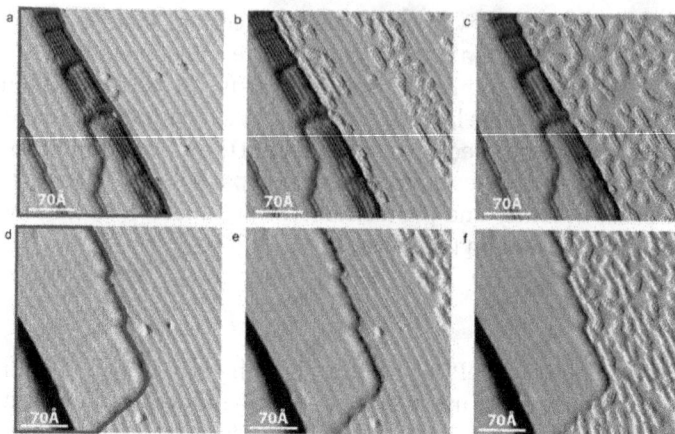

Figure 11.7 STM image of a Pt surface partially coated by graphene (the area at the right side is uncoated, and the rest is coated by graphene). (a-c) Surface exposed to 0, 3, and 63 L of CO, respectively. (d-f) Coated surface exposed to 0, 25, and 40 L of O_2, respectively. Reproduced from Reference 31 with permission from American Chemical Society.

11.6.1 Fabrication

Nishihara *et al.* [33] covered the pore surface of mesoporous silica SBA-15 with 2,3-dihydroxynaphthalene (DN) through a lack of hydration response behavior between the surface silanol aggregates in SBA-15 and the hydroxyl groups of the DN particles. Afterwards, the carbonization of DN in the SBA-15 pores brought about to a great degree thin carbon layer arrangement in the pores including 1-2 graphene sheets. For the synthesis process, the authors added 2.3 g of DN in 5 ml of acetone and further dissolved the mixture into 0.47 g of dried SBA-15 in vacuum. The blend was mixed for a few hours at room temperature which led to the evaporation of acetone. The subsequent strong blend of DN and SBA-15 was warmth treated at 573 K for 1 h under N_2 stream. Amid this progression, fluid state DN was permitted to undergo reaction with SBA-15 through a dehydration reaction between the silanol groups on the pore surface of SBA-15 and the hydroxyl aggregates of DN. The unreacted DN was then washed away with acetone to acquire DN-covered SBA-15 (DN/SBA-15). Lastly, the synthesized DN/SBA-15 was warmth treated under N_2 stream at 1073 K for 4 h to carbonize DN in the silica pores. Thus, graphene was covered/coated on silica particles.

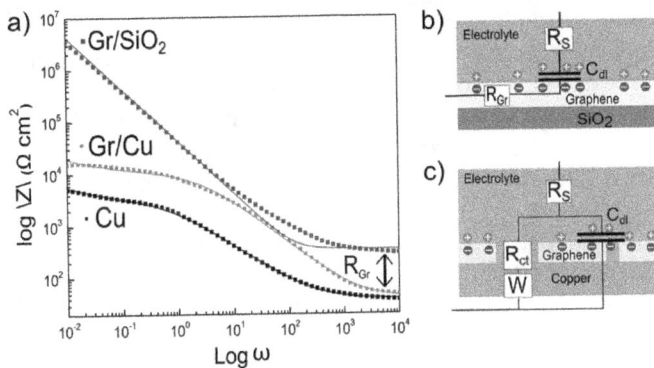

Figure 11.8 (a) Bode plots of Gr/SiO₂, Cu, and Gr/Cu samples (shown as solid symbols), whereas the lines represent best fits to the equivalent circuit models; (b) equivalent circuit model used in modeling Gr/SiO₂ devices, and (c) equivalent circuit model for Cu and Gr/Cu devices. Reproduced from Reference 32 with permission from American Chemical Society.

In another study by Yoshida *et al.* [34], carbon film was generated on SiC filaments in the fabric by electrophoretic deposition technique utilizing industrial colloidal graphite homogeneous mixture and was compared with carbon coating on SiC strands in fiber material synthesized by dip coating or vacuum penetration utilizing the above prepared suspension of colloidal graphite. Lee and Lee [35] synthesized the carbon coating on nano-silica particles. The carbon coating comprised of graphite and coal–tar pitch. For synthesizing oxides of nano-Si, mechano-chemical reduction of SiO was carried out by Al. Later, carbon coating was performed on the ball milled mixture of above oxides with graphite using pyrolysis/heating of coal tar at approx. 900 °C. The procedure was carried out in a glove container with argon gas. Later, size and shape of the composite was controlled by grounding the coated carbon product using rotor mill using variable speeds.

In a recent study by Li and Zhang [36], superhydrophobic coatings were fabricated by using spray-coating technique. The uniform suspension of polysiloxane/multiwalled carbon nanotubes (POS/MWCNTs) nanocomposites was spray coated onto various substrates which was followed by curing for a fixed time span. N-hexadecyltriethoxylsilane and tetraethoxysilane underwent hydrolytic co-condensation with MWCNTs which was activated by acid. In another study, Sevastyanov *et al.* [37] fabricated conducting coatings based on carbon nanotubes and other carbon nanostructures and SnO₂. Vaporized transport method was employed for obtaining the coatings. The accompanying nanomaterials utilized in the study were: carbon nanotubes (CNT), carbon nanofibers (CNF) and carbon nanoflakes (CNFL). For coatings purpose, nanomaterials were used to prepare corresponding suspensions. Carbon nanostructure layers were laid on 10*10 mm glass substrates; cleaned silicon plates were utilized as an extra reference samples. The suspensions with carbon nanomaterials were placed in an ultrasonic tank for 30 min and set into the assembly of a vaporized generator quickly before deposition.

Wang *et al.* [38] also reported the graphene-sulfur composites by wrapping polyethylene glycol (PEG) coated sulfur particles with mildly oxidized graphene oxide sheets, which were decorated by carbon black nanoparticles (Figure 11.9).

11.6.2 Characterization and Testing

Nishihara *et al.* [33] utilized different testing strategies to describe the carbon covered mesoporous silica. TGA was performed to decide the thermal response of DN and carbon stacked on SBA-15. The silanol aggregates on the silica surface were additionally examined with a Fourier transform infrared (FT-IR) spectrometer in the absorbance mode. Powder X-beam diffraction (XRD) was also carried out and the surface areas were ascertained utilizing the Brunauer-Emmet-Teller (BET) technique. The FT-IR spectra of SBA-15 and DN/SBA-15 demonstrated two wide peaks at around 810 and 1030 cm^{-1} which suggested crucial silicon-oxygen vibrations. After reaction with DN, a few new peaks arising from DN were observed, for example, C=C extending vibrations at around 1500 cm^{-1} and C–H out-of-plane disfigurement vibrations at around 750 cm^{-1}. The XRD patterns of SBA-15 and carbon/SBA-15 were observed to be of same nature which demonstrated that the long-run consistency of the silica system in SBA-15 was retained prior and after the carbonization procedure. Additionally, SEM of the initial and carbon-stacked SBA-15 (carbon/SBA-15) demonstrated no morphological change even after the carbon covering, which led to the inference that the carbon was consistently present on the internal surface of the SBA-15 mesopores. In conclusion, the BET analysis concluded that the mesopores were infinitesimally present even on carbon/SBA-15.

Yoshida *et al.* [34] carried out the SEM analysis of carbon-coated SiC fibers generated by various techniques like dip-coating, electrophoretic deposition, and vacuum infiltration. It was observed that the fiber was polycrystalline, and consisted of fine β-SiC particles. In the case of coating with electrophoretic deposition, flaky particles of graphite totally coated the surface of SiC fibers. However, in the case of the coatings by dip coating and vacuum filtration methods,

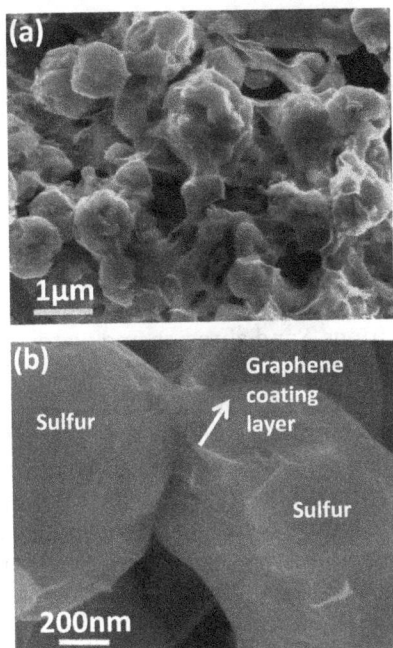

Figure 11.9 SEM images of graphene-sulfur composite at low (a) and high (b) magnifications. Reproduced from Reference 38 with permission from American Chemical Society.

only partial presence of graphite particles on SiC fibers was observed. Later, the analysis of formed composites was performed, especially the bending test and load-crosshead displacement plots of composites were analyzed at room temperature. Very similar load-crosshead curves were obtained for coating with electrophoretic deposition and vacuum filtration methods. Also, it was observed that as crosshead displacement increased after reaching a maximum valued of load, a decrease in the load, though gradually, was observed. However, in the case of coating using dip coating method, the load indicated a linear relationship with crosshead displacement and afterwards a sharp and sudden decrease. The authors concluded that the carbon coating was homogenous.

In the study by Lee and Lee [35], XRD technique was used for investigating the carbon-coated nano-Si dispersed oxides/graphite composite. The SiO-Al-Li_2O_2 powder mixture was milled for different durations, followed by XRD analysis. It was inferred that initially without milling, the pattern had a weak broad signal. Moreover, Si peaks appeared after 6 hours of milling with the disappearance of Li_2O_2 diffraction peaks. It was also seen that as the milling time was increased to 15 hours, there was an increment in the intensity of Si diffraction peaks. The peak of Al_2O_3 was also observed with negligible effect of residual SiO and Al on its electrochemical properties. Calculation of discharge capacity as a function of cycle number for varying milling times was performed for the composite electrode. On the whole, authors inferred that the carbon coating treatment on the composites greatly enhanced their cyclic ability along with minimized irreversible capacity.

Li and Zhang [36] used a wide variety of techniques to characterize the nanocomposites and the coatings. Sand abrasion and water jetting techniques were used for healing capability tests of POS/MWCNTs. In addition, FTIR, SEM, TEM, X-ray photoelectron spectroscopy (XPS) of samples was performed. TEM of nanocomposite exhibited no silica nanoparticles. Moreover, the POS/MWCNTs surface which was damaged because of the artificial sand abrasion and water jetting were healed with the help of toluene and curing. From the FTIR and XRD, it was inferred that the coatings were superhydrophobic, had large contact angles, ultralow sliding angles and were very stable. Furthermore, there was a minute change in CAs of coatings and low SAs were maintained after exposing to severe stability tests of sand abrasion.

Surface of the conducting coatings synthesized by Sevastyanov *et al.* [37] based on different carbon nanomaterials was investigated using various techniques to study composition and structure like X-ray phase analysis (XPA), TEM, SEM, AFM, etc. From SEM, a coating was observed to be deposited on the surface of CNFL sample, along with the presence of voids. However, for CNF-SnO_2, a different coating having polycrystalline tin oxide was observed on its surface which covered the earlier formed layer of carbon nanostructures on the coating surface. Moreover, XRD analysis indicated that a tin dioxide coating was present on all samples.

11.6.3 Applications

The carbon coatings synthesized by Nishihara *et al.* [33] on mesoporous silica find their applications commercially as materials like molecular sieves, adsorbents, etc. Due to the increased hydrophobicity and electrical conductivity, coated silica finds commercial use in solutions which are ionic and organic, and is generally used as supercapacitors electrodes. On the other hand, carbon coatings fabricated by Lee and Lee [35] on nano-silica proved to be a potential anode material for lithium ion batteries and for electronic and ionic conduction.

Hydrophobic coatings fabricated by Li and Zhang [36] are ultrastable, heal-able and are commercially feasible in various applications like separation of oil/water, self-cleaning materials, and controlled drug release. Furthermore, these POS/MWCNTs coatings can be applied onto various substrates like cellulose paper, PTFE, polymeric and aluminum plates along with polymer based textiles, etc. Also, the conductive coatings synthesized by Sevastyanov *et al.* [37] were useful for photovoltaic cells especially organic semiconductors and for the production of solar cells.

11.7 Miscellaneous Carbon Coatings

In the recent past, other carbon based compounds have been used to formulate the coatings like carbon xerogel, microporous carbon, carbides, etc. The below section summarizes the research studies using such materials for the generation of functional coatings.

11.7.1 Fabrication

Cushing and Goodenough *et al.* [39] carried out the carbon coating on the surface of α-NaFeO$_2$-structured LiMn$_{0.5}$Ni$_{0.5}$O$_2$ cathode materials. Carbon xerogel was used as the source of carbon, which was derived from resorcinol-formaldehyde (R-F) polymer. LiMn$_{0.5}$Ni$_{0.5}$O$_2$ particles were carbon-coated by combining as-prepared LiMn$_{0.5}$Ni$_{0.5}$O$_2$ and R-F in an 85:15 weight ratio, compacting the mixture into a 2.5 cm diameter pellet, and heating the pellet to 600 °C for 6 h in air. In a study by Chmiola *et al.* [40], etching of supercapacitor electrodes into conductive titanium carbide substrates was carried out which resulted into the monolithic carbon films. For this purpose, cutting of sintered TiC ceramic plates was carried out and the plates were later polished to reduce the thickness to around 300 nm so that resistance can be minimized before chlorination process starts on each face ranging in a time span of 15 sec to 5 mins. Finally, the obtained coatings had the thickness ranging from 2 to 200 mm. Hence, the technique successfully produced thin carbide derived carbon (CDC) films on bulk TiC ceramic plates.

A recent study by Chung and Manthiram [41] showcased the microporous carbon (MPC) coating on the polyethylene glycol (PEG) as a polysulfide trap. The MPC/PEG slurry was synthesized by mixing 80 wt% MPCs with 20% PEG in 3 mL of isopropyl alcohol (IPA). Later, slurry coating was performed on one side of Celgard separator using tape casting method via an automatic film applicator with blade (standard number 1), followed by drying. Roldan *et al.* [42]

formulated a hydrothermal carbon (HC) layer coating on graphite microfiber felts using one-pot synthesis method. For this purpose, piece of graphite felt was immersed into the carbohydrate solution in an autoclave vessel. Finally, after washing and drying with ethanol and water, the carbon felt got coated with hydrothermal carbon layer denoted by HC-CF. Later, using concentrated sulfuric acid, the treated felt was subjected to sulfonation and the treated sample was denoted as HO_3S-HC-CF.

Zhu *et al.* [43] coated the $Li_4Ti_5O_{12}$ (LTO/C) nanocomposite particles with carbon and proposed a model to illustrate the necessity of carbon content optimization in the composite material. For the coating purpose, varying amount of glucose in $TiOSO_4$ aqueous solution was loaded in an autoclave at 180 °C for 6 h. The resulting precipitate was filtered and washed, followed by drying to generate TiO_2/C precursor. It was later mixed with $LiOH.H_2O$, followed by calcination. In another recent study by Zhang *et al.* [44], carbon coated Fe_3O_4 nanospindles were synthesized by partial reduction of monodispersed hematite nanospindles. For the fabrication purpose, forced hydrolysis of ferric chloride solution along with addition of phosphate ions was carried out which led to the synthesis of hematite particles followed by the coating of carbon layers by pyrolysis of glucose under hydrothermal conditions. Later, drying and heating was performed under N_2 environment so as to achieve carbonization of the layers. Lastly, partial reduction of inner hematite spindles was carried out to generate magnetite, which resulted in final Fe_3O_4-C nanocomposites.

In another study, Wang *et al.* [45] derived interconnected carbon nanosheets from hemp and used for ultrafast supercapacitors with high energy. Figure 11.10 also demonstrates the morphological features of the interconnected partially graphitic carbon nanosheets.

11.7.2 Characterization and Testing

Cushing and Goodenough [39] characterized the products using XRD with a Cu-Kα radiation. Elemental analysis was studied using absorption spectrophotometer by utilizing the method of standard addition. Moreover, the carbon content of $LiMn_{0.5}Ni_{0.5}O2$-C (LMNO-C) was determined by TGA carried out in 30-800 °C range in oxygen atmosphere. XRD exhibited $LiMn_{0.5}Ni_{0.5}O_2$ (LMNO) as O_3 structure of $LiNiO_2$ and the presence of nickel and manganese ions was confirmed. Also, the elemental analysis of LMNO-C indicated the presence of 2.5% (w/w) of carbon. Moreover, SEM images of LMNO and LMNO-C confirmed the presence of small particles in 10-100 nm size range which were bound in the form of agglomerates. On the other hand, EDX carbon map of LMNO-C showed that carbon was uniformly distributed throughout the agglomerates. Also, the discharge capacities were measured by cycling the coated cathodes between 2.75 V and 4.25 V. The values of 87, 101 and 108 mAh/g were obtained corresponding to current densities of 1, 0.5 and 0.25 mA/cm^2.

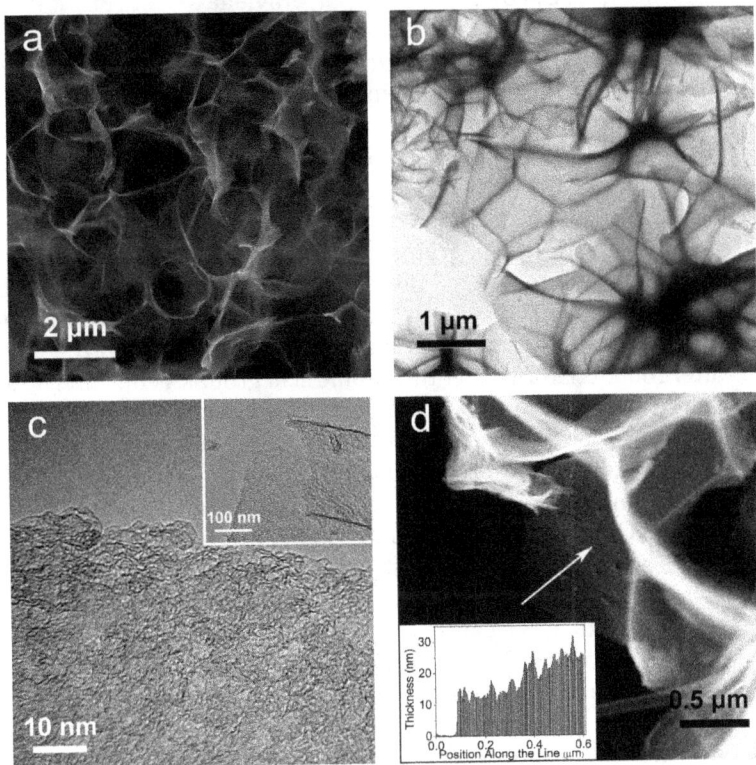

Figure 11.10 (a&b) SEM and TEM micrographs of the interconnected 2D structure; (c) high resolution TEM micrograph indicating the porous and partially ordered structure and (d) ADF TEM micrograph and EELS thickness profile (inset). Reproduced from Reference 45 with permission from American Chemical Society.

Chmiola *et al.* [40] carried out the SEM analysis of the carbide based carbon films on TiC which exhibited an excellent adherence on the surface of TiC along with uniform thickness and some microcracking due to the presence of tensile stress because of Ti removal. Raman spectroscopy further verified the findings of SEM analysis and it was inferred that an increment in the microstructural ordering at the surface of the thicker films was present. Furthermore, volumetric capacitance of TEABF$_4$ and H$_2$SO$_4$ was also calculated for different thickness of the films and it was observed to decrease with increased thickness of the coatings. Chung and Manthiram [41] carried out the SEM analysis of microporous carbon (MPC) coating on the polyethylene glycol (PEG). It was observed that MPCs formed clusters of porous nature with size in microns range and were attached to the separator. Moreover, the MPC nanoparticles had large surface area and high pore volume, along with many micropores. It was also concluded that polysulfides were intercepted by the microporous trapping sites of the MPC/PEG coating. BET analysis, on the other hand, indicated that the coatings were very effective as electrolytes were absorbed in the porous space. Lastly, electrochemical analysis of the Li-S cells was performed with the fabricated MPC/PEG coated separator.

Roldan *et al.* [42] characterized the coated layer of hydrothermal carbon on the graphite felt using SEM, IR-spectroscopy, XPS, etc. and studied its morphology and surface chemistry. Esterification technique was also utilized to test the coatings after functionalization with sulfonic acid groups. From SEM, it was observed that the thickness of the layer increased as the concentration of the glucose enhanced. It was also inferred that hydrothermal carbon synthesized by using HNO$_3$ treated carbon felt yielded higher carbon sphere than the one synthesized by carbon felt without HNO$_3$ treatment. The functional group analysis exhibited various bands after the coating with hydrothermal carbon with most prominent one at 1710 cm^{-1} of C=O and at 2900 and 3000-3700 cm^{-1} representing methyl/methylene and hydroxyl/carboxyl groups respectively [46-49].

Zhu *et al.* [43] studied the carbon coating by *in-situ* Raman analysis. Moreover, structural and morphological studies of the samples were performed by XRD and SEM. Content of carbon in the coatings was studied by elemental analyzer and TGA. Electrochemical impedance spectroscopy (EIS) was also studied at 10 mV AC Volts between 100 kHz to 100 mHz. XRD analysis of TiO$_2$ precursor and the LTO/C samples exhibited no carbon peaks because of the amorphous nature of carbon and its low content. The TEM analysis confirmed the results of XRD and it was further concluded that lower temperature of calcination favors formation of smaller nanoparticles with diameter varying from 70 to 220 nm. Moreover, Raman spectra showed a strong D band, thus, again validating the amorphous nature of carbon with multiple defects. A G band was also noticed corresponding to graphitized carbon.

Zhang *et al.* [44] investigated the properties of carbon coated Fe$_3$O$_4$ nanospindles using SEM, TEM, XRD and various electrochemical techniques. SEM images of nanospindles alone exhibited tiny crystals with a diameter of about 106 nm. Also, SEM images of carbon coated nanospindles showed uniform coating layers. On the other hand, SEM images of final Fe_3O_4-C nanocomposites were smoother with small void spaces. Furthermore, TEM images confirmed the presence of outer layer of carbon and inner magnetite core. To determine the pore structure of Fe_3O_4-C, BET analysis was carried out. The surface area was approximated to be 35.1 m^2 and the Barrett- Joyner-Halenda (BJH) pore size distribution indicated presence of mesopores in the composite, which led to the better cyclic performance for lithium batteries.

11.7.3 Applications

The carbon coating generated by Cushing and Goodenough [39] on the surface of $LiMn_{0.5}Ni_{0.5}O_2$ cathode using xerogel as carbon source is commercially useful in rechargeable lithium ion batteries. On the other hand, the carbon coatings fabricated by Chmiola *et al.* [40] using carbide are useful for micro-supercapacitors as these provide a framework for their integration into a variety of devices because of improved functionality and reduced complexity. The carbon coating led to reducing the abrupt failure and increasing the cycle life of the micro-batteries.

The microporous coating carried out by Chung and Manthiram [41] on the surface of separator having polysulfide trap is beneficial for lithium-sSulfur batteries due to high energy density and discharge capacity. The hydrothermal carbon coating carried out by Roldan *et al.* [42] on the graphite felts was applicable for acid catalyst applications. Also, the graphite felt's electrical conductivity was increased which makes its use more widespread for electrochemical devices. The macroporous structure of felt also made it commercially applicable in flow chemistry studies.

The carbon coating formulated by Zhu *et al.* [43] on the surface of $Li_4Ti_5O_{12}$ (LTO/C) nanocomposite particles paved way for high energy density and power batteries based on lithium ions which made them more efficient for electrochemical energy storage. In a very similar fashion, coatings fabricated by Zhang *et al.* [44] on Fe_3O_4 nanospindles served as a valuable anode material for lithium ion batteries because of enhanced performance and further paved the way for upcoming lithium ion batteries with large energy capacities.

11.8 Conclusions

Carbon based coatings/films have been synthesized in the recent years for various applications like reducing the effect of corrosion on the metallic surfaces, improving properties of lithium ion batteries, etc. Carbon has been used widely because of the multiple properties it exhibits like hybridization, existence in different forms like graphene, nanotubes, amorphous carbon, etc. Major advances have been achieved recently in the synthesis of carbon based films and coatings where different forms of carbon functionalization have been performed, which have further enhanced the commercial potential of these materials.

References

1. Broitman, E., Neidhardt, J., and Hultman, L. (2008) Fullerene-like carbon nitride: A new carbon-based tribological coating. In: *Tribology of Diamond-Like Carbon Films: Fundamentals and Applications*, Donnet, C., and Erdemir A., eds., Springer, USA, pp. 620-653.
2. Broitman, E., Czigany, Z., Greczynski, G., Bohlmark, J., Cremer, R., and Hultman, L. (2010) Industrial-scale deposition of highly adherent CNx films on steel substrates. *Surface and Coatings Technology*, **204**, 3349-3357.
3. Mittal, N., Deva, D., Kumar, R., and Sharma, A. (2015) Exceptionally robust and conductive superhydrophobic free-standing films of mesoporous carbon nanocapsule/polymer composite for multifunctional applications. *Carbon*, **93**, 492-501.
4. Wan, Y., Cui, X., and Wen, Z. (2011) Ordered mesoporous carbon coating on cordierite: Synthesis and application as an efficient adsorbent. *Journal of Hazardous Materials*, **198**, 216-223.
5. Pang, J., Li, X., Wang, D., Wu, Z., John, V. T., Yang, Z., and Lu, Y. (2004) Silica-templated continuous mesoporous carbon films by a spin-coating technique. *Advanced Materials*, **16**, 884-886.
6. Feng, D., Lv, Y., Wu, Z., Dou, Y., Han, L., Sun, Z., Xia, Y., Zheng, G., and Zhao, D. (2011) Free-standing mesoporous carbon thin films with highly ordered pore architectures for nanodevices. *Journal of the American Chemical Society*, **133**, 15148-15156.
7. Mittal, N., Kumar, R., Mishra, G., Deva, D., and Sharma, A. (2016) Mesoporous carbon nanocapsules based coatings with multifunctionalities. *Advanced Materials Interfaces*, **3**(10), 1500708.
8. Yang, Y., Yu, G., Cha, J. J., Wu, H., Vosgueritchian, M., Yao, Y., Bao, Z., and Cui, Y. (2011) Improving the performance of lithium-sulfur batteries by conductive polymer coating. *ACS Nano*, **5**, 9187-9193.
9. Yang, S. J., Nam, S., Kim, T., Im, J. H., Jung, H., Kang, J. H., Wi, S., Park, B., and Park, C. R. (2013) Preparation and exceptional lithium anodic performance of porous carbon-coated ZnO quantum dots derived from a metal-organic framework. *Journal of the American Chemical Society*, **135**(20), 7394-7397.
10. Zhao, X. Y., Tu, J. P., Lu, Y., Cai, J. B., Zhang, Y. J., Wang, X. L., and Gu, C. D. (2013) Graphene-coated mesoporous carbon/sulfur cathode with enhanced cycling stability. *Electrochimica Acta*, **113**, 1 256-262.

11. Ji, L., Guo, G., Sheng, H., Qin, S., Wang, B., Han, D., Li, T., Yang, D., and Dong, A. (2016) Free-standing, ordered mesoporous few-layer graphene framework films derived from nanocrystal superlattices self-assembled at the solid- or liquid-air interface. *Chemistry of Materials*, **28**, 3823-3830.

12. Talapin, D. V., Lee, J. S., Kovalenko, M. V., and Shevchenko, E. V. (2010) Prospects of colloidal nanocrystals for electronic and optoelectronic applications. *Chemical Reviews*, **110**, 389-458.

13. Zhao, D., Huo, Q., Feng, J., Chmelka, B. F., and Stucky, G. D. (2014) Nonionic triblock and star diblock copolymer and oligomeric surfactant syntheses of highly ordered, hydrothermally stable, mesoporous silica structures. *Journal of the American Chemical Society*, **136**, 10546-10546.

14. Noh, M., Kwon, Y., Lee, H., Cho, J., Kim, Y., and Kim, M. G. (2005) Amorphous carbon-coated tin anode material for lithium secondary battery. *Chemistry of Materials*, **17**, 1926-1929.

15. Ng, S. H., Wang, J., Wexler, D., Chew, S. Y., and Liu, H. K. (2007) Amorphous carbon-coated silicon nanocomposites: A low-temperature synthesis via spray pyrolysis and their application as high-capacity anodes for lithium-ion batteries. *Journal of Physical Chemistry C*, **111**, 11131-11138.

16. Kim, H.-S., Kong, M., Kim, K., Kim, I.-J., and Gu, H.-B. (2007) Effect of carbon coating on $LiNi_{1/3}Mn_{1/3}Co_{1/3}O_2$ cathode material for lithium secondary batteries. *Journal of Power Sources*, **171**, 917-921.

17. Alakoski, E., Kiuru, M., Tiainen, V.-M., and Anttila, A. (2003) Adhesion and quality test for tetrahedral amorphous carbon coating process. *Diamond and Related Materials*, **12**, 2115-2118.

18. Cao, Q., Zhang, H. P., Wang, G. J., Xia, Q., Wu, Y. P., and Wu, H. Q. (2007) A novel carbon-coated $LiCoO_2$ as cathode material for lithium ion battery. *Electrochemistry Communications*, **9**, 1228-1232.

19. Das, N. S., Banerjee, D., and Chattopadhyay, K. K. (2011) Enhancement of electron field emission by carbon coating on vertically aligned Si nanowires. *Applied Surface Science*, **257**, 9649-9653.

20. Show, Y. (2007) Electrically conductive amorphous carbon coating on metal bipolar plates for PEFC. *Surface and Coatings Technology*, **202**, 1252-1255.

21. Liu, H., Wang, G., Wang, J., and Wexler, D. (2008) Magnetite/carbon core-shell nanorods as anode materials for lithium-ion batteries. *Electrochemistry Communications*, **10**, 1879-1882.

22. Perez-Huerta, A., and Cusack, M. (2009) Optimizing electron backscatter diffraction of carbonate biominerals-resin type and carbon coating. *Microscopy and Microanalysis*, **15**, 197-203.

23. Kim, B. G., Choi, Y., Lee, J. W., Lee, Y. W., Sohn, D. S., and Kim, G. M. (2000) Multi-layer coating of silicon carbide and pyrolytic carbon on UO_2 pellets by a combustion reaction. *Journal of Nuclear Materials*, **281**, 163-170.

24. Lopez-Honorato, E., Chiritescu, C., Xiao, P., Cahill, D. G., Marsh, G., and Abram, T. J. (2008) Thermal conductivity mapping of pyrolytic carbon and silicon carbide coatings on simulated fuel particles by time-domain thermoreflectance. *Journal of Nuclear Materials*, **378**, 35-39.

25. Kim, J., Kim, B., Lee, J.-G., Cho, J., and Park, B. (2005) Direct carbon-black coating on $LiCoO_2$ cathode using surfactant for high-density Li-ion cell. *Journal of Power Sources*, **139**, 289-294.

26. Ponrouch, A., Goni, A. R., Sougrati, M. T., Ati, M., Tarascon, J.-M., Nava-Avendano, J., and Palacin, M. R. (2013) A new room temperature and solvent free carbon coating procedure for battery electrode materials. *Energy & Environmental Science*, **6**, 3363-3371.

27. Rodriguez-Mirasol, J., Cordero, T., Radovic, L. R., and Rodriguez, J. J. (1998) Structural and textural properties of pyrolytic carbon formed within a microporous zeolite template. *Chemistry of Materials*, **10**, 550-558.

28. Lin, Y., Gao, M. X., Zhu, D., Liu, Y. F., and Pan, H. G. (2008) Effects of carbon coating and iron phosphides on the electrochemical properties of $LiFePO_4/C$. *Journal of Power Sources*, **184**, 444-448.

29. Klebanoff, L. E., Clift, W. M., Malinowski, M. E., Steinhaus, C., Grunow, P., and Bajt, S. (2002) Radiation-induced protective carbon coating for extreme ultraviolet optics. *Journal of Vacuum Science & Technology*, **B20**, 696-703.

30. Wang, J., Yang, J., Tang, Y., Li, R., Liang, G., Sham, T.-K., and Sun, X. (2013) Surface aging at olivine $LiFePO_4$: a direct visual observation of iron dissolution and the protection role of nano-carbon coating. *Journal of Materials Chemistry*, **A1**, 1579-1586.

31. Nilsson, L., Andersen, M., Balog, R., Laegsgaard, E., Hofmaan, P., Basenbacher, F., Hammer, B., Stensgaard, I., and Hornekaer, L. (2012) Graphene coatings: Probing the limits of the one atom thick protection layer. *ACS Nano*, **6**(11), 10258-10266.

32. Prasai, D., Tuberquia, J. C., Harl, R. R., Jennings, G. K., and Bolotin, K. I. (2012) Graphene: Corrosion-inhibiting coating. *ACS Nano*, **6**(2), 1102-1108.

33. Nishihara, H., Fukura, Y., Inde, K., Tsuji, K., Takeuchi, M., and Kyotani, T. (2008) Carbon-coated mesoporous silica with hydrophobicity and electrical conductivity. *Carbon*, **46**, 48-53.

34. Yoshida, K., Matsukawa, K., Imai, M., and Yano, T. (2009) Formation of carbon coating on SiC fiber for two-dimensional SiCf/SiC composites by electrophoretic deposition. *Materials Science and Engineering, Part B*, **161**, 188-192.

35. Lee, H.-Y., and Lee, S.-M. (2004) Carbon-coated nano-Si dispersed oxides/graphite composites as anode material for lithium ion batteries. *Electrochemistry Communications*, **6**, 465-469.

36. Li, B., and Zhang, J. (2015) Polysiloxane/multiwalled carbon nanotubes nanocomposites and their applications as ultrastable, healable and superhydrophobic coatings. *Carbon*, **93**, 648-658.

37. Sevast'yanov, V. G., Kolesnikov, V. A., Desyatov, A. V., Kolesnikov, A. V. (2015) Conducting coatings based on carbon nanomaterials and SnO_2 on glass for photoconverters. *Glass and Ceramics*, **71**, 439-442.

38. Wang, H., Yang, Y., Liang, Y., Robinson, J. T., Li, Y., Jackson, A., Cui, Y., and Dail H. (2011) Graphene-wrapped sulfur particles as a rechargeable lithium-sulfur battery cathode material with high capacity and cycling stability. *Nano Letters*, **11**(7), 2644-2647.

39. Cushing, B. L., and Goodenough, J. B. (2002) Influence of carbon coating on the performance of a $LiMn_{0.5}Ni_{0.5}O_2$ cathode. *Solid State Sciences*, **4**, 1487-1493.

40. Chmiola, J., Largeot, C., Taberna, P. L., Simon, P., and Gogotsi, Y. (2010) Monolithic carbide-derived carbon films for micro-supercapacitors. *Science*, **328**, 480-483.

41. Chung, S.-H., and Manthiram, A. (2014) A polyethylene glycol-supported microporous carbon coating as a polysulfide trap for utilizing pure sulfur cathodes in lithium-sulfur batteries. *Advanced Materials*, **26**, 7352-7357.

42. Roldan, L., Santos, I., Armenise, S., Fraile, J. M., and Garcia-Bordeje, E. (2012) The formation of a hydrothermal carbon coating on graphite microfiber felts for using as structured acid catalyst. *Carbon*, **50**, 1363-1372.

43. Zhu, Z., Cheng, F., and Chen, J. (2013) Investigation of effects of carbon coating on the electrochemical performance of $Li_4Ti_5O_{12}/C$ nanocomposites. *Journal of Materials Chemistry*, **A1**, 9484-9490.

44. Zhang, W.-M., Wu, X.-L., Hu, J.-S., Guo Y.-G., and Wan, L.-J. (2008) Carbon coated Fe_3O_4 nanospindles as a superior anode material for lithium-ion batteries. *Advanced Functional Materials*, **18**, 3941-3946.

45. Wang, H., Xu, Z., Kohandehghan, A., Li, Z., Cui, K., Tan, X., Stephenson, T. J., King'ondu, C. K., Holt, C. M. B., Olsen, B. C., Tak, J. K., Harfield, D., Anyia, A. O., and Mitlin, D. (2013) Interconnected carbon nanosheets derived from hemp for ultrafast supercapacitors with high energy. *ACS Nano*, **7**(6), 5131-5141.

46. Cano-Serrano, E., Blanco-Brieva, G., Campos-Martin, J. M., and Fierro, J. L. G. (2003) Acid-functionalized amorphous silica by chemical grafting-quantitative oxidation of thiol groups. *Langmuir*, **19**, 7621-7627.

47. Figueiredo, J. L., Pereira, M. F. R., Freitas, M. M. A., and Orfao, J. J. M. (1999) Modification of the surface chemistry of activated carbons. *Carbon*, **37**, 1379-1389.

48. Zawadzki, J., Wisniewski, M., Weber, J., Heintz, O., and Azambre, B. (2001) IR study of adsorption and decomposition of propan-2-ol on carbon and carbon-supported catalysts. *Carbon*, **39**, 187-192.

49. Zielke, U., Huttinger, K. J., and Hoffman, W. P. (1996) Surface-oxidized carbon fibers: I. Surface structure and chemistry. *Carbon*, **34**, 983-998.

Performance of Various Adsorbents towards Diverse Gases: A Comparative Study

12.1 Introduction

The ever-increasing consumption of fossil fuels and the resulting combustion products create disastrous impacts on the environments along with the depletion of energy sources. As the combustion of fossil fuels emerges many pollutant gases, namely carbon dioxide (CO_2), methane (CH_4), nitrogen (N_2), nitrous oxide (N_2O), sulfur dioxide (SO_2), etc., into the atmosphere, it directly enhances the accumulation of greenhouse gases responsible for universal climatic alterations and, thereby, causes global warming [1-4]. These growing issues can be rectified using different routes such as enhancement in energy competence, development of other alternative energy sources and making use of pollutant gas capture and storage [5-7].

Among the various techniques for capture and storage of pollutant gases, methods based on adsorption technologies have received considerable attention in terms of efficiency and cost for reducing the concentration of pollutants form the environment. Adsorption is considered as a surface phenomenon caused by the existence of cohesive forces such as van der Waals forces and hydrogen bonding at the interfaces of adsorbent and gas molecules (adsorbate) [8,9]. This phenomenon can be generally categorized into physical adsorption (physisorption) and chemical adsorption (chemisorption) based on the interactions between adsorbent and adsorbate. Physisorption is characterized by the presence of weak interactions such as van der Waals or electrostatic forces while chemisorption is caused by the occupancy of strong chemical bonds [10,11]. In addition, the capturing of exhaust gases by adsorption can be executed using many routes in the process plants. For example, capturing of CO_2, a major constituent of the exhaust gases generated during the combustion of fossil fuels or biomass, can be achieved through any of the following routes such as post-combustion, pre-combustion and oxyfuel combustion. The post-combustion technique is commonly employed for the capture of exhaust CO_2 from gaseous fuels, chemicals, fertilizers and power production, whereas pre-combustion is applicable for mainly power stations, etc.

Anish M. Varghese and Vikas Mittal, The Petroleum Institute (part of Khalifa University of Science and Technology), Abu Dhabi, UAE*
Current address: Bletchington, Wellington County, Australia

Oxyfuel (oxy-firing) combustion is generally implemented in coal fired power plants to yield economic CO_2 capturing [6,7,12-15]

The selection of a suitable adsorbent is the critical parameter which decides the overall performance of gas capture and storage processes. It should follow some requirements including great extent of capability to adsorb gases, superior gas selectivity, swift kinetics, easy regeneration, consistency for substantial adsorption-desorption stages, stability towards moisture and impurities, economical operation and so on [13,16]. Basically, the adsorbent is required to retain its effectiveness during periodic operations consisting of alternative steps of adsorption and desorption, where the desorption can be brought about by means of temperature, pressure or vacuum swing [17]. On the basis of chemical constituents, structural features and operation mechanisms, gas adsorbents may be divided into different categories. Specifically, various kinds of adsorbents have been introduced for the purpose of adsorption of diverse pollutant gas molecules including silica [18], clay [19], activated carbon [20], carbon fibers [21], graphene [22]/graphite oxide [23], nanotubes [24,25], zeolites [26], ion-exchange resins [27], polymers [28], alumina [29], metal oxides [30], metal organic frameworks [31], etc. For instance, Figure 12.1 demonstrates the adsorption isotherms of CH_4 and CO_2 on activated

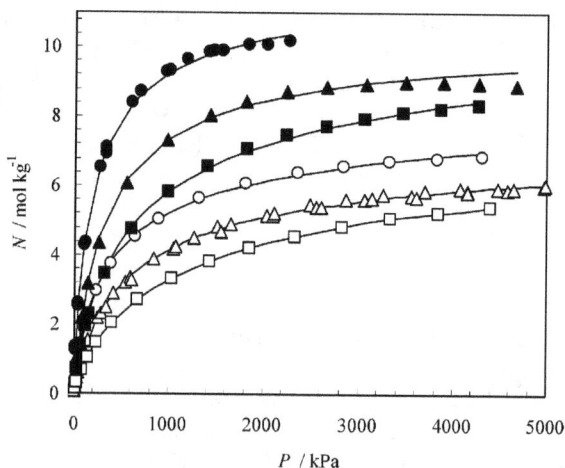

Figure 12.1 Adsorption isotherms of CH_4 and CO_2 on activated carbon. O: methane at 273 K; △: methane at 298 K; □: methane at 323 K; ●: carbon dioxide at 273 K; ▲: carbon dioxide at 298 K; ■: carbon dioxide at 323 K. Reproduced from Reference 20 with permission from American Chemical Society.

carbon [20]. The aim of this work is to overview the performance of various adsorbents other than metal oxides and metal organic frameworks towards the adsorption of gas molecules in the respect of environment and energy applications.

12.2 Silica Based Adsorbents

Silica based materials have received considerable attention as adsorbents for effective capture or removal of greenhouse and other pollutant gases because of their high specific surface area, highly arranged pore structures, ordered distribution of pore diameter, presence of amorphous pore walls, etc. [32].

12.2.1 Adsorption of CO_2

Belmabkhout *et al.* [33] reported the superior volumetric capacity of MCM-41 silica adsorbent for CO_2 storage at ambient temperature and elevated pressure. Also, the adsorption of CO_2 on MCM-41silica was fast and absolutely reversible [33]. In another study, the authors observed the favorable adsorption of CO_2 on MCM-41 silica adsorbent from the binary blend of CO_2 with individual gases N_2, H_2 and CH_4 [34]. Tsai *et al.* [35] accomplished enhancement in the flux and selectivity of CO_2 through the use of dual-layer of microporous silica adsorbents [35]. A study by Himeno *et al.* [36] also observed beneficial adsorption selectivity of CO_2 from a binary mixture of CO_2 and CH_4 on a series of silica Deca-dodecasil 3R zeolites [36]. Bellussi *et al.* [37] studied the bulk adsorption behavior of CO_2 contained in sour natural gas on silica-alumina adsorbent. The reported adsorbent performed well in terms of both specific and working capacity and selectivity [37]. Xomeritakis *et al.* [38] developed nickel doped sol-gel silica adsorbent to enrich the adsorptivity and selectivity of CO_2 in the flue gas. Also, the nickel doping of silica adsorbent was useful to reduce the densification [38].

Leal *et al.* [39] reported chemisorption of CO_2 on amine-surface bonded silica gel at ambient temperature and observed a capability to hold 10 STP cm^3 of dry CO_2 per g. Also, desorption of the stored CO_2 was observed during heating at 100 °C [39]. Instead of post-synthetic grafting of amine moieties on mesoporous silica, a study by Kim *et al.* [40] reported the development of amine modified mesoporous silica adsorbent through anionic surfactant-mediated synthesis. Better adsorption of CO_2 as well as greater stability upon continues cycles of adsorption and desorption was achieved. Yue *et al.* [41] achieved improvement in CO_2 adsorption on amine functionalized silica by the introduction of hydroxyl group. The generation of comparatively thermally weak intermediate at the interface of CO_2 and amine led to simple desorption of CO_2 molecules [41].

Knowles *et al.* [42] reported the existence of notable reversible adsorptivity of CO_2 on aminopropyl-functionalized mesoporous silica even in the influence of water [42]. Hiyoshi *et al.* [43] also reported the application of aminosilane functionalized SBA-15 silica adsorbent for achieving the intensified adsorption capacity of CO_2. The strong adsorption of CO_2 occurred by the generation of alkylammonium carbamate even in the presence of water molecules. Also, the adsorptivity of CO_2 increased with increase in the amine surface density on the adsorbent [43]. In another study, Sakamoto *et al.* [44] accomplished permselectivity of CO_2 from CO_2/N_2 binary mixture on mesoporous silica adsorbent through the grafting of amino-silane [44]. Knofel *et al.* [45] developed diamine modified SBA-16 silica adsorbent with the help of post-synthesis grafting. The positive interactions of CO_2 with the surface of modified silica adsorbent at low pressures caused the adsorbent to exhibit higher enthalpy values [45].

Liu *et al.* [46] modified the surface of mesoporous silica, SBA-15 using triethanolamine with an aim to enrich the selectivity of CO_2 in a mixture containing CO_2 and CH_4. Improvement in CO_2 selectivity was achieved and the modified silica based adsorbent exhibited greater stability during cyclic operation [46]. In another study, Liu *et al.* [47] prepared highly ordered SBA-15 mesoporous silica adsorbent possessing versatile textural characteristics and observed beneficial CO_2 adsorption [47]. Zelenak *et al.* [48] evaluated the influence of amine functionalization of SBA-12 mesoporous silica towards the adsorption of CO_2. Amine functionalization using 3-aminopropyl and 3-(methylamino)propyl resulted in greater capacity of CO_2 adsorption while 3-(phenylamino)propyl presented rapid desorption characteristics [48]. Another study by Jang *et al.* [49] reported the functionalization of mesoporous silica using amine comprising polysilsesquioxane, which resulted in improved permeability and selectivity of CO_2 from the mixture of gases [49]. Hicks *et al.* [50] observed substantial CO_2 adsorptivity on hyperbranched aminosilica adsorbent along with consistency upon cyclic operations [50].

Song *et al.* [51] reported the use of chemically functionalized silica adsorbents for the purpose of CO_2 adsorption. Chemical modification of silica adsorbent was carried out by developing base moieties on the surface through either treating with calcium acetate or synthesis in an alcoholic solution comprising of TEOS and calcium acetate. Improved adsorptivity of CO_2 gas on the resultant adsorbent was observed through chemisorption. The adsorptivity of

CO_2 was enhanced significantly by about 275% at 100 °C and 312% at 250 °C [51].

The impact of polyethyleneimine (PEI) impregnation on a series of silica adsorbents such as MCM-41, MCM-48, SBA-15, SBA-16 and KIT-6 was investigated by Son *et al.* [52]. The appearance of PEI in all silica adsorbents enriched the CO_2 adsorptivity along with stabilized adsorption-desorption reversibility with a significant recovery of more than 99%. Among all studied silica adsorbents, modified KIT-6 silica based adsorbent generated optimal CO_2 adsorption and stability during cyclic operation. Goeppert *et al.* [53] revealed the feasibility of PEI impregnated fumed silica based adsorbent for effective CO_2 capture from the atmospheric air [53]. In another study, Xu *et al.* [54] reported about 24 folds improvement in CO_2 adsorption capacity of MCM-41 silica adsorbent through PEI modification [54]. The authors also suggested the effect of temperature on the absorption process (Figure 12.2), which differed due to

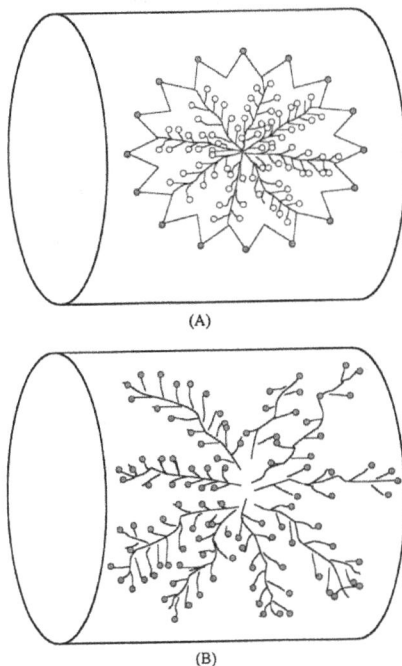

(A)

(B)

Figure 12.2 Schematic of PEI status in MCM-41 at (A) low temperature, and (B) high temperature. (●) Active CO_2 adsorption sites; (O) hidden CO_2 adsorption sites. Reproduced from Reference 54 with permission from American Chemical Society.

the availability of adsorption sites as a function of temperature. In order to further enrich the performance of PEI impregnated silica adsorbent, Rezaei *et al.* [55] introduced aminosilane functionalization, which was useful to improve adsorption of CO_2. Also, PEI impregnated silica adsorbents with and without aminosilane functionalization exhibited decreasing tendency of CO_2 adsorptivity with increase in temperature [55]. A study by Kim *et al.* [18] reported superior rate of CO_2 adsorption on monomeric aminopropyl functionalized MCM-48 silica adsorbent compared to polymeric aminopropyl functionalized one due to the easy approachability of amine modified adsorption spots [18].

Kusakabe *et al.* [56] observed greater permeance of CO_2 on silica/polyimide composite membrane comprising of 60 wt% of silica compared to pure polyimide membrane. The permselectivity of 30 at 30 °C and 13 at 100 °C was observed [56]. In another study, Zornoza *et al.* [57] developed mixed matrix adsorbent comprising of mesoporous silica spheres in respective polyimide and polysulfone matrices. The occupancy of silica on the polymer matrices led to the enrichment of the permeability and selectivity of CO_2 on the adsorbents, which was attributed to the increased interfacial bonding and the mesoporous nature of silica [57]. Another study by Kim *et al.* [58] also reported the existence of advanced permeability and permselectivity of CO_2 over silica/poly(amide-6-b-ethylene oxide) (PEBAX) hybrid owing to the appreciable interaction of CO_2 molecules with SiO_2 groups along with the presence of extra active moieties and organic/inorganic interphase [58].

12.2.2 Permselectivity towards H_2

A study by Morooka *et al.* [59] obtained H_2 permeance of 10^{-8} mol.m^{-2}s^{-1}Pa^{-1} on silica membrane at 600 °C and which was developed through the thermal decomposition of tetraethylorthosilicate at 600-650 °C [59]. In another study, Sea *et al.* [60] reported a permeance of 3×10^{-8} mol.m^{-2}s^{-1}Pa^{-1} for H_2 on silica membranes generated by chemical vapor deposition. Also, the existence of H_2/N_2 selectivity of 7.6 was noted at a temperature of 400 °C [60]. A study by Vos *et al.* [61] also reported silica adsorbent possessing reduced amount of defects and enhanced characteristics using sol-gel method. The existence of much better H_2 permeance of 2×10^{-6} mol m^{-2} s^{-1} Pa^{-1} and permselectivity of more than 700 was noted [61]. Boffa *et al.* [62] reported the existence of H_2 permeance of 6×10^{-7} mol.m^{-2}s^{-1}Pa^{-1} at 473 K on templated mesoporous silica adsorbent [62]. Zornoza *et al.* [63] made use of mixed matrix adsorbents consisting of hollow silicalite spheres in respective polysulfone and polyimide matrices for H_2 adsorption. Adsorbents with 8 wt% of silicalite spheres exhibited maximal H_2 permeability of 38.4 Barrer and H_2/CH_4 selectivity of 180 [63].

12.2.3 Adsorption of H_2S

To remove H_2S from natural gas, Huang *et al.* [64] made use of amine functionalized silica xerogel and MCM-48 silica adsorbents. Both adsorbents were observed to be beneficial with respect to adsorption capacity as well as rate [64] (Figure 12.3). Belmabkhout *et al.* [65] also reported effective removal of H_2S from natural gas and biogas through the use of triamine modified mesoporous silica adsorbent. Better adsorption capacity and selectivity of H_2S on the resultant adsorbent was observed even under the influence of moisture (Figure 12.4). Also, the adsorbent exhibited greater regenerability [65].

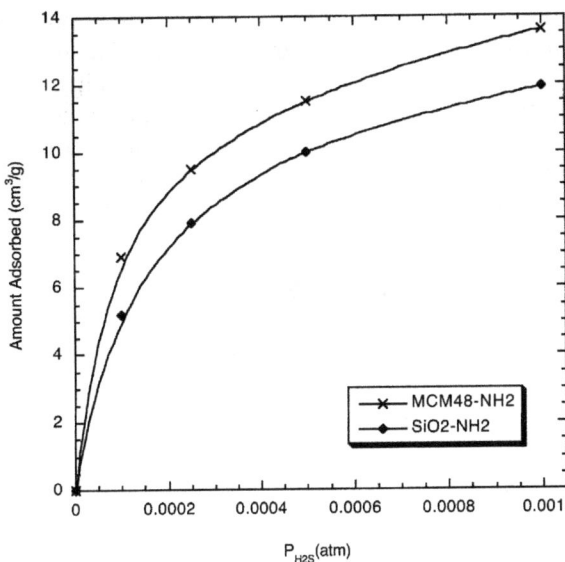

Figure 12.3 Adsorption isotherms of H_2S on amine-modified silica materials at room temperature. Reproduced from Reference 64 with permission from American Chemical Society.

Wang *et al.* [66] examined the behavior of ZnO functionalized aluminum containing mesoporous silica towards the adsorption of H_2S. The resultant adsorbent comprising of 2.1 wt% of ZnO exhibited optimal adsorptivity and therein micropores and mesopores served as energetic adsorption areas for H_2S. The presence of ZnO in the adsorbent composition encouraged the growth of pore structure along with the desulphurization features, which consequently increased the adsorptivity of H_2S [66]. In another study, Wang *et al.* [67] utilized ZnO nanoparticles comprising SBA-15 mesoporous silica adsorbents for H_2S removal and the resultant system comprising of 3.04 wt% ZnO possessed high adsorptivity [67]. Adsorbent based on PEI impregnated SBA-15 mesoporous silica also exhibited higher H_2S capture capacity even under the influence of moisture along with the adequate stability and regenerability [68].

12.2.4 Permselectivity of N_2

Vos *et al.* [69] prepared hydrophobic silica based adsorbent by the application of methyl-tri-ethoxy-silane onto silica sol. The resultant adsorbent presented N_2 permeance in the order of 4×10^{-7} mol.m^{-2}s^{-1}Pa^{-1}. The greater hydrophobicity of the modified silica adsorbent offered good stability even in the humid environments [69]. Sitter *et al.* [70] observed improvement in permeability and reduction in selectivity of N_2 with increase in concentration of silica in the silica/poly(1-trimethylsilyl-1-propyne) nanocomposite based adsorbent [70]. In another study, the authors achieved the stability in N_2 selectivity through

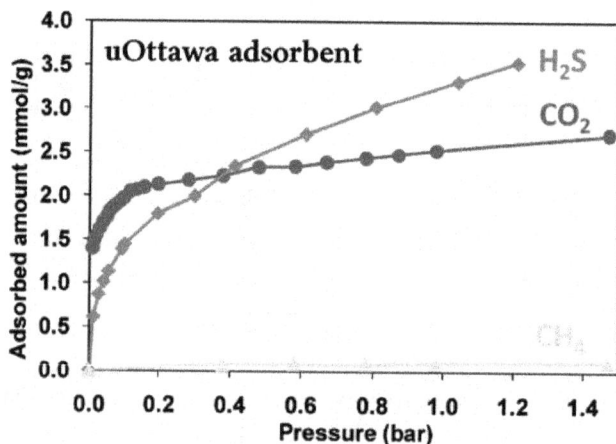

Figure 12.4 Adsorption isotherms of H_2S (and other gases) on the modified adsorbent. Reproduced from Reference 65 with permission from American Chemical Society.

the use of silica/poly(4-methyl-2-pentyne) nanocomposite adsorbent [71].

12.2.5 Adsorption of other gases

Newalkar *et al.* [72] made use of adsorbent based on SBA-15 mesoporous silica to investigate the adsorptivity of methane, ethane, ethylene, acetylene, propane and propylene. Among the studied light hydrocarbons, ethylene and propylene exhibited greater adsorption capacity and acetylene possessed strong adsorption on adsorbent of SBA-15 mesoporous silica [72]. Montanari *et al.* [73] examined the adsorption behavior of silica adsorbents towards hexamethylcyclotrisiloxane (HMCTS), a usual impurity presented in the biogases. Formation of strong hydrogen bond of HMCTS with surface hydroxyl group of silica led to enhanced adsorptivity [73].

12.3 Clay Based Adsorbents

Clay based materials have been used to develop gas adsorbents possessing high adsorption capacities owing to their substantial specific surface area, porous volume, resistance to chemicals, mechanical stability, versatile surface as well as structural features, etc. [74].

12.3.1 Adsorption of H_2S

Stepova *et al.* [75] studied the adsorption of H_2S on adsorbent based on bentonite. The authors also made use of iron and copper chloride functionalization of bentonite to enhance the adsorptivity. Bentonite modified with iron presented pronounced increase in H_2S adsorptivity whereas copper chloride modified bentonite could hold H_2S for long term. Also, the generation of sulfides on the functionalized bentonite surface through the reaction of metal hydroxide with H_2S was observed. The adsorbent could be regenerated when exposed to open air [75]. To accomplish H_2S adsorption, Thanh *et al.* [76] carried out functionalization of montmorillonite (MMT) using iron using different routes such as interchanging of sodium in the sodium rich MMT with iron, incorporation of iron oxocations and iron doping of aluminum-pillared MMTs. Among various iron modifications, iron doping exhibited maximal adsorptivity of H_2S on MMT whereas iron oxocations resulted in lower adsorption capacity. The difference in adsorption behavior of various iron modifications of MMT resulted from the comprehensive effect of material features, dispersion and pore volume [76].

12.3.2 Adsorption of Light Hydrocarbon Gases

Ji *et al.* [77] studied methane adsorption on a series of clay adsorbents, specifically montmorillonite, kaolinite, illite, chlorite and interstratified illite/smectite. Successful methane adsorption on all clay adsorbents was reported *via* physisorption. Adsorbents based on MMT and interstratified illite/smectite clay possessed more volume of micro- and mesopores, which consequently exhibited greater methane adsorption capacity. The methane adsorptivity of various clay adsorbents was noted in the order of MMT > interstratified illite/smectite > kaolinite > chlorite > illite [77]. Pires *et al.* [78] reported the substantial selectivity of methane from CO_2 and ethane on porous smectite clay adsorbent fabricated through gallery templated approach with the help of quaternary ammonium cation and neutral amines [78].

The impact of vapor pressure on the adsorption behavior of C1–C6 hydrocarbon gases on MMT and kaolinite based adsorbents was examined by Cheng *et al.* [79]. An interesting increase was observed in the adsorbed gas concentration with increase in partial pressure of the hydrocarbon gases. Adsorption of methane was higher in kaolinite when compared to MMT, whereas the overall gas adsorption was lower [79]. Choudary *et al.* [80] made use of Ag^+ modified clay based alkene-selective adsorbent to study the adsorption responses of light hydrocarbon gases like ethylene (C_2H_4), ethane (C_2H_6), propylene (C_3H_6) and propane (C_3H_8). Alkenes exhibited greater heats of adsorption whereas alkanes had superior adsorption selectivity on the adsorbent [80]. A study by Pires *et al.* [74] reported beneficial separation of C_2H_6/CH_4 binary mixture on adsorbent based on porous clay heterostructures (PCH) comprising of higher amount of phenyltriethoxysilane in smectite clay.

12.3.3 Adsorption of Volatile Organic Compounds (VOCs)

Zaitan *et al.* [81] revealed the suitability of bentonite based adsorbent to capture volatile organic compound (VOC), namely *o*-xylene. Bentonite exhibited an adsorptivity of 1042 µmol/g for xylene at 300 K and also presented an ability to regenerate about 90% of adsorbed xylene. The authors concluded the use of the adsorbent for effective as well as economic removal of VOCs [81]. In another study, Qu *et al.* [82] developed PCH *via* functionalization of bentonite using cetyltrimethylammonium bromide and dodecylamine to achieve substantial adsorption of VOCs such as acetone, toluene, ethylbenzene, *o*-xylene, *m*-xylene and *p*-xylene. Among the various VOCs, acetone (aliphatic hydrocarbon) presented higher adsorptivity on PCH than aromatic hydrocarbons due to their differences in HOMO energy values [82]. In another study, Zuo *et al.* [83] made comparison of VOCs adsorption behavior on Na-MMT, Al and AlCe pillared Na-MMT adsorbents. Al pillaring of Na-MMT adsorbent led to an increase in the surface area and pore volume, which correspondingly increased the adsorptivity of VOCs [83]. A study by Pires and co-workers [78] reported considerable adsorption of VOCs on adsorbent based on porous smectite clay prepared by gallery templated approach using quaternary ammonium cation and neutral amines . In another study, Nunes *et al.* [84] developed PCH *via* chemical modification of natural clay by bis(triethoxysilyl)benzene and tetraethyl orthosilicate to acquire VOCs adsorption. Resultant adsorbent exhibited effective adsorption of VOCs due to the hydrophobic nature [84].

12.3.4 Adsorption of CO_2

Pires *et al.* [74] examined the adsorption behavior of CO_2 from binary mixtures of CO_2/CH_4 and CO_2/C_2H_6 on PCH developed by impregnating tetraethoxysilane and phenyltriethoxysilane individually in smectite clay. High selectivity and separation of CO_2/CH_4 was observed for PCH adsorbents consisting of tetraethoxysilane and lower concentration of phenyltriethoxysilane [74]. In another study, Pires *et al.* [85] reported optimal adsorption of CO_2 on specific Al_2O_3 and ZrO_2 pillared MMT clay adsorbents compared to other gases such as C_2H_6, CH_4 and N_2. Clay with Al_2O_3 pillaring exhibited greater surface area whereas ZrO_2 pillaring exhibited better separation feasibility.

12.4 Activated Carbon Based Adsorbents

Activated carbons are one of the widely accepted adsorbents for large extent of gas adsorption purposes due to their interactions with the gas molecules resulting from the well-ordered pore structure, along with the tunable surface chemistry [86].

12.4.1 Adsorption of CO_2

A study by Presser *et al.* [87] observed the existence of ultra-high CO_2 adsorptivity of about 7.1mol/kg on carbide derived activated carbon adsorbent at 0 °C and ambient pressure. Also, the adsorbent with pore size smaller than 0.8 nm presented greater CO_2 adsorption feasibility than the larger pores [87]. Valer *et al.* [88] made use of anthracites derived activated carbon for the purpose of CO_2 capture and observed a storage capacity of 65.7 mg-CO_2/g-adsorbent for carbon activated at 800 °C for 2 h. Further improvement in the adsorption features of activated carbon was achieved through PEI infusion and high temperature ammonia treatment [88]. In another study, Goel *et al.* [89] reported optimal CO_2 adsorption capacity for activated mesoporous carbon derived at 700 °C from melamine formaldehyde resin [89]. Siriwardane *et al.* [90] also observed the generation of greater CO_2 adsorptivity on activated carbon compared to molecular sieves at pressure greater than 25 psi.

In another study, Drage *et al.* [91] reported the use of activated carbon based adsorbent in gasification to adsorb CO_2 and CO_2 adsorptivity of about 12 mmol/g under a pressure of 4 MPa was reported. Ganesan *et al.* [92] reported the derivation of activated carbon from graphene exhibiting enormous specific area of about 3240 m^2/g and substantial porous nature. The adsorbent presented high adsorptivity for CO_2 along with swift kinetics. A study by Sui *et al.* [93] observed greater adsorption of CO_2 over carbon based adsorbent derived from graphene aerogels *via* steam activation.

Zhang *et al.* [94] developed KOH activated carbon based adsorbent from polyaniline possessing comparatively smaller micropores favorable for superior CO_2 capture. The resultant adsorbent exhibited adsorption capacities of about 1.86 mmol/g at 75 °C and 1 bar and 1.39 mmol/g at 25 °C and 0.15 bar [94]. Adsorbent based on metal-organic coordination polymers derived activated carbon with a surface area of 2368 m^2/g also exhibited suitability for CO_2 capture with a storage capability of 2.9 mmol/g at ambient pressure and 300 K [95]. Hu *et al.* [96] prepared activated carbon with nanopores from petroleum coke and observed an exceptional adsorptivity of about 15.2 wt% at 1 bar for the material activated at 700 °C [96]. The feasibility of activated carbon sourced from pine nut shell for CO_2 capture was revealed by Deng *et al.* [97]. Alberto *et al.* observed exceptional feasibility of activated carbon molecular sieves to adsorb significant amount of CO_2 at ambient temperature [98]. In another study, Daud *et al.* [99] obtained selectivity of 16 for CO_2 from CH_4 by the amine functionalization of activated carbon molecular sieves [99].

Wickramaratne *et al.* [100] reported the application of activated carbon spheres derived from the carbonization of phenolic resin as adsorbent for CO_2 capture (Figure 12.5). The resultant adsorbent exhibited significantly superior adsorption capacities of 4.55 mmol/g at 25 °C and 8.05 mmol/g at 0 °C [100]. In another study, the authors concluded the CO_2 adsorption capacity of 8.9 mmol/g for phenolic-resin derived activated carbon based adsorbent with a substantial surface area of 2400 m^2/g [101]. Another study by Sevilla *et al.* [102] observed insignificant effect of N-doping of activated carbon microspheres towards the adsorption of CO_2 (Figure 12.6).

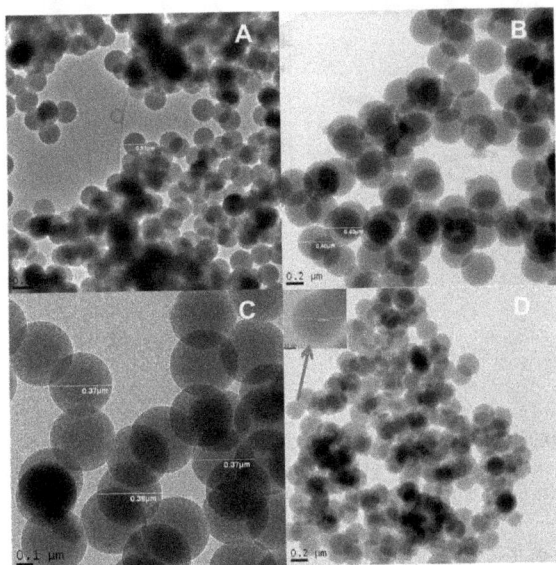

Figure 12.5 TEM images of the activated carbon spheres. Reproduced from Reference 100 with permission from American Chemical Society.

Loser *et al.* [103] reported the use of activated carbon monoliths based adsorbent for CO_2 capture. Volumetric adsorption capacity of 440 g/l was attained in the study. In another study, Ribeiro *et al.* [104] obtained the adsorption capacity of the activated carbon honeycomb monolith as CO_2 > CH_4 > N_2. In another study, Nandi *et al.* [105] accomplished CO_2 adsorption of 5.14 mmol/g at room temperature and pressure by the use of adsorbent based on nitrogen doped activated carbon monoliths derived from the carbonization of polyacrylonitrile. Zhong *et al.* [106] reported the development of polyacrylonitrile based block copolymer templated nitrogen supplemented activated carbon as a better adsorbent for selective CO_2 capture [106].

Xing *et al.* [107] obtained substantial CO_2 adsorption capacity of activated carbon through nitrogen doping. Nitrogen doping of activated carbon was useful to enhance the interactions with CO_2 *via* hydrogen bonding. In another study, Zhao *et al.* [108] made use of activated carbon with high amount (13 wt%) of nitrogen doping to enrich the performance. As a result, the adsorbent system presented an improved adsorption capacity of 10.53 mmol/g at 25 °C and 8 bar. Further, the large amount of nitrogen led to the enhancement of the adsorption selectivity of CO_2/N_2 to about 42 [108]. Saleh *et al.* [109] reported

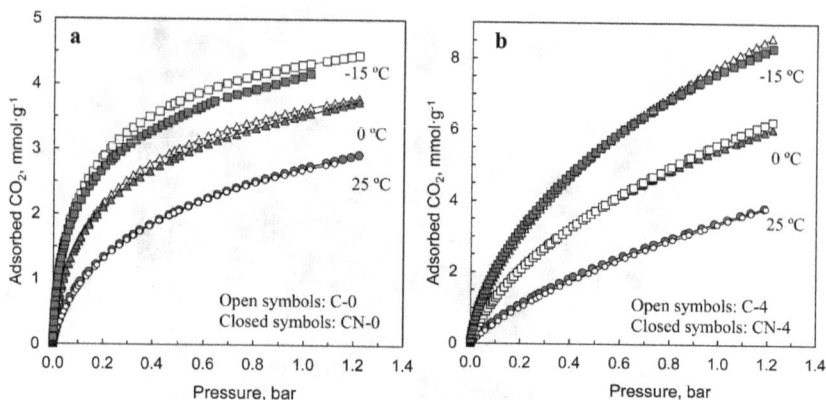

Figure 12.6 Comparison of CO_2 isotherms at several temperatures for (a) non-activated and (b) activated carbons (CN indicates carbon with N-doping). Reproduced from Reference 102 with permission from American Chemical Society.

the generation of nitrogen impregnated activated microporous carbon based adsorbents from polyindole with outstanding CO_2 adsorption selectivity and stability [109]. A study by Sevilla *et al.* [110] reported eminently porous nitrogen doped activated carbon based adsorbent from hydrothermal carbons of algae and glucose. The surface area in the range of 1300–2400 m^2/g and pore volume of about 1.2 cm^3/g for resultant adsorbent was reported. The adsorbent exhibited an adsorption capacity of about 7.4 mmol/g at 0 °C and 1 bar. Zhao *et al.* [111] introduced extra-framework cations onto nitrogen enriched microporous activated carbon with an aim to widen adsorption characteristics. The modification resulted in the improvement in adsorptivity and selectivity of CO_2 on nitrogen doped activated carbon, as shown in Figure 12.7.

The formation of nitrogen impregnated microporous activated carbon from chitosan with the help of K_2CO_3 was also reported by Fan *et al.* [112] to achieve the efficient adsorption of CO_2. An adequate CO_2 adsorption capacity of 3.86 mmol/g was observed at 1 atm and 25 °C for activated carbon prepared at 635 °C with K_2CO_3/chitosan ratio of 2. Also, the system offered significantly improved performance in terms of CO_2 selectivity over N_2, stability and recyclability [112]. Another study by Chen *et al.* [113] observed greater CO_2 adsorptivity, selectivity and recyclability on highly ordered nitrogen doped activated carbon with micro-mesoporous framework. Activated N-doped carbon derived from polypyrrole modified graphene exhibited significant CO_2 uptake of 4.3 mmol/g along with higher selectivity and regeneration [114].

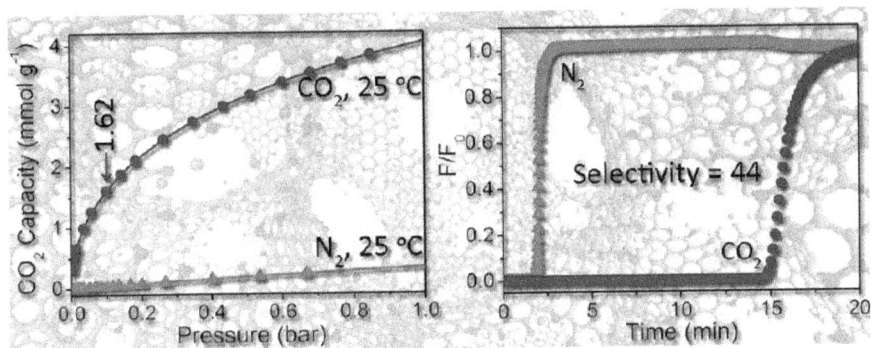

Figure 12.7 Improvement in adsorptivity and selectivity of CO_2 of the adsorbent. Reproduced from Reference 111 with permission from American Chemical Society.

Another study by Xia *et al.* [115] demonstrated the generation of noticeable adsorption energy of 59 KJ/mol and 22 KJ/mol at respective inferior and superior CO_2 coverages on sulfur doped activated microporous carbon based adsorbent [115]. In another study, Seema *et al.* [116] obtained high CO_2 adsorptivity of 4.5 mmol/g at 1 atm and 298 K as well as CO_2 selectivity atop other gases namely, N_2, CH_4 and H_2 on reduced graphene oxide/poly-thiophene composite-derived sulfur impregnated microporous activated carbon. The resultant adsorbent material presented a peak surface area of 1567 m^2/g and exhibited maximal adsorption at 0.6 nm pore gauge [116].

In another study, Builes *et al.* [117] observed superior adsorptivity of CO_2 on adsorbent based on zeolite templated ·activated carbon. Xia *et al.* [118] made use of zeolite templated nitrogen impregnated activated microporous carbon based adsorbent. CO_2 adsorption capacities of 6.9 mmol/g at 273 K and 4.4 mmol/g at room temperature and ambient pressure conditions were reported. Also, the resultant adsorbent exhibited satisfactory adsorption energy along with good selectivity and stability. In another study, Sevilla *et al.* [119] reported the synthesis of cubic and hexagonal mesostructured silica templated activated carbon for the purpose of enriched CO_2 adsorption performances [119].

Pevida *et al.* [120] attained slight enhancement in CO_2 capacity of activated carbon through the surface modification *via* ammonia treatment at 200-800 °C temperature range [120]. Enriched CO_2 adsorptivity of granular activated carbon was accomplished by Shafeeyan *et al.* [121] with the help of oxidation and subsequent ammonia treatment at a high temperature of 800 °C [121]. Zhang *et al.* [122] accomplished an increase of about 28% in CO_2 capture *via* modifications using microwave irradiation under N_2 atmosphere. In another study, Plaza *et al.* [123] analyzed the influence of amine coating over activated carbon towards CO_2 adsorption and observed increment in microporous volume and adsorptivity of activated carbon due to the embellished nitrogen content and basicity. Table 12.1 summarizes the adsorption behavior of selected activated carbon based adsorbents used for CO_2 capture.

12.4.2 Adsorption of H_2S

To accomplish efficient adsorption of H_2S, Guo *et al.* [124] used chemically activated carbon sourced from oil-palm shell. Adsorbents based on chemically activated carbon exhibited greater adsorption degree than thermally activated carbon. Also, the existence of three adsorption mechanisms such as physisorption, chemisorption and H_2S oxidation on the activated carbon was observed on the basis of method and agent used for the activation [124]. Yan *et al.* [125] observed greater capacity of alkaline activated carbon to capture H_2S due to the existence of simultaneous chemisorption and physisorption mechanisms [125].

Seredych *et al.* [126] obtained considerable improvement in H_2S adsorption through nitrogen impregnation on activated carbon procured from titanium carbide. Optimal adsorption capability was observed for 2% nitrogen impregnated activated carbon derived at 800 °C. Also, the strong influence of pore size and volume of the adsorbent on the adsorptivity of H_2S was noted (it was higher for the material with sizeable pore volume). In another study, Bagreev *et al.* [127] reported about 10 folds improvement in H_2S adsorptivity on functionalized bituminous coal derived activated carbon through the combination of nitrogen impregnation and heat treatment processes at 850 °C [127].

Huang *et al.* [128] synthesized copper impregnated activated carbon with an aim to upgrade H_2S adsorption characteristics. About 10 folds improvement in H_2S adsorptivity was achieved through copper impregnation. Also, the detrimental effect of moisture on H_2S adsorptivity was noted. Thanh *et al.* [129] reported the development of metal constituting bentonite binders functionalized activated carbon adsorbent to study the adsorption responses of H_2S. Copper containing activated carbon adsorbent exhibited increase in H_2S adsorptivity and overall performance in contrast to iron and zinc [129]. In another study, Xiao *et al.* [130] observed about 3 folds enhanced adsorption capacity of H_2S by the use of sodium carbonate impregnated activated carbon adsorbent [130].

Table 12.1 Summary of the performance of the selected activated carbon based adsorbents used for CO_2 capture

Authors	activated carbon used	adsorp-tion/uptake capacity	conditions
Presser *et al.* [87]	Carbide derived activated carbon	7.1 mol/kg	0 °C, ambient pressure
Zhang *et al.* [94]	Polyaniline derived activated carbon	1.86 mmol/g	25 °C, 0.15 bar
Deng *et al.* [95]	Metal-organic coordination polymers derived activated carbon	2.9 mmol/g	300 K, ambient pressure
Hu *et al.* [96]	Petroleum coke derived activated nanopores carbon	15.2 wt%	Room temperature, 1 bar
Wickramaratne *et al.* [100]	Phenolic resin derived activated carbon spheres	4.55 m mol/g 8.05 m mol/g	25 °C, ambient pressure 0 °C, ambient pressure
Nandi *et al.* [105]	Polyacrylonitrile derived nitrogen doped activated carbon monoliths	5.14 mmol/g	Room temperature and pressure
Zhao *et al.* [108]	Activated carbon with extremely high nitrogen doping of 13 wt%	10.53 mmol/g	25 °C, 8 bar
Sevilla *et al.* [110]	Eminently porous nitrogen doped activated carbon	7.4 mmol/g	0 °C, 1 bar
Fan *et al.* [112]	Chitosan derived nitrogen impregnated microporous activated carbon	3.86 mmol/g	25 °C, 1 atm
Chen *et al.* [113]	Activated N-doped carbon adsorbent derived from polypyrrole modified graphene sheets	4.3 mmol/g	Room temperature and pressure
Seema *et al.* [116]	Reduced graphene oxide/poly-thiophene composite-derived sulfur impregnated microporous activated carbon	4.5 mmol/g	298 K, 1 atm
Xia *et al.* [118]	Zeolite templated nitrogen impregnated activated microporous carbon based adsorbent	6.9 mmol/g 4.4 mmol/g	273 K, ambient pressure Room temperature and pressure

12.4.3 Adsorption of H₂

Jorda-Beneyto *et al.* [131] used chemically activated carbon to accomplish H_2 adsorption and revealed the dependency of pore volume and adsorption ambience on the H_2 adsorption rate. Peak adsorptivity of 38.8 g H_2/l was observed for the adsorbent based on chemically activated carbon possessing greater specific surface area at 77 K and 4 MPa [131]. Another study by Juan *et al.* [132] reported favorable capability of activated carbon based adsorbent to accumulate more H_2 in volumetric premise than metal organic framework [132]. In another study, Lillo *et al.* [133] obtained H_2 storage capacity of up to 1 wt% at 10 MPa on activated carbon [133]. Kim *et al.* [134] obtained activated carbon based adsorbent from molybdenum carbide and reported about 4.3 wt% H_2 storability. Jord-Beneyto *et al.* [135] also achieved H_2 adsorptivity of 29.7 g H_2/l under the conditions of 77 K and 4 MPa by the application of activated carbon monoliths [135].

Xia *et al.* [115] utilized sulfur impregnated microporous activated carbon possessing highly ordered structure to establish H_2 capture and storage. Sulfur doping of carbon was useful to attain pronounced adsorption heat of 9.2 KJ/mol and adsorption density of 14.3×10^{-3} mmol/m² at -196 °C and 20 bar [115]. Almasoudi *et al.* [136] used zeolite templated activated carbon based adsorbent possessing a surface area of 900–1100 m²/g and a pore volume of about 0.7 cm³ to accomplish effective H_2 adsorption. Significant H_2 storage capacity in the order of 13-15.5 μmol H_2/m² was obtained due to their exceptional microporosity [136].

12.4.4 Adsorption of Hydrocarbon Gases

Sun *et al.* [137] observed substantial methane adsorption on adsorbent based on granular activated carbon obtained from Illinois bituminous coal through KOH activation compared to the one based on steam-activated carbons. This was attributed to the increased pore volume of granular activated carbon. In another study, activated carbon acquired from carbide with a specific surface area of 3360 m²/g exhibited higher methane storage capacity of 18.5 wt% at 60 bar and 25 °C [138]. Prauchner *et al.* [139] also reported about 95 V/v adsorptivity for methane for granular activated carbon based adsorbent. In another study, activated carbon sourced from graphene oxide exhibited greater feasibility for methane uptake owing to the existence of substantial specific surface area of about 1900 m²/g [140]. Yeon *et al.* [141] discussed the use of carbide derived activated carbon monoliths for methane storage purpose and the resultant adsorbent presented a storage capacity of 219 V(STP)/v in volumetric terms under 60 bar pressure and 25 °C temperature [141]. In another study, Castello *et al.* [142] used adsorbents based on activated carbon monoliths to attain methane capture and storage. Comparing to pristine activated carbons, monoliths exhibited reduction in methane adsorptivity, but had no diffusional problems [142].

The application of benzene functionalized activated carbon with a surface area of 1065m²/g for methane capture was reported by Adinata *et al.* [143]. The resultant adsorbent had enhanced methane adsorption capacity and selectivity when compared to CO_2, N_2 and O_2 [143]. Aroua *et al.* [144] generated PEI impregnated activated carbon with enhanced methane adsorption capacity, which was higher for the adsorbent containing 0.26 wt% of PEI.

12.4.5 Adsorption of Volatile Organic Compounds (VOCs)

Chiang *et al.* [145] observed adsorption of benzene on activated carbon compared to other VOCs like carbon tetrachloride, chloroform and methylene chloride. This was ascribed to the superior heats of adsorption and entropy change of benzene on activated carbon [145]. Das *et al.* [146] obtained considerable feasibility of activated carbon fiber based adsorbent towards adsorption and desorption of VOC under dynamic conditions of temperature, gas concentration, gas flow rate and adsorbent weight [146]. Kim *et al.* [147] reported advantageous utility of 1 wt% H_3PO_4 impregnated activated carbon based adsorbent for repeated VOCs capture operations [147].

12.5 Graphene/Graphite Oxide (GO) Based Adsorbents

Adsorbents based on graphene or GO have acquired great research interest in the last few years, as compared to other adsorbents because of their outstanding adsorption performances, which results from the distinctive features such as peculiar molecular structure, vast surface area, adjustable porosity, outstanding mechanical properties, high chemical permanence, good thermal conductivity, etc. [22,23,148].

12.5.1 Adsorption of CO_2 on Graphene

Mishra *et al.* [149] showcased the capability of graphene based adsorbent to capture CO_2 *via* physisorption. The existence of peak adsorption capacity of 21.6 mmol/g for the resultant graphene based adsorbent at 25 °C and 11 bar was observed. Meng *et al.* [150] reported the generation of low temperature (150-400 °C) exfoliated graphene nanoplates based adsorbent for CO_2 capture. The resultant graphene nanoplates exhibited predominant interlayer spacing and extensive pore volume, which consequently led to greater adsorption capacity of 248 wt% at 30 bar and 25 °C [150]. Hierarchically porous graphene derived through the CO_2 activation of GO at 850 °C exhibited an adsorptivity of 1.76 mmol/g for CO_2 at 1 bar and 273 K [151]. In another study, Ning *et al.* [152] obtained favorable selectivity of CO_2 on corrugated graphene porous structure. Adsorbent based on 3-dimensional graphene generated a consequential adsorption capacity of 2.98 wt% for CO_2 under the conditions of 273 K and 106.6 kPa [153].

To enrich the CO_2 adsorption features of porous graphene based adsorbent, Shan *et al.* [154] carried out chemical functionalization and obtained improvement in uptake capacity and selectivity compared to N_2 owing to better electrostatic interactions [154]. Carrillo *et al.* [155] observed adsorption of CO_2 on titanium enriched graphene structure and the corresponding dissociation into CO and O [155]. Mishra *et al.* [156] reported the use of polyaniline impregnated graphene nanosheets as adsorbent for CO_2 capture and displayed an ultra-high adsorption capacity of up to 75 mmol/g at 25 °C and 11 bar [156]. In another study, Dasgupta *et al.* [157] reported the improvement in CO_2 adsorption and selectivity on graphene nanoribbons through COOH functionalization.

Kumar *et al.* [158] detailed the fabrication of porous graphene framework through iodobenzene treatment of reduced GO and subsequent C-C coupling. The resultant adsorbent exhibited up to 112 wt% of storage capacity for CO_2 under the conditions of 1 atm and 195 K due to the characteristic large surface area and porosity [158]. In another study, Ding *et al.* [159] synthesized graphene/manganese oxide composite *via* hydrothermal treatment of GO/MnO(OH)$_2$ blend with an aim to achieve significant CO_2 uptake. The resultant adsorbent had a specific surface area of up to 680 m²/g and which consequently created an optimal storage capacity of 11 wt% [159].

Kemp *et al.* [160] utilized nitrogen doped graphene/polyaniline composite with notable surface area of up to 1336 m²/g for effective capture of CO_2. At 1 atm pressure and 273 K temperature, resultant adsorbent exhibited an adsorption capacity of 2.7 mmol/g along with the adequate recyclability and high selectivity over H_2, N_2, Ar and CH_4 [160]. In another study, Saleh *et al.* [161] reported the application of nitrogen doped graphene/polyindole hybrid as a better CO_2 adsorbent and observed a high storage capacity of 3.0 mmol/g at 1 atm and 25 °C. Also, the system exhibited excellent stability up to 10 cycles and substantial selectivity compared to H_2, N_2, and CH_4 [161]. Table 12.2 summarizes the adsorption behavior of selected graphene based adsorbents used for CO_2 capture.

12.5.2 Adsorption of CO_2 on GO

Sui *et al.* [162] yielded hydrothermally reduced GO based adsorbent manifesting three dimensional porous network structure along with sizeable surface area and pore volume. The excellent properties of GO led to considerable adsorption of CO_2 with an adsorption capacity of 2.4 mmol/g at 273K and 1 bar [162]. Chen *et al.* [163] demonstrated enhanced CO_2 capture on GO through the impregnation of light metals like Li and Al [163]. Tsoufis *et al.* [164] claimed pronounced adsorptivity of CO_2 even under the wet conditions on GO by the intercalation of poly(propylene imine) dendrimer, which led to strong crosslinking between vicinal sheets of GO [164]. In another study, Sui *et al.* [165] used PEI impregnated GO based adsorbent, as represented schematically in Figure 12.8, for CO_2 uptake and exhibited a high storage capacity of 11.2 wt% at 1 bar and 273 K [165].

Table 12.2 Summary of selected graphene based adsorbents used for CO_2 capture

Authors	adsorbent used	Adsorption / uptake	conditions
Mishra *et al.* [149]	Graphene sheets	21.6 mmol/g	25 °C, 11 bar
Meng *et al.* [150]	Low temperature (150-400°C) exfoliated graphene nanoplates	248 wt%	25 °C, 30 bar
Xia *et al.* [151]	Hierarchically porous graphene	1.76 mmol/g	273 K, 1 bar
Wang *et al.* [153]	3-dimensional graphene	2.98 wt%	273 K, 106.6 kPa
Mishra *et al.* [156]	Polyaniline impregnated graphene nanosheets	75 mmol/g	25 °C; 11 bar
Kumar *et al.* [158]	Porous graphene framework	112 wt%	195 K; 1 atm
Kemp *et al.* [160]	Nitrogen doped graphene/polyaniline hybrid	2.7 mmol/g	273 K; 1 atm
Saleh *et al.* [161]	Nitrogen doped graphene/polyindole hybrid	3.0 mmol/g	25 °C; 1 atm

In order to upgrade the CO_2 adsorption features, Zhao *et al.* [166] carried out amine intercalation of GO using various amines such as ethylenediamine, diethylenetriamine and triethylenetetramine. Aminated GO presented better adsorption of CO_2, especially the GO with ethylenediamine intercalation exhibited an optimal adsorptivity of 53.62 mg CO_2/g-adsorbent [166]. In another study, Zhao *et al.* [167] achieved notable CO_2 adsorption capacity of 1.06 mmol/g on aminated GO/metal-organic framework composite based adsorbent, even under the influence of moisture. The observed behavior resulted from the combined effect of the features of the composites such as substantial surface area and pore volume, presence of additional active moieties after functionalization and compatibility of pore dimensions with adsorbate [167].

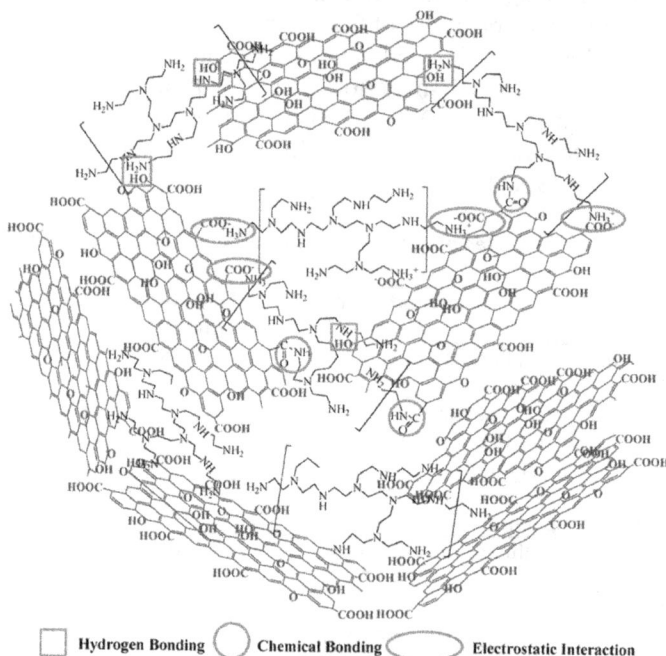

Hydrogen Bonding   Chemical Bonding   Electrostatic Interaction

Figure 12.8 Schematic representation of GO-PEI hydrogel, exhibiting various interactions like hydrogen bonding, chemical bonding and electrostatic interactions. Reproduced from Reference 165 with permission from American Chemical Society.

Cao *et al.* [168] reported the fabrication of GO/zirconium metal organic framework composite with an ultimate objective to enhance the CO_2 adsorption capacity. Under the conditions of 298 K and 1 bar, GO based composite exhibited an optimal CO_2 adsorption capacity of 3.37 mmol/g, which was about 48% greater than zirconium metal organic framework [168]. In another study, Chowdhury et al. [169] fabricated GO/TiO_2 composite based adsorbent and obtained a satisfactory adsorptivity of 1.88 mmol/g for CO_2 under ambient temperature along with significantly high selectivity from exhaust gases, thus, confirming the functional nature of the adsorbent.

12.5.3 Adsorption of H_2S

With an aim to achieve adsorption of H_2S, Mabayoje et al. [170] developed GO/cobalt oxide composite based adsorbent. The existence of meaningful adsorption of H_2S through the generation of sulfites on the composite surface was noted even in the presence of moisture [170]. In another study, Seredych et al. [171] also obtained appreciable H_2S adsorption by the use of adsorbent based on GO/zirconium hydroxide composite owing to the substantial basicity and the newly generated active centers on the composite structure. The enhanced H_2S adsorption of composite resulted from the existing reactions of H_2S with epoxy and carboxylic functional groups of GO. Adsorbent based on GO/metal-organic framework composite also exhibited beneficial H_2S uptake capacity through the creation of miniature pores [172]. Seredych et al. [173] also used GO based ammonia adsorbent as H_2S adsorbent using high temperature treatment at 950 °C. The existence of nitrogen functionalities on the treated GO surface led to enhanced H_2S adsorptivity. These studies confirmed the high potential of graphene and modified GO based adsorbents for the selective adsorption of H_2S from the gas mixtures.

12.5.4 Adsorption of Ammonia

Seredych et al. [174] revealed the possibility of successful ammonia capture on graphene through the sequential intercalation and the reactive adsorption mechanisms [174]. In another study, the authors revealed greater ammonia capture for GO adsorbent derived from the amorphous carbon rich graphite [175]. Seredych et al. [176] examined the adsorption behavior of ammonia on exfoliated GO/bentonite hybrid based adsorbent under dynamic conditions. The layered composite structure offered the consumption of considerable concentration of ammonia *via* Brønsted and Lewis active centers on the composite framework as well as interlayered hybrid structures [176]. Composites of GO and manganese oxide were also developed as adsorbent for ammonia and appreciable adsorption capacity for composite composed of immensely oxidized GO was observed [177].

In another study, Seredych *et al.* [178] reported the functionalization of GO using aluminum and zirconium-aluminum polyoxycations separately and the subsequent calcination at 300 °C to enrich ammonia adsorption characteristics. Enhanced adsorption performance was reported for zirconium-aluminum polyoxycation functionalized GO owing to the superior acidity [178]. The intensity of ammonia adsorption degree on the zirconium-aluminum polyoxycation functionalized GO based adsorbent was improved by pre-humidification [179]. The reduction in adsorptivity for ammonia on GO/aluminum polyoxycation composite under humid environment was also reported [180].

12.5.5 Adsorption of H_2

Srinivas *et al.* [181] obtained H_2 adsorptivies of 1.2 and 0.1 wt% at temperatures corresponding to 77 K and 298 K by the use of graphene sheets as an adsorbent with a surface area of 640 m^2/g. Favorable adsorption heats of up to 5.9 KJ/mol on graphene sheets owing to their strong interaction with H_2 were also reported [181]. Yuan *et al.* [182] reported the development of graphene sheets through the reduction of GO in the presence of glucose for the purpose of effective H_2 adsorption. The resultant graphene was characterized with a surface area of 1205.8 m^2/g, which consequently led to an effective H_2 storage capacity of 2.7 wt% at 25 bar and 298 K [182]. In another study, Xia *et al.* [151] also reported an adsorptivity of about 3.76 mmol/g for H_2 at 1 bar and 77 K on highly ordered micro-meso-macroporous graphene obtained from the CO_2 activation of GO at a high temperature of 850 °C. The adsorbent exhibited a high surface area of about 532 m^2/g along with a sizeable pore volume of around 1.67 cm^3/g [151].

Ghosh *et al.* [183] obtained H_2 capture of 3.1 wt% at 298 K and 100 atm on graphene based adsorbent derived from the exfoliation of GO. An interesting improvement in adsorption performance was reported by decreasing the mean graphene layers, a salso suggested in Figure 12.9. Ma *et al.* [184] reported poor adsorption of H_2 on single layer graphene sheets owing to their inadequate surface area and poor interaction with H_2 [184], which underlined the need of high surface area and interfacial intercaitos for better adsorbent performance. Wang *et al.* [153] revealed the use of 3-dimensional graphene with a specific surface area of 477 m^2/g along with the mesoporosity for effective H_2 capture and observed a storage capacity of 1.40 wt% at 106.6 kPa and

77 K [153]. In another study, Lyth *et al.* [185] achieved H_2 storage capacity of 2.1 wt% on graphene foam based adsorbent derived from sodium ethoxide with a specific surface area of $1200 \ m^2/g$ [185].

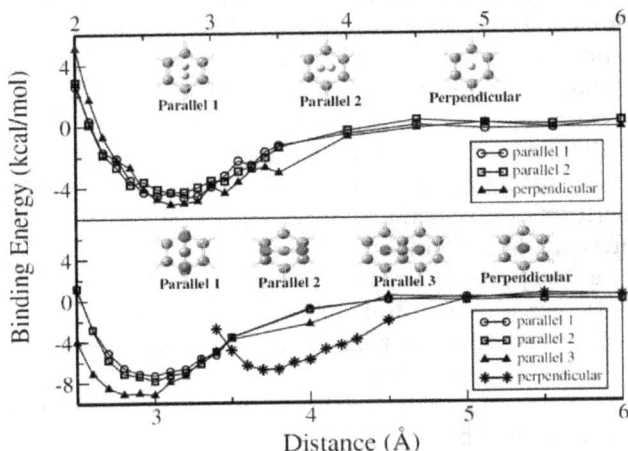

Figure 12.9 Representation of the binding energies of a single molecule on a graphene supercell as a function of distance with different adsorption orientations (shown in insets) for hydrogen (upper panel) and carbon dioxide (lower panel). Reproduced from Reference 183 with permission from American Chemical Society.

Studies of Burress *et al.* [23] and Srinivas *et al.* [186] reported the development of GO framework based adsorbent with the aid of diboronic acid for enhanced H_2 capture. Adsorption performance such as adsorptivity and heat of adsorption of the adsorbents were about two folds greater when compared to carbon based adsorbents and competitive when compared to metal organic frameworks. Also, GO frameworks exhibited higher thermal stability than GO [186]. Petit *et al.* [187] observed an improved H_2 capture on GO/copper metal organic framework composite due to the presence of miniature pores [187].

12.6 Nanotubes Based Adsorbents

Nanotubes are one-dimensional porous nanomaterials possessing tubular honeycomb like network structure. Better adsorption performance of nanotubes is derived from their porous structure and these materials provide five possible active adsorption sites such as cavities, interstitial channels, two ends and exterior surfaces for successful gas capture [188-190].

12.6.1 Adsorption of CO_2

Babu *et al.* [191] used ordered as well as vertically positioned double-walled carbon nanotubes (double-walled CNTs) based adsorbent exhibiting substantial surface area and purity for CO_2 uptake. The adsorbent exhibited competitive adsorption properties when compared to other conventional adsorbents such as mesoporous molecular sieves and metal organic frameworks under elevated pressure. Also, the impact of oxygen plasma induced chemical modifications and diameter reduced structural modifications on the CO_2 adsorption features of these materials were investigated. The former modification caused significant performance at nether pressure through C-O moieties generation, while the latter modification resulted in greater capture capacity under elevated pressure [191]. Cinke *et al.* [192] observed CO_2 capture on single-walled CNTs (SWCNTs) *via* physisorption route and noted up to two folds greater storage capacity than adsorbent based on activated carbon [192]. Hu *et al.* [193] reported the CO_2 adsorption on SWCNTs through interior as well as exterior surfaces. In another study, Golkhar *et al.* [194] also achieved high rate of CO_2 uptake on adsorbent based on nanofluids of CNTs than nanosilica.

Casco *et al.* [195] made comparison of adsorption characteristics of multi-walled CNTs (MWCNTs), acidified MWCNTs and unzipped MWCNTs of carbon and N-doped carbon. Significant influence of surface roughness and curved regions of MWCNTs on CO_2 adsorption was revealed, with the materials rich in these parameters exhibiting superior adsorption capacities. In the study, adsorbent based on unzipped MWCNTs of N-doped carbon exhibited optimal adsorption features for CO_2 [195]. Adsorbent based on boron CNTs exhibited appreciable adsorption for CO_2 over N_2, H_2 and CH_4, which was attributed to the creation of strong interactions. For successful CO_2 adsorption, Upender *et al.* [197] used sodium and potassium titanate nanotubes as adsorbent. Comparing with pristine titanate, these materials exhibited higher extent of adsorption degree and recyclability. Also, an interesting improvement in CO_2 adsorptivity was observed in the presence of moisture [197].

Hsu *et al.* [198] revealed the usefulness of 3-aminopropyl-triethoxysilane (APTS) functionalized CNTs for continuous operations without affecting the performance. The decreasing tendency of adsorption intensity was observed with increasing temperature. In another study, Su *et al.* [199] reported the generation of better CO_2 adsorption features of APTS functionalized MWCNTs over conventional silica or activated carbon based adsorbents at 20 °C. Also, the workable regeneration of captured CO_2 under low temperature conditions was obtained. Lu *et al.* [200] also reported greater CO_2 adsorption capacity of

APTS modified CNTs over granular activated carbon and zeolite based adsorbents. The authors also revealed accomplishable regeneration of adsorbed CO_2 under low temperature conditions for APTS modified CNTs. Figure 12.10 shows morphological changes of CNTs before and after APTS modification [200].

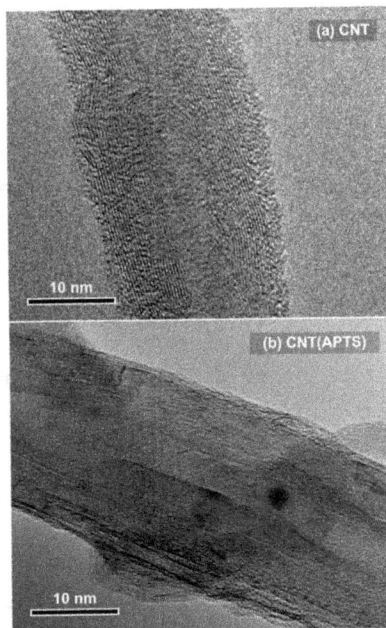

Figure 12.10 TEM images of CNTs (up) before and (down) after APTS modification. Reproduced from Reference 200 with permission from American Chemical Society.

To achieve efficient adsorption of CO_2 from atmosphere, Jana *et al.* [201] evaluated the feasibility of halloysite and APTS-functionalized halloysite nanotubes. Amine modification of halloysite nanotubes was useful to upgrade the CO_2 uptake and successful regeneration under ambient air conditions. APTS- functionalized halloysite nanotubes also exhibited superior recyclability along with outstanding stability, even under the oxidative surroundings [201]. Effectiveness of amine impregnated TiO_2 nanotubes towards CO_2 capture even under the wet conditions was demonstrated by Song *et al.* [202]. At 60 °C, TiO_2 nanotubes based adsorbent containing tetraethylenepentamine exhibited maximal storage capacity of 4.37 mmol/g for CO_2, which was improved to 5.24 mmol/g in the presence of moisture. Also, the adsorbent exhibited greater recycling stability under the dry conditions. In another study, Ko et al. [203] obtained significant CO_2 capture on adsorbent based on amine functionalized double walled silica nanotubes. A reduction in efficiency of the resultant adsorbent was also observed with the corresponding increase in temperature.

Liu et al. [204] reported the use of PEI impregnated protonated titanate nanotubes as a beneficial adsorbent for CO_2 capture. Optimal adsorptivity of 130.8 mg/g for modified protonated titanate nanotubes with 50 wt% of PEI at 100 °C was observed. Also, adsorbent based on the modified nanotubes exhibited greater recycling stability. In another study, Dillon et al. [205] developed PEI impregnated SWCNTs for CO_2 capture. A peak adsorption capacity of 9.2 wt% for CO_2 on modified SWCNTs at 27 °C was observed. Moreover, the adsorbed CO_2 was easily desorbed from the modified SWCNTs by heating at 75 °C [205]. Cong et al. [206] reported the application of composite composed of CNTs and brominated poly(2,6-diphenyl-1,4-phenylene oxide) as CO_2 adsorbent, which presented a greater adsorption degree than the pure polymer. The adsorption degree was further enhanced through the use of carboxylic acid treated SWCNTs [206].

12.6.2 Adsorption of H_2

Zuttel et al. [207] employed SWCNTs for H_2 adsorption and observed good adsorption and desorption characteristics. Successful adsorption of H_2 on SWCNTs through either cavities or monolayered structure was observed. The intensity of H_2 adsorption on SWCNTs enhanced on increasing the CNT diameter, with the CNTs with 2.2 nm diameter reported to have the potential to hold about 5 mass% of H_2 [207]. In another study, Ma et al. [208] examined the feasibility of multiwalled boron nitride nanotubes for H_2 capture. Adsorption capacity of up to 2.6 wt% for H_2 under ambient temperature and 10 MPa was observed. Nour et al. [209] attained adsorption selectivity of H_2 from CH_4 on MWCNT/polydimethylsiloxane composite based adsorbent and an improvement in effectiveness was observed with increase in concentration of MWCNTs in the composite.

12.6.3 Adsorption of Sulfur Gases

Wang et al. [210] investigated the adsorption behavior of sulfur gases such as H_2S and SO_2 from gas mixtures on SWCNTs arrays based adsorbent. SWCNTs with larger diameter exhibited optimal adsorption capacities of 16.31 mmol/g for H_2S and 16.03 mmol/g for SO_2 at a temperature of 303 K. At the same time, substantial selectivity of sulfur gases from gas mixtures was reported for SWCNTs possessing smaller diameter. Further, the functionalization of SWCNTs surfaces with carbonyl moieties enhanced the adsorption selectivity of sulfur gases from the gas mixture containing CH_4 to about two folds [210]. Mohamadalizadeh *et al.* [211] carried out a comparative study concerning the adsorption behavior of H_2S on various MWCNTs based adsorbents such as pristine MWCNTs, tungsten impregnated MWCNTs and amino/amido functionalized MWCNTs at the conditions of 20 °C and 1-10 bar. The adsorptivity of MWCNTs was enhanced through amine functionalization, whereas the impregnation of tungsten nanoparticles reduced it.

12.6.4 Adsorption of VOCs

Hussain *et al.* [212] reported strong adsorption and desorption characteristics of VOCs on both single and multiwalled carbon nanotubes based adsorbent owing to high aspect ratio. Also, the materials offered mass transfer resistance free conditions in consequence to their non-porous nature. In another study, Li *et al.* [213] further confirmed the potential of MWCNTs for effective VOCs capture and subsequent successful recovery. For obtaining improved VOCs adsorption characteristics of MWCNTs, Wang *et al.* [214] reported the use of MWCNTs/porous silica gel composites. In another study, Zheng *et al.* [215] reported the application of SWCNTs in the paper form for effective adsorption and subsequent desorption of VOCs like methyl ethyl ketone, toluene and dimethyl methylphosphonate.

12.6.5 Adsorption of Other Gases

Delavar *et al.* [216] used MWCNTs with a specific surface area of 294 m^2/g and a mean pore volume of 0.62 cm^3/g for methane capture and attained adsorption capacity of about 34 wt% at 283 K and 50 bar. In another study, Cao *et al.* [217] theoretically evaluated the adsorption characteristics of methane on triangular arrays of SWCNTs and obtained a volumetric adsorptivity of 216 V/V and a gravimetric adsorptivity of 215 g CH_4/kg at ambient temperature and 4.1 MPa.

Kim *et al.* [218] reported the development of CNTs/poly(imide siloxane) composite based adsorbent for capture of O_2, N_2 and CH_4. The occupancy of CNTs on copolymer matrix caused to enrich the adsorption performance, which was higher for the material containing more amount of CNTs. Wang *et al.* [219] fabricated titania nanotubes *via* hydrothermal route and applied successively for mercury capture from exhaust gases. In another study, Ding *et al.* [220] also detailed the feasibility of CNTs for advantageous adsorption of SF_4 and SOF_2.

Foroutan *et al.* [221] achieved higher capability of CNT bundles to capture considerable amount of noble gas mixture comprising of argon, krypton and xenon. Jalili *et al.* [222] also observed substantial adsorption of xenon and krypton through internal and external surfaces of SWCNTs. In another study, the adsorptivity of neon on open ended SWCNTs exhibited decreasing tendency with increase in temperature [223]. Foroutan *et al.* [224] also reported the adsorption degree of heavier noble gas on SWCNTs based adsorbent to be greater as compared to the lighter one.

12.7 Zeolite Based Adsorbents

Zeolites are crystalline microporous materials with three dimensional structure comprising of hydrated aluminosilicate minerals. These materials offer wide range of gas adsorption and storage applications owing to their structural features [225,226].

12.7.1 Adsorption of CO_2

Montanari *et al.* [227] investigated the adsorption of CO_2 from synthetic biogas on adsorbents based on 4A and 13X zeolites. Significant adsorptivity of CO_2 on 13X zeolites was reported and the adsorption intensity was additionally improved under wet conditions. Adsorbent based on 4A zeolites exhibited poor adsorptivity of CO_2 even under wet conditions due to the methane co-adsorption. 4A zeolites presented swift regeneration compared to 13X zeolites under nitrogen environment at ambient temperature. In another study, adsorbent based on Y-type zeolite was observed to exhibit greater permeance for CO_2 than N_2 owing to the substantial adsorption selectivity [228]. The permeance of CO_2 on Y-type zeolite adsorbent was more than 10^{-7} mol.m^{-2}s^{-1}Pa^{-1}. Also, a good permselectivity of 20-100 for CO_2 to N_2 was reported. Saha *et al.* [229] obtained sizeable adsorption of CO_2 and N_2O from atmospheric air as well as significant CO_2 uptake from methane on zeolite 5A based adsorbent compared to metal organic frameworks (Figure 12.11).

Chatti *et al.* [230] applied 13X zeolite and its amine functionalized versions for successful CO_2 capture. Adsorbent based on 13X zeolite had an adsorption capacity of 16.01mg/g, which enhanced to 22.78 mg/g and 19.98 mg/g after amine functionalization using isopropanol amine and monoethanol amine, respectively [230]. In another study, Zhao *et al.* [226] achieved improved performance of 13X zeolite towards CO_2 capture through alkali (NaOH) modification. The creation of enlarged active adsorption surfaces and the reduced diffusion barrier after alkali treatment of 13X zeolite resulted in intensifying CO_2 adsorption rate as well as capacity. Tagliabue *et al.* [231] used sodium impregnated FAU zeolite based adsorbent for the mass adsorption of CO_2 from natural gas. Mixed matrix membrane based adsorbent consisting of 50 wt% of zeolite 4A and poly(vinyl acetate) was also developed by Adams *et al.* [232] for efficient removal of CO_2 from CH_4.

Li *et al.* [233] reported considerable adsorption of CO_2 over β-zeolite based adsorbents compared to CH_4, N_2 and O_2 [233]. Xu *et al.* [234] also executed

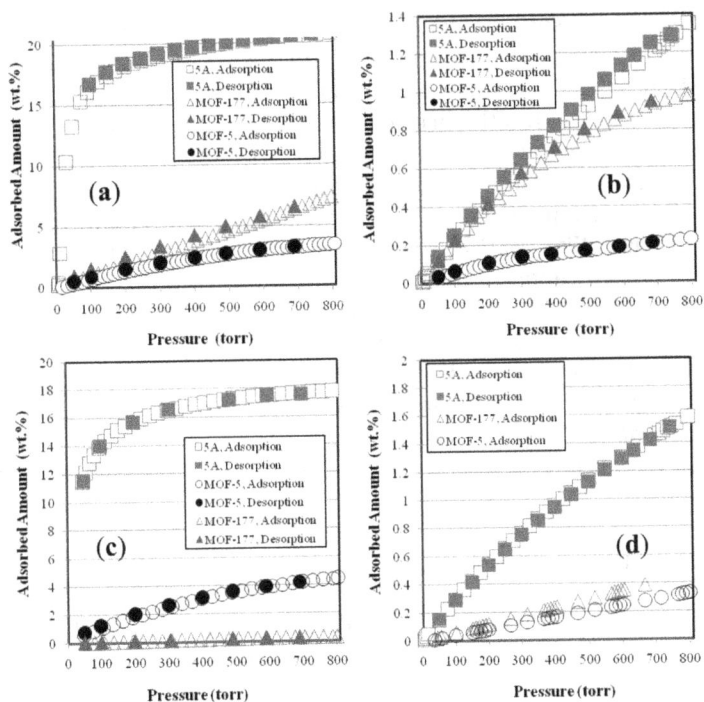

Figure 12.11 Adsorption isotherms of (a) CO_2, (b) CH_4, (c) N_2O, and (d) N_2 on MOF-5, MOF-177, and zeolite 5A at 298 K and pressures up to 800 Torr. Reproduced form reference 229 with permission from American Chemical Society.

monoethanol amine functionalization of β-zeolite in order to enhance the adsorption performance. Amine functionalization of β-zeolite improved the adsorption selectivity of CO_2, with the material having 40 wt% of monoethanol amine exhibiting the maximum enhancement. Maximal selectivity of 25.67 for CO_2/N_2 and 7.70 for CO_2/CH_4 over amine functionalized β-zeolite was reported due to the strong steric effect and, thus, improved adsorbent-adsorbate interaction. In another study, Xu *et al.* [235] carried out Na impregnation of β-zeolite, which correspondingly enhanced the adsorption features. Thus, Na impregnated β-zeolite exhibited appreciable CO_2 adsorptivity and selectivity than CH_4 and H_2 resulting from the enhanced electrostatic interaction of CO_2 with Na^+ ions.

12.7.2 Adsorption of Other Gases

Diaz *et al.* [236] observed superior adsorption capacity of Zeolite 13X over Zeolite 5A for various VOCs such as several alkanes, cyclic hydrocarbons, aromatic hydrocarbons and chlorinated compounds owing to the extensive pore dimensions. Sen *et al.* [237] fabricated zeolite 4A/polycarbonate mixed matrix adsorbent for achieving efficient capture of gas molecules like N_2, H_2, O_2, CO_2 and CH_4. Increase in zeolite 4A concentration in the composition increased the selectivity of all gas molecules at the cost of adsorption permeabilities. For adsorption of gases like O_2, CO_2, N_2 and He as well as gas mixture of CO_2/CH_4, Husain *et al.* [238] reported the use of functionalized HSSZ-13 zeolite/PEI hybrid based adsorbents. Functionalization of zeolite was carried out through two different routes, namely PEI sizing and Grignard treatment. PEI sizing had no significant effect on the adsorption selectivity of gases whereas Grignard treatment was advantageous for improved adsorption selectivity. In another study, Sen *et al.* [239] also achieved improvement in adsorption selectivities of H_2 and CO_2 on polycarbonate/*p*-nitroaniline blend through the incorporation of zeolite 4A.

12.8 Polymer Based Adsorbents

In this section, only polymer based adsorbents not mentioned in the earlier sections have been reviewed. Many polymeric materials offer effective gas adsorption and storage feasibility based on their structural characteristics. Rolker *et al.* [240] revealed the usefulness of branched polymers of polyethers, polyesters and polyamines as adsorbent for CO_2 uptake. Branched polymers exhibited large CO_2 adsorption capacity along with CO_2/N_2 selectivity.

Bali *et al.* [241] used polyamines such as PEI and poly(allylamine) along with alumina as adsorbent for CO_2 capture. The oxidation stability of adsorbents is a critical parameter which decides their reusability. In this study, the adsorbent based on poly(allylamine) exhibited greater oxidation stability than PEI due to the absence of secondary amines and the presence of primary amines.

An *et al.* [242] generated an ionic polymer namely, poly(1,1,3,3-tetramethylguanidine acrylate) based adsorbent to achieve effective SO_2 adsorption and successive desorption. The ionic polymer exhibited high adsorption capacity, adsorption rate and selectivity for SO_2 compared to monomer based adsorbents. Successful desorption of SO_2 with the aid of elevated temperature or vacuum was reported, which correspondingly offered great potential of copolymer based adsorbent for reuse. Kikkinides *et al.* [243] reported adsorption of SO_2 molecules over nitrogen monoxide (NO) modified polystyrene based adsorbent. The hydrophobic NO-modified polystyrene exhibited excellent SO_2 storage capacity along with high diffusion rate and SO_2/CO_2 selectivity. Up to 90% removal of SO_2 from the flue gases was accomplished using the adsorbent. In another study, Deshmukh *et al.* [244] genertaed Hofmann-type porous coordination polymer $\{Fe(Pz)[Pt(CN)_4]\}_n$ (PCP) with three dimensional structure (Figure 12.12) to examine the adsorption behavior of CO_2 and CS_2. Compared to CO_2, gas molecules of CS_2 exhibited efficient adsorption on PCP owing to superior dispersion interactions generated through Pt and Pz moieties of PCP with CS_2. Noro *et al.* [245] also reported an outstanding adsorption density of 0.21 gm/L for methane on porous coordination polymer $\{CuSiF_6(4,4'-bipyridine)_2\}_n$ based adsorbent.

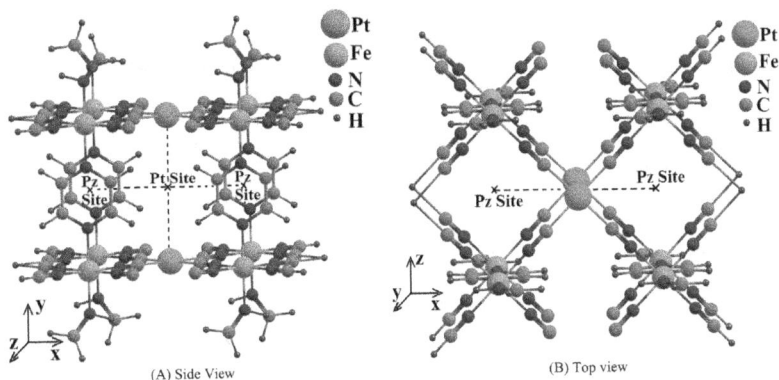

Figure 12.12 Large realistic model (LRM) of the Hofmann-type PCP. Reproduced from Reference 244 with permission from American Chemical Society.

12.9 Conclusions

Adsorption of diverse pollutant gases with the help of miscellaneous adsorbents is immensely important research field in terms of environmental and energy aspects. The selection of suitable adsorbent is the critical parameter which decides the performance during the adsorption of gas molecules. In this study, the adsorption performance of various adsorbents based on silica, clay, activated carbon, graphene, graphite oxide, nanotubes, zeolites and polymers have been thoroughly compared. The adsorption potential of each adsorbent for different gas molecules including CO_2, N_2, H_2, sulfur gases, hydrocarbon gases, ammonia, volatile organic compounds and other gases has been evaluated with respect to adsorption capacity, selectivity, stability, recyclability/reusability, heat of adsorption, etc. The modifications of adsorbent materials using chemical treatment, physical methods, elemental doping, polymer impregnation, metal/metal oxide infusion, *etc.* are useful to upgrade the ultimate performance of the adsorbents, which has also been thoroughly reviewed.

References

1. Yu, K. M. K., Curcic, I., Gabriel, J., and Tsang, S. C. E. (2008) Recent advances in CO_2 capture and utilization. *ChemSusChem,* **1**, 893-899.
2. Basic Research Needs for Carbon Capture: Beyond 2020 (2010) *Report of the Basic Energy Sciences Workshop for Carbon Capture: Beyond 2020,* USA. Online: https://science.energy.gov/~/media/bes/pdf/reports/files/Basic_Research_Needs_for_Carbon_Capture_rpt.pdf [assessed 21st April 2017].
3. Griggs, D. J., and Noguer, M. (2002) Climate change 2001: the scientific basis. Contribution of working group I to the third assessment report of the intergovernmental panel on climate change. *Weather,* **57**, 267-269.
4. Lee, S.-Y. and Park, S.-J. (2015) A review on solid adsorbents for carbon dioxide capture. *Journal of Industrial and Engineering Chemistry,* **23**, 1-11.
5. Wang, Q., Luo, J., Zhong, Z., and Borgna, A. (2011) CO_2 capture by solid adsorbents and their applications: current status and new trends. *Energy & Environmental Science,* **4**, 42-55.
6. Olivares-Marin, M., and Maroto-Valer, M. M. (2012) Development of adsorbents for CO_2 capture from waste materials: a review. *Greenhouse Gases: Science and Technology,* **2**, 20-35.
7. D'Alessandro, D. M., Smit, B., and Long, J. R. (2010) Carbon dioxide capture: prospects for new materials. *Angewandte Chemie International Edition,* **49**, 6058-6082.

8. Srivastava, N., and Eames, I. (1998) A review of adsorbents and adsorbates in solid-vapour adsorption heat pump systems. *Applied Thermal Engineering,* **18**, 707-714.
9. Suzuk, M. (1990) Adsorption engineering. *Kodansha,* **551**, 128.
10. Yang, R. T. (2013) *Gas Separation by Adsorption Processes*, Butterworth-Heinemann, USA.
11. Bhown, A. S., and Freeman, B. C. (2011) Analysis and status of post-combustion carbon dioxide capture technologies. *Environmental Science & Technology,* **45**, 8624-8632.
12. Ben, T., Li, Y., Zhu, L., Zhang, D., Cao, D., Xiang, Z., Yao, X., and Qiu, S. (2012) Selective adsorption of carbon dioxide by carbonized porous aromatic framework (PAF). *Energy & Environmental Science,* **5**, 8370-8376.
13. Sayari, A., Belmabkhout, Y., and Serna-Guerrero, R. (2011) Flue gas treatment via CO_2 adsorption. *Chemical Engineering Journal,* **171**, 760-774.
14. Choi, S., Drese, J. H., and Jones, C. W. (2009) Adsorbent materials for carbon dioxide capture from large anthropogenic point sources. *ChemSusChem,* **2**, 796-854.
15. Florin, N. H., Blamey, J., and Fennell, P. S. (2010) Synthetic CaO-based sorbent for CO_2 capture from large-point sources. *Energy & Fuels,* **24**, 4598-4604.
16. Drage, T., Arenillas, A., Smith, K. M., and Snape, C. E. (2008) Thermal stability of polyethylenimine based carbon dioxide adsorbents and its influence on selection of regeneration strategies. *Microporous and Mesoporous Materials,* **116**, 504-512.
17. Bae, Y. S., and Snurr, R. Q. (2011) Development and evaluation of porous materials for carbon dioxide separation and capture. *Angewandte Chemie International Edition,* **50**, 11586-11596.
18. Kim, S., Ida, J., Guliants, V. V., and Lin, Y. (2005) Tailoring pore properties of MCM-48 silica for selective adsorption of CO_2. *The Journal of Physical Chemistry B,* **109**, 6287-6293.
19. Pereira, P., Pires, J., and Brotas de Carvalho, M. (1998) Zirconium pillared clays for carbon dioxide/methane separation. 1. Preparation of adsorbent materials and pure gas adsorption. *Langmuir,* **14**, 4584-4588.
20. Himeno, S., Komatsu, T., and Fujita, S. (2005) High-pressure adsorption equilibria of methane and carbon dioxide on several activated carbons. *Journal of Chemical & Engineering Data,* **50**, 369-376.
21. Thiruvenkatachari, R., Su, S., An, H., and Yu, X. X. (2009) Post combustion CO_2 capture by carbon fibre monolithic adsorbents. *Progress in Energy and Combustion Science,* **35**, 438-455.
22. Balasubramanian, R., and Chowdhury, S. (2015) Recent advances and progress in the development of graphene-based adsorbents for CO_2 capture. *Journal of Materials Chemistry A,* **3**, 21968-21989.
23. Burress, J. W., Gadipelli, S., Ford, J., Simmons, J. M., Zhou, W., and Yildirim, T. (2010) Graphene oxide framework materials: theoretical predictions and experimental results. *Angewandte Chemie International Edition,* **49**, 8902-8904.
24. Zhao, J., Buldum, A., Han, J., and Lu, J. P. (2002) Gas molecule adsorption in carbon nanotubes and nanotube bundles. *Nanotechnology,* **13**, 195.
25. Ren, X., Chen, C., Nagatsu, M., and Wang, X. (2011) Carbon nanotubes as adsorbents in environmental pollution management: A review. *Chemical Engineering Journal,* **170**, 395-410.

26. Rouquerol, J., Rouquerol, F., Llewellyn, P., Maurin, G., and Sing, K. S. (2013) *Adsorption by Powders and Porous Solids: Principles, Methodology and Applications.* 2nd edition, Academic Press, USA.

27. LeVan, M. D., Carta, G., and Yon, C. M. (1997) Adsorption and ion exchange. *Energy,* **16**, 17.

28. Farha, O. K., Spokoyny, A. M., Hauser, B. G., Bae, Y.-S., Brown, S. E., Snurr, R. Q., Mirkin, C. A., and Hupp, J. T. (2009) Synthesis, properties, and gas separation studies of a robust diimide-based microporous organic polymer. *Chemistry of Materials,* **21**, 3033-3035.

29. Yong, Z., Mata, V., and Rodrigues, A. E. (2000) Adsorption of carbon dioxide on basic alumina at high temperatures. *Journal of Chemical & Engineering Data,* **45**, 1093-1095.

30. Bhatta, L. K. G., Subramanyam, S., Chengala, M. D., Olivera, S., and Venkatesh, K. (2015) Progress in hydrotalcite like compounds and metal-based oxides for CO_2 capture: A review. *Journal of Cleaner Production,* **103**, 171-196.

31. Liu, J., Thallapally, P. K., McGrail, B. P., Brown, D. R., and Liu, J. (2012) Progress in adsorption-based CO_2 capture by metal-organic frameworks. *Chemical Society Reviews,* **41**, 2308-2322.

32. Hoffmann, F., Cornelius, M., Morell, J., and Froba, M. (2006) Silica-based mesoporous organic-inorganic hybrid materials. *Angewandte Chemie International Edition,* **45**, 3216-3251.

33. Belmabkhout, Y., Serna-Guerrero, R., and Sayari, A. (2009) Adsorption of CO_2 from dry gases on MCM-41 silica at ambient temperature and high pressure. 1: Pure CO_2 adsorption. *Chemical Engineering Science,* **64**, 3721-3728.

34. Belmabkhout, Y., and Sayari, A. (2009) Adsorption of CO_2 from dry gases on MCM-41 silica at ambient temperature and high pressure. 2: Adsorption of CO_2/N_2, CO_2/CH_4 and CO_2/H_2 binary mixtures. *Chemical Engineering Science,* **64**, 3729-3735.

35. Tsai, C.-Y., Tam, S.-Y., Lu, Y., and Brinker, C. J. (2000) Dual-layer asymmetric microporous silica membranes. *Journal of Membrane Science,* **169**, 255-268.

36. Himeno, S., Tomita, T., Suzuki, K., and Yoshida, S. (2007) Characterization and selectivity for methane and carbon dioxide adsorption on the all-silica DD3R zeolite. *Microporous and Mesoporous Materials,* **98**, 62-69.

37. Bellussi, G., Broccia, P., Carati, A., Millini, R., Pollesel, P., Rizzo, C., and Tagliabue, M. (2011) Silica-aluminas for carbon dioxide bulk removal from sour natural gas. *Microporous and Mesoporous Materials,* **146**, 134-140.

38. Xomeritakis, G., Tsai, C., Jiang, Y., and Brinker, C. (2009) Tubular ceramic-supported sol–gel silica-based membranes for flue gas carbon dioxide capture and sequestration. *Journal of Membrane Science,* **341**, 30-36.

39. Leal, O., Bolivar, C., Ovalles, C., Garcia, J. J., and Espidel, Y. (1995) Reversible adsorption of carbon dioxide on amine surface-bonded silica gel. *Inorganica Chimica Acta,* **240**, 183-189.

40. Kim, S.-N., Son, W.-J., Choi, J.-S., and Ahn, W.-S. (2008) CO_2 adsorption using amine-functionalized mesoporous silica prepared via anionic surfactant-mediated synthesis. *Microporous and Mesoporous Materials,* **115**, 497-503.

41. Yue, M. B., Sun, L. B., Cao, Y., Wang, Z. J., Wang, Y., Yu, Q., and Zhu, J. H. (2008) Promoting the CO_2 adsorption in the amine-containing SBA-15 by hydroxyl group. *Microporous and Mesoporous Materials,* **114**, 74-81.
42. Knowles, G. P., Graham, J. V., Delaney, S. W., and Chaffee, A. L. (2005) Aminopropyl-functionalized mesoporous silicas as CO_2 adsorbents. *Fuel Processing Technology,* **86**, 1435-1448.
43. Hiyoshi, N., Yogo, K., and Yashima, T. (2005) Adsorption characteristics of carbon dioxide on organically functionalized SBA-15. *Microporous and Mesoporous Materials,* **84**, 357-365.
44. Sakamoto, Y., Nagata, K., Yogo, K., and Yamada, K. (2007) Preparation and CO_2 separation properties of amine-modified mesoporous silica membranes. *Microporous and Mesoporous Materials,* **101**, 303-311.
45. Knofel, C., Descarpentries, J., Benzaouia, A., Zelenak, V., Mornet, S., Llewellyn, P., and Hornebecq, V. (2007) Functionalised micro-/mesoporous silica for the adsorption of carbon dioxide. *Microporous and Mesoporous Materials,* **99**, 79-85.
46. Liu, X., Zhou, L., Fu, X., Sun, Y., Su, W., and Zhou, Y. (2007) Adsorption and regeneration study of the mesoporous adsorbent SBA-15 adapted to the capture/separation of CO_2 and CH_4. *Chemical Engineering Science,* **62**, 1101-1110.
47. Liu, X., Li, J., Zhou, L., Huang, D., and Zhou, Y. (2005) Adsorption of CO_2, CH_4 and N_2 on ordered mesoporous silica molecular sieve. *Chemical Physics Letters,* **415**, 198-201.
48. Zelenak, V., Halamova, D., Gaberova, L., Bloch, E., and Llewellyn, P. (2008) Amine-modified SBA-12 mesoporous silica for carbon dioxide capture: Effect of amine basicity on sorption properties. *Microporous and Mesoporous Materials,* **116**, 358-364.
49. Jang, K.-S., Kim, H.-J., Johnson, J., Kim, W.-g., Koros, W. J., Jones, C. W., and Nair, S. (2011) Modified mesoporous silica gas separation membranes on polymeric hollow fibers. *Chemistry of Materials,* **23**, 3025-3028.
50. Hicks, J. C., Drese, J. H., Fauth, D. J., Gray, M. L., Qi, G., and Jones, C. W. (2008) Designing adsorbents for CO_2 capture from flue gas-hyperbranched aminosilicas capable of capturing CO_2 reversibly. *Journal of the American Chemical Society,* **130**, 2902-2903.
51. Song, H.-K., Cho, K. W., and Lee, K.-H. (1998) Adsorption of carbon dioxide on the chemically modified silica adsorbents. *Journal of Non-crystalline Solids,* **242**, 69-80.
52. Son, W.-J., Choi, J.-S., and Ahn, W.-S. (2008) Adsorptive removal of carbon dioxide using polyethyleneimine-loaded mesoporous silica materials. *Microporous and Mesoporous Materials,* **113**, 31-40.
53. Goeppert, A., Czaun, M., May, R. B., Prakash, G. S., Olah, G. A., and Narayanan, S. (2011) Carbon dioxide capture from the air using a polyamine based regenerable solid adsorbent. *Journal of the American Chemical Society,* **133**, 20164-20167.
54. Xu, X., Song, C., Andresen, J. M., Miller, B. G., and Scaroni, A. W. (2002) Novel polyethylenimine-modified mesoporous molecular sieve of MCM-41 type as high-capacity adsorbent for CO_2 capture. *Energy & Fuels,* **16**, 1463-1469.
55. Rezaei, F., Lively, R. P., Labreche, Y., Chen, G., Fan, Y., Koros, W. J., and Jones, C. W. (2013) Aminosilane-grafted polymer/silica hollow fiber adsorbents for CO_2 capture from flue gas. *ACS Applied Materials & Interfaces,* **5**, 3921-3931.

56. Kusakabe, K., Ichiki, K., Hayashi, J.-i., Maeda, H., and Morooka, S. (1996) Preparation and characterization of silica-polyimide composite membranes coated on porous tubes for CO_2 separation. *Journal of Membrane Science,* **115**, 65-75.

57. Zornoza, B., Tellez, C., and Coronas, J. (2011) Mixed matrix membranes comprising glassy polymers and dispersed mesoporous silica spheres for gas separation. *Journal of Membrane Science,* **368**, 100-109.

58. Kim, J. H., and Lee, Y. M. (2001) Gas permeation properties of poly (amide-6-b-ethylene oxide)-silica hybrid membranes. *Journal of Membrane Science,* **193**, 209-225.

59. Morooka, S., Yan, S., Kusakabe, K., and Akiyama, Y. (1995) Formation of hydrogen-permselective SiO_2 membrane in macropores of α-alumina support tube by thermal decomposition of TEOS. *Journal of Membrane Science,* **101**, 89-98.

60. Sea, B.-K., Watanabe, M., Kusakabe, K., Morooka, S., and Kim, S.-S. (1996) Formation of hydrogen permselective silica membrane for elevated temperature hydrogen recovery from a mixture containing steam. *Gas Separation & Purification,* **10**, 187-195.

61. De Vos, R. M., and Verweij, H. (1998) Improved performance of silica membranes for gas separation. *Journal of Membrane Science,* **143**, 37-51.

62. Boffa, V., Ten Elshof, J., and Blank, D. (2007) Preparation of templated mesoporous silica membranes on macroporous α-alumina supports via direct coating of thixotropic polymeric sols. *Microporous and Mesoporous Materials,* **100**, 173-182.

63. Zornoza, B., Esekhile, O., J. Koros, W., Tellez, C., and Coronas, J. (2011) Hollow silicalite-1 sphere-polymer mixed matrix membranes for gas separation. *Separation and Purification Technology,* **77**, 137-145.

64. Huang, H. Y., Yang, R. T., Chinn, D., and Munson, C. L. (2003) Amine-grafted MCM-48 and silica xerogel as superior sorbents for acidic gas removal from natural gas. *Industrial & Engineering Chemistry Research,* **42**, 2427-2433.

65. Belmabkhout, Y., De Weireld, G., and Sayari, A. (2009) Amine-bearing mesoporous silica for CO_2 and H_2S removal from natural gas and biogas. *Langmuir,* **25**, 13275-13278.

66. Wang, X., Jia, J., Zhao, L., and Sun, T. (2008) Chemisorption of hydrogen sulphide on zinc oxide modified aluminum-substituted SBA-15. *Applied Surface Science,* **254**, 5445-5451.

67. Wang, X., Sun, T., Yang, J., Zhao, L., and Jia, J. (2008) Low-temperature H_2S removal from gas streams with SBA-15 supported ZnO nanoparticles. *Chemical Engineering Journal,* **142**, 48-55.

68. Wang, X., Ma, X., Sun, L., and Song, C. (2007) A nanoporous polymeric sorbent for deep removal of H_2S from gas mixtures for hydrogen purification. *Green Chemistry,* **9**, 695-702.

69. De Vos, R. M., Maier, W. F., and Verweij, H. (1999) Hydrophobic silica membranes for gas separation. *Journal of Membrane Science,* **158**, 277-288.

70. De Sitter, K., Winberg, P., D'Haen, J., Dotremont, C., Leysen, R. Martens, J. A., Mullens, S., Maurer, F. H., and Vankelecom, I. F. (2006) Silica filled poly (1-trimethylsilyl-1-propyne) nanocomposite membranes: relation between the transport of gases and structural characteristics. *Journal of Membrane Science,* **278**, 83-91.

71. De Sitter, K., Andersson, A., D'Haen, J., Leysen, R., Mullens, S., Maurer, F. H., and Vankelecom, I. F. (2008) Silica filled poly (4-methyl-2-pentyne) nanocomposite membranes: Similarities and differences with poly (1-trimethylsilyl-1-propyne)-silica systems. *Journal of Membrane Science*, **321**, 284-292.

72. Newalkar, B. L., Choudary, N. V., Kumar, P., Komarneni, S., and Bhat, T. S. (2002) Exploring the potential of mesoporous silica, SBA-15, as an adsorbent for light hydrocarbon separation. *Chemistry of Materials*, **14**, 304-309.

73. Montanari, T., Finocchio, E., Bozzano, I., Garuti, G., Giordano, A., Pistarino, C., and Busca, G. (2010) Purification of landfill biogases from siloxanes by adsorption: A study of silica and 13X zeolite adsorbents on hexamethylcyclotrisiloxane separation. *Chemical Engineering Journal*, **165**, 859-863.

74. Pires, J., Bestilleiro, M., Pinto, M., and Gil, A. (2008) Selective adsorption of carbon dioxide, methane and ethane by porous clays heterostructures. *Separation and Purification Technology*, **61**, 161-167.

75. Stepova, K. V., Maquarrie, D. J., and Krip, I. M. (2009) Modified bentonites as adsorbents of hydrogen sulfide gases. *Applied Clay Science*, **42**, 625-628.

76. Nguyen-Thanh, D., Block, K., and Bandosz, T. J. (2005) Adsorption of hydrogen sulfide on montmorillonites modified with iron. *Chemosphere*, **59**, 343-353.

77. Ji, L., Zhang, T., Milliken, K. L., Qu, J., and Zhang, X. (2012) Experimental investigation of main controls to methane adsorption in clay-rich rocks. *Applied Geochemistry*, **27**, 2533-2545.

78. Pires, J., Araujo, A., Carvalho, A., Pinto, M., Gonzalez-Calbet, J., and Ramırez-Castellanos, J. (2004) Porous materials from clays by the gallery template approach: synthesis, characterization and adsorption properties. *Microporous and Mesoporous Materials*, **73**, 175-180.

79. Cheng, A.-L., and Huang, W.-L. (2004) Selective adsorption of hydrocarbon gases on clays and organic matter. *Organic Geochemistry*, **35**, 413-423.

80. Choudary, N. V., Kumar, P., Bhat, T. S., Cho, S. H., Han, S. S., and Kim, J. N. (2002) Adsorption of light hydrocarbon gases on alkene-selective adsorbent. *Industrial & Engineering Chemistry Research*, **41**, 2728-2734.

81. Zaitan, H., Bianchi, D., Achak, O., and Chafik, T. (2008) A comparative study of the adsorption and desorption of o-xylene onto bentonite clay and alumina. *Journal of Hazardous Materials*, **153**, 852-859.

82. Qu, F., Zhu, L., and Yang, K. (2009) Adsorption behaviors of volatile organic compounds (VOCs) on porous clay heterostructures (PCH). *Journal of Hazardous Materials*, **170**, 7-12.

83. Zuo, S., Liu, F., Zhou, R., and Qi, C. (2012) Adsorption/desorption and catalytic oxidation of VOCs on montmorillonite and pillared clays. *Catalysis Communications*, **22**, 1-5.

84. Nunes, C. D., Pires, J., Carvalho, A. P., Calhorda, M. J., and Ferreira, P. (2008) Synthesis and characterisation of organo-silica hydrophobic clay heterostructures for volatile organic compounds removal. *Microporous and Mesoporous Materials*, **111**, 612-619.

85. Pires, J., Saini, V. K., and Pinto, M. s. L. (2008) Studies on selective adsorption of biogas components on pillared clays: approach for biogas improvement. *Environmental Science & Technology*, **42**, 8727-8732.

86. Sircar, S., Golden, T., and Rao, M. (1996) Activated carbon for gas separation and storage. *Carbon,* **34**, 1-12.

87. Presser, V., McDonough, J., Yeon, S.-H., and Gogotsi, Y. (2011) Effect of pore size on carbon dioxide sorption by carbide derived carbon. *Energy & Environmental Science,* **4**, 3059-3066.

88. Maroto-Valer, M. M., Tang, Z., and Zhang, Y. (2005) CO_2 capture by activated and impregnated anthracites. *Fuel Processing Technology,* **86**, 1487-1502.

89. Goel, C., Bhunia, H., and Bajpai, P. K. (2015) Mesoporous carbon adsorbents from melamine-formaldehyde resin using nanocasting technique for CO_2 adsorption. *Journal of Environmental Sciences,* **32**, 238-248.

90. Siriwardane, R. V., Shen, M.-S., Fisher, E. P., and Poston J. A. (2001) Adsorption of CO_2 on molecular sieves and activated carbon. *Energy & Fuels,* **15**, 279-284.

91. Drage, T. C., Blackman, J. M., Pevida, C., and Snape, C. E. (2009) Evaluation of activated carbon adsorbents for CO_2 capture in gasification. *Energy & Fuels,* **23**, 2790-2796.

92. Ganesan, A., and Shaijumon, M. M. (2016) Activated graphene-derived porous carbon with exceptional gas adsorption properties. *Microporous and Mesoporous Materials,* **220**, 21-27.

93. Sui, Z.-Y., Meng, Q.-H., Li, J.-T., Zhu, J.-H., Cui, Y., and Han, B.-H. (2014) High surface area porous carbons produced by steam activation of graphene aerogels. *Journal of Materials Chemistry A,* **2**, 9891-9898.

94. Zhang, Z., Zhou, J., Xing, W., Xue, Q., Yan, Z., Zhuo, S., and Qiao, S. Z. (2013) Critical role of small micropores in high CO_2 uptake. *Physical Chemistry Chemical Physics,* **15**, 2523-2529.

95. Deng, H.-g., Jin, S.-l., Liang, Z., Wang, Y.-l., Lu, B.-h., Qiao, W.-m., and Ling, L.-c. (2012) Synthesis of porous carbons derived from metal-organic coordination polymers and their adsorption performance for carbon dioxide. *New Carbon Materials,* **27**, 194-199.

96. Hu, X., Radosz, M., Cychosz, K. A., and Thommes, M. (2011) CO_2-filling capacity and selectivity of carbon nanopores: synthesis, texture, and pore-size distribution from quenched-solid density functional theory (QSDFT). *Environmental Science & Technology,* **45**, 7068-7074.

97. Deng, S., Wei, H., Chen, T., Wang, B., Huang, J., and Yu, G. (2014) Superior CO_2 adsorption on pine nut shell-derived activated carbons and the effective micropores at different temperatures. *Chemical Engineering Journal,* **253**, 46-54.

98. Silvestre-Albero, J., Wahby, A., Sepulveda-Escribano, A., Martinez-Escandell, M., Kaneko, K., and Rodriguez-Reinoso, F. (2011) Ultrahigh CO_2 adsorption capacity on carbon molecular sieves at room temperature. *Chemical Communications,* **47**, 6840-6842.

99. Daud, W. W., Ahmad, M., and Aroua, M. (2007) Carbon molecular sieves from palm shell: Effect of the benzene deposition times on gas separation properties. *Separation and Purification Technology,* **57**, 289-293.

100. Wickramaratne, N. P., and Jaroniec, M. (2013) Activated carbon spheres for CO_2 adsorption. *ACS Applied Materials & Interfaces,* **5**, 1849-1855.

101. Wickramaratne, N. P., and Jaroniec, M. (2013) Importance of small micropores in CO_2 capture by phenolic resin-based activated carbon spheres. *Journal of Materials Chemistry A*, **1**, 112-116.
102. Sevilla, M., Parra, J. B., and Fuertes, A. B. (2013) Assessment of the role of micropore size and N-doping in CO_2 capture by porous carbons. *ACS Applied Materials & Interfaces*, **5**, 6360-6368.
103. Marco-Lozar, J., Kunowsky, M., Suarez-Garcia, F., Carruthers, J., and Linares-Solano, A. (2012) Activated carbon monoliths for gas storage at room temperature. *Energy & Environmental Science*, **5**, 9833-9842.
104. Ribeiro, R. P., Sauer, T. P., Lopes, F. V., Moreira, R. F., Grande, C. A., and Rodrigues, A. r. E. (2008) Adsorption of CO_2, CH_4, and N_2 in activated carbon honeycomb monolith. *Journal of Chemical & Engineering Data*, **53**, 2311-2317.
105. Nandi, M., Okada, K., Dutta, A., Bhaumik, A., Maruyama, J., Derks, D., and Uyama, H. (2012) Unprecedented CO_2 uptake over highly porous N-doped activated carbon monoliths prepared by physical activation. *Chemical Communications*, **48**, 10283-10285.
106. Zhong, M., Natesakhawat, S., Baltrus, J. P., Luebke, D., Nulwala, H., Matyjaszewski, K., and Kowalewski, T. (2012) Copolymer-templated nitrogen-enriched porous nanocarbons for CO_2 capture. *Chemical Communications*, **48**, 11516-11518.
107. Xing, W., Liu, C., Zhou, Z., Zhang, L., Zhou, J., Zhuo, S., Yan, Z., Gao, H., Wang, G., and Qiao, S. Z. (2012) Superior CO_2 uptake of N-doped activated carbon through hydrogen-bonding interaction. *Energy & Environmental Science*, **5**, 7323-7327.
108. Zhao, Y., Zhao, L., Yao, K. X., Yang, Y., Zhang, Q., and Han, Y. (2012) Novel porous carbon materials with ultrahigh nitrogen contents for selective CO_2 capture. *Journal of Materials Chemistry*, **22**, 19726-19731.
109. Saleh, M., Tiwari, J. N., Kemp, K. C., Yousuf, M., and Kim, K. S. (2013) Highly selective and stable carbon dioxide uptake in polyindole-derived microporous carbon materials. *Environmental Science & Technology*, **47**, 5467-5473.
110. Sevilla, M., Falco, C., Titirici, M.-M., and Fuertes, A. B. (2012) High-performance CO_2 sorbents from algae. *RSC Advances*, **2**, 12792-12797.
111. Zhao, Y., Liu, X., Yao, K. X., Zhao, L., and Han, Y. (2012) Superior capture of CO_2 achieved by introducing extra-framework cations into N-doped microporous carbon. *Chemistry of Materials*, **24**, 4725-4734.
112. Fan, X., Zhang, L., Zhang, G., Shu, Z., and Shi, J. (2013) Chitosan derived nitrogen-doped microporous carbons for high performance CO_2 capture. *Carbon*, **61**, 423-430.
113. Chen, C., Kim, J., and Ahn, W.-S. (2012) Efficient carbon dioxide capture over a nitrogen-rich carbon having a hierarchical micro-mesopore structure. *Fuel*, **95**, 360-364.
114. Chandra, V., Yu, S. U., Kim, S. H., Yoon, Y. S., Kim, D. Y., Kwon, A. H., Meyyappan, M., and Kim, K. S. (2012) Highly selective CO_2 capture on N-doped carbon produced by chemical activation of polypyrrole functionalized graphene sheets. *Chemical Communications*, **48**, 735-737.
115. Xia, Y., Zhu, Y., and Tang, Y. (2012) Preparation of sulfur-doped microporous carbons for the storage of hydrogen and carbon dioxide. *Carbon*, **50**, 5543-5553.

116. Seema, H., Kemp, K. C., Le, N. H., Park, S.-W., Chandra, V., Lee, J. W., and Kim, K. S. (2014) Highly selective CO_2 capture by S-doped microporous carbon materials. *Carbon*, **66**, 320-326.

117. Builes, S., Roussel, T., Ghimbeu, C. M., Parmentier, J., Gadiou, R., Vix-Guterl, C., and Vega, L. F. (2011) Microporous carbon adsorbents with high CO_2 capacities for industrial applications. *Physical Chemistry Chemical Physics*, **13**, 16063-16070.

118. Xia, Y., Mokaya, R., Walker, G. S., and Zhu, Y. (2011) Superior CO_2 adsorption capacity on N-doped, high-surface-area, microporous carbons templated from zeolite. *Advanced Energy Materials*, **1**, 678-683.

119. Sevilla, M., and Fuertes, A. B. (2012) CO_2 adsorption by activated templated carbons. *Journal of Colloid and Interface Science*, **366**, 147-154.

120. Pevida, C., Plaza, M., Arias, B., Fermoso, J., Rubiera, F., and Pis, J. (2008) Surface modification of activated carbons for CO_2 capture. *Applied Surface Science*, **254**, 7165-7172.

121. Shafeeyan, M. S., Daud, W. M. A. W., Houshmand, A., and Arami-Niya, A. (2011) Ammonia modification of activated carbon to enhance carbon dioxide adsorption: Effect of pre-oxidation. *Applied Surface Science*, **257**, 3936-3942.

122. Zhang, Z., Xu, M., Wang, H., and Li, Z. (2010) Enhancement of CO_2 adsorption on high surface area activated carbon modified by N_2, H_2 and ammonia. *Chemical Engineering Journal*, **160**, 571-577.

123. Plaza, M., Pevida, C., Arenillas, A., Rubiera, F., and Pis, J. (2007) CO_2 capture by adsorption with nitrogen enriched carbons. *Fuel*, **86**, 2204-2212.

124. Guo, J., Luo, Y., Lua, A. C., Chi, R.-a., Chen, Y.-l., Bao, X.-t., and Xiang, S.-x. (2007) Adsorption of hydrogen sulphide (H_2S) by activated carbons derived from oil-palm shell. *Carbon*, **45**, 330-336.

125. Yan, R., Liang, D. T., Tsen, L., and Tay, J. H. (2002) Kinetics and mechanisms of H_2S adsorption by alkaline activated carbon. *Environmental Science & Technology*, **36**, 4460-4466.

126. Seredych, M., Portet, C., Gogotsi, Y., and Bandosz, T. J. (2009) Nitrogen modified carbide-derived carbons as adsorbents of hydrogen sulfide. *Journal of Colloid and interface Science*, **330**, 60-66.

127. Bagreev, A., Menendez, J. A., Dukhno, I., Tarasenko, Y., and Bandosz, T. J. (2004) Bituminous coal-based activated carbons modified with nitrogen as adsorbents of hydrogen sulfide. *Carbon*, **42**, 469-476.

128. Huang, C.-C., Chen, C.-H., and Chu, S.-M. (2006) Effect of moisture on H_2S adsorption by copper impregnated activated carbon. *Journal of Hazardous Materials*, **136**, 866-873.

129. Nguyen-Thanh, D., and Bandosz, T. J. (2005) Activated carbons with metal containing bentonite binders as adsorbents of hydrogen sulfide. *Carbon*, **43**, 359-367.

130. Xiao, Y., Wang, S., Wu, D., and Yuan, Q. (2008) Experimental and simulation study of hydrogen sulfide adsorption on impregnated activated carbon under anaerobic conditions. *Journal of Hazardous Materials*, **153**, 1193-1200.

131. Jorda-Beneyto, M., Suarez-Garcia, F., Lozano-Castello, D., Cazorla-Amoros, D., and Linares-Solano, A. (2007) Hydrogen storage on chemically activated carbons and carbon nanomaterials at high pressures. *Carbon*, **45**, 293-303.

132. Juan-Juan, J., Marco-Lozar, J., Suarez-Garcia, F., Cazorla-Amoros, D., and Linares-Solano, A. (2010) A comparison of hydrogen storage in activated carbons and a metal-organic framework (MOF-5). *Carbon,* **48**, 2906-2909.

133. De la Casa-Lillo, M., Lamari-Darkrim, F., Cazorla-Amoros, D., and Linares-Solano, A. (2002) Hydrogen storage in activated carbons and activated carbon fibers. *The Journal of Physical Chemistry B,* **106**, 10930-10934.

134. Kim, H. S., Singer, J. P., Gogotsi, Y., and Fischer, J. E. (2009) Molybdenum carbide-derived carbon for hydrogen storage. *Microporous and Mesoporous Materials,* **120**, 267-271.

135. Jorda-Beneyto, M., Lozano-Castello, D., Suarez-Garcia, F., Cazorla-Amoros, D., and Linares-Solano, A. (2008) Advanced activated carbon monoliths and activated carbons for hydrogen storage. *Microporous and Mesoporous Materials,* **112**, 235-242.

136. Almasoudi, A., and Mokaya, R. (2012) Preparation and hydrogen storage capacity of templated and activated carbons nanocast from commercially available zeolitic imidazolate framework. *Journal of Materials Chemistry,* **22**, 146-152.

137. Sun, J., Rood, M. J., Rostam-Abadi, M., and Lizzio, A. A. (1996) Natural gas storage with activated carbon from a bituminous coal. *Gas Separation & Purification,* **10**, 91-96.

138. Yeon, S.-H., Osswald, S., Gogotsi, Y., Singer, J. P., Simmons, J. M., Fischer, J. E., Lillo-Rodenas, M. A., and Linares-Solano, A. (2009) Enhanced methane storage of chemically and physically activated carbide-derived carbon. *Journal of Power Sources,* **191**, 560-567.

139. Prauchner, M. J., and Rodriguez-Reinoso, F. (2008) Preparation of granular activated carbons for adsorption of natural gas. *Microporous and Mesoporous Materials,* **109**, 581-584.

140. Srinivas, G., Burress, J., and Yildirim, T. (2012) Graphene oxide derived carbons (GODCs): Synthesis and gas adsorption properties. *Energy & Environmental Science,* **5**, 6453-6459.

141. Yeon, S.-H., Knoke, I., Gogotsi, Y., and Fischer, J. E. (2010) Enhanced volumetric hydrogen and methane storage capacity of monolithic carbide-derived carbon. *Microporous and Mesoporous Materials,* **131**, 423-428.

142. Lozano-Castello, D., Cazorla-Amoros, D., Linares-Solano, A., and Quinn, D. (2002) Activated carbon monoliths for methane storage: Influence of binder. *Carbon,* **40**, 2817-2825.

143. Adinata, D., Daud, W. M. A. W., and Aroua, M. K. (2007) Production of carbon molecular sieves from palm shell based activated carbon by pore sizes modification with benzene for methane selective separation. *Fuel Processing Technology,* **88**, 599-605.

144. Aroua, M. K., Daud, W. M. A. W., Yin, C. Y., and Adinata, D. (2008) Adsorption capacities of carbon dioxide, oxygen, nitrogen and methane on carbon molecular basket derived from polyethyleneimine impregnation on microporous palm shell activated carbon. *Separation and Purification Technology,* **62**, 609-613.

145. Chiang, Y.-C., Chiang, P.-C., and Huang, C.-P. (2001) Effects of pore structure and temperature on VOC adsorption on activated carbon. *Carbon,* **39**, 523-534.

146. Das, D., Gaur, V., and Verma, N. (2004) Removal of volatile organic compound by activated carbon fiber. *Carbon,* **42**, 2949-2962.

147. Kim, K.-J., Kang, C.-S., You, Y.-J., Chung, M.-C., Woo, M.-W., Jeong, W.-J., Park, N.-C., and Ahn, H.-G. (2006) Adsorption-desorption characteristics of VOCs over impregnated activated carbons. *Catalysis Today,* **111**, 223-228.
148. Kemp, K. C., Seema, H., Saleh, M., Le, N. H., Mahesh, K., Chandra, V., and Kim, K. S. (2013) Environmental applications using graphene composites: water remediation and gas adsorption. *Nanoscale,* **5**, 3149-3171.
149. Mishra, A. K., and Ramaprabhu, S. (2011) Carbon dioxide adsorption in graphene sheets. *AIP Advances,* **1**, 032152.
150. Meng, L.-Y., and Park, S.-J. (2012) Effect of exfoliation temperature on carbon dioxide capture of graphene nanoplates. *Journal of Colloid and interface Science,* **386**, 285-290.
151. Xia, K., Tian, X., Fei, S., and You, K. (2014) Hierarchical porous graphene-based carbons prepared by carbon dioxide activation and their gas adsorption properties. *International Journal of Hydrogen Energy,* **39**, 11047-11054.
152. Ning, G., Xu, C., Mu, L., Chen, G., Wang, G., Gao, J., Fan, Z., Qian, W., and Wei, F. (2012) High capacity gas storage in corrugated porous graphene with a specific surface area-lossless tightly stacking manner. *Chemical Communications,* **48**, 6815-6817.
153. Wang, Y., Guan, C., Wang, K., Guo, C. X., and Li, C. M. (2011) Nitrogen, hydrogen, carbon dioxide, and water vapor sorption properties of three-dimensional graphene. *Journal of Chemical & Engineering Data,* **56**, 642-645.
154. Shan, M., Xue, Q., Jing, N., Ling, C., Zhang, T., Yan, Z., and Zheng, J. (2012) Influence of chemical functionalization on the CO_2/N_2 separation performance of porous graphene membranes. *Nanoscale,* **4**, 5477-5482.
155. Carrillo, I., Rangel, E., and Magana, L. (2009) Adsorption of carbon dioxide and methane on graphene with a high titanium coverage. *Carbon,* **47**, 2758-2760.
156. Mishra, A. K., and Ramaprabhu, S. (2012) Nanostructured polyaniline decorated graphene sheets for reversible CO_2 capture. *Journal of Materials Chemistry,* **22**, 3708-3712.
157. Dasgupta, T., Punnathanam, S. N., and Ayappa, K. (2015) Effect of functional groups on separating carbon dioxide from CO_2/N_2 gas mixtures using edge functionalized graphene nanoribbons. *Chemical Engineering Science,* **121**, 279-291.
158. Kumar, R., Suresh, V. M., Maji, T. K., and Rao, C. (2014) Porous graphene frameworks pillared by organic linkers with tunable surface area and gas storage properties. *Chemical Communications,* **50**, 2015-2017.
159. Zhou, D., Liu, Q., Cheng, Q., Zhao, Y., Cui, Y., Wang, T., and Han, B. (2012) Graphene-manganese oxide hybrid porous material and its application in carbon dioxide adsorption. *Chinese Science Bulletin,* **57**, 3059-3064.
160. Kemp, K. C., Chandra, V., Saleh, M., and Kim, K. S. (2013) Reversible CO_2 adsorption by an activated nitrogen doped graphene/polyaniline material. *Nanotechnology, 24,* 235703.
161. Saleh, M., Chandra, V., Kemp, K. C., and Kim, K. S. (2013) Synthesis of N-doped microporous carbon via chemical activation of polyindole-modified graphene oxide sheets for selective carbon dioxide adsorption. *Nanotechnology,* **24**, 255702.
162. Sui, Z.-Y., and Han, B.-H. (2015) Effect of surface chemistry and textural properties on carbon dioxide uptake in hydrothermally reduced graphene oxide. *Carbon,* **82**, 590-598.

163. Chen, C., Xu, K., Ji, X., Miao, L., and Jiang, J. (2014) Enhanced adsorption of acidic gases (CO_2, NO_2 and SO_2) on light metal decorated graphene oxide. *Physical Chemistry Chemical Physics*, **16**, 11031-11036.

164. Tsoufis, T., Katsaros, F., Sideratou, Z., Romanos, G., Ivashenko, O., Rudolf, P., Kooi, B., Papageorgiou, S., and Karakassides, M. (2014) Tailor-made graphite oxide-DAB poly (propylene imine) dendrimer intercalated hybrids and their potential for efficient CO_2 adsorption. *Chemical Communications*, **50**, 10967-10970.

165. Sui, Z.-Y., Cui, Y., Zhu, J.-H., and Han, B.-H. (2013) Preparation of three-dimensional graphene oxide-polyethylenimine porous materials as dye and gas adsorbents. *ACS Applied Materials & Interfaces*, **5**, 9172-9179.

166. Zhao, Y., Ding, H., and Zhong, Q. (2012) Preparation and characterization of aminated graphite oxide for CO_2 capture. *Applied Surface Science*, **258**, 4301-4307.

167. Zhao, Y., Ding, H., and Zhong, Q. (2013) Synthesis and characterization of MOF-aminated graphite oxide composites for CO_2 capture. *Applied Surface Science*, **284**, 138-144.

168. Cao, Y., Zhao, Y., Lv, Z., Song, F., and Zhong, Q. (2015) Preparation and enhanced CO_2 adsorption capacity of UiO-66/graphene oxide composites. *Journal of Industrial and Engineering Chemistry*, **27**, 102-107.

169. Chowdhury, S., Parshetti, G. K., and Balasubramanian, R. (2015) Post-combustion CO_2 capture using mesoporous TiO_2/graphene oxide nanocomposites. *Chemical Engineering Journal*, **263**, 374-384.

170. Mabayoje, O., Seredych, M., and Bandosz, T. J. (2012) Cobalt (hydr) oxide/graphite oxide composites: Importance of surface chemical heterogeneity for reactive adsorption of hydrogen sulfide. *Journal of Colloid and Interface Science*, **378**, 1-9.

171. Seredych, M. and Bandosz, T. J. (2011) Reactive adsorption of hydrogen sulfide on graphite oxide/$Zr(OH)_4$ composites. *Chemical Engineering Journal*, **166**, 1032-1038.

172. Petit, C., Levasseur, B., Mendoza, B., and Bandosz, T. J. (2012) Reactive adsorption of acidic gases on MOF/graphite oxide composites. *Microporous and Mesoporous Materials*, **154**, 107-112.

173. Seredych, M., and Bandosz, T. J. (2009) Adsorption of hydrogen sulfide on graphite derived materials modified by incorporation of nitrogen. *Materials Chemistry and Physics*, **113**, 946-952.

174. Seredych, M., and Bandosz, T. J. (2007) Removal of ammonia by graphite oxide via its intercalation and reactive adsorption. *Carbon*, **45**, 2130-2132.

175. Seredych, M., Petit, C., Tamashausky, A. V., and Bandosz, T. J. (2009) Role of graphite precursor in the performance of graphite oxides as ammonia adsorbents. *Carbon*, **47**, 445-456.

176. Seredych, M., Tamashausky, A. V., and Bandosz, T. J. (2008) Surface features of exfoliated graphite/bentonite composites and their importance for ammonia adsorption. *Carbon*, **46**, 1241-1252.

177. Seredych, M. and Bandosz, T. J. (2012) Manganese oxide and graphite oxide/MnO_2 composites as reactive adsorbents of ammonia at ambient conditions. *Microporous and Mesoporous Materials*, **150**, 55-63.

178. Seredych. M. and Bandosz, T. J. (2008) Adsorption of ammonia on graphite oxide/aluminium polycation and graphite oxide/zirconium-aluminium polyoxycation composites. *Journal of Colloid and Interface Science*, **324**, 25-35.

179. Seredych, M. and Bandosz, T. J. (2009) Graphite oxide/AlZr polycation composites: Surface characterization and performance as adsorbents of ammonia. *Materials Chemistry and Physics*, **117**, 99-106.

180. Seredych, M., and Bandosz, T. J. (2010) Adsorption of ammonia on graphite oxide/Al 13 composites. *Colloids and Surfaces A: Physicochemical and Engineering Aspects*, **353**, 30-36.

181. Srinivas, G., Zhu, Y., Piner, R., Skipper, N., Ellerby, M., and Ruoff, R. (2010) Synthesis of graphene-like nanosheets and their hydrogen adsorption capacity. *Carbon*, **48**, 630-635.

182. Yuan, W., Li, B., and Li, L. (2011) A green synthetic approach to graphene nanosheets for hydrogen adsorption. *Applied Surface Science*, **257**, 10183-10187.

183. Ghosh, A., Subrahmanyam, K., Krishna, K. S., Datta, S., Govindaraj, A., Pati, S. K., and Rao, C. (2008) Uptake of H_2 and CO_2 by graphene. *The Journal of Physical Chemistry C*, **112**, 15704-15707.

184. Ma, L.-P., Wu, Z.-S., Li, J., Wu, E.-D., Ren, W.-C., and Cheng, H.-M. (2009) Hydrogen adsorption behavior of graphene above critical temperature. *International Journal of Hydrogen Energy*, **34**, 2329-2332.

185. Lyth, S. M., Shao, H., Liu, J., Sasaki, K., and Akiba, E. (2014) Hydrogen adsorption on graphene foam synthesized by combustion of sodium ethoxide. *International Journal of Hydrogen Energy*, **39**, 376-380.

186. Srinivas, G., Burress, J. W., Ford, J., and Yildirim, T. (2011) Porous graphene oxide frameworks: synthesis and gas sorption properties. *Journal of Materials Chemistry*, **21**, 11323-11329.

187. Petit, C., Burress, J., and Bandosz, T. J. (2011) The synthesis and characterization of copper-based metal–organic framework/graphite oxide composites. *Carbon*, **49**, 563-572.

188. Krungleviciute, V., Heroux, L., Migone, A. D., Kingston, C. T., and Simard, B. (2005) Isosteric heat of argon adsorbed on single-walled carbon nanotubes prepared by laser ablation. *The Journal of Physical Chemistry B*, **109**, 9317-9320.

189. Valcarcel, M., Cardenas, S., Simonet, B. M., Moliner-Martinez, Y., and Lucena, R. (2008) Carbon nanostructures as sorbent materials in analytical processes. *TrAC Trends in Analytical Chemistry*, **27**, 34-43.

190. Fujiwara, A., Ishii, K., Suematsu, H., Kataura, H., Maniwa, Y., Suzuki, S., and Achiba, Y. (2001) Gas adsorption in the inside and outside of single-walled carbon nanotubes. *Chemical Physics Letters*, **336**, 205-211.

191. Babu, D. J., Lange, M., Cherkashinin, G., Issanin, A., Staudt, R., and Schneider, J. J. (2013) Gas adsorption studies of CO_2 and N_2 in spatially aligned double-walled carbon nanotube arrays. *Carbon*, **61**, 616-623.

192. Cinke, M., Li, J., Bauschlicher, C. W., Ricca, A., and Meyyappan, M. (2003) CO_2 adsorption in single-walled carbon nanotubes. *Chemical Physics Letters*, **376**, 761-766.

193. Hu, Y. H., and Ruckenstein, E. (2006) Applicability of Dubinin–Astakhov equation to CO_2 adsorption on single-walled carbon nanotubes. *Chemical Physics Letters*, **425**, 306-310.

194. Golkha A., Keshavar P., and Mowla, D. (2013) Investigation of CO_2 removal by silica and CNT nanofluids in microporous hollow fiber membrane contactors. *Journal of Membrane Science,* **433,** 17-24.

195. Casco, M. E., Morelos-Gomez, A., Vega-Diaz, S. M., Cruz-Silva, R., Tristan-Lopez. F., Muramatsu, H., Hayashi, T., Martinez-Escandell, M., Terrones, M., and Endo, M. (2014) CO_2 adsorption on crystalline graphitic nanostructures. *Journal of CO_2 Utilization,* **5,** 60-65.

196. Sun, Q., Wang, M., Li, Z., Ma, Y., and Du, A. (2013) CO_2 capture and gas separation on boron carbon nanotubes. *Chemical Physics Letters,* **575,** 59-66.

197. Upendar, K., Kumar, A. S. H., Lingaiah, N., Rao, K. R., and Prasad, P. S. (2012) Low-temperature CO_2 adsorption on alkali metal titanate nanotubes. *International Journal of Greenhouse Gas Control,* **10,** 191-198.

198. Hsu, S.-C., Lu, C., Su, F., Zeng, W., and Chen, W. (2010) Thermodynamics and regeneration studies of CO_2 adsorption on multiwalled carbon nanotubes. *Chemical Engineering Science,* **65,** 1354-1361.

199. Su, F., Lu, C., Cnen, W., Bai, H., and Hwang, J. F. (2009) Capture of CO_2 from flue gas via multiwalled carbon nanotubes. *Science of the Total Environment,* **407,** 3017-3023.

200. Lu, C., Bai, H., Wu, B., Su, F., and Hwang, J. F. (2008) Comparative study of CO_2 capture by carbon nanotubes, activated carbons, and zeolites. *Energy & Fuels,* **22,** 3050-3056.

201. Jana, S., Das, S., Ghosh, C., Maity, A., and Pradhan, M. (2015) Halloysite nanotubes capturing isotope selective atmospheric CO_2. *Scientific Reports,* **5,** 2015.

202. Song, F., Zhao, Y., Cao, Y., Ding, J., Bu, Y., and Zhong, Q. (2013) Capture of carbon dioxide from flue gases by amine-functionalized TiO_2 nanotubes. *Applied Surface Science,* **268,** 124-128.

203. Ko, Y. G., Lee, H. J., Oh, H. C., and Choi, U. S. (2013) Amines immobilized double-walled silica nanotubes for CO_2 capture. *Journal of Hazardous Materials,* **250,** 53-60.

204. Liu, J., Liu, Y., Wu, Z., Chen, X., Wang, H., and Weng, X. (2012) Polyethyleneimine functionalized protonated titanate nanotubes as superior carbon dioxide adsorbents. *Journal of Colloid and Interface Science,* **386,** 392-397.

205. Dillon, E. P., Crouse, C. A., and Barron, A. R. (2008) Synthesis, characterization, and carbon dioxide adsorption of covalently attached polyethyleneimine-functionalized single-wall carbon nanotubes. *ACS Nano,* **2,** 156-164.

206. Cong, H., Zhang, J., Radosz, M., and Shen, Y. (2007) Carbon nanotube composite membranes of brominated poly (2, 6-diphenyl-1, 4-phenylene oxide) for gas separation. *Journal of Membrane Science,* **294,** 178-185.

207. Zuttel, A., Sudan, P., Mauron, P., Kiyobayashi, T., Emmenegger, C., and Schlapbach, L. (2002) Hydrogen storage in carbon nanostructures. *International Journal of Hydrogen Energy,* **27,** 203-212.

208. Ma, R., Bando, Y., Zhu, H., Sato, T., Xu, C., and Wu, D. (2002) Hydrogen uptake in boron nitride nanotubes at room temperature. *Journal of the American Chemical Society,* **124,** 7672-7673.

209. Nour, M., Berean, K., Balendhran, S., Ou, J. Z., Du Plessis, J., McSweeney, C., Bhaskaran, M., Sriram, S., and Kalantar-zadeh, K. (2013) CNT/PDMS composite membranes for

H$_2$ and CH$_4$ gas separation. *International Journal of Hydrogen Energy*, **38**, 10494-10501.

210. Wang, W., Peng, X., and Cao, D. (2011) Capture of trace sulfur gases from binary mixtures by single-walled carbon nanotube arrays: a molecular simulation study. *Environmental Science & Technology*, **45**, 4832-4838.

211. Mohamadalizadeh, A., Towfighi, J., Rashidi, A., Mohajeri, A., and Golkar, M. (2011) Modification of carbon nanotubes for H$_2$S sorption. *Industrial & Engineering Chemistry Research*, **50**, 8050-8057.

212. Hussain, C. M., Saridara, C., and Mitra, S. (2008) Microtrapping characteristics of single and multi-walled carbon nanotubes. *Journal of Chromatography A*, **1185**, 161-166.

213. Li, Q.-L., Yuan, D.-X., and Lin, Q.-M. (2004) Evaluation of multi-walled carbon nanotubes as an adsorbent for trapping volatile organic compounds from environmental samples. *Journal of Chromatography A*, **1026**, 283-288.

214. Wang, L., Liu, J., Zhao, P., Ning, Z., and Fan, H. (2010) Novel adsorbent based on multi-walled carbon nanotubes bonding on the external surface of porous silica gel particulates for trapping volatile organic compounds. *Journal of Chromatography A*, **1217**, 5741-5745.

215. Zheng, F., Baldwin, D. L., Fifield, L. S., Anheier, N. C., Aardahl, C. L., and Grate, J. W. (2006) Single-walled carbon nanotube paper as a sorbent for organic vapor preconcentration. *Analytical Chemistry*, **78**, 2442-2446.

216. Delavar, M., Ghoreyshi, A. A., Jahanshahi, M., Khalili, S., and Nabian, N. (2012) Equilibria and kinetics of natural gas adsorption on multi-walled carbon nanotube material. *RSC Advances*, **2**, 4490-4497.

217. Cao, D., Zhang, X., Chen, J., Wang, W., and Yun, J. (2003) Optimization of single-walled carbon nanotube arrays for methane storage at room temperature. *The Journal of Physical Chemistry B*, **107**, 13286-13292.

218. Kim, S., Pechar, T. W., and Marand, E. (2006) Poly (imide siloxane) and carbon nanotube mixed matrix membranes for gas separation. *Desalination*, **192**, 330-339.

219. Wang, H., Zhou, S., Xiao, L., Wang, Y., Liu, Y., and Wu, Z. (2011) Titania nanotubes - A unique photocatalyst and adsorbent for elemental mercury removal. *Catalysis Today*, **175**, 202-208.

220. Ding, W., Hayashi, R., Ochi, K., Suehiro, J., Imasaka, K., Hara, M., Sano, N., Nagao, E., and Minagawa, T. (2006) Analysis of PD-generated SF 6 decomposition gases adsorbed on carbon nanotubes. *IEEE Transactions on Dielectrics and Electrical Insulation*, **13**, 1200-1207.

221. Foroutan, M., and Nasrabadi, A. T. (2010) Adsorption behavior of ternary mixtures of noble gases inside single-walled carbon nanotube bundles. *Chemical Physics Letters*, **497**, 213-217.

222. Jalili, S., and Majidi, R. (2007) Study of Xe and Kr adsorption on open single-walled carbon nanotubes using molecular dynamics simulations. *Physica E: Low-dimensional Systems and Nanostructures*, **39**, 166-170.

223. Foroutan, M., and Nasrabadi, A. T. (2010) Molecular dynamics simulation study of neon adsorption on single-walled carbon nanotubes. *Physica E: Low-dimensional Systems and Nanostructures*, **43**, 261-265.

224. Foroutan, M., and Nasrabadi, A. T. (2011) Adsorption and separation of binary mixtures of noble gases on single-walled carbon nanotube bundles. *Physica E: Low-dimensional Systems and Nanostructures,* **43**, 851-856.

225. Kiselev, A. V., and Yashin, Y. I. (2013) *Gas-adsorption Chromatography*, Springer, USA.

226. Zhao, Z., Cui, X., Ma, J., and Li, R. (2007) Adsorption of carbon dioxide on alkali-modified zeolite 13X adsorbents. *International Journal of Greenhouse Gas Control,* **1**, 355-359.

227. Montanari, T., Finocchio, E., Salvatore, E., Garuti, G., Giordano, A., Pistarino, C., and Busca, G. (2011) CO_2 separation and landfill biogas upgrading: a comparison of 4A and 13X zeolite adsorbents. *Energy,* **36**, 314-319.

228. Kusakabe, K., Kuroda, T., Murata, A., and Morooka, S. (1997) Formation of a Y-type zeolite membrane on a porous α-alumina tube for gas separation. *Industrial & Engineering Chemistry Research,* **36**, 649-655.

229. Saha, D., Bao, Z., Jia, F., and Deng, S. (2010) Adsorption of CO_2, CH_4, N_2O, and N_2 on MOF-5, MOF-177, and zeolite 5A. *Environmental Science & Technology,* **44**, 1820-1826.

230. Chatti, R., Bansiwal, A. K., Thote, J. A., Kumar, V., Jadhav, P., Lokhande, S. K., Biniwale, R. B., Labhsetwar, N. K., and Rayalu, S. S. (2009) Amine loaded zeolites for carbon dioxide capture: Amine loading and adsorption studies. *Microporous and Mesoporous Materials,* **121**, 84-89.

231. Tagliabue, M., Rizzo, C., Onorati, N. B., Gambarotta, E. F., Carati, A., and Bazzano, F. (2012) Regenerability of zeolites as adsorbents for natural gas sweetening: A case-study. *Fuel,* **93**, 238-244.

232. Adams, R. T., Lee, J. S., Bae, T.-H., Ward, J. K., Johnson, J., Jones, C. W., Nair, S., and Koros, W. J. (2011) CO_2-CH_4 permeation in high zeolite 4A loading mixed matrix membranes. *Journal of Membrane Science,* **367**, 197-203.

233. Li, P., and Tezel, F. H. (2007) Adsorption separation of N_2, O_2, CO_2 and CH_4 gases by β-zeolite. *Microporous and Mesoporous Materials,* **98**, 94-101.

234. Xu, X., Zhao, X., Sun, L., and Liu, X. (2009) Adsorption separation of carbon dioxide, methane and nitrogen on monoethanol amine modified β-zeolite. *Journal of Natural Gas Chemistry,* **18**, 167-172.

235. Xu, X., Zhao, X., Sun, L., and Liu, X. (2008) Adsorption separation of carbon dioxide, methane, and nitrogen on Hβ and Na-exchanged β-zeolite. *Journal of Natural Gas Chemistry,* **17**, 391-396.

236. Diaz, E., Ordonez, S., Vega, A., and Coca, J. (2004) Adsorption characterisation of different volatile organic compounds over alumina, zeolites and activated carbon using inverse gas chromatography. *Journal of Chromatography A,* **1049**, 139-146.

237. Sen, D., Kalipcilar, H., and Yilmaz, L. (2006) Development of zeolite filled polycarbonate mixed matrix gas separation membranes. *Desalination,* **200**, 222-224.

238. Husain, S., and Koros, W. J. (2007) Mixed matrix hollow fiber membranes made with modified HSSZ-13 zeolite in polyetherimide polymer matrix for gas separation. *Journal of Membrane Science,* **288**, 195-207.

239. Sen, D., Kalipcilar, H., and Yilmaz, L. (2007) Development of polycarbonate based zeolite 4A filled mixed matrix gas separation membranes. *Journal of Membrane Science,* **303**, 194-203.

240. Rolker, J., Seiler, M., Mokrushina, L., and Arlt, W. (2007) Potential of branched polymers in the field of gas absorption: Experimental gas solubilities and modeling. *Industrial & Engineering Chemistry Research*, **46**, 6572-6583.

241. Bali, S., Chen, T. T., Chaikittisilp, W., and Jones, C. W. (2013) Oxidative stability of amino polymer-alumina hybrid adsorbents for carbon dioxide capture. *Energy & Fuels*, **27**, 1547-1554.

242. An, D., Wu, L., Li, B.-G., and Zhu, S. (2007) Synthesis and SO_2 absorption/desorption properties of poly (1,1,3,3-tetramethylguanidine acrylate). *Macromolecules*, **40**, 3388-3393.

243. Kikkinides, E., and Yang, R. (1993) Gas separation and purification by polymeric adsorbents: Flue gas desulfurization and sulfur dioxide recovery with styrenic polymers. *Industrial & Engineering Chemistry Research*, **32**, 2365-2372.

244. Deshmukh, M. M., Ohba, M., Kitagawa, S., and Sakaki, S. (2013) Absorption of CO_2 and CS_2 into the Hofmann-type porous coordination polymer: Electrostatic versus dispersion interactions. *Journal of the American Chemical Society*, **135**, 4840-4849.

245. Noro, S. i., Kitagawa, S., Kondo, M., and Seki, K. (2000) A new, methane adsorbent, porous coordination polymer [{CuSiF6 (4,4'-bipyridine)2}n]. *Angewandte Chemie International Edition*, **39**, 2081-2084.

Index

A

B

C

www.ingramcontent.com/pod-product-compliance
Lightning Source LLC
Chambersburg PA
CBHW061614220326
41598CB00026BA/3764